Stream Ecology

Stream Ecology

Structure and function of running waters

J. David Allan

School of Natural Resources and Environment, University of Michigan, USA

CHAPMAN & HALL

London · Glasgow · Weinheim · New York · Tokyo · Melbourne · Madras

Published by Chapman & Hall, 2–6 Boundary Row, London SE1 8HN

Chapman & Hall, 2–6 Boundary Row, London SE1 8HN, UK

Blackie Academic & Professional, Wester Cleddens Road, Bishopbriggs, Glasgow G64 2NZ, UK

Chapman & Hall GmbH, Pappelallee 3, 69469 Weinheim, Germany

Chapman & Hall USA, 115 Fifth Avenue, New York, NY 10003, USA

Chapman & Hall Japan, ITP-Japan, Kyowa Building, 3F, 2-2-1 Hirakawacho, Chiyoda-ku, Tokyo 102, Japan

Chapman & Hall Australia, 102 Dodds Street, South Melbourne, Victoria 3205, Australia

Chapman & Hall India, R. Seshadri, 32 Second Main Road, CIT East, Madras 600 035, India

First edition 1995
Reprinted 1995

Typeset in 10.5/12.5pt Sabon by Witwell Ltd, Southport
Printed in Great Britain by Alden Press, Oxford

ISBN 0 412 29430 3 (HB) 0 412 35530 2 (PB)

A catalogue record for this book is available from the British Library

♾ Printed on acid-free text paper, manufactured in accordance with ANSI/NISO Z39.48-1992 and ANSI/NISO Z39.48-1984 (Permanence of Paper).

For my family

Contents

Preface

When my general interest in aquatic ecology began to focus increasingly on streams, immediately after completing my PhD, it was my good fortune that H.B.N. Hynes classic, *The Ecology of Running Waters*, had just appeared. Many of the current generation of stream ecologists benefitted from that masterful survey of the field. During the 1970s and 1980s, ecological studies of streams and rivers made many advances. Major new paradigms emerged, such as the river continuum concept and nutrient spiralling. Community ecologists made impressive advances in documenting the occurrence of predation, competition and other species interactions. The importance of physical processes in rivers drew increasing attention, particularly the areas of hydrology and geomorphology, and the interrelationships between physical and biological factors became better understood. While each advance in knowledge has opened new questions, the progress made in the past 20 years has been considerable, and has formed the basis for continuing and future work. In this volume I attempt to summarize these developments for the current generation of students.

The diversity of running water environments is enormously broad. When one considers torrential mountain brooks, large rivers of lowlands, and great rivers whose basins occupy subcontinents, it is apparent how location-specific environmental factors contribute to the sense of uniqueness and diversity of running waters. At the same time, however, our improved understanding of ecological, hydrological and geomorphological processes provides insight into the functional and structural characteristics of river systems that brings a unifying framework to this area of study. The inputs and transformation of energy and materials are important in all river systems, regional species richness and local species interactions influence the structure of all riverine communities, and the interaction of physical and biological forces is important to virtually every question that is asked. It appears to be the processes acting in running waters that are general, and the settings that often are unique.

As is true for every area of ecology during the closing years of the twentieth century, the study of streams and rivers cannot be addressed exclusive of the role of human activities, nor can we ignore the urgency of the need for conservation. This is a two-way street. Ecologists who study streams without considering how past or present human modifications of the stream or its valley might have contributed to their observations, do so at the risk of incomplete understanding. Conservation efforts that lack an adequate scientific basis are less likely to succeed. One trend that seems safe to forecast in stream ecology is toward a greater emphasis on understanding human impacts. Fortunately, signs of this trend are already apparent.

I cannot pretend to do justice to the entire range of ecological topics pertaining to streams and rivers. The diversity of running water environments, the richness of taxa from microbes to higher vertebrates, and the range of topic areas all are too great. The reader will surely detect biases towards studies based in small streams, at middle or higher latitudes, and reported in the English-language literature, although I have done my best to resist these tendencies. I greatly appreciate the constructive suggestions of many colleagues who were generous in their advice on draft chapters of the book. They have done much to clarify and correct what I have written, but the responsibility for choice of topics as well as any errors or omissions is mine. Scott Cooper, Bobbi Peckarsky and Scott Wissinger used an early draft of the entire manuscript in their stream ecology courses and provided valuable feedback from their students as well as their own reactions. Individuals whose efforts improved individual chapters include Fred Benfield, Art Benke, Art Brown, Scott Cooper, Stuart Findlay, Alex Flecker, Nancy Grimm, David Hart, Chuck Hawkins, Bob Hughes, Steve Kohler, Gary Lamberti, Rex Lowe, Rich Merritt, Diane McKnight, Judy Meyer, Bobbi Peckarsky, Pete Ode, Walt Osterkamp, M.L. Ostrofsky, Margaret Palmer, LeRoy Poff, Karen Prestergaard, Ike Schlosser, Len Smock, Al Steinman, Scott Wissinger and Jack Webster. I thank Ken Cummins, Alen Flecker, Jim Gore and Al Steinman for providing drawings for figures. The stream group at the University of Maryland was a continual source of ideas during my tenure there. I began writing this book while a sabbatical visitor at the University of Lund, and benefitted greatly from my interactions with Bjorn Mamlqvist, Per Sjöström, Christian Otto and other Swedish colleagues. To all of these individuals I extend my deepest appreciation. I also wish to thank those people who helped with many of the details of manuscript preparation. Tara Burke, Teresa Petersen and Cindy Smith helped to compile and prepare figures and Mary Lammert and Robin Abell were most careful and efficient editorial assistants. Thanks to Gary Williams for his photograph of the Maple River. I thank my editors at Chapman & Hall, Bob Carling and Clem Earle, for their combination of patience and prodding.

Lastly, I wish to express my deepest appreciation to my extremely supportive and tolerant family, Susan, Jenni and Brian. Their encouragement has been a constant throughout my career, and I especially appreciate their help and understanding during the time this book was being written.

Chapter one

Channels and flow

Our first impressions, when we gaze upon a river, are of the strength of the current, the dimensions of the channel, and perhaps the boulders in the streambed or the shape of the banks. Mentally we catalog this flowing water system as a stream or a river, in flood or in repose, and perhaps take note of bends, pools and meanders. Very likely we have encountered other streams with similar features, and some that are very different. We might also note whether this waterway is suitable for various activities, such as the passage of boats, recreational uses and whether it poses a hazard to humans. With a moment's further reflection it is apparent that these same channel and flow characteristics likely influence the functioning of running water ecosystems, and the biota found therein.

Physical processes acting on river systems traditionally have been the domain of hydrologists and geomorphologists. It is increasingly apparent, however, that our understanding of the ecology of rivers will benefit from an appreciation of flow variability, channel structure and other physical aspects of river ecosystems. Running waters are enormously diverse. They range from small streams to great rivers, and occur under widely differing conditions of climate, vegetation, topography and geology. In order to make sense of biological findings from such disparate settings, it is important to have a framework that reflects the physical dimensions of the study system.

1.1 Hydrology

1.1.1 The water cycle

Until the sixteenth century, oceans were thought to be the source of rivers and springs via underground seepage. Palissy and others suggested that storage of rain-water was the real source, based on several lines of reasoning. It was suggested that springs would not dry up in summer if oceans were the source, since the oceans do not decrease noticeably. Springs should be more common at low elevations if they derive from oceanic water. However springs often dry up in summer, they are more common on mountain slopes, and finally, springs are fresh. In 1674, measurements by Perrault showed that precipitation into the Seine basin was six times greater than discharge (Morisawa, 1968). This finding changed the focus of interest from studying whether rainfall is sufficient to provision rivers, to studying where the rest of the rainfall goes.

The hydrologic cycle describes the continuous cycling of water from atmosphere to earth and oceans and back again (Figure 1.1). Conceptually this cycle can be viewed as a series of storage places and transfer processes, although water in rivers is both a storage place, however temporary, and a transfer between land and sea. The hydrologic cycle is powered by solar energy. This drives evaporation and evapotranspiration, transferring water from the surface of

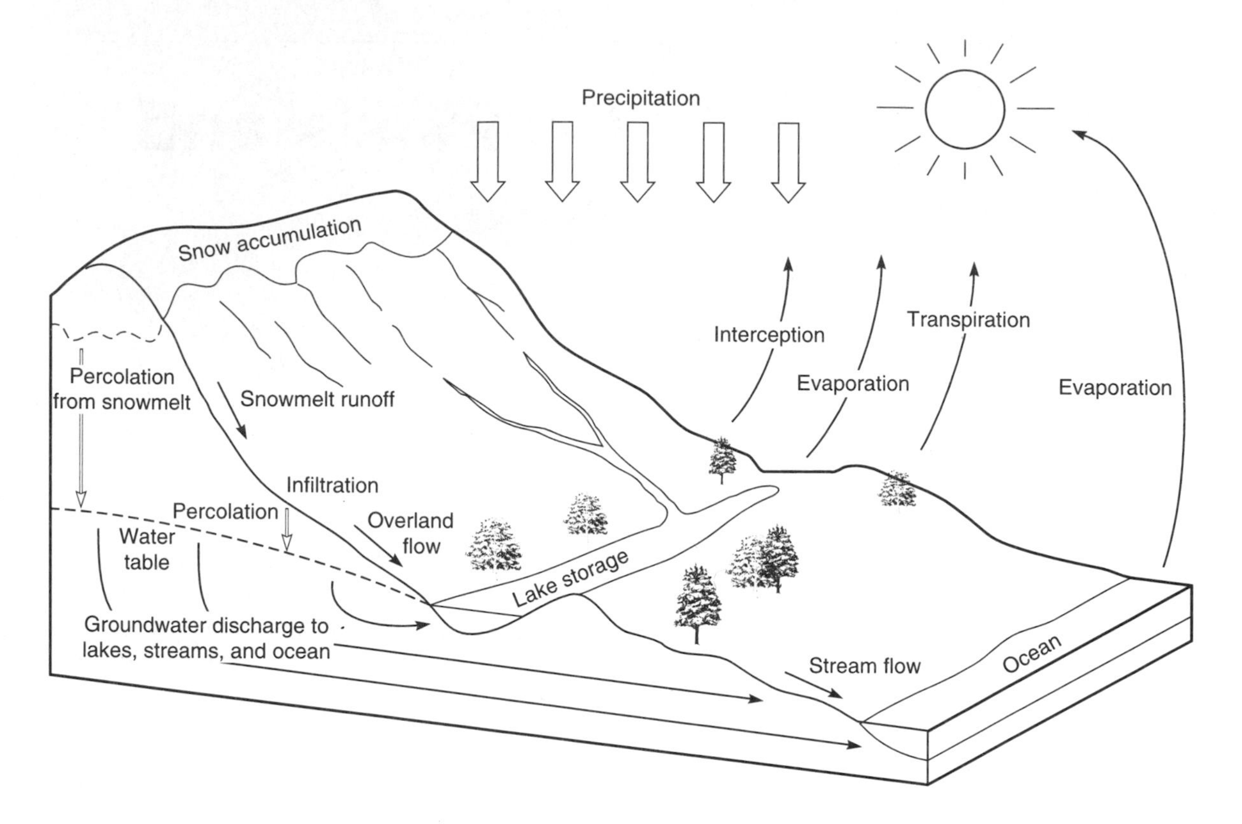

FIGURE 1.1 Schematic drawing of the hydrologic cycle. (Redrawn from Dunne and Leopold, 1978.)

the land, from plant tissue, and especially from the oceans, into the atmosphere. Precipitation, primarily as rain and snow, transfers water from the atmosphere to the land surface. These inputs immediately run off as surface water, or follow a number of alternative pathways, some of which (e.g. groundwater) release to the stream channel much more slowly and so are, in effect, storage places as well.

Despite the enormous significance of rivers in the development of civilizations and the shaping of land masses, the amount of water in rivers at any one time is tiny in comparison to other stores. Only 2.8% of the world's total water occurs on land. Ice caps and glaciers comprise the majority (2.24%), and groundwater (0.61%) is also a sizeable percentage. Only 0.009% of the total water is stored in lakes, about 0.001% is stored in the atmosphere, and rivers contain ten times less, 0.0001% of the world's water. Because the volume in the atmosphere and river water at any instant in time is small, the average water molecule cycles through them rapidly, residing only days to weeks, compared with much longer residence times of water in other compartments.

Estimates of the amount of water discharged by rivers to the world's oceans range between 32 and 37 × $10^3\,km^{-3}\,yr^{-1}$ (Milliman, 1990). The ten largest rivers in terms of runoff volume account for nearly 40% of the total, and the Amazon alone contributes more than 15%. In the United States, the Mississippi contributes some 40% of total discharge, while the Columbia, Mobile and Susquehana together contribute an additional 20%. Globally, the

TABLE 1.1 Latitudinal distribution of runoff (From Milliman, 1990)

Latitude	*Volume of runoff (km^3)*	*Runoff (%)*	*Land area (%)*
60–90 N	3551	8.9	11.6
30–60	8252	20.8	31.4
0–30	12597	31.8	24.5
0–30	11746	29.6	19.6
30–60	1567	3.9	9.4
60–90 S	1987	5.0	9.4

greatest runoff occurs in tropical and subtropical areas, because these latitudes also receive the greatest rainfall (Table 1.1). This is especially apparent if one compares specific discharge (discharge divided by drainage basin area).

(a) Pathways of water from land to sea

Water moves downhill by various routes (Figures 1.1 and 1.2). Climate, vegetation, topography, geology, land use and soil characteristics determine how much surface runoff occurs compared with other pathways (Dunne and Leopold, 1978). This in turn affects the rate and the chemical composition (chapter 2) of the runoff.

Before tracing the pathways of water over or through the land, it should be noted that a substantial fraction of precipitation inputs return directly to the atmosphere by evapotranspiration. First, some rain-water evaporates from the surface of vegetation immediately during and after a rainstorm, never reaching the ground or being absorbed by plants. This is referred to as interception. Second, water on the surface of the ground and in lakes and streams evaporates directly back to the atmosphere. Third, plants lose water to the atmosphere during the exchange of gases necessary for photosynthesis. Water loss by transpiration constitutes a major flux back to the atmosphere. Plant leaves display special adaptations

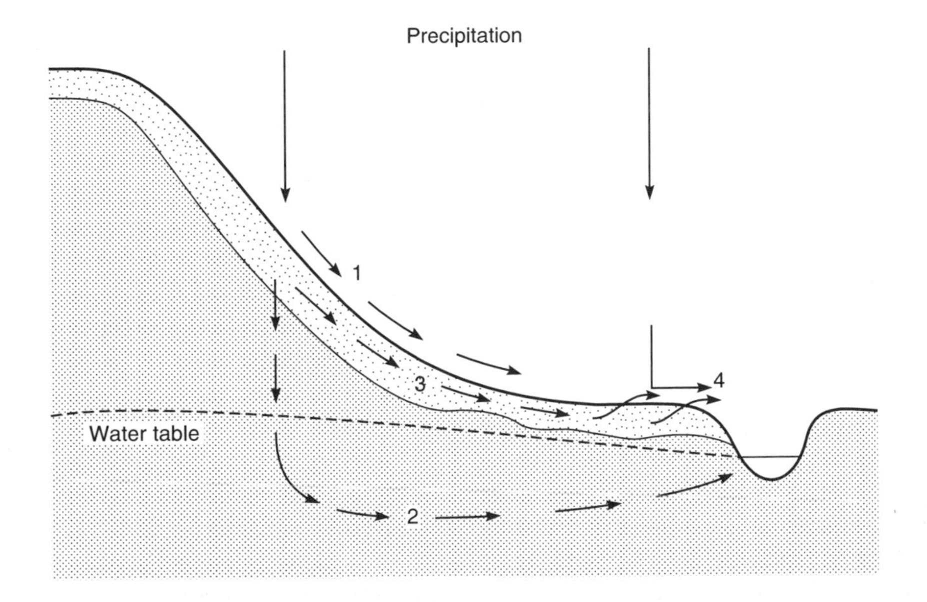

FIGURE 1.2 Pathways of water moving downhill. Overland flow (1) occurs when precipitation exceeds the infiltration capacity of the soil. Water that enters the soil adds to groundwater flow (2) and usually reaches streams, lakes, or the oceans. A relatively impermeable layer will cause water to move laterally through the soil (3) as shallow sub-surface stormflow. Saturation of the soil can force sub-surface water to rise to the surface where, along with direct precipitation, it forms saturation overland flow (4). The stippled area is relatively permeable topsoil. (Redrawn from Dunne and Leopold, 1978.)

to minimize transpiration, and the geography of plant species and leaf forms reflects the importance of this water loss. When an experimental forest in New Hampshire was clear-cut and subsequent re-growth suppressed with herbicides, stream runoff increased 40% on an annual basis, 400% during the summer (Likens, 1984). This represents water that would have returned to the atmosphere primarily via transpiration in an intact forest.

Once rain- or meltwater encounters the ground surface, it follows several pathways in reaching a stream channel or groundwater (Figure 1.2). Soil can absorb rainfall at some maximum rate, termed the infiltration capacity. This capacity declines during a rain event, normally approaching a constant some 0.5 to 2 h into the storm (Free, Browning and Musgrave, 1940). Any rainfall in excess of infiltration capacity accumulates on the surface, and any surface water in excess of depression storage capacity will move as an irregular sheet of overland flow. Overland flow is likely in arid areas due to the low permeability of their soils, when the soil surface is frozen, and where human activities make the land surface less permeable (Likens, 1984). In extreme cases, 50–100% of the rainfall can travel as overland flow (Horton, 1945), attaining velocities of 10–500 $m\,h^{-1}$. However, overland flow rarely occurs in undisturbed humid regions because their soils have high infiltration capacities.

Rain that penetrates the soil, particularly less intense rain that does not exceed the infiltration capacity, reaches the groundwater, from which it discharges to the stream slowly and over a long period of time. Baseflow or dry-weather flow in a river is due to groundwater entering the stream channel. However, if the soil structure includes a relatively impermeable layer underlying highly permeable topsoil, water will accumulate at that layer and move downhill, reaching streams laterally through their banks. This movement, termed shallow sub-surface stormflow, is slower than Horton overland flow but faster than groundwater flow, moving at up to 11 $m\,day^{-1}$ through sandy loam on a steep hill (Linsley, Kohler and Paulhus, 1958). Finally, when there is a large enough rainstorm or a shallow enough water table, the water table will rise to the ground surface (at certain elevations), causing sub-surface water to escape from the saturated soil as saturation overland flow. This is composed of return flow forced up from the soil and direct precipitation onto the saturated soil (Dunne and Leopold, 1978). Velocities are similar to the lower range of Horton overland flow.

Most rivers continue to flow during periods of no rainfall. These are called perennial, as opposed to intermittent, and much of the water in the channel comes from groundwater. In humid regions the water table slopes toward the stream channel, with the consequence that groundwater discharges into the channel. Discharge from the water table into the stream accounts for baseflow during periods without precipitation, and also explains why baseflow increases as one proceeds downstream, even without tributary input. Such streams are called gaining or effluent (Figure 1.3(a)). Streams originating at high altitudes often flow into drier areas where the local water table is below the bottom of the stream channel. Depending on permeability of materials underlying the stream bed, the stream may lose water into the ground. This is referred to as a losing or influent stream (Figure 1.3(b)). The same stream can shift between gaining and losing conditions along its course due to changes in underlying lithology and local climate, or temporally due to alternation of baseflow and stormflow conditions. The exchange of water between the channel and groundwater will turn out to be important to the dynamics of nutrients and the ecology of the biota that dwells within the substrate of the stream bottom.

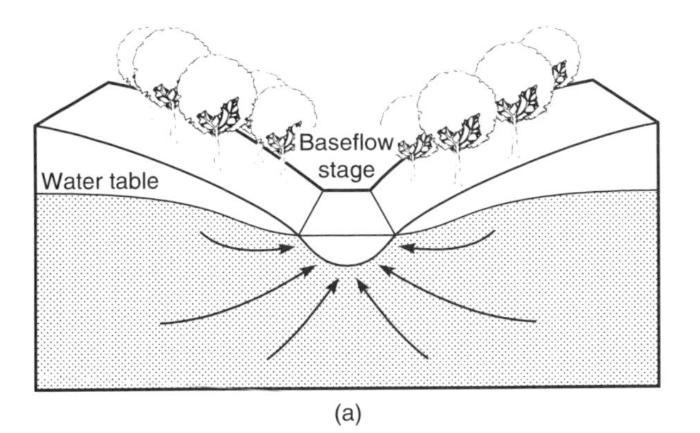

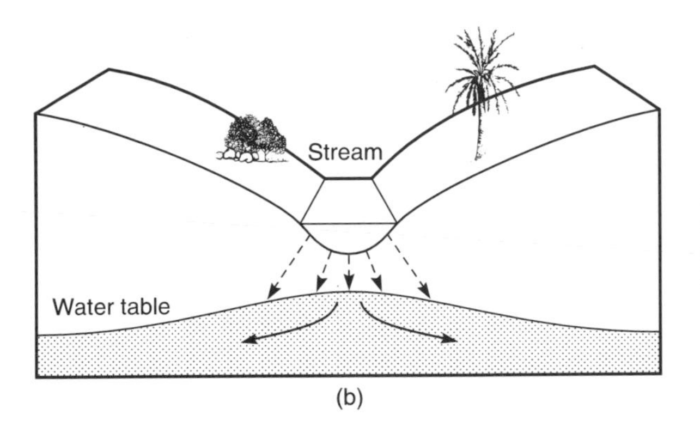

FIGURE 1.3 (a) Cross-section of a gaining stream, typical of humid regions, where groundwater recharges the stream. (b) Cross-section of a losing stream, typical of arid regions, where streams can recharge groundwater. The same stream can be gaining during low flows and losing when in flood. (After Fetter, 1988.)

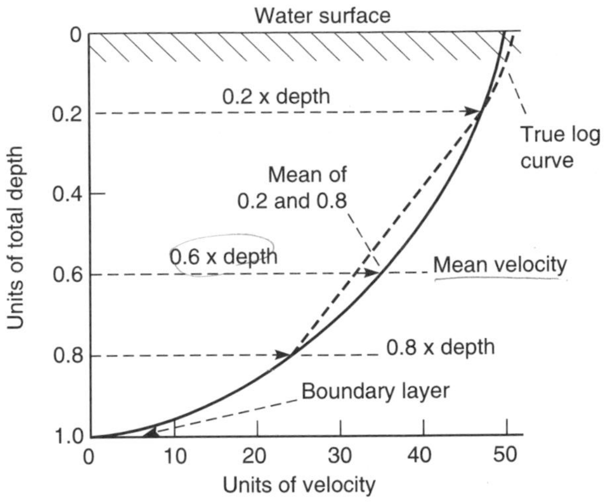

FIGURE 1.4 The velocity of water flow as a function of depth in an open channel. Mean velocity is obtained at a depth of 0.6 from surface to bottom. (Redrawn from Hynes, 1970.)

1.1.2 Discharge

The speed of water in a channel, also referred to as current velocity (U, usually expressed in $m\,s^{-1}$), varies considerably within a stream's cross-section owing to friction with the bottom and sides, and to sinuosity and obstructions. This makes it very difficult to ascertain what current velocity an organism experiences, especially one on the streambed, and we will return to this issue in chapter 3. Highest velocities are found where friction is least, generally at or near the surface and near the center of the channel. In shallow streams U is greatest at the surface due to friction with the bed; in deeper rivers U is greatest just below the surface due to friction with the atmosphere (Gordon, McMahon and Finlayson, 1992). Velocity then decreases as a function of the logarithm of depth (Figure 1.4), approaching zero at the substrate surface. In streams with logarithmic velocity profiles, one can obtain an average value fairly easily by measuring current speed at 0.6 of the depth from the surface to the bottom. In the absence of an appropriate current meter, floats give a rough measure of surface velocity, which multiplied by 0.8 provides an estimate of mean velocity.

The cross-sectional area of a stream is calculated from width (W) multiplied by mean depth (D) and is measured in m^2. Thus the total volume of water moving past a point, or discharge (Q), is:

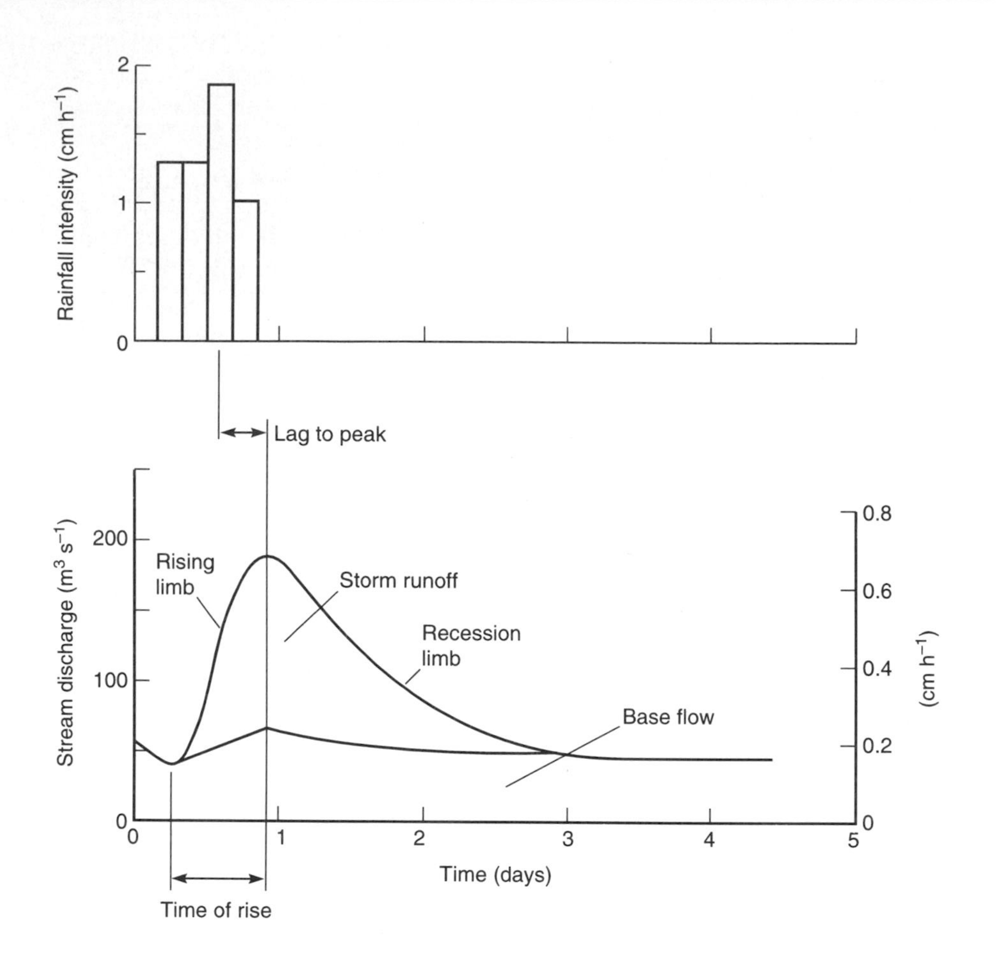

FIGURE 1.5 Streamflow hydrograph resulting from a rainstorm. (Redrawn from Dunne and Leopold, 1978.)

$$Q = WDU \qquad (1.1)$$

measured in m^3 s^{-1}. In practice, discharge is obtained by dividing the stream cross-section into segments, measuring velocity and area for each, and summing the discharge estimates for the segments. An alternative approach for small streams requires the installation of notched weirs, and estimates then are made based on a known relationship between discharge and height of water at the notch (Gordon, McMahon and Finlayson, 1992). In larger rivers, well-like structures on the streambank measure river height, or stage, and discharge is estimated from equation 1.1 over a range of flow conditions. A stage–discharge rating curve is then developed for that location, and discharge is estimated thereafter by monitoring stage.

(a) The hydrograph

A continuous record of discharge plotted against time is called a hydrograph. It can depict in detail the passage of a flood event over several days (Figure 1.5), or the discharge pattern over a year or more.

A hydrograph has a number of characteristics

that reflect the pathways and rapidity with which precipitation inputs reach the stream or river (Dunne and Leopold, 1978). Baseflow represents groundwater input to river flow. Rainstorms result in increases above baseflow, called stormflow or quickflow. There is a rising limb, the timing and shape of which is of interest. The peak of the hydrograph has a magnitude determined by the severity of the storm event and the relative importance of various pathways by which rain-water enters the stream (Figure 1.2). The lag to peak measures the time between the center of mass of rainfall and the peak of the hydrograph. The recession limb describes the return to baseflow conditions.

Some consideration of the various pathways by which rain-water enters stream channels will indicate how hydrographs can vary, and also why river outflow does not equal precipitation input, as Perrault first calculated for the Seine. It has been estimated that, world-wide, precipitation averages about $100\,cm\,yr^{-1}$, most of which returns to the atmosphere as evaporation and transpiration, leaving about 20% to become runoff (either immediate or via groundwater). For the United States, comparable figures are 75 cm and about 30% (Leopold, 1962).

Hydrographs exhibit wide variation over time, at scales of hours, days, months and years; and over space, from small stream to large river; and among geographic regions. Much of this variation makes sense based on amount and distribution of precipitation throughout the year, its storage as snow, the size and topography of the basin, and soil and vegetation characteristics. Substantial overland flow causes a rapid and pronounced rising limb to the hydrograph and can result in significant sediment erosion from the land surface (discussed below). Such streams are called 'flashy'. In contrast, little or no overland flow occurred in a humid New Hampshire forest even after deforestation (Likens, 1984), demonstrating that soil permeability, enhanced by litter, other organic matter, and root structure, provides an infiltration capacity that is rarely exceeded. Because flow is slower in sub-surface pathways, this should result in a less pronounced rising limb of the hydrograph. Furthermore, the likelihood of sediment transport from the landscape is reduced, while the transport of dissolved materials is enhanced. Urbanization and agricultural land use lead to increases in the amount of overland flow in human-altered catchments.

Another general pattern is for flood hydrographs to become broader and less sharp as a river gathers tributaries in a downstream direction. This is due to differences among tributary sub-basins in the amount and intensity of precipitation received, causing the sum of tributary inputs to a larger river to be less well defined than the individual events. In addition, a flood peak attenuates as it travels downstream. As a consequence, a flood peak can pass in minutes in a small stream and persist for days in a large river.

(b) Flood frequency computations

Hydrograph data can be analyzed to determine baseflow, total flow, highs and lows, and the probability of occurrence of a given discharge event. These have considerable practical application in engineering, flood control, agriculture and perhaps even in biology, though more work needs to be done to investigate the last assertion. The likelihood of floods has received the most attention and can be analyzed in several ways. Typically one estimates the probability of a '1-in-N-year' event of a flood of a given size or larger. Thus a 1-in-100-year flood has a probability of 0.01 (1%) of occurring in any year and the average recurrence interval is 100 years between two floods of that magnitude or larger (Gordon, McMahon and Finlayson, 1992). Flood probability (P) and average recurrence (T) are reciprocals:

$$P = \frac{1}{T} \quad (1.2)$$

Given a record of annual maximum flows or other measures of flood events, a number of methods can be used to estimate P and T (Dunne and Leopold, 1978; Interagency Advisory Committee on Water Data, 1982; Gordon, McMahon and Finlayson, 1992). Flood-frequency analysis is generally performed on a data series consisting of the single highest flow of each year, and can be graphed on special probability paper or fitted to one of several probability distributions. The Weibull method (equation 1.3) provides a useful illustration of this approach. Other techniques include the partial-duration series, which uses all events greater than some arbitrary value (usually the smallest observed annual maximum) and the application of regional data to an ungauged site.

One begins with a list of annual maximum floods, preferably based on the peak of the flood hydrograph rather than average daily discharge. This is especially important in small rivers where the peak flow passes in hours and will be under-estimated by the daily average, although this may not be critical for large rivers. By fitting a probability distribution to the data set, it is possible to predict the average recurrence interval for floods of a given magnitude or, conversely, the magnitude of the flood that occurs with a given frequency. The recurrence interval (T) for an individual flood is calculated as:

$$T = \frac{n + 1}{m} \quad (1.3)$$

where n is the years of record and m is the rank magnitude of that flood. The largest event is scored as $m = 1$. Figure 1.6 illustrates flood-frequency curves for Sycamore Creek, Arizona, which experiences irregular flash floods, and the upper reaches of the Colorado, with a highly repeatable snowmelt-driven flow regime. The recurrence interval for floods of a given magnitude is read directly from the graph. One can also determine exceedance probability ($P = 1/T$), which is the likelihood that the annual maximum flood for a given year will equal or exceed the value of a 10 year, 20 year or 50 year flood event.

Estimating the likelihood of rare events is obviously risky, and becomes more so when only a short hydrologic record is available for analysis. Dunne and Leopold (1978) describe methods for placing 90% confidence bands around the flood-frequency curve, and Costa (1978) discusses the use of geological and radiocarbon methods for estimating the recurrence time of outstanding large floods. In addition to the possibility that any given string of years can include an individual flood the recurrence interval of which is much longer than the record, changes in land use or climate can result in a heterogeneous data set. For instance, the flood-frequency curves for a river before and after construction of a major dam, or large-scale deforestation, likely will be very different. Note also that a 50 year flood is not one that occurs like clockwork, every 50 years. It is a flood that has a 2% chance of occurring in any given year and, if one occurred last year, there is still a 2% chance of another such flood occurring this year.

1.1.3 The transport of material

Water flowing in a channel may exhibit laminar flow, in which parallel layers of water shear over one another vertically, or turbulent flow with the complex mixing that implies. Laminar flow is uncommon, except at channel boundaries where flow is very low and in groundwater (flow characteristics are discussed in greater detail in chapter 3). Thus the flow of water in rivers generally is turbulent, and exerts a shearing force that causes particles to move along the bed by pushing, rolling and skipping, referred to as the bedload. This same shear causes turbulent eddies that entrain particles

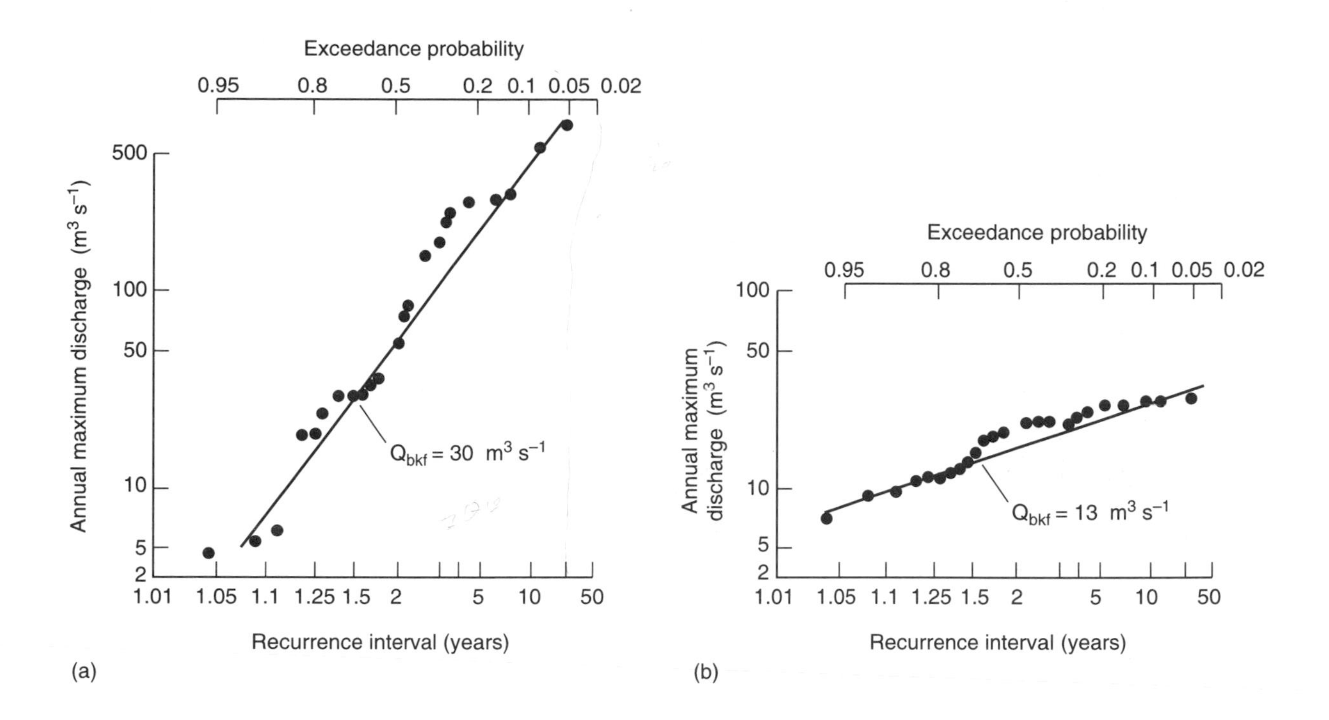

FIGURE 1.6 Example of a flood-frequency analysis for two rivers, based on annual peak events from a gauge record of over 20 years. The bankfull flood (Q_{bkf}) is estimated using $T = 1.5$ years, and the probability or recurrence interval for more extreme events (e.g. 20- and 50-year floods) can be read from the graph. Lines are fitted by eye. (a) Sycamore Creek, Arizona, is an aridland stream subject to flash floods. (b) The Colorado River in its upper reaches, near Grand Lake, Colorado, has a highly regular snowmelt-driven hydrograph. Note the steeper slope of the graph for Sycamore Creek.

into suspension, called the suspended load. The suspended load also includes fine sediments, primarily clays, silts and fine sands, that require only low velocities and minor turbulence to remain in suspension. These are referred to as washload, because this load is 'washed' into the stream from banks and upland areas (Gordon, McMahon and Finlayson, 1992). Thus the suspended load includes the washload plus coarser materials that may be part of the bedload at lower flows. The suspended load and bedload together constitute the solid load. In practice, the bedload is extremely difficult to measure and is generally less than 5–10% of the total load. The bedload will be a lower fraction in bedrock streams than in alluvial streams where channels are composed of easily transported material.

All sediments ultimately derive from erosion of basin slopes, but the immediate supply usually derives from the river channel and banks, while the bedload comes from the streambed itself and is replaced by erosion of bank regions (Richards, 1982). This is illustrated by the experimental deforestation of a New Hampshire stream. Annual export of eroded material increased some six- to tenfold (Bormann *et al.*, 1974). Because overland flow was never observed, most of this transported material must have come from the stream channel itself, probably mainly from debris dams (Bilby, 1981; Likens, 1984).

In addition, a substantial amount of material is transported as the dissolved load. Solutes generally are derived from chemical weathering of bedrock and soils, and their contribution is greatest in non-flashy hydrologic regimes where most flow is sub-surface, and in regions of limestone geology (Richards, 1982; Likens, 1984). The dissolved constituents of river water are discussed more fully in chapter 2.

The relative amount of material transported as solute rather than solid load depends on basin characteristics, lithology and hydrologic pathways. In dry regions, sediments make up as much as 90% of the total load, whereas the contribution of solutes approaches or exceeds sediment load in areas of very high runoff (Richards, 1982). World-wide, it is estimated that rivers carry approximately 15–20 billion tonnes of suspended materials annually to the oceans. This is roughly five times the dissolved load (Holeman, 1968; Martin and Meybeck, 1979).

Sediment flux from rivers is estimated from fluvial sediment loads and water volume, though it is likely that storage of sediments in floodplains, lakes, and the lower sections of rivers substantially reduces the actual delivery to the ocean. The world average sediment yield (sediment discharge per unit area of drainage basin) is estimated to be about 150 tonnes yr^{-1}, with great variation from place to place (Milliman, 1990). Discharge volume alone is a poor predictor of sediment load except within a region. Rivers in just 10% of the world's drainage basins account for more than 60% of sediment discharge. The greatest sediment export occurs in southern Asia and Oceania, where the amount of rainfall, steepness of slopes, lithology and intensity of land cultivation result in very high yields. The Hwang Ho (Yellow River) of northern China is believed to carry the highest suspended load of any river, as much as 40% sand, silt and clay by weight, during high discharge (Cressey, 1963). The great rivers of South America make a significant but nonetheless much smaller contribution to the world sediment flux, and large northern rivers account for considerably less.

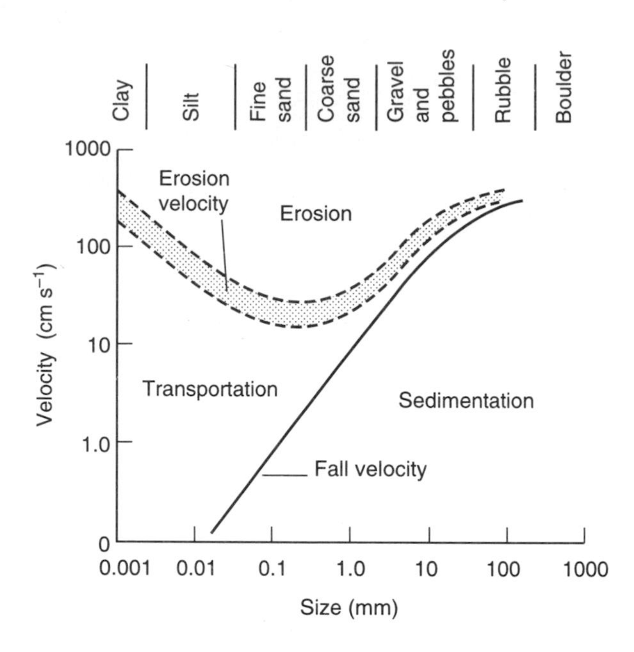

FIGURE 1.7 Relation of mean current velocity in water at least 1 m deep to the size of mineral grains that can be eroded from a bed of material of similar size. Below the velocity sufficient for erosion of grains of a given size (shown as a band), grains can continue to be transported. Deposition occurs at lower velocities than required for erosion of a particle of a given size. (Redrawn from Morisawa, 1968.)

Not surprisingly, the size of particle that can be eroded and transported is a function of current velocity (Figure 1.7). The competence of a stream refers to the largest particle that can be moved as bedload, and the critical erosion (competent) velocity is the lowest velocity at which a particle resting on the streambed will move (Morisawa, 1968). Sand particles are the most easily eroded, having a critical erosion

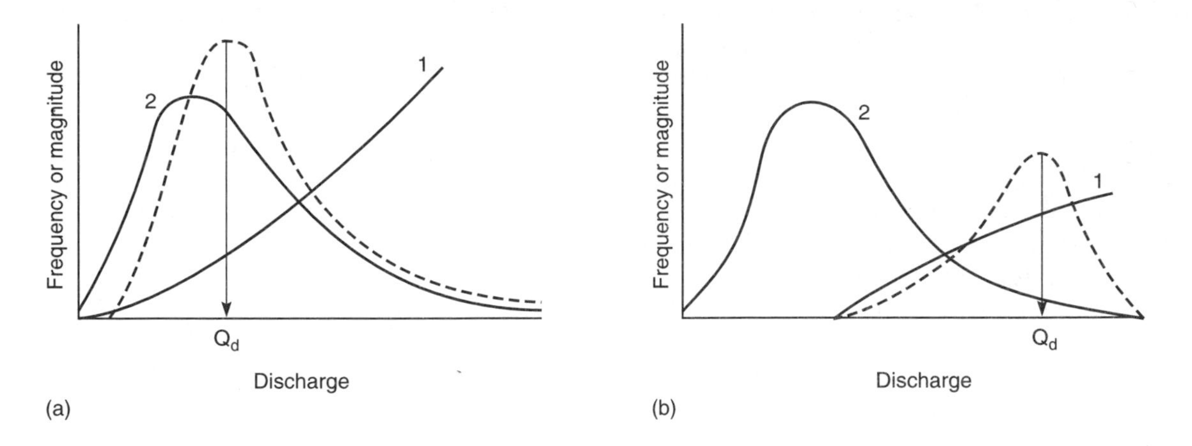

FIGURE 1.8 The relationship between frequency and magnitude of discharge events causing sediment transport. (a) Suspended load, (b) bedload. Curve 1 depicts the increase of sediment load with increasing magnitude of discharge, and curve 2 describes the frequency of discharge events of a given magnitude. Their product (dashed curve) is the total sediment transported. Q_d, the dominant discharge, is approximately Q_{bkf} for suspended sediments, and is in the range $Q_{1.5}$–Q_{10} for bedload. (Redrawn from Richards, 1982.)

velocity of about 20 cm s^{-1}. The greater mass of larger particles requires higher current velocities to initiate movement, for example 1 m s^{-1} or more for coarse gravel. However, grains smaller than sands, including silts and clays, have greater critical erosion velocities because of their cohesiveness. Once in transport, particles will continue in motion at somewhat slower velocities than were necessary to initiate movement, and will settle at still lower velocities (Figure 1.7).

The concentration and amount of transported solids increases with discharge unless the supply of sediment becomes depleted (Richards, 1982). Throughout most of the year, discharge usually is too low to scour, shape channels, or move significant quantities of sediment, though sand-bed streams can experience change much more readily. Extreme events produce the greatest scour, and the amount of material removed is impressive. The depth of scour can be as great as half of the increase in water level. A flood in the Colorado River at Lees Ferry increased bed depth by approximately 1.5 m. Redeposition of sediment as the flood receded re-established bed elevation at very close to its previous value (Leopold, 1962), further evidence of the dynamic equilibrium between erosion and deposition.

While one might suppose that extreme events also account for the greatest proportion of total sediment transport, Wolman and Miller (1960) argue that channel formation and maintenance are due primarily to events of intermediate frequency and strength (Figure 1.8). The usual flow of water lacks the competence to fashion changes, while the extreme event is too infrequent. Thus events of intermediate magnitude (near the bankfull flood) are particularly important in bedload movement and bedform construction, while more extreme events define channel capacity. Together these create channel form. 'Flashy' streams tend to be influenced by more frequent events, while rarer events control channel form in baseflow-dominated streams (Richards, 1982). Bankside vegetation and sediment composition both influence the stability of channel banks.

Human activities can increase or reduce sediment flux (chapter 14). Deforestation and poor agricultural practices greatly increase erosion, perhaps as much as 5-fold in Asia and Oceania. On the other hand, sediment flux is greatly reduced in rivers that are dammed and diverted. Both the Nile and the Colorado have experienced a complete cessation of sediment export, and the Rhône is estimated to export about 5% of its load of a century ago. Thus in a number of large rivers we have the apparent paradox of increased erosion within the drainage basin coupled with reduced export to the ocean, a combination that bodes ill for the long-term effectiveness of dams. It is noteworthy that most of the world's major rivers are in developing countries, and continuing pressure for economic development seems likely to bring increased demand for river regulation along with increased erosion.

1.2 Characteristics of river channels

Before the close of the eighteenth century, the erosive capabilities of running water were not appreciated. Streams were believed to flow in valleys because the valleys were already there, not because the stream cut the valley. Catastrophism, with its emphasis on the biblical flood as the final stage in shaping of the earth's surface, obviously influenced this perspective (Morisawa, 1968). By the late 1700s, however, geologists reasoned that the dendritic pattern of drainage nets gave evidence of erosion, as did the observation that valleys in headwaters are smaller than valleys downriver.

A recurring theme in fluvial geomorphology today is that rivers determine their own channels, and these channels exhibit geometric relationships attesting to the operation of physical laws that are yet to be fully understood (Leopold, Wolman and Miller, 1964; Richards, 1982). The development of stream channels and entire drainage networks, and the existence of various regular patterns in the shape of channels, indicate that rivers are in dynamic equilibrium between erosion and deposition, and governed by common hydraulic processes. However, because channel geometry is three dimensional with a long profile, a cross-section, and a plan view (what one would see from the air), and because these mutually adjust over a time scale of years to centuries, cause and effect are difficult to establish. Leopold and Maddock (1953) argue that discharge, sediment load and ultimate base level (approximately sea level, which together with the altitude of precipitation determines the altitudinal extent of the stream) are variables of hydraulic geometry that the river cannot control, and therefore must adjust to. For a particular reach, slope exerts significant control over channel characteristics because it adjusts more slowly than other hydraulic variables. Channel width and depth, velocity, grain size of sediment load, bed roughness, and the degree of sinuosity and braiding are other variables presumed to interact as the river achieves its graded state.

1.2.1 The importance of scale

The physical characteristics of river systems can be investigated at spatial scales ranging from the individual particle to the entire drainage basin and over an equally broad temporal scale (Frissell *et al.*, 1986, Figure 1.9). Richards (1982) distinguishes present time (1–10 yr), modern time (10–100 + yr) and geologic time (1000 + yr), and notes that the interdependence of river channel variables changes across time scales. In the short term, channel form is unvarying. Daily fluctuations in discharge result simply in changes in width, depth, velocity and sediment transport. Over the intermediate, or modern time frame, channel form approaches its state of dynamic equilibrium between sediment and water transport on the one hand, and the longer-term inherited conditions of gradient, sedimentology, and paleochannel features on the other.

This spatial–temporal framework likely has meaning for the biota as well. The permanence

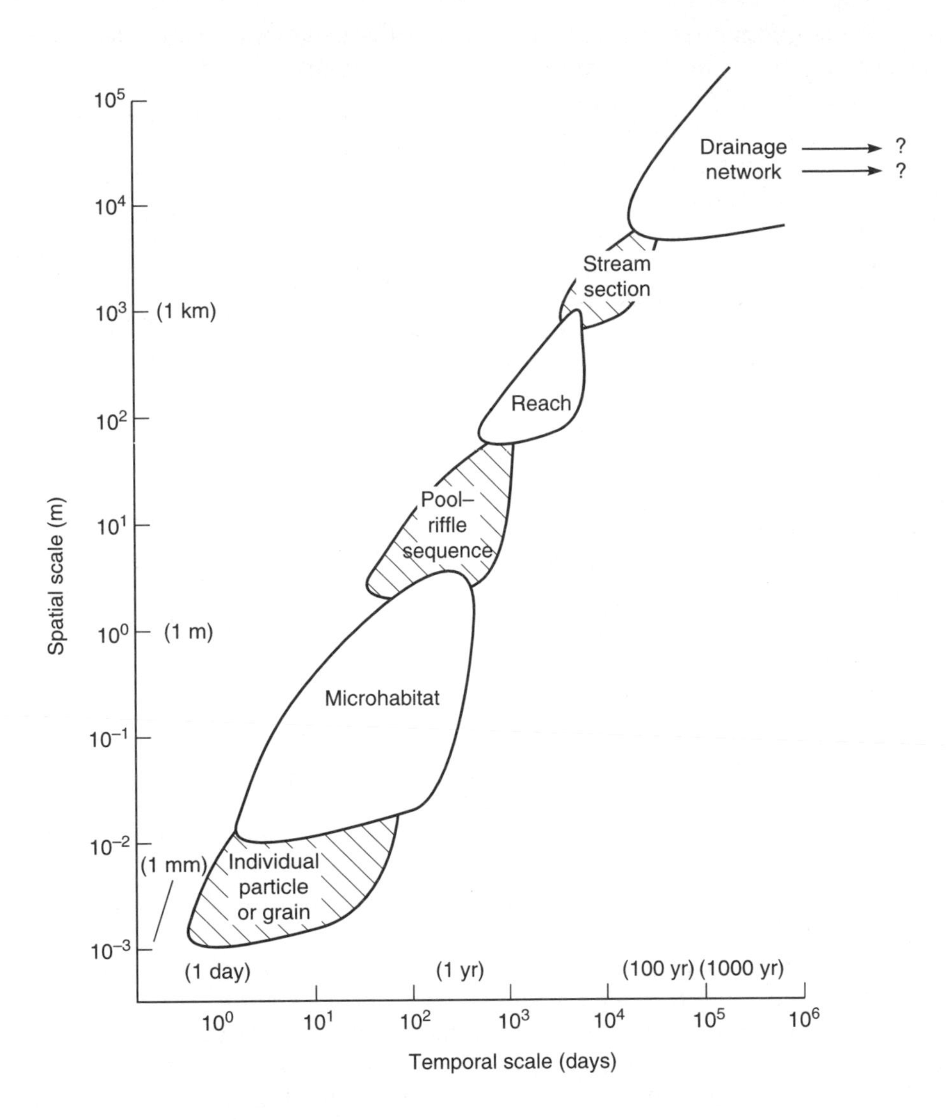

FIGURE 1.9 An approximate spatial and temporal scale over which physical change takes place in rivers. (From Frissell *et al.*, 1986.)

of structure at finer spatial scale is fairly brief: small particles on the riverbed are moved by scour many times per year. Microhabitat structure, on the scale of centimeters or less for a sessile insect larva and perhaps several meters for a fish, can persist for weeks to years. Longer persistence might be associated with the upper limit of the microhabitat designation. Pool–riffle sequences, meters to hundreds of meters in length, persist in their present location for perhaps tens to hundreds of years, but the existence of the floodplain, side channels and meanders attests to lateral wanderings of the channel itself. As spatial scale is expanded to encompass the broader units of the reach and stream section, it is likely that these units have existed in something like their present state for hundreds to thousands of years. (Bear in mind, however, that human activities have altered channel form for nearly as long, a topic we will return to in chapter 14.) An entire drainage basin, some tens to thousands or more of square kilometers, has a history intertwined with the geological history of the region.

While most north temperate lakes date from the glacial retreat of 10 000–12 000 years ago, and some lakes of tectonic origin date from the Tertiary (Hutchinson, 1957), the age of rivers is difficult to specify. Existing rivers derive from earlier rivers, which in turn derive from still earlier progenitors, analogous to the biological evolution of a species (Leopold, 1962). This sequence extends backwards through glaciation into the geological history of the landscape. Changes in elevation of the continental mass, along with changes in climate, runoff and vegetation, have influenced the river basin over its history. The oldest large rivers have been carving their channels over millions of years, and some rivers have been able to maintain their valleys across rising tectonic ridges. The New River of North Carolina is such an antecedent river, cutting its channel though the rising Appalachian Mountains. Other watersheds are known to have been grandly influenced by recent geological events. The greatest floods known in earth history occurred when large lakes which had formed behind glacial dams broke through in the late Pleistocene to carve major portions of the upper basins of the Columbia and the Snake Rivers in the American west (Malde, 1968). Indeed, most major stream valleys in the glaciated midcontinent of North America originated from or were significantly altered by torrential discharges from glacial-lake outbursts during the retreat of the Laurentide Ice Sheet (Kehew and Lord, 1987). In brief, unlike the age of lakes which can be known reasonably accurately and refers to the existence of that water body in the same location (allowing for expansion and contraction), the age of rivers is a messier business.

1.2.2 Discharge relationships and hydraulic geometry

Rivers generally increase in size as one proceeds downstream, because tributaries and groundwater add to the flow. Since discharge $Q = WDU$, any increase in discharge must result in an increase in width, depth, velocity or some combination of these. Similarly, fluctuations in the hydrograph at a single station can come about due to changes in any of W, D or U. Hydraulic geometry, as defined by Leopold and Maddock (1953), describes the relationship among hydraulic characteristics, chiefly width, depth, velocity and discharge. Power equations of the form:

$$W = aQ^b \quad (1.4)$$

$$D = cQ^f \quad (1.5)$$

$$U = kQ^m \quad (1.6)$$

provide good fits to empirical data. Because $Q = WDU$, the constants a+c+k, and b+f+m, must sum to unity. Early work suggested that a fixed set of coefficients related changes in width, depth and velocity to discharge 'at a station', while another set of coefficients held for the 'downstream' case along a river's length. This turns out to be an over-generalization because coefficients vary with the nature of the material forming the channel perimeter, as Osterkamp, Lane and Foster (1983) show using channel data from the western United States. Bearing in mind that statements about the interrelationships among Q, W, D and U describe general trends rather than invariant relationships, some broad patterns are discernible.

At locations above the floodplain, fluctuations in discharge at a station usually result in little increase in width, while depth changes more, and velocity most of all. At flows greater than bankflow, width can increase greatly with discharge, especially in floodplains.

As one proceeds downstream and a river enlarges, width, depth and velocity all increase log-linearly with mean annual discharge. The increase of width with discharge is greater than the increase of depth, while velocity increases least with discharge and can remain almost constant (Leopold, 1962). This is in contrast to the commonly held view that velocity should decrease downstream, which we might expect if velocity depended solely on gradient.

In fact, water velocity does vary with gradient, but also with depth and inversely with the roughness of the bottom. The Manning equation for velocity of flow in a channel (in metric units) is:

$$U = \frac{1}{n} R^{2/3} S^{1/2} \quad (1.7)$$

where R is the hydraulic radius, about equal to mean depth for most channels, S is the energy gradient, approximately the slope of the water surface, and n is the Manning resistance coefficient. Published photographs of rivers, with measured values of n, provide a guide to the application of this formula (Barns, 1967).

This equation helps to explain why velocity usually increases along a stream's length, even as gradient decreases. Channel depth generally is greater and substrates are finer as one proceeds downstream, hence resistance decreases longitudinally and this offsets the effects of reduction in slope. The River Tweed in Scotland illustrates this nicely (Ledger, 1981). At most flow levels, the highest velocities are found at the lower and flatter end of the river system. Only in some situations involving floods does mean velocity not exhibit an increase in the downstream direction.

(a) River profiles

Most streams and rivers exhibit a downstream decrease in gradient along their length. Slopes are steep in the headwaters, and become less so as one proceeds downstream, resulting in a concave longitudinal profile. Bearing in mind that diverse geography provides for almost unlimited variation, a lengthy river that originates in a mountainous area typically comes into existence as a series of springs and rivulets; these coalesce into a fast-flowing, turbulent mountain stream, and the addition of tributaries results in a large and smoothly flowing river that winds through the lowlands to the sea.

Almost everything about a river varies with position along its length. Discharge increases, resulting in changes in width, depth and velocity as just described. As will be discussed more fully in later chapters, a host of biological variables correlate with stream size and distance downstream. In terms of physical appearance, however, the most striking changes are in steepness of slope and in the transition from a shallow stream with boulders and a stony substrate to a deep river with a preponderance of sand.

Over the length of a river, the particle size of bed material usually shifts from an abundance of coarser material upstream to mainly finer material in downstream areas. The causes for this are still uncertain. When it was believed that velocity decreased downstream, the well-known sorting effect of current speed (Figure 1.7) was thought to be responsible for changes in the size of material in the riverbed. However, the demonstration that velocity remains constant or increases as one progresses downstream forced geomorphologists to search for other explanations for this usually observed decline in average grain size along a river's length. Mechanical abrasion, sorting, weathering in place during temporary storage and underlying lithology may all contribute to the observed downstream decrease in grain size. However, a fully satisfactory answer seems to be lacking (Leopold, Wolman and Miller, 1964; Richards, 1982).

(b) Sinuosity

Flowing water will follow a sinuous course, whether it is melt water on the surface of a glacier, the Gulf Stream which is unconstrained by banks, or a river channel. The most commonly used measure is the sinuosity index (SI).

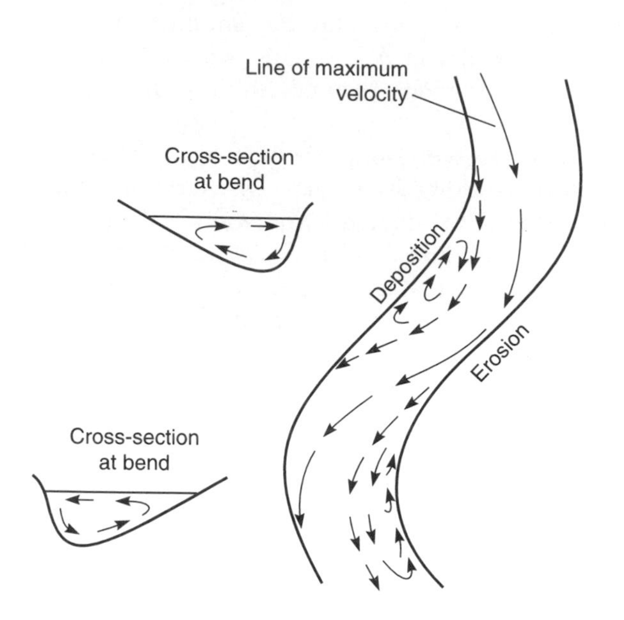

FIGURE 1.10 A meandering reach, showing the line of maximum velocity and the separation of flow that produces areas of deposition and erosion. Cross-sections show the lateral movements of water at the bends. (Redrawn from Morisawa, 1968.)

$$SI = \frac{\text{Channel (thalweg) distance}}{\text{Downvalley distance}} \quad (1.8)$$

Meandering is usually defined as an arbitrarily extreme level of sinuosity, typically an SI greater than 1.5 (Gordon, McMahon and Finlayson, 1992). If one draws the channels of rivers of very different size, scaled to fit on the same page, the similarity in sinuosity is striking. Small channels wind in small curves and large channels wind in large curves (Leopold, 1962). Many variables affect the degree of sinuosity, however, and so SI values range from near unity in simple, well-defined channels to four in highly meandering channels (Gordon, McMahon and Finlayson, 1992).

Flow through a meander stretch follows a predictable pattern and causes regular regions of erosion and deposition. The streamlines of maximum velocity and the deepest part of the channel, or thalweg, lie close to the outer side of each bend and cross over near the point of inflection between the banks (Figure 1.10). A super-elevation of water at the outside of a bend causes a helical flow of water towards the opposite bank. In addition, a separation of surface flow causes a back eddy. The result is zones of erosion and deposition, and explains why point bars develop in a downstream direction in depositional zones.

(c) The floodplain

River channels, self-formed and self-adjusting, also influence the shape of the valley floor through which they course. A generalized cross-section is shown in Figure 1.11. The flat area near to the stream is the modern floodplain, constructed by the river in the present climate and inundated at times of high discharge. Because of lateral movement of the streambanks due to erosion and deposition, the channel may have occupied positions throughout the entire valley flat in the recent past. Thus, channel movement and valley flooding are regular and natural behaviors of the river. The bankfull level of a river can be recognized by a sharp leveling of the bank, gravel deposits, or scoured vegetation; or by directly observing the flood where the river just overflows its banks. In practice, however, this is not always easily done. Hydrologists often estimate Q_{bkf} as the flood magnitude that occurs in two years out of three (the 1.5 year recurrence event in Figure 1.6).

Owing to changes in climate or basin conditions, a river can change its level upward (aggradation) or downward (degradation). In the latter circumstance, the old floodplain, abandoned as the river cuts downward, remains as a terrace (Figure 1.11).

(d) Pool–Riffle sequences

At a smaller scale than discussed previously, the reach (perhaps a few hundred meters in length)

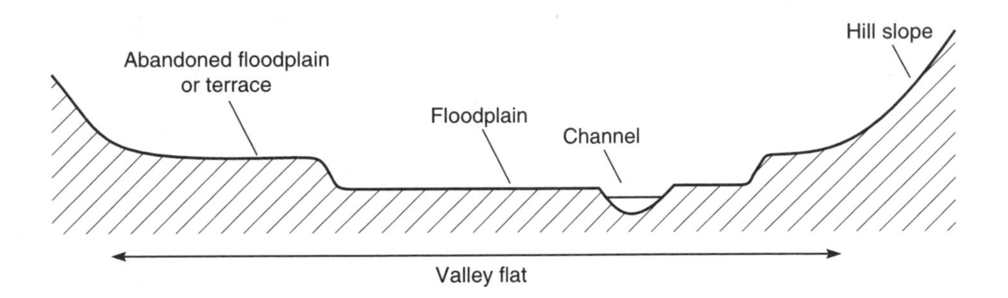

FIGURE 1.11 Diagrammatic cross-section of a valley showing present channel, the floodplain occupied in modern time and a terrace representing a previous floodplain. (Redrawn from Dunne and Leopold, 1978.)

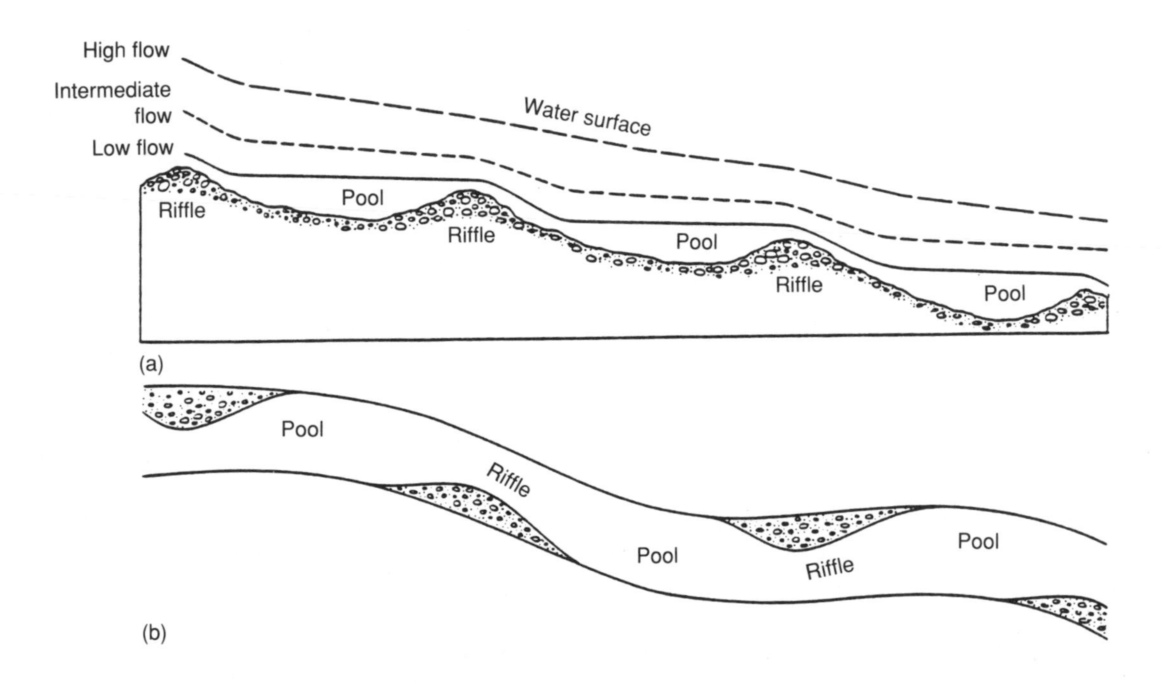

FIGURE 1.12 A longitudinal profile (a) and a plan view (b) of a riffle–pool sequence. Water surface profiles in (a) depict high-, intermediate- and low-flow conditions. (Redrawn from Dunne and Leopold, 1978.)

displays a more or less regular alternation between shallow areas of higher velocity and mixed gravel–cobble substrate, called riffles, and deeper areas of slower velocity and finer substrate, called pools (Figure 1.12). The riffle is a topographical hillock, the pool a depression. Riffles are formed by the deposition of gravel bars in a characteristic alternation from one side of the channel to the other, at a distance of approximately 5–7 channel widths in gravel-bed rivers. Pool–riffle sequences are the result of particle sorting and require a range

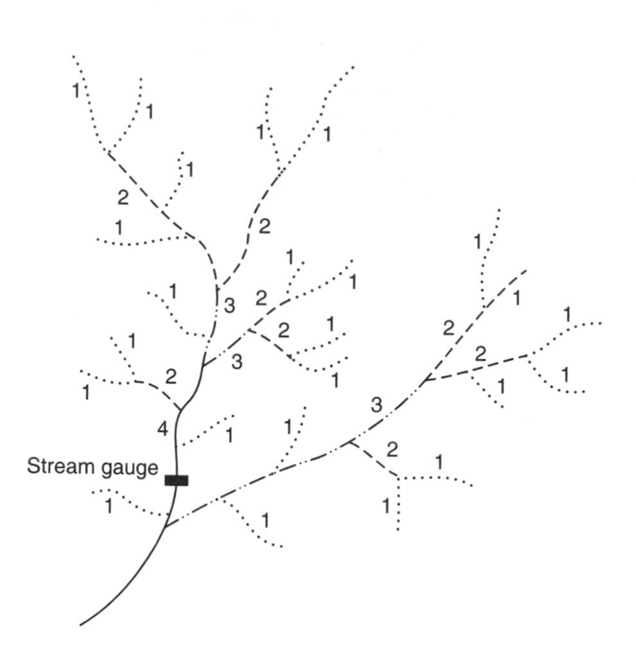

FIGURE 1.13 A drainage network illustrating stream order classification for a fourth-order watershed.

of sediment sizes to develop. As a consequence, regular riffle–pool alternation may not be apparent in sandy bottom streams.

(e) Depositional features

Rivers experience a decrease in sediment transporting ability when the gradient or discharge lessens, load increases, or channel obstructions occur. The point bars that form riffles are one example of the consequences of deposition. Braided streams have mid-channel bars, perhaps islands if the bar is large enough and vegetation becomes established. Highly variable discharge and easily eroded banks favor channel braiding, whereas lower gradients and less variable discharge favor meandering. When a river enters standing or more slowly moving water, its debris load is dropped forming a delta under water or an alluvial fan if sufficient material accumulates.

1.2.3 Watershed characteristics

The drainage basin is the total area drained by the multitude of tributaries that feed the main channel or set of channels. (Drainage basin and catchment are synonymous, though the use of catchment generally is reserved for small basins. The term 'watershed' describes the boundary separating two drainages (in British usage) and is a synonym for catchment (in American usage).) The drainage basin exhibits a branching or dendritic pattern, but detailed shape varies greatly, and the relative importance of terrain, physical laws and chance in determining network shape is not yet understood. Nevertheless, it is clear that drainage nets exhibit striking regularity in certain characteristics.

Horton (1945) developed a hierarchical classification system, subsequently modified by Strahler (1952, 1964), that remains in wide use (Figure 1.13). The smallest, permanently flowing stream is termed first order, and the union of two streams of order n creates a stream of order $n + 1$. This provides a convenient system in which rivers increase in order as they increase in size. In addition to its simplicity, stream order classification is useful as a correlate of other watershed variables. Horton proposed two laws, one describing stream order's negative relationship with the logarithm of the number of streams, and another for its positive relationship with the logarithm of stream length (Figure 1.14).

Stream order also correlates with basin area, stream gradient and basin relief. There are usually some three to four times as many streams of order $n - 1$ as of order n, each of which is roughly less than half as long, and drains somewhat more than one-fifth of the area. The river continuum concept (chapter 12) uses stream order as its physical template, and the term has found wide use among stream ecologists.

In practice, however, stream order has several drawbacks (Richards, 1982; Hughes

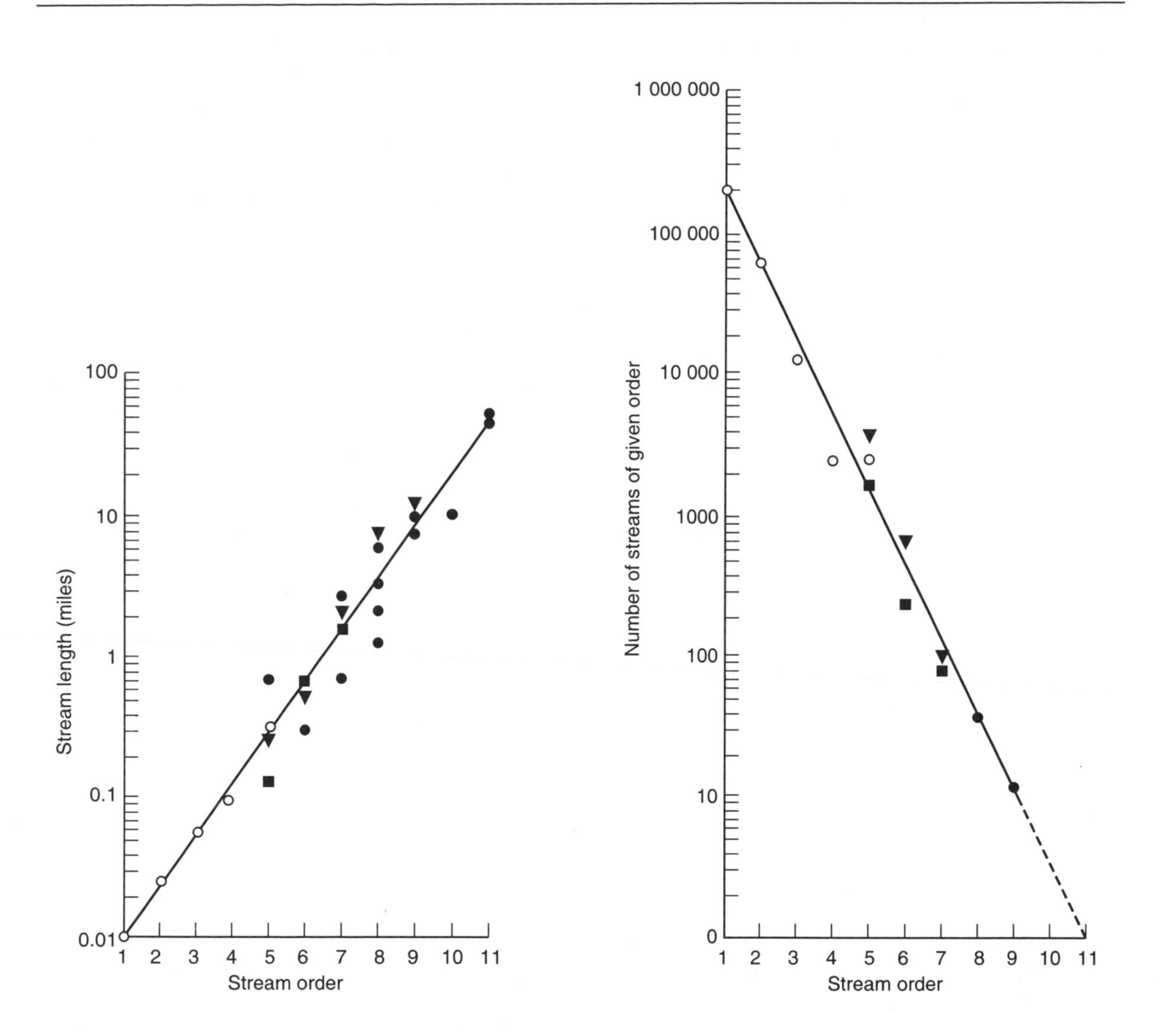

FIGURE 1.14 The relationship of stream order to stream length and to number of streams of a given order, from a network of arroyos near Sante Fe, New Mexico. ○ = Average values, Arroyo Caliente, ■ = Arroya de los Frijoles, ▼ = Arroya de las Trampas to Rio Santa Fe, ● = Rio Santa Fe, Rio Galisteo and Cañada Ancha (del oriente). (Redrawn from Leopold, Wolman and Miller, 1964.)

and Omernik, 1983). First-order streams are difficult to identify because of unevenness in the accuracy of maps, the influence of choice of map scale (1:24 000 or 1:25 000 is recommended) and variation between wet and dry years. In addition, this approach ignores the entry of streams of order n into order $n + 1$. Link classification (Strahler, 1952; Shreve, 1966, 1967) incorporates the addition of first-order streams into higher-order branches of a drainage net, but shares the other failings. As an alternative, Hughes and Omernik (1983) advocate use of drainage area and discharge to estimate mean annual runoff. They argue that drainage area better correlates with mean annual discharge than does stream order, and their approach allows further analysis of hydrologic variables. It requires less accurate topographic maps than does order analysis, and the necessary data usually can be found in local or national governmental agencies' files.

Further evidence that running waters conform to regular, physical laws is provided by a number of interrelations of river dimensions. River length (L) increases with drainage area (A) according to:

$$L = 1.4\,A^{0.6} \tag{1.9}$$

Bankfull discharge also tends to increase log-linearly with drainage area, but the coefficients of the equations vary according to a number of basin characteristics.

1.3 Summary

Rivers contain a tiny fraction of the world's freshwaters. Yet they are a vital component of the hydrologic cycle, annually transporting 32–37 $km^3\,yr^{-1}$ of water to the world's oceans. Most of the precipitation that falls within a river basin returns directly to the atmosphere by evaporation and plant transpiration. Some water may reach stream channels by overland flow, but this is usually a small fraction, except in very intense storms or in areas of little vegetation cover and compacted soils. In mesic vegetated areas, virtually all water enters stream channels via sub-surface pathways. Erosion of the land and transport of sediments are greater with overland flow, whereas more dissolved materials are transported by sub-surface flows. Baseflow or dry weather flow in a river is due to groundwater entering the stream channel.

The hydrograph, which is a record of discharge plotted against time, depicts the daily, seasonal and annual variation in total flow. Flow patterns can be analyzed to determine the size of the bankfull flood, the 50 year flood, and other measures of magnitude and frequency of extremes of high and low discharge.

Rivers transport an enormous amount of sediment to the oceans, some 15–20 billion tonnes annually. The rivers of southern Asia and Oceania carry the greatest sediment loads. World-wide, differences in lithology, land use, precipitation and steepness of slope are more important than are differences in discharge in determining regional patterns of sediment export. Most particulate material is carried in suspension, and the minimum particle size that can be eroded and transported increases with current velocity. A lesser amount of material is transported by skipping and rolling along the bottom, termed the bedload.

River channels and the catchment basin exhibit a number of regular relationships with regard to discharge, longitudinal changes, channel morphology and the size of tributaries in the drainage network. Discharge fluctuates at a location because of daily, seasonal and annual variation in rainfall. Discharge increases as one proceeds downstream due to tributary inputs and the addition of groundwater. At locations above the floodplain, fluctuations in discharge at a location usually result in a substantial increase in velocity, some increase in depth and little increase in width. However, width can increase greatly in response to floods above bankfull, particularly within the floodplain. As one proceeds downstream and a river enlarges, width, depth and velocity all increase log-linearly with mean annual discharge. The increase of width with Q is greater than the increase of depth, while velocity increases least with discharge and may remain almost constant.

Other changes can be seen in a river's appearance as one proceeds from source to mouth. Slopes are steep in the headwaters and become less so as one proceeds downstream. Coarser particles including gravel and boulders are typical of upland streams, while a finer and softer substrate often is found in large lowland rivers. Particle sorting by the current produces riffles at regular intervals, approximately five to seven times the channel width, but only where a range of sediment sizes is available. Rivers meander, especially when the gradient is low and topographic features do not constrain this natural tendency. The river channel itself changes location over time. In downstream locations of low relief, rivers tend to flow through wide, flat valleys. This is the flood-plain, inundated at times of high discharge.

Some of the most regular features of rivers are apparent at the scale of the entire basin and drainage net. As a rule there are three to four times as many streams of order $n - 1$ as of order n, each of which is less than half as long, and drains somewhat more than one-fifth of the area. Both river length and bankfull discharge increase log-linearly with basin area. Although the theoretical basis for many of these relationships remains to be elucidated, their existence is clear evidence of processes that produce regular patterns in the physical characteristics of river systems.

Chapter two

Streamwater chemistry

We all have an intuitive appreciation that river water contains a variety of dissolved and suspended constituents. Mountain streams appear pure, farm creeks are often muddy with sediments, and drainages in limestone-rich regions are fertile while those containing only granitic rocks are usually less so. The rivers of heavily populated areas, and even the rain, are polluted by human activities.

Many factors influence the composition of river water, causing variation from place to place. Rain is one source of chemical inputs to rivers, and a stream flowing through a region of relatively insoluble rocks can be chemically very similar to rain-water in its composition. Most streams and rivers contain much more suspended and dissolved material than is found in rain-water, however. Ultimately, all of the constituents of river water originate from dissolution of the earth's rocks. The dissolving of rocks is commonly the major determinant of river water chemistry locally as well, but this varies with geology, and with the magnitude of inputs via other pathways including rain-water, volcanic activity and pollution. Materials are concentrated by evaporation and altered by chemical and biological interactions within the stream. Unlike sea water, which is quite constant everywhere and can be approximated with an artificial standard, river water varies considerably in its chemical composition (Livingstone, 1963).

The materials transported in river water can be subdivided according to whether they are dissolved or suspended, organic or inorganic and by chemical description. Following Berner and Berner (1987), a useful breakdown includes:

- water
- suspended inorganic matter (including such major elements as Al, Fe, Si, Ca, K, Mg, Na and P)
- dissolved major ions (Ca^{2+}, Na^+, Mg^{2+}, K^+, HCO_3^-, SO_4^{2-}, Cl^-)
- dissolved nutrients (N, P and to some extent Si)
- suspended and dissolved organic matter
- gases (N_2, CO_2, O_2)
- trace metals, both dissolved and suspended.

In this chapter we shall concentrate on the major dissolved constituents and the gases. The remaining components are discussed in other chapters. We briefly considered sediment transport in chapter 1. Organic matter and dissolved nutrients are strongly intertwined with biological processes and are the subject of two chapters (12 and 13) delving into ecosystem dynamics. Readers wishing for more detailed discussions of aquatic chemistry and geochemistry should consult Berner and Berner (1987), Stumm and Morgan (1981) or other specialized volumes.

TABLE 2.1 Concentration of dissolved oxygen and carbon dioxide in saturated pure water for atmospheric partial pressure at sea level

Temperature (°C)	O_2 *(mg l^{-1})*	CO_2 *(mg l^{-1})*
0	14.2	1.1
15	9.8	0.6
30	7.5	0.4

2.1 Dissolved gases

Oxygen, carbon dioxide and nitrogen occur as dissolved gases in river water in significant amounts. Biologically, N_2 gas is the least important of the three. Although nitrogen gas takes part in nitrogen cycling within stream ecosystems, the concentration of dissolved N_2 is of little biological importance. Transformations affecting nitrogen in its various forms are discussed in detail in chapter 13 from the perspective of nutrient dynamics. In this section we shall consider only O_2 and CO_2.

Both oxygen and carbon dioxide gas occur in the atmosphere and dissolve into water according to partial pressure and temperature (Table 2.1). Air is nearly 21% oxygen by volume and just 0.03% carbon dioxide, but the latter is more soluble in water. Hence, although saturated freshwater has higher concentrations of oxygen than carbon dioxide, the difference is not nearly as great as is found in air.

Groundwater can be very low in dissolved oxygen and enriched in carbon dioxide due to microbial processing of organic matter as water passes through soil. Localities that receive substantial groundwater inputs and have had little opportunity for equilibration with the atmosphere may reflect this.

In small, turbulent streams that have received only limited pollution, diffusion maintains oxygen and carbon dioxide near saturation. Should biological or chemical processes create a demand for or an excess of either within the water column, exchange with the atmosphere maintains concentrations very near to equilibrium. Concentrations will change seasonally and daily in response to shifts in temperature, but saturation will remain near 100%. However, diffusion plays a reduced role in large rivers, because of the smaller surface area relative to volume; and in more smoothly flowing rivers, because of less turbulence. In these circumstances high naturally occurring biological activity can alter the concentrations of oxygen and carbon dioxide, organic pollution can greatly increase respiratory demand for oxygen, and acid precipitation can alter the carbonate buffer system, which influences the concentration of free carbon dioxide in solution.

Photosynthesis and respiration are the two important biological processes that alter the concentration of oxygen and carbon dioxide. In highly productive waters, such as slow-moving rivers with abundant macrophytes, oxygen is elevated and carbon dioxide is reduced during the daytime, while the reverse occurs at night. Such changes are evidence of strong biological control over the concentration of these dissolved gases, because in the absence of this biological activity, shifts in temperature between day and night would cause oxygen to exhibit just the opposite pattern.

Diel (24 h) changes in oxygen concentration provide a means of estimating photosynthesis and respiration of the total ecosystem (Odum, 1956). This approach is useful only if biological production is high relative to diffusion. The basic idea is that the rate of change of dissolved oxygen per unit area of surface between an upstream and downstream station equals $PS - R \pm D$, where PS is photosynthesis, R is respiration and D is diffusion. Figure 2.1(a) shows diel oxygen curves for two stations of an English chalk stream rich in macrophytes (Edwards and Owens, 1962a, b). Diffusion is estimated from temperature and oxygen concentration, and respiration is separated from photosynthesis by using night data to estimate R. In this highly productive system the daily changes in PS, R and D are pronounced (Figure 2.1(b)). Limitations of this method, especially due to the indirect estimation of D, are discussed by Owens (1965).

Carbon dioxide likewise tends to deviate from atmospheric equilibrium in highly productive lowland streams where luxuriant growths of macrophytes and microbenthic algae can result in diel shifts in dissolved CO_2 (Rebsdorf, Thyssen and Erlandsen, 1991). Because of the interdependence of CO_2 concentration and pH (discussed below), mid-day pH can increase by as much as 0.5 units. In larger rivers receiving a substantial organic load, degassing (diffusion of CO_2 out of river water) is unable to compensate for excess CO_2 generated by microbial respiration. As a consequence the partial pressure of carbon dioxide (pCO_2) in the water column can exceed the pCO_2 in the atmosphere by as much as two to five times, and occasionally by even more (Small and Sutton, 1986; Rebsdorf, Thyssen and Erlandsen, 1991). The CO_2 concentration of the Rhine is a good example. Water leaving the source, Lake Constance, in summer is lower than the atmospheric partial pressure due to the productivity of lake phytoplankton. In winter, however, water leaves the lacustrine source at about twice the atmospheric partial pressure. Because organic pollution increases as one proceeds downriver, the pCO_2 increases also. High summer temperatures permit high microbial respiration, with the result that the downstream average pCO_2 is about twenty times the atmospheric value (Kempe, Fettine and Cauwet, 1991).

The River Thames is a good example of the effect of organic pollution on dissolved oxygen concentrations (Gameson and Wheeler, 1977). Human and animal wastes have been a documented source of foulness since at least 1620, and 1858 was known as the 'Year of the Great Stink', but the volume of untreated human sewage reduced water quality to an all-time low in the mid-1950s. Parts of the Thames around London became anaerobic from microbial respiration driven by organic waste. These several cycles of pollution and recovery of the Thames are discussed more fully in chapter 14. The impact of high oxygen demand due to pollution can be exacerbated by high summer temperatures, which reduce the solubility of oxygen in water, and by ice cover, which minimizes diffusion.

2.2 Major dissolved components of river water

The total dissolved solids (TDS) content of fresh water is the sum of the concentrations of the dissolved major ions. The world average is about $100\,mg\,l^{-1}$ (Table 2.2). Both the total and the concentration of the constituents vary considerably from place to place, due to variability in natural and anthropogenic inputs. However, the vast majority of the world's rivers have TDS that contain more than 50% HCO_3^-, and 10–30% ($Cl^- + SO_4^{2-}$). This reflects the dominance of sedimentary rock weathering, and especially of carbonate minerals (Berner and Berner, 1987). Salinity is sometimes used interchangeably with total dissolved solids. Generally, salinity has the broader meaning of all anions and cations dissolved in water, and is a synonym for total dissolved salts. Total dissolved solids refers just to the major ions listed in Table 2.2, or additional ions if specified.

The ionic concentration of rain-water (Table 2.3) is much more dilute than most river water, with an average value of a few milligrams per liter (Berner and Berner, 1987). Na^+, K^+, Ca^{2+}, Mg^{2+} and Cl^- are derived primarily from particles in the air, whereas SO_4^{2-}, NH_4^+, and NO_3^- are derived mainly from atmospheric gases. Marine salts (NaCl) are especially important near the oceans, and a transition to rain dominated by $CaSO_4$ or $Ca(HCO_3)_2$ occurs as one proceeds inland. The relative importance of these various inputs can vary seasonally and over quite short distances, as Sutcliffe and

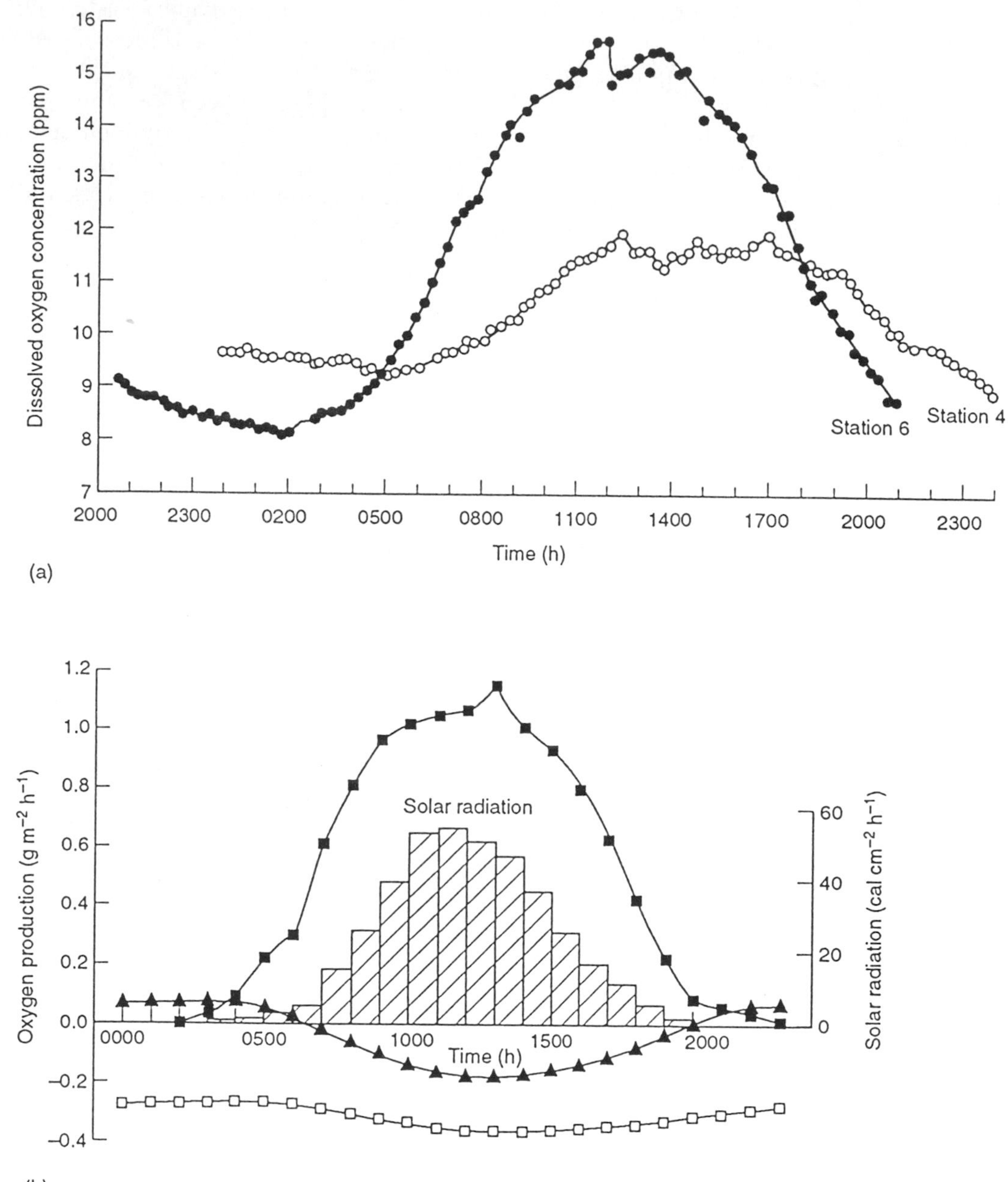

FIGURE 2.1 (a) Oxygen concentrations measured over 24 h at two sites in an English chalk stream. ● = Downstream station; ○ = upstream station. The plot of the downstream station has been displaced backwards to compensate for the differences between sites in retention time. (b) Estimated daily changes in photosynthesis (■), respiration (□) and diffusion (▲) based on the data from Figure 2.1(a). (From Edwards and Owens, 1962b.)

TABLE 2.2 Chemical composition of river water ($mg\,l^{-1}$) (From Berner and Berner, 1987)[a]

	Total dissolved solids	*Ca^{2+}*	*Mg^{2+}*	*Na^+*	*K^+*	*Cl^-*	*SO_4^{2-}*	*HCO_3^-*	*SiO_2*	*Discharge ($km^3\,yr^{-1}$)*	*Runoff ratio[b]*
World average											
Actual	110.1	14.7	3.7	7.4	1.4	8.3	11.5	53.0	10.4	37.4	0.46
Natural	99.6	13.4	3.4	5.2	1.3	5.8	8.3	52.0	10.4		
North America											
Actual	142.6	21.2	4.9	8.4	1.5	9.2	18.0	72.3	7.2	5.5	0.38
Natural	133.5	20.1	4.9	6.5	1.5	7.0	14.9	71.4	7.2		
South America											
Actual	54.6	6.3	1.4	3.3	1.0	4.1	3.8	24.4	10.3	11.0	0.41
Natural	54.3	6.3	1.4	3.3	1.0	4.1	3.5	24.4	10.3		
Europe											
Actual	212.8	31.7	6.7	16.5	1.8	20.0	35.5	86.0	6.8	2.6	0.42
Natural	140.3	24.2	5.2	3.2	1.1	4.7	15.1	80.1	6.8		
Africa											
Actual	60.5	5.7	2.2	4.4	1.4	4.1	4.2	26.9	12.0	3.4	0.28
Natural	57.8	5.3	2.2	3.8	1.4	3.4	3.2	26.7	12.0		
Asia											
Actual	134.6	17.8	4.6	8.7	1.7	10.0	13.3	67.1	11.0	12.5	0.54
Natural	123.5	16.6	4.3	6.6	1.6	7.6	9.7	66.2	11.0		
Oceania											
Actual	125.3	15.2	3.8	7.6	1.1	6.8	7.7	65.6	16.3	2.4	–
Natural	120.6	15.0	3.8	7.0	1.1	5.9	6.5	65.1	16.3		

[a] Actual concentrations include inputs from human activity. Natural values are corrected to exclude pollution.
[b] Runoff ratio = average runoff per unit area/average rainfall.

TABLE 2.3 Typical concentrations of major ions in rainfall ($mg\,l^{-1}$) (From Berner and Berner, 1987)

Ion	*Continental rain*	*Marine and coastal rain*
Na^+	0.2–1	1–5
Mg^{2+}	0.05–0.5	0.4–1.5
K^+	0.1–0.5	0.2–0.6
Ca^{2+}	0.2–4	0.2–1.5
NH_4^+	0.1–0.5	0.01–0.05
H^+	pH 4–6	pH 5–6
Cl^-	0.2–2	1–10
SO_4^{2-}	1–3	1–3
NO_3^-	0.4–1.3	0.1–0.5

TABLE 2.4 Factors to convert units of common ions from mg l^{-1} to meq l^{-1}, and vice versa

Ion	*meq* l^{-1} = *mg* l^{-1} *	*mg* l^{-1} = *meq* l^{-1} *
Ca^{2+}	0.04990	20.04
Mg^{2+}	0.08224	12.16
Na^{+}	0.04350	22.99
K^{+}	0.02558	39.10
HCO_3^-	0.01639	61.02
CO_3^{2-}	0.03333	30.01
SO_4^{2-}	0.02082	48.03
Cl^-	0.02820	35.46

Carrick (1983) document for streams of the English Lake District.

Ions are reported as units of mass, mg l^{-1} (equal to ppm), or as chemical equivalents. In the latter case, milliequivalents per liter are calculated from mg l^{-1}, by dividing by the equivalent weight of the ion (its ionic weight divided by its ionic charge). Table 2.4 gives conversion factors from mg l^{-1} to meq l^{-1} for principal ions.

Of course, river water will be more concentrated than rain-water simply because of evaporation. Using the world average runoff ratio of 0.46, which means that 46% of precipitation becomes runoff (Table 2.2), the concentration of ions in river water should be 2.2 times greater than the concentration in rain. Because the true differential is roughly 20-fold, rock weathering, other natural sources, and anthropogenic inputs must account for the majority of dissolved ions in river water (Berner and Berner, 1987). Figure 2.2 illustrates this discrepancy for North America. Roughly 10–15% of the Ca, Na and Cl in US river water comes from rain, compared to a quarter of the K and almost half of the sulphate. In contrast, almost none of the SiO_2 or HCO_3^- comes from rain. This illustrates the need to examine the origins of each of these major cations and anions in order to understand what influences their concentrations.

Calcium is the most abundant cation in the world's rivers. It originates almost entirely from the weathering of sedimentary carbonate rocks,

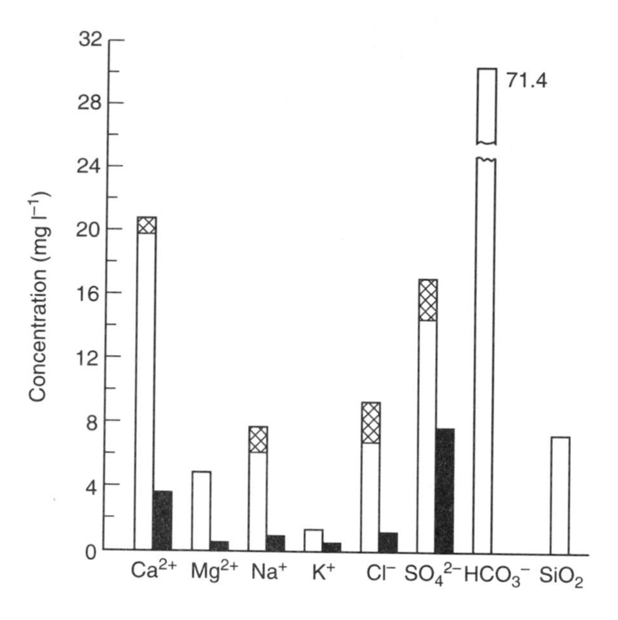

FIGURE 2.2 Dissolved ionic concentrations for natural river water (□) from North America, and rain-water (■) from the USA. Rain-water concentrations are multiplied by 2.6 to correct for evaporation. ▩ shows the anthropogenic contribution to river water values. (From Berner and Berner, 1987.)

although pollution and atmospheric inputs constitute small sources. Its concentration (along with magnesium) is used to characterize soft and hard waters, which are discussed more fully below. Magnesium likewise originates almost entirely from the weathering of rocks, particularly Mg-silicate minerals and dolomite. Atmospheric inputs are minimal, and pollution contributes only slightly.

Sodium is generally found in association with chloride, indicating their common origin. Weathering of NaCl-containing rocks accounts for most of the Na^+ found in river water. However, rain-water inputs from sea salts can contribute significantly, especially near coasts. Pollution, due to domestic sewage, fertilizers and road salt, is an especially important factor. Berner and Berner (1987) estimate that, world-wide, about 28% of the sodium in rivers is anthropogenic.

Potassium is the least abundant of the major cations in river water, and the least variable. Roughly 90% originates from the weathering of silicate materials, especially potassium feldspar and mica.

Silica also derives almost exclusively from the weathering of silicate rocks. Concentrations thus vary with the underlying geology, and also increase substantially from polar latitudes toward the tropics, apparently due to more complete chemical weathering at higher temperatures. Silica is used by diatoms in the formation of their external cell wall and thus affects biological productivity (discussed further in chapter 13).

Bicarbonate (HCO_3^-) derives almost entirely from the weathering of carbonate minerals. However, the immediate source of the majority of bicarbonate is CO_2 dissolved in soil and groundwater, which is produced by bacterial decomposition of organic matter, and derives in turn from the photosynthetic fixation of atmospheric CO_2. Bicarbonate is a biologically important anion. High concentrations are reflected in measures of alkalinity and are indicative of fertile waters. The carbonate buffer system, alkalinity and hardness are interrelated, as will be discussed more fully below. Anthropogenic increases in acidity, caused by acid precipitation or mining, reduce bicarbonate levels through the formation of H_2CO_3. While this can be locally important, Berner and Berner (1987) find no evidence that increased acid precipitation over the past century has yet caused reduced bicarbonate on a world-wide scale.

The origins of chloride are essentially the same as for sodium: mostly from weathering of rocks, but inputs of sea salts and pollution can be locally important. Chloride is chemically and biologically unreactive, and so is useful as a tracer in nutrient release experiments.

Sulphate has many sources, especially the weathering of sedimentary rocks and pollution (from fertilizers, wastes, mining activities); but biogenically derived sulphate in rain, and volcanic activity, are additional inputs. In areas of sulphuric acid rain, such as Hubbard Brook, New Hampshire, sulphate concentrations are high relative to overall ionic concentrations (Likens *et al.*, 1977). Sulphate and bicarbonate concentrations tend to be inversely correlated in streamwater, especially in low alkalinity areas.

Conductivity is a measure of electrical conductance of water, and an approximate predictor of total dissolved ions. Distilled water has a very high resistance to electron flow, and the presence of ions in the water reduces that resistance. In many instances an excellent relationship can be established between total dissolved salts and specific conductance (SC) (Figure 2.3).

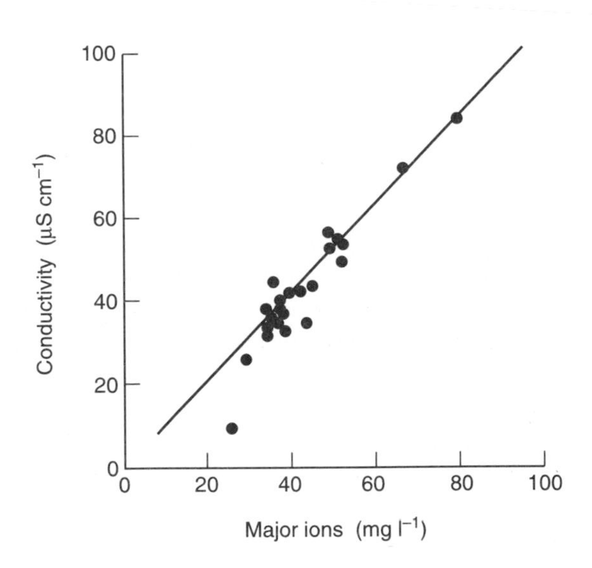

FIGURE 2.3 Conductance versus concentration of major ions ($Ca^{2+} + Mg^{2+} + K^+ + Na^+ + Cl^- + HCO_3^- + SiO_2$) for some West African rivers, illustrating the usefulness of conductance as an approximate measure of total ions. (From Grove, 1972.)

The relationship typically is linear (TDS = k*SC) with a value of k between 0.55 and 0.75. However, the value of the constant varies with location and must be established empirically (Walling, 1984). Differences in conductivity result mainly from the concentration of the charged ions in solution, and to a lesser degree from ionic composition and temperature. Values are reported as $\mu S\,cm^{-1}$ at 20 or 25°C, and in the older literature as $mho\,cm^{-1}$ (Golterman, Clyno and Ohnstad, 1978).

2.2.1 Quantification of transported constituents

The terms concentration, load and yield are widely used in descriptions of materials transported by river water. Concentration refers to the amount in a volume of water, and commonly is expressed as $mg\,l^{-1}$. For suspended particulates this can be as simple as collecting a known volume of water, which is then filtered, dried and weighed. Ionic concentrations require chemical assay procedures, which are described in *Standard methods of water and wastewater analysis* (American Public Health Association, 1989).

By combining estimates of concentration ($mg\,l^{-1}$) and discharge ($m^3\,s^{-1}$), the total amount of material in transport (mass per unit time) can be determined. This is the load, and usually it is summed over some time period, often a year. Of course, concentration and discharge might each vary considerably over time, and so many estimates of both are needed to calculate an annual load of material delivered to a lake or ocean. Walling (1984) provides a useful discussion of dissolved load calculations.

Finally, to compare the dissolved load of two rivers, it is useful to correct for differences in basin size. This is accomplished by dividing by basin area, giving an estimate of yield (mass per area per year).

2.2.2 Variation in time and space

As already stated, the chemistry of freshwaters is quite variable, rivers usually more so than lakes. Natural spatial variation is determined mainly by the type of rocks available for weathering, how wet or dry is the climate, and by the composition of rain, which in turn is influenced by proximity to the sea. The ionic concentration of rivers draining igneous and metamorphic terrains is roughly half that of rivers draining sedimentary terrain, because of the differential resistance of rocks to weathering. All of these factors provide the opportunity for substantial local variation in river chemistry. As a consequence the concentration of total dissolved ions can vary considerably amongst the headwater branches of a large drainage. However, these heterogeneities tend to average out, and concentrations tend to increase, as one proceeds downstream (Livingstone, 1963). Comparison of the Rio Negro and the Solimões (Amazon mainstem) dramatically illustrates the differences between tributaries. The Negro is much lower in ions, and the unique chemical signatures of these two mighty rivers can be detected as far as 100 km below their confluence. In smaller rivers, turbulence quickly eliminates heterogeneities from tributary inputs.

Even at the level of continental averages, substantial differences can be seen between, say, Africa and North America (Table 2.2). Moreover, for various reasons these continental averages do not portray the full range of natural variation. They are dominated by the contributions of a few, very large rivers, measured at downriver locations. In addition, large-scale averages of river chemistry are determined by the overwhelming contribution of sedimentary rocks. Just under a quarter of the earth's land surface is covered by igneous and metamorphic rocks, with three-quarters covered by sedimentary rocks. Combining this information with the known differences in ionic concentrations associated with different geologies, it is evident that dissolved materials from sedimentary rocks contribute by far the greatest amount (over 80%, Berner and Berner, 1987) of the total dissolved load of rivers, and thus dominate the composition of 'average' river water.

Studies of the chemical composition of rivers across a gradient from arid to humid conditions establish a general inverse relation between annual precipitation and total solute concentration. High concentrations of total dissolved salts are found in rivers draining arid areas due to the small volumes of precipitation and runoff, salt accumulation in the soil and evaporation (Walling, 1984). Concentrations decline as annual precipitation and runoff increase, first gradually and then more rapidly, according to Langbein and Dawdy (1964).

In addition to this broad trend relating ion concentrations to annual runoff, local concentration differences can be found over quite small distances. Within a series of small streams in southwest England, Walling and Webb (1975) reported a concentration range of total ions from 25 to 650 $mg\,l^{-1}$, resulting from small-scale shifts between igneous and sedimentary rocks.

River chemistry also varies over time, due to the influence of seasonal changes in discharge regime, precipitation inputs and biological activity. Flow variation has especially strong effects on ionic concentrations. Rivers are fed by a combination of groundwater and surface water, depending upon local geology and rainfall. Because of its longer association with rocks, the chemistry of groundwater typically is both more concentrated and less variable than that of surface waters. As a consequence, increases in flow typically dilute streamwater, though it is not a simple relationship (Livingstone, 1963). Walling (1984) provides a nice summary of some of the different dilution patterns that hydrologists have described statistically. Golterman (1975) states that there are two common patterns. Total dissolved salts may decline with increasing discharge, which is expected when the input of materials is constant. Alternatively, ion concentrations might not change greatly with fluctuations in discharge. This is expected when water chemistry reaches an equilibrium with the soil through which it percolates, or when concentrations approach saturation values. In addition to these two common patterns, however, some ions have been found to increase in concentration with rising discharge.

Long-term studies of streamwater draining a hardwood forest in New Hampshire illustrate how ionic concentrations can change in response to seasonal variation in precipitation inputs, discharge, and the cycle of growth of the terrestrial vegetation (Likens *et al.*, 1977). The most significant point is the relative constancy in stream chemistry, which probably is typical of intact, undisturbed ecosystems (Figure 2.4). Most dissolved substances vary within a narrow range (less than twofold), whereas streamflow can vary as much as four orders of magnitude over an annual cycle. In the Hubbard Brook Experimental Forest virtually all drainage water must pass through its mature and highly permeable podzolic soils. This affords considerable buffering capacity, and accounts for the considerable chemical stability of streamwater (Likens *et al.*, 1967).

As can be seen from Figure 2.4, concentration trends varied among cations. Both magnesium and calcium showed no significant correlation with discharge, although the latter was the more variable of the two. Sodium concentrations exhibited a significant inverse relationship with discharge, presumably because of its low availability, and so rising discharge caused dilution. With the exception of some very high values during summer drought, potassium concentrations generally increased with increasing discharge. The explanation for this is complicated, and apparently includes biological activity as well as soil buffering. Stream discharge is low during the summer, higher throughout the winter, and highest at snowmelt. Plant growth during the summer corresponds with low potassium

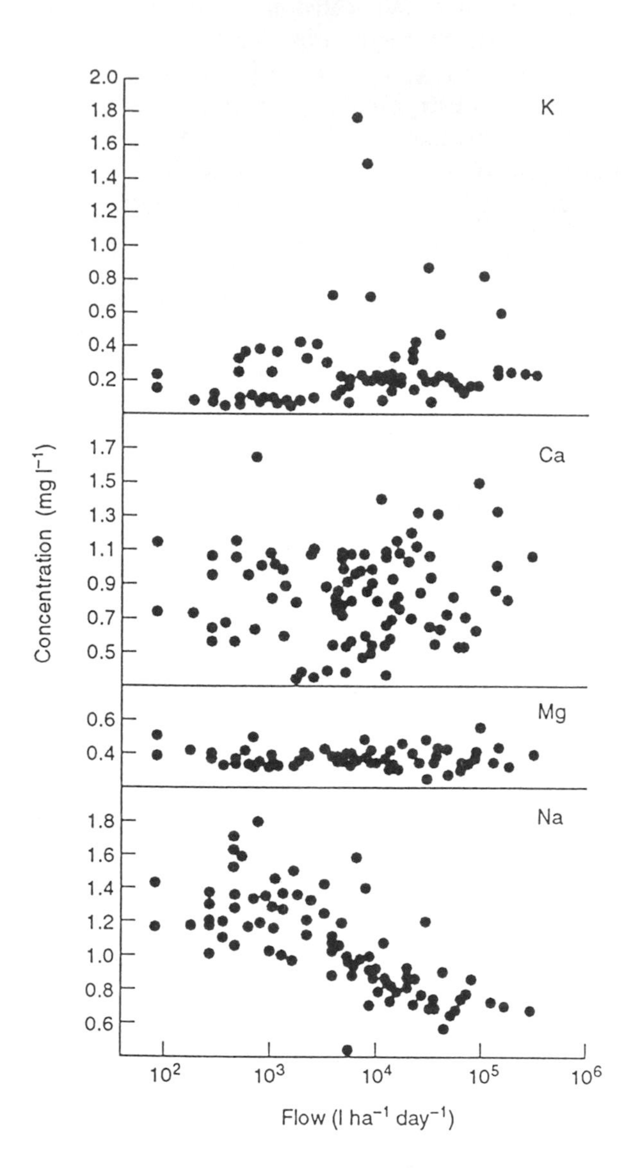

FIGURE 2.4 The concentration of major ions in relation to stream discharge in a small forested watershed in the Hubbard Brook Experimental Forest, New Hampshire, from 1963 to 1965. (From Likens *et al.*, 1970.)

concentrations, and so it appears that seasonal changes in biological demand correlate with seasonal changes in flow conditions.

It should be noted that trends in streamwater chemistry differ among studies. In mountain streams of California, Johnson and Needham (1966) found strong inverse relationships between discharge and the concentration of calcium and magnesium, as well as sodium. In a southern Appalachian watershed of grasses and shrubs, sodium concentration was variable, but not clearly related to discharge (Johnson and Swank, 1973). The importance of vegetation cover can be seen in Johnson and Swank's comparison of four watersheds: a cut forest planted with fescue grass and experiencing some regrowth of shrubs; a 7-year-old hardwood regrowth; a 13-year-old stand of white pine; and a mature hardwood forest. Compared with the mature forest, magnesium levels were twice as great in the grass–shrub watershed; calcium, magnesium and sodium were 20–50% greater. Streams draining the young hardwood and white pine stands exhibited reduced cation concentrations relative to streams draining the mature forest.

The profound effect of disturbance has been replicated experimentally on watersheds within the Hubbard Brook Experimental Forest (Likens *et al.*, 1970). Following deforestation and suppression of regrowth by herbicides, most major ions exhibited large increases in streamwater concentration, and total output increased sixfold. Only ammonium and carbonate remained low and constant, and sulphate declined because of reductions in sulphate generation by sources internal to the ecosystem. The average values for calcium and magnesium increased by over 400%, sodium by 177% and potassium increased over 18-fold. Altered ion concentrations were attributed to increased discharge, changes in the nitrogen cycle within the ecosystem, and higher temperatures.

The dissolved load of a river is the product of concentration and discharge, which as previously described, are often inversely related. As a consequence, the range of the dissolved load of salts varies over only two orders of magnitude, from around 3 to around 300 $t\,m^{-2}\,yr^{-1}$ (with a

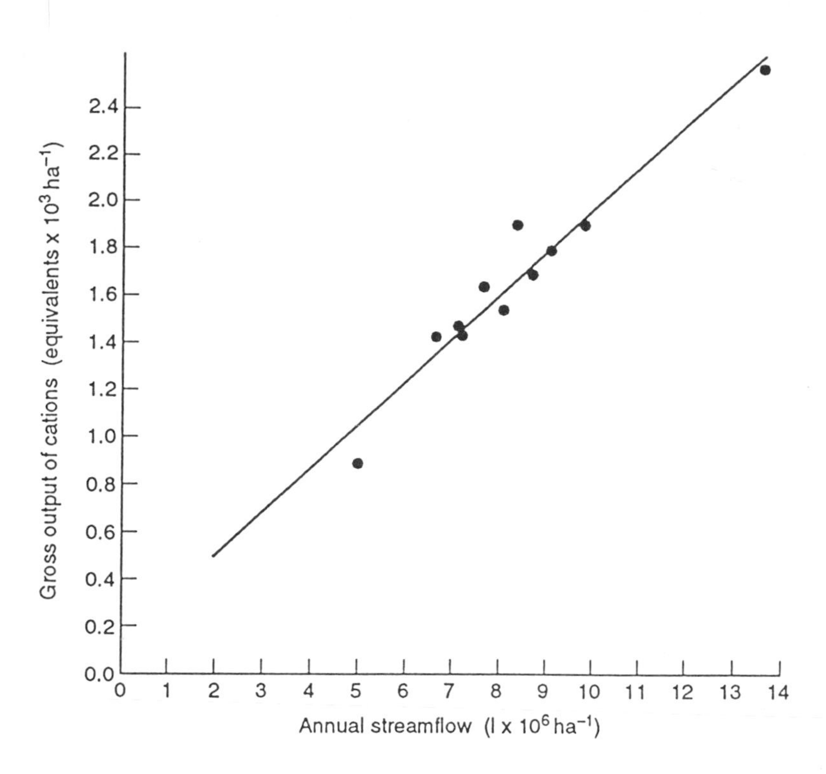

FIGURE 2.5 Between-year variation in the gross output of calcium, sodium, magnesium and potassium depends mainly on between-year variation in discharge for undisturbed watersheds of the Hubbard Brook Experimental Forest. Data span 1963–1974. (From Likens *et al.*, 1977.)

mean of $32\,t\,m^{-2}\,yr^{-1}$, Meybeck, 1976). However, the greater discharge of rivers in humid areas more than compensates for lower ionic concentrations, and so the dissolved load is least in arid areas and greater in areas of higher runoff.

Year-to-year variation in streamflow likewise influences the amount of dissolved material exported from a watershed. Because the concentration of most ions in Hubbard Brook streamwater is relatively constant, the amount exported is determined largely by streamflow, and between-year variation in export of ions depends strongly on inter-annual variation in discharge (Figure 2.5).

2.2.3 Chemical classification of river water

If one graphs the relative proportions of principle anions and cations in the world's surface waters against total dissolved salts, a curve with two arms emerges (Figure 2.6). Gibbs (1970) interprets this as evidence that three major mechanisms control surface water chemistry. At the left side of the 'boomerang', lie systems where the rocks and soils of river basins are the predominant source of their dissolved materials. Relief, climate, and age and hardness of rocks determine the positions of rivers within this grouping. Proceeding along the lower arm to the right of the figure we encounter waters

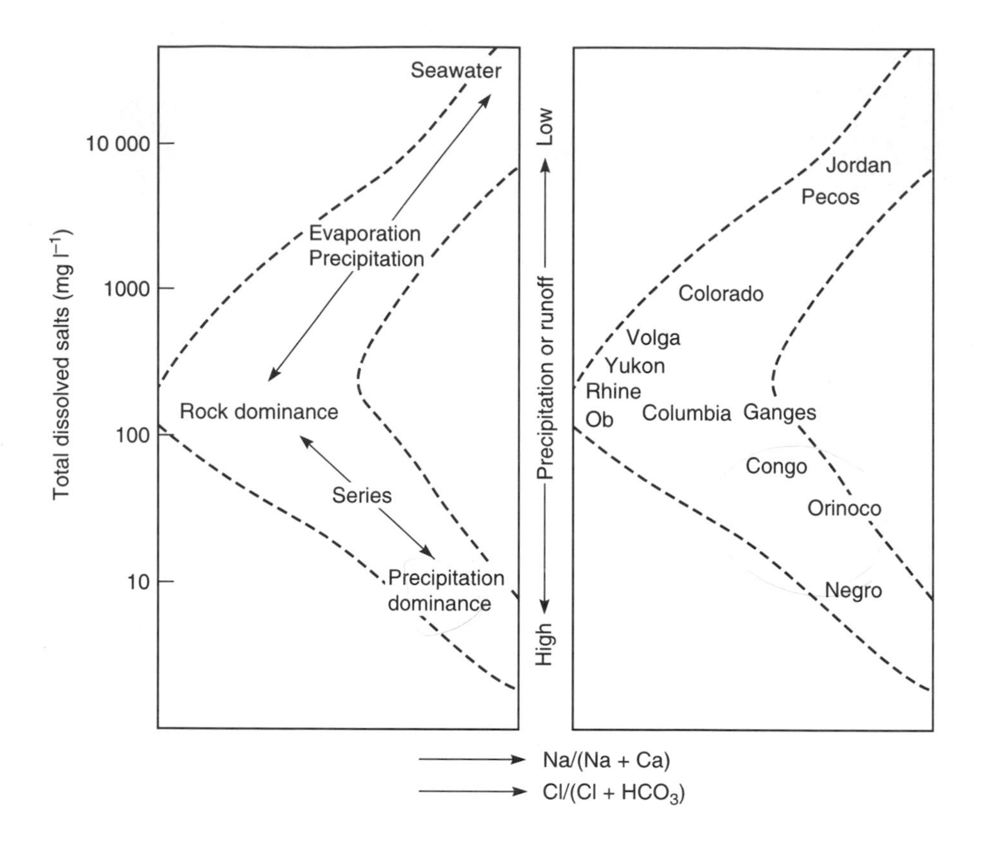

FIGURE 2.6 A classification of surface waters of the world based on ratios of sodium to calcium and chloride to bicarbonate, in relation to total dissolved salts. As one proceeds from left to right along the lower arm, inputs shift from a dominance of rock dissolution to a dominance of precipitation. The majority of large tropical rivers are found to the lower right. As one proceeds from left to right along the upper arm, sodium and chloride increase. These high salinity rivers lie in arid regions where evaporation is great. Note the vertical axis also reflects a gradient from high precipitation and runoff at the base to arid regions at the top. (Modified from Gibbs (1970) and Payne (1986).)

that are lower in salts and whose chemical composition most closely resembles rain-water. These are mainly the tropical rivers of Africa and South America with their sources in highly leached areas of low relief. In these humid regions, precipitation with a composition similar to sea water is of major importance as a source of salts. Average river water concentrations are less than 30 mg l^{-1}. Proceeding along the upper arm to the right, we encounter systems with high concentrations of dissolved salts and again a relative predominance of Na and Cl. These are rivers of hot, arid regions, and the combined influence of evaporation and precipitation of $CaCO_3$ from solution accounts for their chemistry. Total dissolved salts exceed 1000 mg l^{-1}, and can be as high as 6–7 g l^{-1}, as in Kazakhstan. Thus three mechanisms – atmospheric precipitation, dissolution of rocks, and the evaporation–crystallization process – are considered to account for the principal trends of dissolved salts in the world's surface waters. Other factors, including relief, vegetation and the composition of rocks and soils then can be invoked to explain differences within these major groupings (Gibbs, 1970).

Critics of this scheme question the interpretation of control at the ends of the boomerang. The Rio Negro's chemistry is equally a consequence of its long history of intense weathering, and as a basin of mainly silicious rocks (Stallard and Edmond, 1983). Similarly, saline rivers might be strongly influenced by near-surface halite deposits (Feth, 1971; Kilham, 1990). These arguments play down the roles of precipitation and evaporation, and suggest that local geology is of primary importance in determining river chemistry over all extremes. Whatever the resolution of this debate, most of the world's rivers are closer to the middle than the ends of this diagram, are low in $Na^+/(Na^+ + Ca^{2+})$, and are dominated by Ca^{2+} and HCO_3^- from carbonate dissolution. This accords with the view that the weathering of sedimentary rocks provides most of the dissolved ions in most of the world's major rivers (Berner and Berner, 1987).

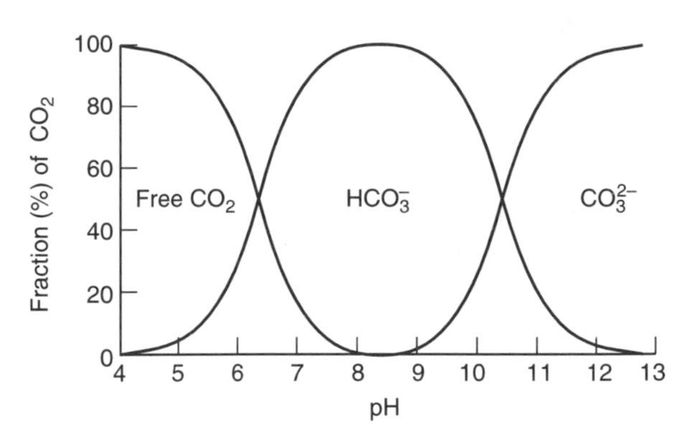

FIGURE 2.7 Influence of pH on the relative proportions of inorganic carbon species, CO_2 (+ H_2CO_3), HCO_3^-, and CO_3^{2-} in solution. (From Wetzel, 1983.)

2.3 The bicarbonate buffer system, alkalinity and hardness

Most natural waters contain various bicarbonate and carbonate compounds, originating from dissolution of sedimentary rocks. The calcium bicarbonate content of freshwaters determines the pH or acidity/alkalinity balance. When CO_2 dissolves in pure water, a small fraction is hydrated to form carbonic acid. However, streamwater usually contains bicarbonates and carbonates, and H_2CO_3 readily dissolves calcium carbonate rocks, neutralizing the soil and river water, and forming calcium bicarbonate. The resulting streamwater is a solution of carbon dioxide, carbonic acid, and bicarbonate and carbonate ions forming an effective buffer system that resists changes in pH (Wetzel, 1983).

The relative proportions of CO_2, HCO_3^- and CO_3^{2-} are pH dependent (Figure 2.7). At a pH below 4.5 only CO_2 and H_2CO_3 are present, and almost no bicarbonate or carbonate can be found. Indeed, bicarbonate concentration is commonly measured by titration with a strong acid until reaching a pH of about 4.3. At higher pH values dissociation of carbonic acid occurs, bicarbonate and carbonate are present, and CO_2 and H_2CO_3 are no longer detectable. At intermediate pH values, HCO_3^- predominates. Above a pH of about 8.3, bicarbonate also declines. These dissociation dynamics are influenced by both temperature and ionic concentrations, and the relationships shown in Figure 2.7 may not be valid for water of very high ionic concentration.

Freshwaters can vary widely in acidity and alkalinity due to natural causes as well as anthropogenic inputs. Extreme pH values, generally those much below 5 or above 9, are harmful to most organisms, and so the buffering capacity of water is critical to the maintenance of life. In freshwater, the CO_2–HCO_3^-–CO_3^{2-}

equilibrium serves as the major buffering mechanism.

pH is a measure of the concentration of hydrogen ions, hence the strength and amount of acid present. A value of 7 is neutral, and because the scale is logarithmic to the base 10, a decline of one pH unit represents a tenfold increase in hydrogen ion concentration.

In natural waters, carbonic acid is the main source of hydrogen ions, resulting in a pH of 5.7. Rain-water normally is acid because of its carbon dioxide content, and also due to naturally occurring sulphate. Normally these acids are neutralized as rain-water passes through the soil. However, in catchments of hard rocks, little buffering capacity, and high surface water (as opposed to groundwater) inputs, streamwater will be acid even if pollution is absent. Organic acids also contribute to low pH values. Where decaying plant matter is abundant, especially in swamps, bogs, and peaty areas, humic acids result in 'brown' or 'black' waters, and a pH in the range of 4–5. In addition, volcanic fumes and local seepage from sulphurous or soda springs can produce natural extremes of pH.

Over the past few decades, industrial activity has contributed to acid precipitation in many areas. The strong inorganic acids H_2SO_4 and HNO_3, formed in the atmosphere from oxides of sulphur and nitrogen, have seriously lowered surface water pH in large areas of Europe and North America, especially in granitic drainages with poor buffering capacity. These anthropogenic inputs initially produce only slow declines in pH, but as the bicarbonate buffering capacity becomes exhausted further acid inputs cause pH to decline rapidly. The biological consequences can be severe, and will be described later in this chapter.

Alkalinity refers to the quantity and kinds of compounds which collectively shift the pH into the alkaline range (pH over 7). It is more appropriately thought of as an index of alkaline compounds in water than as the opposite of acidity (Cole, 1979). Bicarbonates, carbonates and hydroxides are mainly responsible for alkalinity. It is measured by titration with a strong acid, and expressed as milliequivalents per liter or $mg\,l^{-1}$. Often alkalinity is measured as $mg\,l^{-1}$ of $CaCO_3$, which assumes that alkalinity is due solely to carbonate and bicarbonate. In some highly alkaline waters, however, this is not the case.

Hardness is another commonly used water quality term. It is determined by cations that form insoluble compounds with soap, and so primarily is a measure of the amount of calcium and magnesium salts. Ca and Mg occur mainly in combination with bicarbonate, sulphate and chloride, and the common co-occurrence of calcium and bicarbonate has led some to equate hardness with alkalinity. However, it is possible to find very high alkalinity with very little calcium or magnesium, and so it can be incorrect to equate these terms (Cole, 1979).

2.4 Influence of chemical factors on the biota

The biological consequences of chemical variation within freshwaters appear not to be very significant when conditions are reasonably close to the average. However, when one encounters the extremes, due to any of the natural causes already discussed or because of human influence, chemical variation from place to place can be of some biological significance.

Unfortunately, even when there is good evidence of a correlation between chemical variation and the distribution of the biota, the primary cause is difficult to pin down. For example, the finding that certain species are found in hard but not in soft waters might be due to differences in the amounts of calcium and magnesium present. However if, as seems likely, there is also a correlation with bicarbonate, alkalinity and total dissolved salts, any could be the cause. While it is plausible that the inability of mollusks to obtain sufficient calcium from soft waters for shell formation restricts the distribution of some species, it is not clear whether this explanation extends to other organisms. Even pollution-caused acidification can influence the biota via multiple pathways, and species differ in how they are affected. However, the study of acid streams has taken on considerable urgency in recent years, and substantial progress is being made toward understanding the resultant alterations in water chemistry, and their impact upon the biota.

2.4.1 Variation in ionic concentration

If one attempts to develop an entirely artificial medium for the culture of freshwater invertebrates, as has been done successfully for planktonic microcrustaceans, a long list of chemicals must be included (D'Agostino and Provasoli, 1970). If one holds mayfly larvae in distilled water, 50% mortality is exceeded within a few days (Willoughby and Mappin, 1988). Unquestionably, stream-dwelling organisms require water of some minimal ionic concentration. Unfortunately, there has been little effort to establish these requirements experimentally. The majority of the evidence linking the ionic content of water to the stream biota comes from surveys. Often such studies establish that streamwater of very low ionic concentration has a restricted flora and fauna, in both abundance and species richness. Many such studies are reported in terms of water hardness, but others have utilized alkalinity, conductivity and measurements of specific ions. Hynes (1970) describes a number of examples where the species of algae, mosses and higher plants differ between soft and hard waters. The growth rate and upper size limit of trout apparently are greater in hard waters, as McFadden and Cooper (1962) found in a comparison of fish populations from three soft-water and three hard-water streams in Pennsylvania. The difference might be due to differences in the invertebrate abundance between such streams, but this has not been clearly established.

Among the invertebrates, it appears that mollusks, crustaceans and leeches are more responsive to the range of ionic concentrations than are aquatic insects (Hynes, 1970; Macan 1974). The amphipod *Gammarus* apparently is common in streams of the English Lake District that have over $3\,mg\,l^{-1}$ calcium, and rare in streams of lower concentrations. A number of molluskan surveys have described the particular species that occupy soft waters, and reported a positive correlation between hardness and species richness. According to Russell-Hunter *et al.* (1967), roughly 5% of the molluskan species of a region will occur in extremely soft waters (less than $3\,mg\,l^{-1}\,Ca^{2+}$). Moderately soft waters (less than $10\,mg\,l^{-1}$) will support perhaps 40% of the species of a region, intermediate waters ($10–25\,mg\,l^{-1}$) will support up to 55%, and hard waters (over $25\,mg\,l^{-1}$) are needed to include all the molluskan species of an area. While those species found in water of high calcium content apparently cannot maintain populations in soft water, there is no indication that molluskan species of soft waters are excluded by water chemistry from regions of higher calcium content.

Such a dependence of mollusks on calcium availability makes very good sense. Aquatic mollusks derive a large fraction of their considerable calcium needs by absorption directly from the external medium. $CaCO_3$ is necessary for shell deposition and growth, and calcium is important in general fluid and electrolyte balance. Despite the evidence and the plausibility of this interpretation, however, our understanding is still far from complete. As Macan (1974) points out, there is no universal dividing line between ion-poor and ion-rich waters. Based on a survey of many Norwegian lakes, Økland (1983) reported that 18 (67%) of the total of 27 gastropod species tolerated water as soft as about $5\,mg\,l^{-1}\,Ca^{2+}$, and 13 species (48%) were found in water as soft as about

$3\,mg\,l^{-1}\,Ca^{2+}$. These observations indicate greater tolerance of water of low calcium content than claimed by Russell-Hunter *et al.* (1967).

Some of the variation in reported tolerance of calcium content might be due to differing physiological capabilities in ion extraction. Although molluskan shell thickness and mass often correlate with the dissolved calcium content of the surrounding waters, Russell-Hunter *et al.* (1967) report an example of strong inter-population differences in shell thickness and mass within the stream limpet *Ferrissia rivularis* unrelated to calcium availability. The authors speculate that populations differ genetically in their ability to extract calcium from dilute media, which would also imply differences in energetic cost. Harrison, Williams and Greig (1970) did indeed demonstrate a greater potential for population growth in a pulmonate snail reared at high rather than low calcium levels. Presumably the greater energetic cost associated with calcium metabolism in very soft water accounts for this finding.

Two surveys of aquatic insect distributions illustrate some of the difficulties of making sense of correlations with ionic concentrations. In their study of 52 streams in the Scottish Highlands, Egglishaw and Morgan (1965) found fewer total cations (less than $400\,\mu eq\,l^{-1}$) in streams draining rocks of granite and schist compared with areas of basalt, limestone or sandstone. The bottom fauna also was less diverse and abundant than in streams of greater ionic content. Although the relationship between the biota and water chemistry was expressed in terms of total cations, at least half of this was (Ca + Mg), and so in all likelihood the authors could have used hardness instead.

A survey of the plecopterans (stoneflies) in Irish streams is also instructive. Conductivity and hardness were both high in limestone-rich locales, and much lower in another region where harder rocks accounted for the softer water (Costello, McCarthy and O'Farrell, 1984). Stoneflies were more common at the latter sites, but it is unlikely that they prefer or are physiologically restricted to streams of low ionic content. A more plausible explanation is that soft-water sites were at higher elevations with cooler water temperatures, and this survey simply accords with the observation that plecopterans are most diverse in montane and high latitude locales of clean, cool running waters (Hynes, 1970).

2.4.2 Salinization

This describes the situation where total dissolved salts are unnaturally high, rather than naturally low. Salinization refers to either the process or the result of the buildup of dissolved salts in freshwaters. The natural range of salinity in inland waters is very great, but when referring to the result of human activities, we often are concerned with changes from relatively low background concentrations. Clearly this is subjective, but $300\,mg\,l^{-1}$ is a useful boundary value. From the human perspective, salinities greater than $500\,mg\,l^{-1}$ are undesirable for drinking water, and detrimental effects on crops begin at levels over 0.5–$1\,g\,l^{-1}$ (Williams, 1987).

Salinization is a particular problem in arid and semi-arid areas, due to irrigation and dryland farming. Irrigation concentrates salts first because of evaporation, and also because the remaining, more concentrated solution leaches soil salts. Ultimately this more concentrated water returns to the stream, via surface or subsurface flows. In Australia, where salinization is widespread in semi-arid agricultural areas, a number of faunal changes have been attributed to this problem. Species have disappeared from large areas where they once were common, and brackish-water forms have replaced freshwater taxa (Williams, 1987). In North America, the elevated salinity of the Colorado (over $0.8\,g\,l^{-1}$ at its delta) is the recent consequence of irrigation and impoundments. Although the biological effects of this are not well known, river

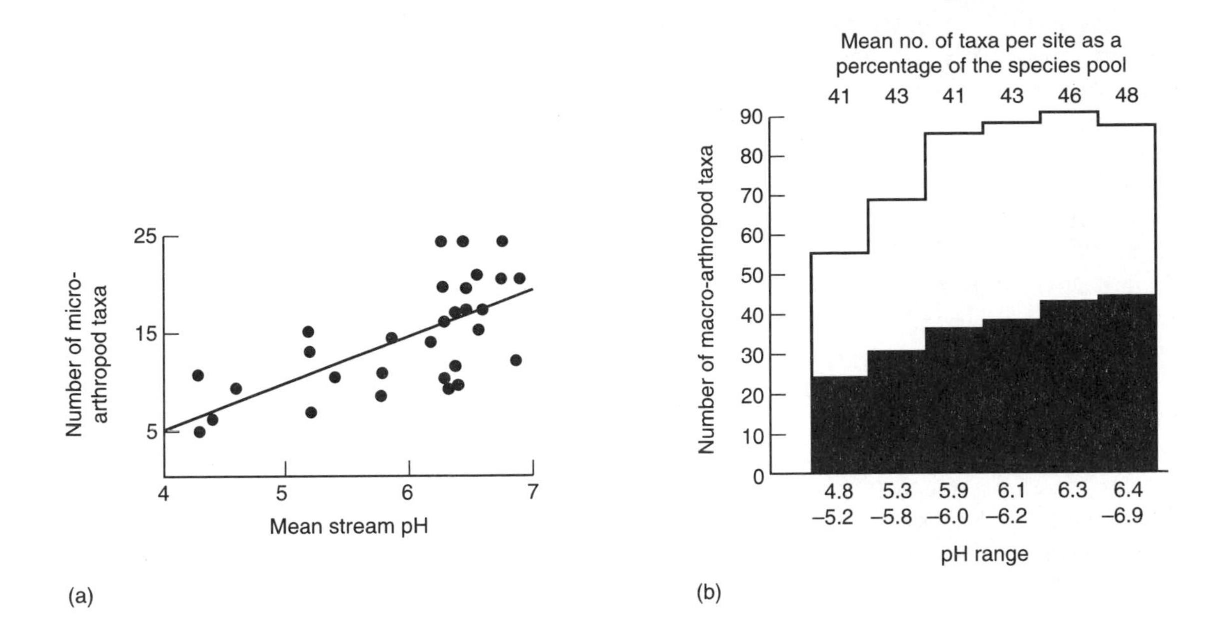

FIGURE 2.8 Influence of pH on the number of species occurring in streams of the Ashdown Forest, southern England. (a) Microarthropods: mainly mites (Hydrachnellae), copepods (Harpacticoida and Cyclopoida) and Cladocera, from Rundle and Omerod (1991). (b) Macroarthropods: mainly aquatic insects, from Townsend, Hildrew and Francis (1983). Note that the number of species of macroarthropods found at a site was a fairly constant 40–50% of the species known to occur under the conditions of water chemistry found at that site.

water in the lowermost sections is unsuitable for agriculture without expensive desalinization facilities. It is doubtful that the river biota is completely unharmed by such conditions.

The use of salt to clear highways of ice is another source of elevated streamwater concentrations of NaCl, since almost all of the deicing salts applied to roads eventually enter streams and rivers. In some cases values are of the order of $10\,g\,l^{-1}\,Cl^{-}$, and the fauna is greatly impoverished. However, according to Crowther and Hynes (1977), macroinvertebrates experience few deleterious effects until Cl^{-} concentrations surpass $1\,g\,l^{-1}$, which does not appear to happen commonly in North American streams.

2.4.3 Effects of acidity on stream ecosystems

The deleterious effects of acidic streamwater are well established, primarily in terms of reduced numbers of species and individuals (Figure 2.8), but there is also evidence of altered ecosystem processes. At present the underlying mechanisms are only partially understood, and a number of variables appear to determine how acidic streamwater influences the biota. Naturally acid streams seem less affected than those acidified by atmospheric pollutants. The degree of acidification is of course very important, and depends upon both inputs and buffering capacity. Organisms evidently are harmed via diverse pathways, and taxa differ in their susceptibility. Consequently, while one can assert that anthropogenic acidification is generally harmful once pH falls much below 5.0, the details depend on many factors.

Streamwater can be acid for a variety of reasons. A comparison of two regions, the English Lake District, and the west coast of the South Island, New Zealand, illustrates that pH alone provides an incomplete story.

Streams and lakes of the English Lake District, and particularly the River Duddon, clearly show how extremes of pH can influence the biota (Sutcliffe and Carrick, 1973, 1988). The industrialization of Europe has resulted in acidic precipitation in the region for probably the past 100 years, and prior to this a long and gradual period of post-glacial acidification is thought to have occurred. The range of pH in this region is approximately 4–7 and varies with geology, upstream or downstream location and season. Based on pH and invertebrate distributions, Sutcliffe and Carrick (1973) recognize three categories of streams: pH greater than 5.7, independent of season or rainfall; pH less than 5.7, independent of season or rainfall; pH less than 5.7 during winter and wet periods; pH greater than 5.7 during summer and dry periods.

The permanently acid streams have a characteristic and restricted macroinvertebrate fauna, consisting of six plecopterans, four caddis and three dipterans. Locations with pH permanently greater than 5.7 have these plus additional taxa, including a number of mayflies, two other species of trichopterans, the limpet, *Ancylus*, and the amphipod, *Gammarus*. Permanently acid streams are found in areas of relatively hard, slow-weathering rocks, often in the headwaters. Locations where pH invariably is greater than 5.7 occur in regions of sedimentary slates and exposed veins of calcite. They have not changed in acidity since at least the 1950s, evidently because bicarbonate ion availability counters the continued inputs from acid rain.

The New Zealand streams studied by Winterbourn and Collier (1987) were acidic due to high concentrations of humic substances, resulting in a natural pH range 4.1–8.1. A survey of 34 sites showed no correlation between pH and taxonomic diversity. Indeed, similar numbers of taxa were obtained from streams with pH 5.5 or less, between 5.6 and 6.9, and 7.0 or more. Only at a pH below 4.5 was there any evidence of faunal exclusion, in marked contrast to the Duddon example.

These contrasting cases suggest that the mechanism must involve more than acidity alone. Indeed, there is growing evidence that lowered pH is accompanied by a number of chemical changes, and also that organism response is due to various physiological, behavioral and indirect effects.

Leaching of metals from soils has only recently been identified as an important consequence of hydrogen ion deposition. In particular, aluminum has been found in elevated concentrations in acidic waters, and also been shown to increase in response to experimental acidification (Hall *et al.*, 1980). Possible effects of increased aluminium, in addition to direct toxicity, include reduced buffering capacity and altered concentrations of trace metals, orthophosphate, and dissolved organic carbon (Hall, Driscoll and Likens, 1987). Using separate and combined additions of aluminum compounds and inorganic acids to stream channels, several studies have attempted to distinguish the direct influence of hydrogen ion concentration from the effects of elevated aluminum. In a short-term (24 h) manipulation of a soft-water stream in upland Wales, two salmonid species exhibited far greater susceptibility to the combined effects of acid and aluminum compared with sulphuric acid alone, apparently because of respiratory inhibition (Omerod *et al.*, 1987). However, aquatic insects showed little difference in response to the two treatments. Only the mayfly, *Baetis rhodani*, showed a greater density reduction in the combined treatment. In contrast, in a similar study, Hall, Driscoll and Likens (1987) found the drift behavior of aquatic invertebrates did respond to the combined treatments, but not to reduced pH alone.

Clearly, more must be learned of the changes in stream chemistry that accompany or are induced by acid deposition. However, it is apparent that elevated aluminum concentrations are a contributing factor. Moreover, this evidence provides a valuable clue regarding the difference between naturally acidic brown-water streams and more recently acidified clearwater streams. Because of the chelating abilities of humic acids, which bind metal ions mobilized at low pH, aluminum toxicity likely is less important in the streams studied by Winterbourn and Collier (1987).

In addition to the mobilization of toxic metals, there are numerous other pathways by which acidity is detrimental to the stream biota. Direct physiological effects of acidity are implicated by field and laboratory demonstration of increased mortality or failure of eggs to develop as pH is reduced (e. g. Carrick, 1979; Burton, Stanford and Allan, 1985; Willoughby and Mappin, 1988). Inability to regulate ions apparently is responsible, including loss of body sodium and failure to obtain sufficient calcium from surrounding waters (Økland and Økland, 1986). There is some experimental evidence that the concentration of other ions in the medium is important, and also that species differ in their susceptibility to ion loss (Willoughby and Mappin, 1988). Field collections generally show stoneflies and caddisflies to tolerate waters of lower pH than do mayflies and some dipterans. Hall, Driscoll and Likens (1987) speculate that differences in life cycle and respiratory style of these groups account for their differential susceptibility to acid stress.

Behavioral avoidance of acidic waters is a sublethal mechanism that might account for the rareness or absence of many taxa. The ability to avoid exposure obviously is a useful adaptation when one considers that exposures as short as 15 min cause death in some taxa (Hall, Driscoll and Likens, 1987). Following experimental acidification, many species almost immediately enter the water column, to be transported downstream in the drift (Hall *et al.*, 1980; Hall, Driscoll and Likens, 1987). Adults also might avoid depositing their eggs in acidic waters, as has been suggested for the mayfly, *Baetis*, which enters the stream to oviposit on the surface of stones (Sutcliffe and Carrick, 1973). Species that oviposit on the water surface would not be expected to show such avoidance behavior, however.

Evidence that stream acidification affects the biota indirectly via alteration of food availabilities is somewhat weak, but nonetheless suggestive. The mayfly, *Ephemerella ignita*, survives well in the laboratory at a pH of 4.8, although it does not occur in sections of the River Duddon with such pH values. From this and the apparent scarcity of algae in low pH streams, Willoughby and Mappin (1988) inferred that food availability, rather than water chemistry, accounts for the limited distribution of *E. ignita* within the Duddon catchment. It also is likely that acidity reduces microbial populations that are important in organic matter decomposition, such as converting freshly fallen leaves into an energy source available to consumer organisms (chapters 5 and 6). By measuring loss of strength in strips of cellulose test cloth, Hildrew, Townsend and Francis (1984) found a significant negative effect of acid streamwater upon decomposition rate.

Of course, the susceptibility of the streams of a region to acidification depends greatly on geology and soils. The most dramatic examples generally are found in areas of igneous rocks and low alkalinity, which have succumbed to decades or more of anthropogenic acidic deposition because of their limited buffering capacity. Scandinavia and the eastern parts of the USA and Canada have been particularly impacted. Where alkalinities are appreciably higher than about 0.1–0.2 $mmol\,l^{-1}$, surface waters are believed to be resistant to acidification (Schindler, Kasian and Hesslein, 1989).

Trends of pH in different regions of Denmark nicely illustrate how vulnerability to human-caused acid deposition varies with local geology (Rebsdorf, Thyssen and Erlandsen, 1991). Streams on leached, sandy soils in western Denmark have low alkalinities (mean of

$0.59 \ mmol \ l^{-1}$), while those to the east have soils that are mostly calcareous and clayey, with alkalinities averaging $2.24 \ mmol \ l^{-1}$. Water samples taken over the past 15 years in streams with alkalinities above $1.5 \ mmol \ l^{-1}$ in eastern Denmark show no acidification trend, while several western streams with mean alkalinities of $0.05–0.79 \ mmol \ l^{-1}$ underwent an annual decrease of 0.027 pH units. Thus, vulnerability to human-caused acid deposition generally requires a combination of substantial inputs, along with little natural alkalinity to act as a buffer.

2.5 Summary

The constituents of river water include suspended inorganic matter, dissolved major ions, dissolved nutrients, suspended and dissolved organic matter, gases and trace metals. The dissolved gases of particular interest are oxygen and carbon dioxide. Exchange with the atmosphere maintains the concentrations of both at close to the equilibrium determined by temperature and atmospheric partial pressure, especially in streams that are small and turbulent. Photosynthetic activity in highly productive settings, usually involving filamentous algae or macrophytes, can elevate oxygen to super-saturated levels and result in strong fluctuations between day and night. Respiration has the opposite effect, reducing oxygen and elevating CO_2. High levels of organic waste can reduce oxygen concentrations below life-sustaining levels, and elevate carbon dioxide to many times atmospheric pCO2.

Many factors influence the composition of river water, and as a consequence it is highly variable in its chemical composition. The concentration of the dissolved major ions (Ca^{2+}, Na^+, Mg^{2+}, K^+, HCO_3^-, SO_4^{2-}, Cl^-) is roughly $100 \ mg \ l^{-1}$ on a world average. However, river water is highly variable, ranging from a few $mg \ l^{-1}$ where rain-water collects in catchments of very hard rocks, to some thousands of $mg \ l^{-1}$ in arid areas.

Variation from place to place is determined mainly by the type of rocks available for weathering, by the amount of precipitation, and by the composition of rain, which in turn is influenced by proximity to the sea. The concentration of total dissolved salts is roughly twice as great in rivers draining sedimentary terrain, compared with igneous and metamorphic terrains, due to differential resistance of rocks to weathering. Areas of high rainfall and surface water runoff usually have less concentrated streamwater compared with arid areas where evaporation is greater and dilution is less. Precipitation inputs typically are of lesser importance to streamwater chemistry, except in areas of very hard rocks and high surface runoff. Human pollutants enter river water through, for example, precipitation and dry deposition, by stormwater transport of fertilizers and road salt and by direct disposal.

River chemistry changes temporally under the multiple influences of seasonal changes in discharge regime, precipitation inputs and biological activity. Groundwater typically is both more concentrated and less variable than surface waters, because of its longer association with rocks. In undisturbed catchments some ions are remarkably immune to discharge fluctuations spanning several orders of magnitude. However, because rainfall increases the surface water contribution, ion concentrations often are diluted by increases in flow.

Natural waters contain a solution of carbon dioxide, carbonic acid, and bicarbonate and carbonate ions in an equilibrium that serves as the major determinant of the acidity/alkalinity balance of freshwaters. Freshwaters can vary widely in acidity and alkalinity, and extreme pH values (much below 5 or above 9) are harmful to most organisms. The bicarbonate buffer system, consisting of the CO_2–HCO_3^-–CO_3^{2-} equilibrium, provides the buffering capacity that is critical to the health of the freshwater biota.

Although freshwater is highly variable in its

chemical composition, and rivers more so than lakes, the biological importance of such variation is only evident at the extremes, and where human pollutants are substantial. Water of very low ionic concentration appears to support a reduced fauna, particularly of crustaceans and mollusks. The number of species commonly increases with hardness, and many taxa are distinctly soft-water or hard-water forms. Anthropogenic additions of strong inorganic acids set off a number of changes in water chemistry, and at a pH much below 5.0, the biological consequences are serious.

Chapter three

Physical factors of importance to the biota

The physical environment of running waters has a number of characteristic features that pose special challenges to the organisms that dwell there. Current is the defining feature that unites all rivers and streams. It conveys benefits, such as transport of resources to the organism and removal of wastes; and also risks, of which being swept away is the most obvious. The substrate of running waters differs greatly from place to place, and is important to many insects as the surface on which they dwell, and to many fishes as the structure near which they find shelter from current or enemies. Temperature affects all life processes, and as most stream-dwelling organisms are ectothermic, growth rates, life cycles and the productivity of the entire system are strongly under its influence. Thus current, substrate and temperature are the three physical variables that we should understand in order to appreciate the functioning of a lotic ecosystem and the adaptations of its denizens. Oxygen, a chemical variable introduced in the previous chapter, is usually of minor significance in unpolluted running waters, but sometimes it is important, and those circumstances are governed by the three main physical factors we examine in this chapter.

The following sections will describe the main features of current, substrate, temperature and oxygen: the range of conditions encountered, how these variables affect the biota, and some of the many adaptations that organisms possess to cope with their physical surroundings. A focus on these four parameters also provides an opportunity to describe many of the plants and animals found in running waters, and that is a secondary purpose of this chapter.

3.1 Current

Water velocity and the associated physical forces collectively represent perhaps the most important environmental factor affecting the organisms of running waters. The speed of the current influences the size of particles of the substrate. Current affects food resources via the delivery and removal of nutrients and food items. And, of course, current velocity presents a direct physical force that organisms experience within the water column as well as at the substrate surface. It is this latter topic, concerning what fluid forces are experienced and how organisms are affected, that is our main concern here. The complexities of flow around obstructions and near the streambed are of particular importance, because most organisms of running waters live under these complicated nearbed conditions, and not in the middle of the water column.

Biologists have long believed that water as a

medium, and current as a force, strongly determine ecological distributions and shape anatomical and behavioral adaptations. Many species are found mainly in fast flowing sections of a stream, or in slow water, but not in both. The body shapes of animals and growth forms of plants exhibit a number of morphological adaptations that are viewed as adaptations to move about in current or avoid being swept away. A low vertical profile, streamlined shape, and attachment devices are frequently observed in the running water biota. However, because of difficulties in characterizing flow in biologically meaningful ways and the complexity of interacting factors, it is apparent that we are still very far from a complete understanding of the effects of current on organisms and processes in river ecosystems.

3.1.1 Effects of flow on organisms

The effects of flowing water on the biota are complex. Not only is flow itself difficult to understand and measure, but size, shape and lifestyle of organisms profoundly influence how they will be affected by hydrodynamic forces. Clearly we do not expect the same forces to be encountered by a salmon swimming upstream, a sculpin in contact with the stream bottom, an insect clinging to a stone, or an algal cell attached to the substrate by its gelatinous sheath. Current is important in every instance, as the following examples will illustrate, but not necessarily in the same manner.

The segregation of two species of caddis larvae between fast- and slowwater habitats is a classic example of the specialization of organisms to flow conditions (Edington, 1968). Careful measurements of water velocity 1.5 cm above the substrate indicated that *Hydropsyche instabilis* was found where currents ranged from 15 to 100 cm s^{-1}, whereas *Plectrocnemia conspersa* occurred primarily in the 0–20 cm s^{-1} range. These insects are suspension feeders, capturing food particles from the water column using nets

TABLE 3.1 Percentage of individuals of two caddis larvae that constructed nets in laboratory streams maintained at a given current velocity

	Current velocity (cm s^{-1})		
	10	*15*	*20*
Hydropsyche instabilis	20%	48%	73%
Plectrocnemia conspersa	72%	50%	4%

spun of silk (chapter 6). When larvae of each species were placed in a laboratory stream at several current speeds, the percentage of larvae that constructed nets was indicative of their different behavioral preferences (Table 3.1).

Additionally, Edington conducted an experimental manipulation that demonstrated the rapid adjustment of larvae of *H. instabilis* to altered current regime. After carefully mapping current and net location, he removed nets (but not the larvae, which remained in their retreats within a mossy layer) and diverted current with a baffle. A second mapping less than 2 days later revealed that nets were reconstructed only at higher current velocities, and most larvae had deserted the low velocity locales created by the baffle. Finally, a comparison of net structure provides further evidence of adaptation to fast versus slow water in these two species. The nets of the hydropsychid were rigidly supported, streamlined and not damaged by fast currents. In contrast the polycentropodid's net was flimsier, and ruptured in the laboratory when current velocity was increased from 10 to 25 cm s^{-1}.

Aquatic invertebrates exhibit a number of anatomical features that apparently enhance their ability to move about, or minimize their likelihood of being swept away (Figure 3.1). In some instances the functional benefits of the features are clear, while others are less certain (Hynes, 1970). Direct attachment devices including silk and other sticky secretions, hooks and suckers seem appropriately identified as adaptations that aid their owners in maintaining position against the current. Blepharocerid larvae occur

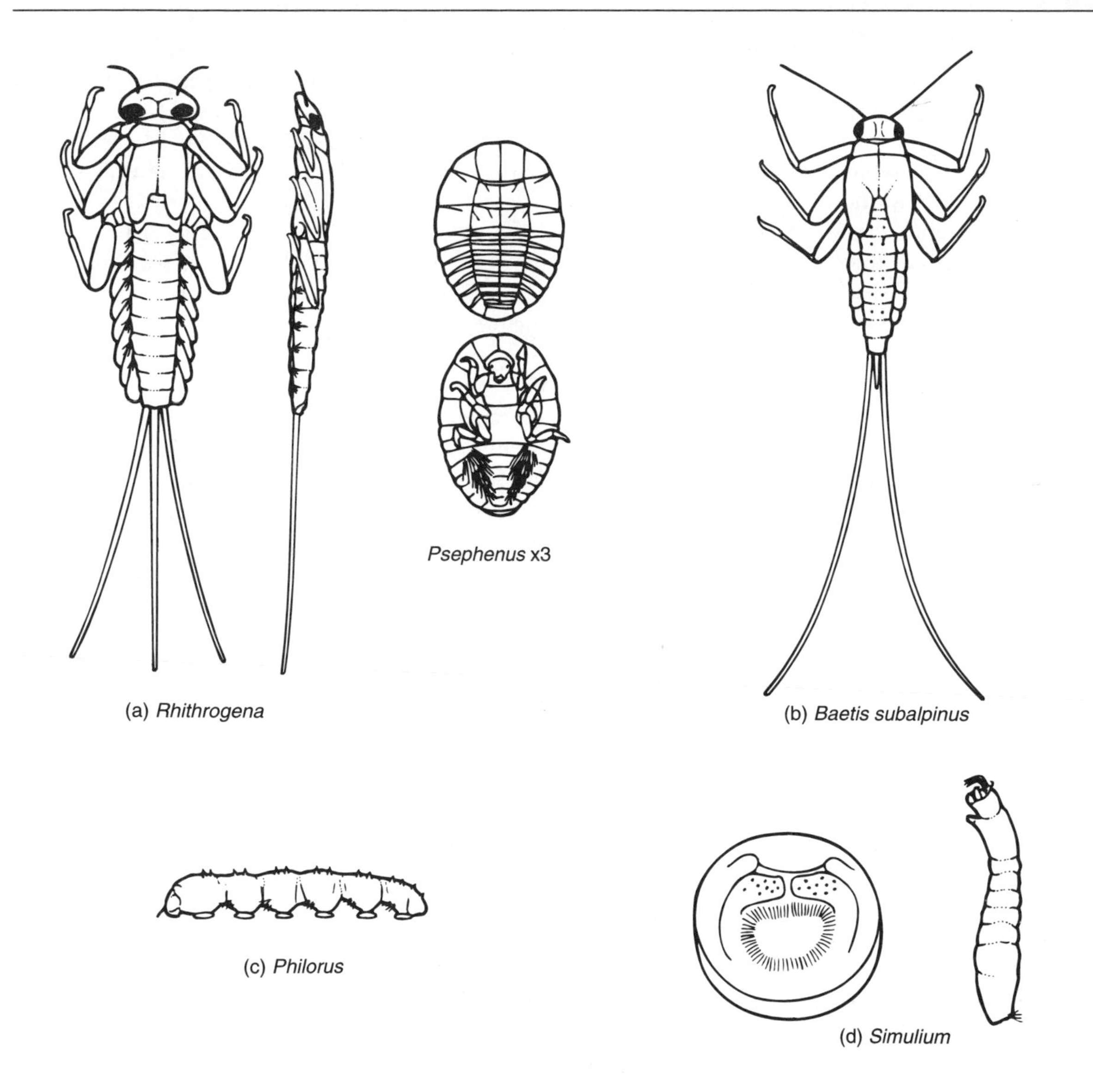

FIGURE 3.1 Examples of invertebrate body shapes thought to be adaptive to life in running waters. (a) Dorsoventral compression in *Rhithrogena* and *Psephenus*, (b) streamlining in *Baetis*, (c) the blepharocerid *Philorus* with sucker-like pseudopods, (d) circlet of hooks on the posterior proleg of *Simulium*, which attach to a silken mat produced by the larva. (From Hynes, 1970.)

on smooth rocks in fast water, where their row of six ventral suckers allows them to move against very high current velocities. Black fly larvae are another example of extreme adaptation to withstand current. They are able to occupy high-velocity habitats by spinning a mat of silk onto a stone surface, to which they attach with specialized prolegs. Circlets of outwardly directed hooks on both anterior and posterior prolegs aid the larva in attachment and movement. Should it be dislodged, an additional line of silk connects the larva to the substrate, allowing the animal to climb down the thread and re-attach. Silk is also used for attachment in other dipterans, especially midges, and by caddis larvae to attach their pupal cases to stones while they molt. One last example of a specialized adaptation for attachment is provided by the zebra mussel, *Dreissena polymorpha*. It is unusual among freshwater clams in possessing byssal threads in the adult form, and this permits it to live in flowing waters provided a hard substrate is available. Zebra mussels have long been a nuisance in Europe, where dense aggregations occasionally block water-intake pipes. They are now spreading throughout the waterways of the Great Lakes and are likely to invade an even larger area of North America (Strayer, 1991), with uncertain consequences for the native biota.

Some other anatomical features have been suggested to be adaptive, or at least beneficial, to life in streams, but they are less cogently argued to be specializations primarily to withstand current. Body shapes that are dorsoventrally flattened or streamlined, such as of the mayflies *Rhithrogena* and *Baetis*, respectively, might be adaptations to avoid or resist the pressures of flow (Figure 3.1). Attractive as this explanation may be, related species of the same general body plans (e. g. *Ecdyonurus, Caenis*) can be found in slow water and lakes, where a flattened shape is useful for life under stones, and good swimming ability is useful simply for moving about. Ballast is beneficial in an erosive environment, and so the stone cases of some caddis larvae can act to resist dislodgement. However, caddis cases also provide camouflage and direct protection against enemies. We should be careful not to make current the explanation for every feature of the biota of flowing waters.

The adaptations of fishes to life in water, be it still or moving, have received much study. Current is of course only one factor to consider. The much greater resistance of water relative to air is of primary importance. In addition, a variety of ecological considerations such as maneuverability and acceleration reflect the continual struggle between predator and prey to improve the tactics of attack and escape. Nevertheless, fishes, like invertebrates, exhibit a number of anatomical adaptations that reflect life in flowing water.

Fast-swimming fishes and fishes that swim in fast current generally are streamlined and rounded in cross-section (Figure 3.2). Examples include the salmonids, the trout-like *Galaxias* of Australia and New Zealand, the dace *Rhinichthys*, and many more (Hynes, 1970). Laterally flattened fishes, of which members of the sunfish family, Centrarchidae, are excellent examples, are less well suited to speed than to maneuvering. Of course, many stream-dwelling fishes live near the bottom or take shelter behind obstructions, and these taxa display another suite of adaptations. Many are dorso-ventrally flattened, with eyes that are dorsally situated, and their paired fins are muscular and lateral. Examples include the darters (*Etheostoma*) and sculpins (*Cottus*) of North America, along with many catfishes, loaches, gobies and some suckers (Hynes, 1970). Another frequent adaptation is reduction of the buoyancy-inducing swim bladder, which has occurred independently in a number of lineages. This can be seen in the comparison of two close relatives, the leopard dace *Rhinichthys falcatus* and the longnose dace *R. cataractae*. The former is found in slow currents along river banks, and has a greater swim bladder volume than the current-dwelling longnose dace (Gee and Northcote, 1963). Suckers are yet another adaptation to facilitate movement in rapid water. Tadpoles of the Asian frog *Rana hainensis* have a hydraulic sucker formed by papillation of the front lip, and this enables them to crawl leech-like over stones. Hynes (1970) remarks that stones of five times the weight of a tadpole can be lifted from the water, using the animal as a handle. The lamprey's sucker is an adaptation to an ectoparasitic existence, but it can be employed to climb the sheer rock walls of waterfalls during upstream migrations.

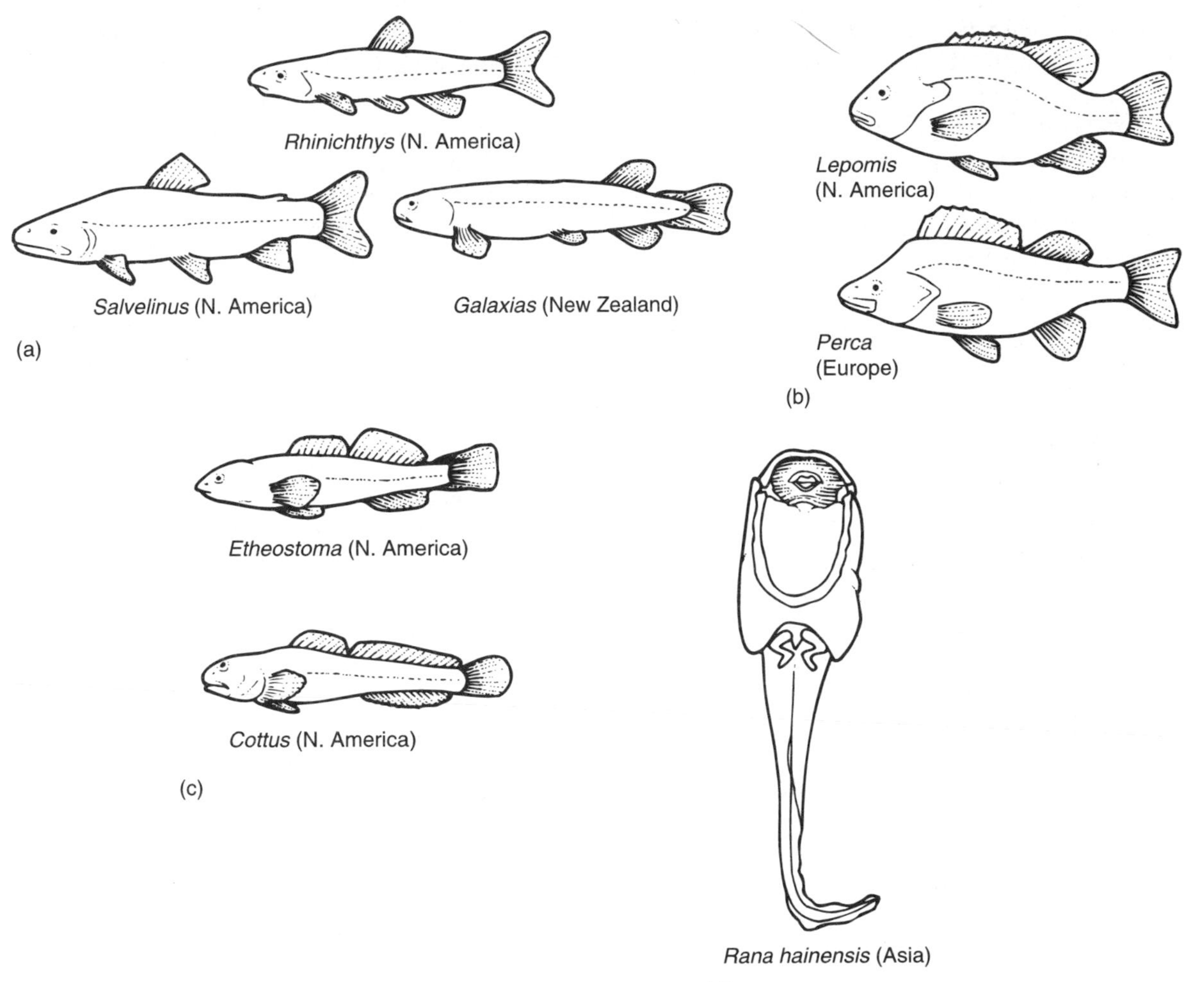

FIGURE 3.2 Examples of vertebrate body shapes thought to be adaptive to life in running waters. (a) Streamlined fishes of the water column, (b) deeper body shape of slow-water fishes, (c) note enlarged pectorals and dorsal eyes of benthic fishes, (d) oral sucker of tadpole of Asian frog. (From Hynes, 1970.) 가슴지느러미

Fishes display behavioral as well as anatomical adaptations to withstand the force of current. Bottom-dwelling fishes behaviorally alter the position of their bodies in order to prevent displacement by the current. Outspread pectoral fins help to hold the fish in place both by friction and by acting as hydrofoils, using the force of current to press the animal against the substrate. Benthic fishes respond to increasing current in laboratory flumes by altering their body posture to reduce slippage (Webb, 1989). This includes arching of the back, and pressing the body close against the substrate.

What about algae and other microorganisms that live on surfaces in flowing water? It is very difficult to specify what flow conditions a microorganism actually experiences, and to separate the direct from the indirect effects of current. It might even be possible to be too sheltered from turbulence and mixing, which benefits microorganisms living on substrate surfaces by augmenting the supply of gases, nutrients and organic material.

Whatever the underlying mechanism, surveys of the distribution of diatoms reveal the existence of distinctive taxa in fast versus slow water. Moreover, attached algae seem to be most abundant in fast water with hard substrate, while rooted vascular plants are found mainly in locations of slow water and soft bottoms (Hynes, 1970). Thus, the ecological distribution of plants is indicative of the general importance of current as a habitat characteristic. A closer look at the diatoms, which are the most widespread periphyton of running waters, strongly suggests that their distribution with regard to current velocity is based on differences in structure. Diatoms have an external skeleton or test composed of two valves that fit together like the top and bottom of a Petri dish. In shape they resemble either a pie plate or a cigar. Most species in fast water are elongate and attach to surfaces of stones or other plants. Some diatoms, like *Gomphonema* and *Cymbella*, stand erect on a stalk composed of mucilaginous material. Others, such as *Achnanthes* and *Cocconeis*, glue their entire valve directly to stone surfaces with mucilage (Patrick, 1948). Inspection of the micro-distribution of diatoms in a small Tennessee stream established a close correspondence between algal growth form and current (Keithan and Lowe, 1985). Erect and colonial taxa predominated in slow-moving water, whereas most diatoms found in rapid currents were prostrate and tightly adherent. *Cladophera glomerata*, a green alga the filaments of which can exceed a meter in length under some circumstances, has different growth forms depending on current (Whitton, 1975a). Thick plumose growths occur in slow water; long, tough, rope-like strands are seen in faster water. The diatom *Desmogonium* likewise is longer in fast water, shorter and broader in slow water (Patrick, 1948).

Although the above examples leave little doubt that current has a direct effect on the distribution of microscopic algae and that body plans differ in their suitability for particular microenvironments, experimental confirmation is essentially lacking. Moreover, there is a significant alternative explanation for the importance of current to microorganisms: resupply of necessary materials, and removal of potentially harmful metabolites. As discussed more fully in chapter 4, a number of investigators have monitored the formation of algal mats in laboratory flumes at different current speeds. Temporal succession, species composition and mat height typically change with velocity (e. g. McIntire, 1968), but the relative contributions of attachment ability *versus* physiological response to flow regime are difficult to distinguish.

From these varied examples it is apparent that some features of stream-dwelling organisms represent remarkable specialization to move about or remain in place in currents that can exceed 200 cm s^{-1}. However, for several reasons the influence of current on the biota remains poorly understood. Adaptive explanations of morphology and behavior must recognize that the principal features of any organism are inherited within a taxonomic lineage and subject to multiple selective forces. Current has indirect as well as direct effects on organisms. Water velocity affects substrate size composition, the delivery of gases and food items, and other environmental factors, making causation difficult to ascertain even with imaginative experimentation. Lastly and very importantly, we are just beginning to understand the actual forces that organisms experience, and this of course limits our interpretations. Toward this end, we need a better understanding of the complexities of flow, of velocities near the streambed and around obstructions, and of the fluid forces that organisms actually experience.

3.1.2 Channel and nearbed flow environments

The biota of running waters dwells in a highly variable environment from the standpoint of current regime. This is apparent to anyone mesmerized by the delicate swirls on the surface of the smallest stream, or the awesome power of a storm-swollen river. Three fundamental types of flow characterize moving fluids: laminar, turbulent and transitional. In laminar flow, fluid particle movement is regular and smooth, and particles can be thought of as 'sliding' in parallel layers with little mixing. Turbulent flow is characterized by irregular movement with considerable mixing. Intermediate conditions are described as transitional. In fact, laminar flow conditions are so rare in aquatic environments that they are relevant primarily as a theoretical reference point. Although laminar flow can occur in pipes, and over smooth mud surfaces, even beds of sand produce complex flow. This complexity increases with increasing roughness of the channel bottom and with mean velocity. As a consequence, quantification of flow conditions is problematic, and this is especially true for organisms dwelling on or near the substrate.

At the interface between a fluid and a solid, the velocity of the two is identical (the 'no-slip' condition, Vogel, 1981), which means that water in contact with non-eroding substrate has zero velocity. Because surface water can move quite rapidly, there must be a gradient in velocity as one approaches the bottom and sides of streams (recall Figure 1.4). This decrease of velocity with depth produces a region of shear, known as the boundary layer. The upper limit of the boundary layer occurs where the speed of the current is no longer influenced by the presence of the stream bottom. In a shallow stream, the boundary layer may extend to the surface. Very close to the stream bottom, there can be a laminar (or viscous) sublayer where shear stress is zero and flow is greatly reduced.

The possibility that a thin layer of low flow exists very near the stream bottom, perhaps functioning as a refuge from the turbulence and high velocities of the water column just above, has attracted the attention of lotic ecologists since at least the turn of the century (e. g. Steinmann, 1908). This was based on dorsally compressed body shapes exemplified by the water penny *Psephenus* and a number of mayflies, and the expectation that current must be reduced at the water–substrate interface. An imaginative study by Ambühl (1959) greatly heightened interest in this topic by providing a first glimpse of a viscous sublayer. By photographing particles of acetyl cellulose flowing around and over obstructions in an artificial stream channel, Ambühl was able to demonstrate a viscous sublayer some 1–3 mm thick. Although best results were obtained under rather special conditions (smooth, plaster-of-Paris substrates and low flows), Ambühl's work reinforced the expectation that viscous sublayers of perhaps several millimeters in height provided shelter for invertebrates.

This view has changed substantially in recent years. As Vogel (1981) says, "most biologists . . . have the fuzzy notion that [the boundary layer] is a discrete region rather than the discrete notion that it's a fuzzy region." Moreover, the terms boundary layer and viscous sublayer should not be used interchangeably. Strictly speaking, the region of greatly reduced flow is the viscous sublayer, which is found very close to the streambed or other surface. It now appears that as flow becomes more turbulent and more typical of natural streams, the viscous layer is thinned to the point that most benthic invertebrates likely experience a turbulent, three-dimensional flow microenvironment (Nowell and Jumars, 1984). This shift in perspective heightens the need for a better understanding of the hydrodynamic conditions that organisms actually experience.

TABLE 3.2 Some terms and equations useful in describing streamflow (Adapted from Davis and Barmuta, 1989; and Carling, 1992)

Terms			
$\bar{U}$	Mean velocity	Measured at 0.6 depth from surface or from velocity profile	
U_*	Shear velocity	Estimated from fine-scale velocity *versus* log depth profile at nearbed depths	
D	Water depth	Total depth, surface to bottom	
k	Height of surface roughness elements	Difficult to quantify; methods described in text	
ν	Kinematic viscosity	$1.004 \times 10^{-6}\,m^2\,s^{-1}$ at 20°C	
g	Acceleration due to gravity	$9.8\,m^2\,s^{-1}$	
Equations			
Re	Bulk flow Reynolds number		
	$Re = \bar{U}D/\nu$	$Re < 500$	$\Longrightarrow$ laminar flow
		$500 < Re < 10^3\text{–}10^4$	$\Longrightarrow$ transitional flow
		$Re > 10^3\text{–}10^4$	$\Longrightarrow$ turbulent flow
Fr	Froude number		
	$Fr = \bar{U}\sqrt{(gD)}$	$Fr < 1$	$\Longrightarrow$ sub-critical flow
		$Fr = 1$	$\Longrightarrow$ critical flow
		$Fr > 1$	$\Longrightarrow$ super-critical flow
D/k	Relative roughness	Height of roughness elements relative to water depth; influences flow type	
Re_*	Roughness Reynolds number	Describes flow near streambed	
	$Re_* = U_*k/\nu$	$Re_* < 5$	$\Longrightarrow$ hydraulically smooth flow
		$5 < Re_* < 70$	$\Longrightarrow$ transitional flow
		$Re_* > 70$	$\Longrightarrow$ hydraulically rough flow
δ	Thickness of laminar sublayer	Describes region of viscous flow	
	$\delta = 11.5\nu/U_*$	$\delta/k < 1$	$\Longrightarrow$ hydraulically smooth flow
		$\delta/k > 1$	$\Longrightarrow$ hydraulically rough flow

(a) Quantification of flow conditions

For biologists interested in life in moving water, the subject of fluid dynamics provides an extensive if complicated framework of theory and empirical evidence that promises greatly to enrich our understanding of the physical conditions under which stream-dwelling organisms function. Traditionally, this has been the province of engineers. However, the biological applications of this subject are increasing, and recent books by Vogel (1981), Denny (1988), and Gordon, McMahon and Finlayson (1992) provide excellent treatments of fluid mechanics for biologists. Reviews by Davis and Barmuta (1989) and Carling (1992) are especially useful from the perspective of life on the streambed. Webb (1988, also see Webb and Weihs 1986) nicely summarizes the hydrodynamic conditions experienced by vertebrates moving through

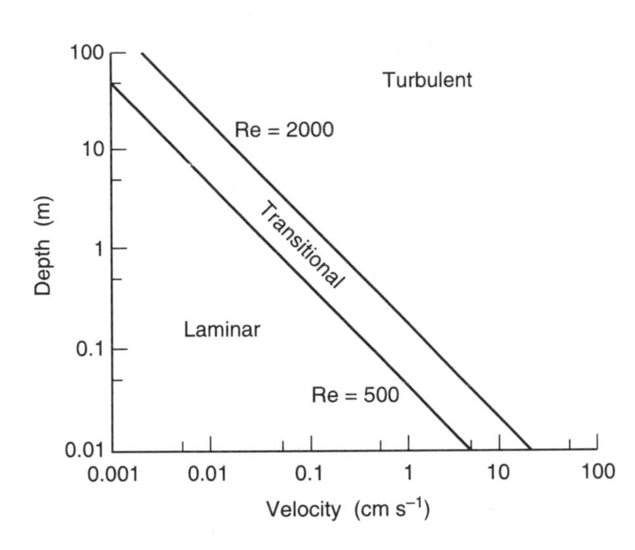

FIGURE 3.3 Reynolds number conditions for the occurrence of laminar, transitional and turbulent flow in stream channels. Note that turbulent conditions are the norm. (Redrawn from Davis and Barmuta (1989), after Smith (1975).)

water. Some basic applications of fluid dynamics are set out in Table 3.2, and discussed below.

The most useful single quantifier in biological fluid dynamics is the dimensionless Reynolds number:

$$\mathrm{Re} = \frac{\bar{U}L}{\nu} \tag{3.1}$$

where $\bar{U}$ is the velocity of the fluid (m s^{-1}), L is a characteristic length scale (m), and ν is the kinematic viscosity (1.004×10^{-6} $m^2 s^{-1}$ for freshwater at 20°C). The units must be chosen to be internally consistent, and depend on the scale of interest.

Re can be estimated for the stream channel, the nearbed region, or an individual organism. Re for the river channel (often called the bulk flow Reynolds number) is easily calculated from the mean velocity (in m s^{-1}) and water depth (in m, Table 3.2). Depending on circumstances, mean velocity is estimated from the average of measurements over a depth profile or a single measurement at 0.6 depth. The hydraulic radius R, which equals the cross-sectional area divided by the wetted perimeter, is used instead of depth for some applications. As Figure 3.3 illustrates, turbulent flow is the norm in the channels of rivers and streams. Laminar flow usually requires current velocities well below 10 cm s^{-1}, especially if depth exceeds 0.1 m; in short, quite shallow and slow-moving water.

Estimation of Re for a fish or an insect larva is conceptually straightforward, but a technical challenge. For a solid exposed to a current, characteristic length is usually approximated by the length of the organism in the direction of flow or movement. One measures current velocity at the front of the animal, or swimming speed in the case of propulsion through still water. In general, small organisms close to the substrate where velocity is low have low Re values and large organisms experiencing greater velocities have high Re values.

The Reynolds number quantifies the ratio of inertial to viscous forces within a fluid. Re can be used to distinguish types of flow, and what forces are experienced by an organism. At low Re, viscous forces predominate and flow is laminar, whereas at high Re, inertial forces predominate and turbulence occurs. Physical conditions between the extremes of low and high Reynolds number differ profoundly (Vogel, 1981). At high Re values, pressure drag is the important force and streamlining is an adaptive counter-measure. An airfoil, a trout and a *Baetis*, each with blunt front and tapered rear, are ideal shapes to minimize turbulent drag due to flow separation downstream of the object. By minimizing wake turbulence, streamlined shapes reduce the pressure differential between front and rear, which creates the drag we experience on our legs as we wade a swift stream. At low Re, water is more viscous, flow is much more laminar, and the force exerted as layers of water slide over one another is greater. This last force, due to the 'no-slip condition', creates a shear region near the object's surface called skin friction. It is minimized by reduction in surface area, and so stubby or rotund shapes might be advantageous. Streamlining will be of little benefit due to the reduced role of pressure drag and the increased surface area that streamlining entails. These are only generalizations, however; at Re between 10^2 and 10^4, the best shapes to minimize total drag are not known (Vogel, 1981).

The Froude number, Fr, is another useful descriptor of main channel flow that is easily calculated from velocity and depth.

$$Fr = \frac{\bar{U}}{\sqrt{gD}} \qquad (3.2)$$

where $\bar{U}$ is again mean current velocity, D is total water depth, and g is the acceleration due to gravity (9.8 $m^2\,s^{-1}$). Fr represents the ratio of inertial forces to gravitational forces, and differentiates tranquil flow from broken and turbulent flow (Davis and Barmuta, 1989). Like Re, the Froude number is dimensionless.

Estimates of bulk flow Re and Fr are easily obtained, as each requires only routine measurement of depth and velocity. In fast-flowing streams, the result almost invariably is turbulent, super-critical flow (see ranges given in Table 3.2). For these reasons, channel Re and Fr are of perhaps only modest value to biologists, and have little direct bearing on conditions near the streambed. Moreover, these measurements are not likely to differentiate effectively among environmental conditions, because most streams will be found to have turbulent rough flow. As a consequence we need to redirect our attention to the channel bottom, and attempt to quantify conditions near the streambed. The roughness Reynolds number remains useful, because it can be applied to flow near the bottom or to flow through the axis of the organism.

(b) Laboratory studies of boundary layers

Benthic boundary layers are most easily studied in the laboratory, where current and bed roughness are easily manipulated, flow profiles can be measured precisely, and the response of organisms can be closely observed. Although laboratory settings do not reproduce the full complexity of nature, they nonetheless provide extremely useful insights provided that flume design meets necessary geometric constraints. Based on recent studies, the region of greatly reduced flow usually is quite thin. As a consequence, this region likely is insufficient to shelter the larger invertebrates, whether or not they are dorso-ventrally compressed. However, the smallest invertebrates and microorganisms live within a layer of greatly reduced flow. The drag forces that organisms experience thus depend on size because size determines whether they are fully within the viscous sublayer, and also because size determines the relative importance of viscous forces, which create friction drag, *versus* inertial forces, which create pressure drag.

The boundary layer is due to friction between a moving fluid and a stationary surface. Its vertical extent is defined experimentally as the point where flow reaches 90% (or higher) of the mean velocity. This can be calculated in the laboratory for flow over flat surfaces such as glass plates, and should be relevant to organisms living on flat bottoms or the leaves of large plants such as *Potamogeton*. At velocities of 20 $cm\,s^{-1}$ or less, viscous sublayers of a few millimeters thickness are likely, perhaps as great as 5 mm (Silvester and Sleigh, 1985). Under these circumstances, algae and microorganisms would experience current velocities less than 1% of the main current. However, at flow speeds above 20 $cm\,s^{-1}$, turbulence is enhanced and viscous sublayers shrink dramatically. Theoretical expectations give a δ of perhaps 0.5 to 1 mm at velocities of 20–50 $cm\,s^{-1}$ (Figure 3.4). Roughness of the surface would further shrink δ, but can not be calculated by the methods available.

Flow visualization provides another way to describe viscous sublayers. Using laser Doppler anenometry, which measures light scatter from

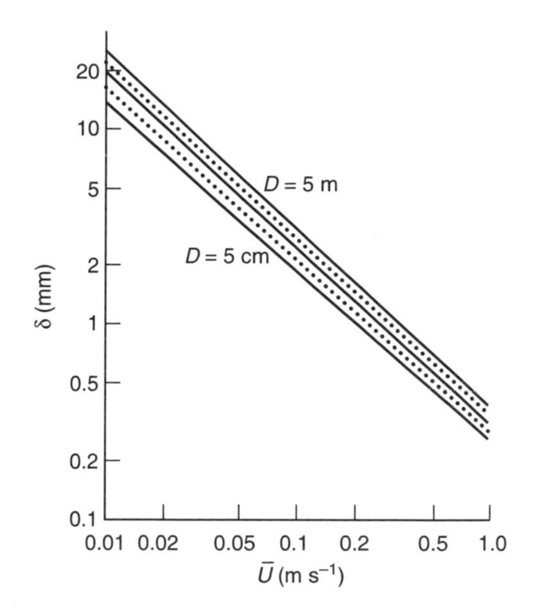

FIGURE 3.4 Depth of viscous sublayer (δ) in turbulent flow over a smooth surface. The central continuous line is for a water depth of 50 cm and the dashed lines correspond to 15 cm and 1.5 m. (From Silvester and Sleigh, 1985.)

particles ranging from 0.1 to 10 μm in size, Statzner and Holm (1982) were able to quantify flow conditions around dead invertebrates glued to the surface of a flume. The dorsally compressed mayfly *Ecdyonurus venosus*, and the stream limpet *Ancylus fluviatilis*, are considered well adapted to minimize the effects of flow. However, at 16 and 38 cm s^{-1}, the flow gradient encountering the frontal aspect of both organisms was compressed, resulting in a dead-water space behind the animal, a steep velocity gradient over the body, and thus substantial friction force (Figure 3.5). The authors concluded that the viscous sublayer was approximately 0.5 mm thick at 16 cm s^{-1}, and perhaps only 0.1 mm thick at 38 cm s^{-1}. In comparison, the vertical height of the test organisms was roughly 2 mm.

These results directly contradict the inference, drawn from Ambühl's (1959) pioneering work, that dorsally compressed insect larvae are not subject to high shear stress and turbulence. A careful reconsideration of Ambühl's experiment suggests that his results are valid for hydraulically smooth flow (Carling, 1992). However, Ambühl's results have limited application to natural rivers, where hydraulically rough flow is common. In general, if bed material consists of 8-mm size gravel or larger, flow is likely to be rough-turbulent, with no intact viscous sublayer.

The growing recognition that viscous sublayers can be quite limited in extent has led to greater efforts to measure forces experienced by the organism. Once again, the Reynolds number is useful. The force that current exerts on organisms can be estimated from the Reynolds number obtained using velocity through the animal's frontal aspect and body length in equation 3.1. This is denoted as R_L. Friction drag predominates at low R_L, the environment can be likened to what we might experience swimming in honey, and a bluff or hemispheric body shape best minimizes total drag. In contrast, pressure drag predominates at high R_L, the environment is experienced as more watery, and a streamlined shape most effectively minimizes total drag.

Using several species in a flume with simulated roughness, Statzner (1988) demonstrated that R_L changes greatly with organism size as well as with nearbed current velocity (Figure 3.6). Laser Doppler anenometry was used to estimate velocity through the frontal axis of a mud snail, a limpet and an amphipod, under flume current speeds of 5–37 cm s^{-1}. Very small invertebrates, including the newly hatched larvae of many taxa, are approximately 0.5–1 mm in length. Their R_L values should vary between 1 and 10, depending on current. In larger individuals, say 10 mm or more length, R_L is considerably greater, in the 10^2–10^3 range. Thus, a growing larva initially is subject mainly to friction drag, but subsequently experiences mainly pressure drag. This analysis implies that animals

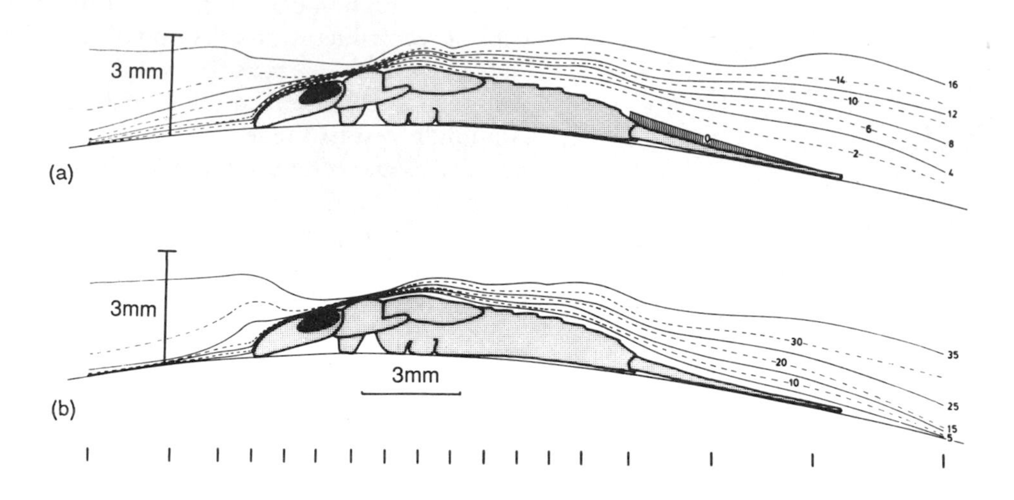

FIGURE 3.5 Lines of equal flow (iso-vels) around a dead mayfly nymph glued to a surface in a laboratory flume, measured using a flow visualization technique. Note that all iso-vels are compressed over the mayfly's dorsal surface. (a) Maximum velocity 16 cm s^{-1}, (b) 32 cm s^{-1}. (After Statzner and Holm, 1982.)

might be expected to be more rotund when small, and become more streamlined as they grow. Since this does not appear to be true, either the analysis is flawed, or body shape is constrained by other agents of selection.

Although each of these laboratory studies has its limitations, collectively they undermine the idea that animals on substrate surfaces routinely find shelter from current within a viscous sublayer. Except under low flows over flat surfaces, layers of greatly reduced flow appear to be less than 1 mm in height, and perhaps less than 200–300 μm (Silvester and Sleigh, 1985; Statzner and Müller, 1989). Only the smaller invertebrates, and of course microorganisms (except when in mats or on stalks), would truly lie within a viscous sublayer. Many invertebrate taxa, including those whose flattened shape has long been viewed as an adaptation to dwell within the viscous sublayer, in fact experience complex flows and relatively high shear. For these organisms, size and shape are important mainly because they influence the ratio of inertial to viscous forces that the organism experiences, and not because of sheltering from the current.

(c) Field studies of nearbed flow conditions

Because direct measurement of nearbed conditions at a very fine scale is not yet possible in natural rivers, it is necessary to estimate nearbed conditions from water column measurements. This requires a bed roughness estimate and measurement of current velocity at several depths very close to the substrate. Linear regression of velocity *versus* the logarithm of distance above the streambed is used to estimate shear velocity, U_*, provided the regression relationship is strong. Bed roughness (k) is difficult to estimate. Usually it is obtained from the diameter of streambed materials and tabulations of experimentally derived values (Chow, 1981), but it also can be derived from the slope of the log-linear velocity profile through the boundary layer (Carling, 1992). Gore (1978) developed a surface profiler with a six-by-six matrix of steel rods on a 0.1 m^2 frame. The frame is leveled and the rods pushed gently down until an

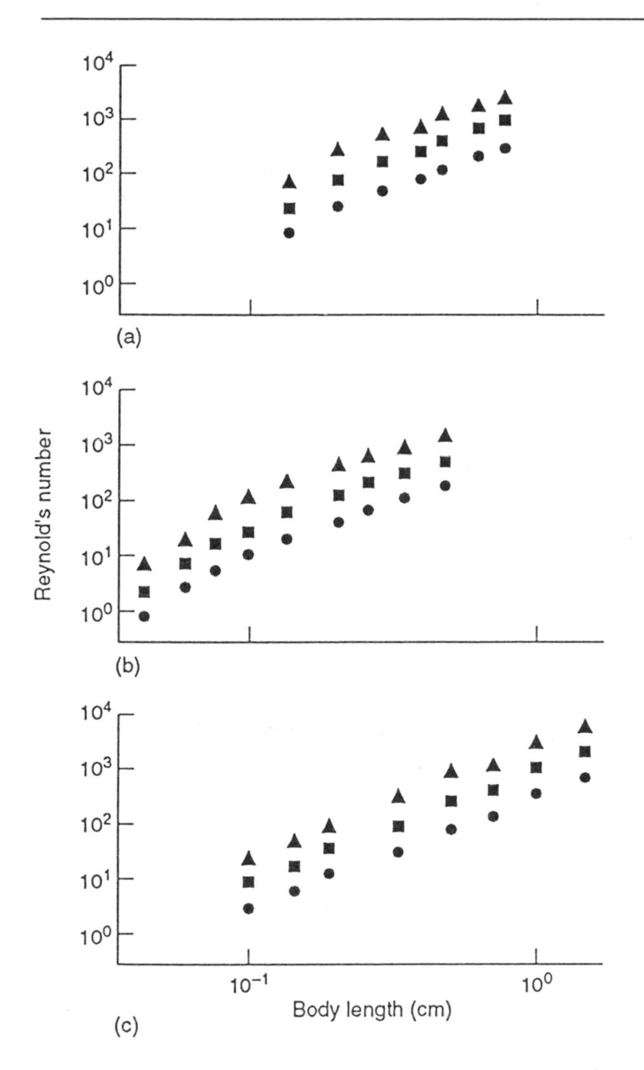

FIGURE 3.6 Reynolds number through the long axis (R_L) of aquatic invertebrates in a laboratory flume: (a) freshwater limpet *Ancylus fluviatilis*, (b) mud snail *Potamopyrgus jenkinsi*, (c) amphipod *Gammarus fossarum*. Current velocities measured 2 mm above the bottom are ● = 5, ■ = 16 and ▲ = 37 cm s^{-1}. As is evident from equation 3.1, R_L increases linearly with current velocity and body length. (After Statzner, 1988.)

obstacle is reached, producing a map of the substrate profile. Other, more approximate techniques based on substrate size classification, including visual methods, are described by Statzner, Gore and Resh (1988).

Inserting estimates of the nearbed velocity profile and of substrate roughness into equation 3.1 results in a roughness Reynolds number (Re_*), useful for comparing nearbed flow conditions. Actual detection of a viscous sublayer of greatly reduced flow requires a finer-scale velocity profile than present techniques permit. However, a laminar sublayer is likely to be present when Re_* is less than 5, disrupted in the range 5–70, and absent when Re_* is greater than 70 (Table 3.2).

A number of indirect methods have been suggested as surrogates for direct measurement of nearbed flow conditions. Smith (1975) suggested that the relationship between mean velocity and U_* could be estimated from the ratio of D to k, where D is the total depth and k is grain length or height of a projection (Table 3.2). This allows U_* and Re_* to be estimated without fine-scale velocity data (Davis, 1986). First, one finds U_* from the formula:

$$\bar{U}/U_* = (5.75 \log [12D/k]) \quad (3.3)$$

The thickness of the viscous sublayer (δ) is then estimated as:

$$\delta = 11.5v/U_* \quad (3.4)$$

However, this value of δ is an average for a stream section, and thus says little about δ at a point, and is inappropriate under conditions of rough flow (Carling, 1992).

Another approach, developed by Statzner (1981), results in a hydraulic stress index (HSI). This is a dimensionless number calculated from easily obtained measurements, namely depth and current in the main channel, hydraulic radius (R), channel slope (I), and a roughness coefficient (k_s). The HSI, a dimensionless number, is calculated from these measurements as:

$$\mathrm{HSI} = \frac{\langle \bar{U} \rangle^2}{\langle k_s [D/(2+D)]^{2/3} \rangle} \frac{\langle 1 \rangle}{\langle 1000 \rangle} \quad (3.5)$$

where k_s, is obtained by solution of the Manning equation:

$$\bar{U} = k_s R^{2/3} I^{1/2} \quad (3.6)$$

Statzner suggests that the HSI is an estimate of streambed hydraulic stress experienced by benthic organisms; in essence, an indirect estimate of the thickness of the laminar sublayer.

Subsequently, Statzner and Müller (1989) developed an apparatus to provide quick and standard estimates of nearbed flows. Hemispheres of the same size but of different densities are placed on a horizontal plane of Plexiglas resting on the streambed. The heaviest hemisphere moved by the current is used as an indicator of flow characteristics near the streambed. Their data in several streams comparing hemisphere density, flow statistics such as Re, Fr and Re*, and measurement variables such as $\bar{U}$, D and k, establish that all are highly correlated. The potential benefits of such an approach are several. By eliminating the need for more labor-intensive approaches, Statzner and Müller's standard hemispheres allow detailed mapping of the mosaic of hydraulic environments on the stream-bed. Their method can be used under high-flow conditions, and very likely can be left in place to quantify episodic events by tethering hemispheres with fish line. However, because the accuracy of this method in estimating shear stress at the streambed is uncertain, it is probably best used as a relative measure.

Despite much recent progress in this area, it is best to end on a cautionary note. Roughness is difficult to measure, and results are affected by how one arrives at an estimate. Nearbed flow conditions in nature cannot be measured at a fine enough scale to verify actual flow conditions near the substrate. All of the estimates of nearbed flow conditions, viscous sublayer depth and 'hydraulic stress' are based on untested assumptions of the applicability of engineering equations to nearbed conditions in real streams. Novel methods such as the hemispheres of varying density are themselves surrogates for these unconfirmed estimators. For these reasons, further development of fine-scale current meters and shear-stress sensors is critical to a full understanding of nearbed flow conditions.

(d) Classification of nearbed flows

Davis and Barmuta (1989) suggest a classification scheme for nearbed flows based on earlier work of Morris (1955). Their approach will require further testing to determine its practical value, but at least it serves to integrate many of the ideas already discussed. Shear velocity, height of roughness elements, and their spacing are considered to be the main components governing nearbed flow conditions.

Hydraulically smooth flow (a subdivision of turbulent, not laminar flow (Carling, 1992)) occurs over smooth streambeds, requires a low Re*, and allows a viscous sublayer to form. This is most likely under conditions of fine-grained beds and slow currents. More commonly, flow will be hydraulically rough, and Davis and Barmuta recognize four subcategories. When channel depth is shallow relative to substrate roughness ($D < 3k$), such as in riffles and broken water, flow will be very complex. They call this 'chaotic'. Between the extremes of hydraulically smooth and chaotic flow, at water depths greater than three times the height of roughness elements, Davis and Barmuta recognize three additional categories (Figure 3.7). When substrate elements are separated by sufficient distance, the wake behind each element dissipates before the next element is encountered. This is called isolated roughness flow. When spacing between roughness elements is less, their wakes interfere with one another, producing turbulence and high local velocities, termed wake interference flow. Lastly, skimming flow describes the circumstance when roughness elements are very closely spaced, which allows flow to skim across the tops of elements and produces a relatively smooth flow environment and slow eddies in the intervening spaces.

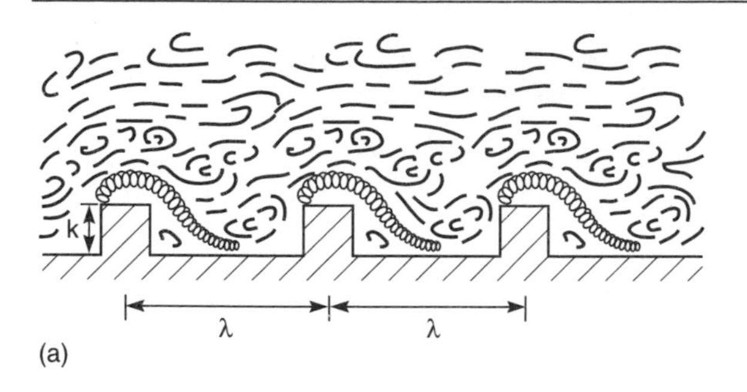

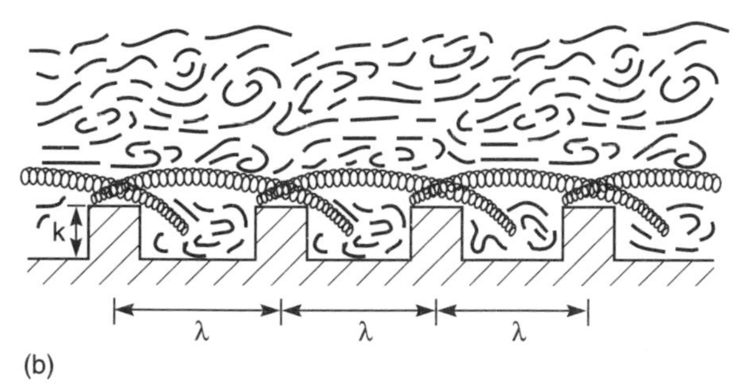

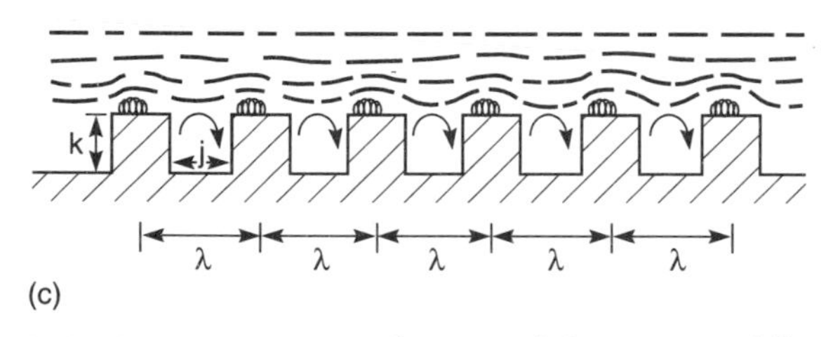

FIGURE 3.7 Conceptualization of three types of flow occurring over a rough surface, depending upon differences in relative roughness and longitudinal spacing between roughness elements. (a) Isolated roughness flow, (b) wake interference flow, (c) quasi-smooth flow. The spatial arrangement of substrate elements is defined by roughness height, k; groove width, j; and the longitudinal distance between roughness elements, λ. (Redrawn from Davis and Barmuta (1989), after Chow (1981).)

3.2 Substrate

Substrate is a complex aspect of the physical environment. What comes to mind first are the cobbles and boulders in the bed of a mountain stream, and the silts and sands that are more typical of lowland rivers. Current, together with available parent material, determines mineral substrate composition. However, even in this seemingly straightforward situation, organic detritus is found in conjunction with mineral material, and can strongly influence the organism's response to substrate. There are, additionally, many kinds of organic substrates in running waters, from minute organic fragments up to fallen trees in size, along with rooted plants, filamentous algae, even other animals. In essence, the substrate includes everything on the bottom or sides of streams or projecting out into the stream, not excluding a variety of human artifacts and debris, on which organisms reside (Minshall, 1984). There are instances where substrate is relatively uniform, as in sandy bottoms of low-gradient rivers, but usually it is very heterogeneous.

Determination of the role of substrate is further complicated by its tendency to interact with other environmental factors. For example, slower currents, finer substrate particle size and (possibly) lower oxygen are often correlated. In addition, the size and amount of organic matter, which affect algal and microbial growth, vary with substrate. This natural covariation of environmental factors makes it very difficult to ascribe causality from field surveys; moreover, the experimental manipulation of single variables is not always practical. However, while it is unwise to claim that substrate, and only substrate, accounts for observed distributions of organisms, there is no doubt of its importance. We will begin with an examination of the types of substrates found in running waters, and the evidence that this influences the distribution and abundance of the biota. Finally, we will review some of the studies that attempt to dissect the different ways that substrate affects organisms. Excellent reviews of organism–substrate relationships in streams can be found in Cummins (1962), Hynes (1970) and Minshall (1984).

3.2.1 Different kinds of substrates

Because substrate is so diverse, it is not easily categorized on a linear scale as can be done with current, temperature and other physical

TABLE 3.3 The classification of mineral substrates by particle size, according to the Wentworth Scale (After Cummins, 1962; Minshall, 1984)

Size Category	*Particle Diameter (range in mm)*	*Phi (ϕ) Value ($-\log_2$ smallest diameter)*
Boulder	>256	≤−8
Cobble		
Large	128–256	−7
Small	64–128	−6
Pebble		
Large	32–64	−5
Small	16–32	−4
Gravel		
Coarse	8–16	−3
Medium	4–8	−2
Fine	2–4	−1
Sand		
Very coarse	1–2	0
Coarse	0.5–1	1
Medium	0.25–0.5	2
Fine	0.125–0.25	3
Very fine	0.063–0.125	4
Silt	<0.063	≥5

variables. As we shall see, particle size lends itself to rigorous measurement and can be reduced to a single term such as median size. However, one cannot average observations of, say, moss, rocks and submerged wood; as a consequence, descriptions of substrate often consist of the relative amount of various categories. What feature of the substrate should be measured can be a difficult decision. Is it the size of a stone that matters, its surface area, or the interstitial spaces and how tightly stones are packed together? Substrate typically is highly variable from place to place, exhibiting small-scale patchiness both vertically and horizontally within the streambed, and changing over time in response to fluctuations in flow. For simplicity we will separate inorganic from organic substrates, but this distinction is mainly a convenience.

(a) Inorganic substrates

Probably no substrate is totally lacking in organic matter, but the bed material of many streams consists primarily of inorganic particles ranging in size from clays and silts to boulders and bedrock. Particle size, factors relating to the mix of particles, surface texture and the temporal persistence of bed material are among the most critical physical aspects of mineral substrates.

Particle size, usually expressed as particle diameter obtained by direct measurement, from settling velocities, or by sieving, is the most easily quantified of these measures. Cummins (1962) introduced to river ecology the Wentworth (1922) scale of particle size classification (Table 3.3). Phi (ϕ) is the negative $\log_2$ of the smallest diameter in each size group, in millimeters. Larger particles are easily sieved, but silts and clays require elutriation and settling. The material in each size category is dried and weighed. Then one can calculate, from the size composition based on mass, such terms as median particle size (MPS), standard deviation, and so on. Table 3.3 presents the more precise descriptive terms used in classifying particular substrates. However, there are some inorganic substrates for which it is not helpful. These include bedrock, where the streambed consists entirely of exposed and unfragmented bed material; and marl, which is a precipitate of calcite that forms crusty deposits in limestone-rich areas.

Some of the more obvious and large-scale patterns in substrate composition were discussed in chapter 1. Substrate of course depends on the parent material available, but there is a general tendency for particle size to decrease as one proceeds downstream. In many regions one finds larger stones and boulders in mountainous areas, and sandy bottoms in lowland rivers. Even 'muddy' rivers have mainly sand and fine gravel in their substratum, and silts are found primarily in backwaters or during periods of greatly reduced flow (Hynes, 1970). Riverbeds of mud and silt are unusual because the size of the smallest particle entrained depends on current velocity, and currents usually remain constant or increase somewhat as one proceeds down river (chapter 1).

At a more local scale than the length of a river, we also can recall from chapter 1 the topics of riffle–pool alternation, meandering and point bar development, which contribute to small-scale horizontal variation in substrate composition. Vertical heterogeneity is observed in substrate due to the ability of coarse material to protect finer materials beneath them, referred to as armoring. Flood events, with their potential to sort, remove, and re-deposit substrate, add temporal variability.

Mineral substrates have characteristics beyond their average size. The surface area of individual particles and the amount of texture are rarely quantified in field measurements. Surface area is at least amenable to measurement, whereas texture requires some arbitrary ranking scale. Some effort has been made to experiment with the effects of these variables on substrate choice by invertebrates, but little can be said about the range of conditions that occur in natural streams.

The stability of the substrate is one of its most important features. This depends on the spatial and temporal scale being considered (Figure 1.9), and on the frequency of extreme hydrological events. Sand beds are active at flows below bankfull. Large rivers in alluvial plains have beds of shifting sand that are transported as slowly moving dunes, as high as 6–8 m in the Amazon River (Sioli, 1984). In general, the larger the particle, the longer its expected residence in place. Given that movement of the bed load is difficult for geomorphologists to measure, it is not surprising that river ecologists have little direct information concerning how substrate stability affects organisms.

(b) Organic substrates

Very small organic particles (less than 1 mm) usually serve as food rather than as substrate, except perhaps for the smallest invertebrates and microorganisms. Larger organic material, from plant stems to submerged logs, generally functions as substrate rather than food. However, there are plenty of examples that blur this distinction. Autumn-shed leaves on the streambed are a substrate to insects that graze algae from their surfaces, and food to insects that eat the leaves themselves. Even logs meet the nutritional needs of some invertebrates (chapter 6). More commonly, however, large organic substrates serve as perches from which to capture food items transported in the water column, as sites where fine detrital material accumulates, and as surfaces for algal growth.

Size categories also are useful for organic matter, although mainly for investigating the dynamics of the detritus itself (chapter 5). From the substrate perspective, the main categories are fine particles, autumn-shed leaves, submerged wood, moss and the surfaces of higher plants. Clearly these are not amenable to the statistical averaging one does with mineral substrates.

Fine-scale heterogeneity in current and mineral substrate affects the distribution of organic

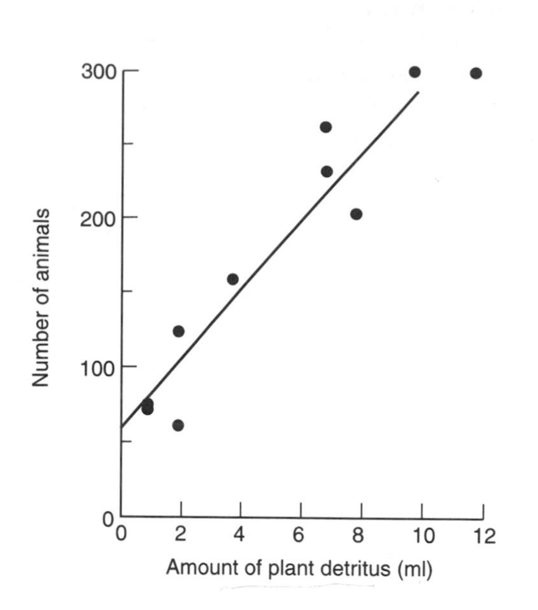

FIGURE 3.8 The number of animals collected from a small area of stony substrate using a handnet, *versus* the volume of fine plant detritus collected at the same time. Samples collected from a single riffle in a Scottish stream in April. (From Egglishaw, 1964.)

detrital particles, and the availability of detritus influences the distribution of organisms within the substrate. Egglishaw (1964) collected macroinvertebrates in a handnet at a number of sites in a riffle in the River Almond, Scotland, and also quantified the volume of fine plant detritus in the net. The number of animals collected depended strongly on the amount of plant detritus (Figure 3.8). To test whether the detritus acted as food or was itself a substrate, Egglishaw compared insect colonization of mineral substrates containing a standard volume of plant detritus, *versus* substrates containing an artificial detritus made from the inner-tubing of a car tire. After 17 days, the number of invertebrates that colonized stones with the artificial substrate was similar to controls lacking any detritus. Substantially more insects colonized mineral substrates supplemented with fine organic detritus, and these were organisms that routinely used fine organic matter as food.

As we progress to larger organic particles, their role as substrate becomes increasingly apparent. Autumn-shed leaves are a significant feature of woodland streams during at least part of the year. Aggregations of leaves on the stream bottom usually support the greatest diversity and abundance of invertebrates (Mackay and Kalff, 1969), and the addition of leaves to mineral substrates results in higher densities of animals (Reice, 1980). On the other hand, mosses and some other plants that are macroscopic but relatively small maintain very high local densities of animals without themselves serving as food. Vegetated substrates in Doe Run, Kentucky, supported far greater populations of the amphipod *Gammarus* and isopod *Asellus* than did bare riffles and shifting sand regions (Minckley, 1963). Based on Hynes' (1970) re-calculations of Minckley's data, the ratio of total numbers on a moss (*Fissidens*), watercress (*Nasturtium*), bare riffles and sandy areas was 47:4.8:1.4:1 at one site, 22:7.5:1.6:1 at another. The plants serve as a refuge, and a trap for silt and organic matter, but provide little or no direct nourishment. Larger plants such as *Potamogeton* support very high densities of midge larvae that graze upon algae growing epiphytically on their leaves and other surfaces (Tokeshi and Pinder, 1985). Submerged wood is yet another category of organic substrate, variable in size but as large as fallen trees. Here we might expect the material to serve almost exclusively as substrate, and as a surface on which more edible material might be found. For many taxa this is the case, but no organic substrate is entirely free of consumers, as we shall see below.

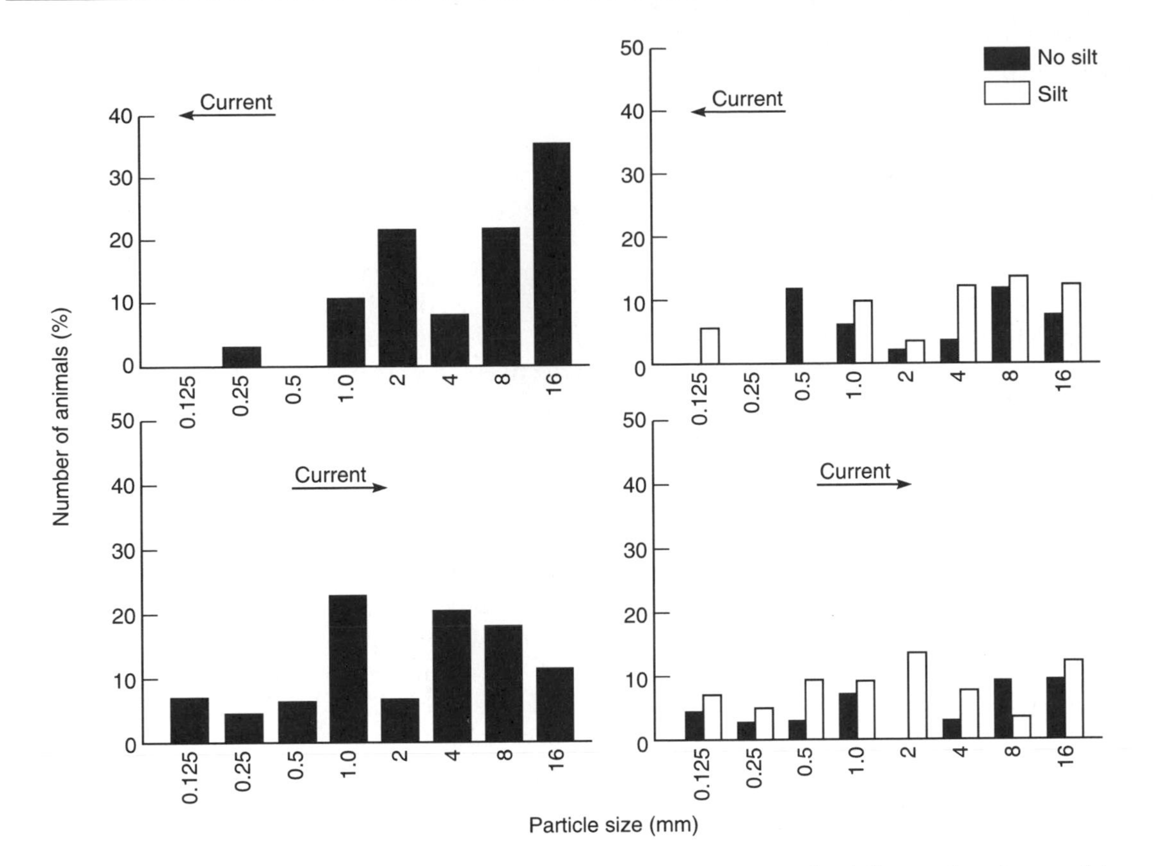

FIGURE 3.9 Preference of the mayfly *Caenis latipennis* tested among eight different size classes of inorganic substrates in a laboratory stream. *Caenis latipennis* preferred coarse sand and larger particles, but because it tended to move upstream, this size preference was weakened when test substrates were arranged so that largest substrates were at the downstream end of the test chamber. Addition of a 1 mm film of silt increased the suitability of all substrates. (From Cummins and Lauff, 1969.)

3.2.2 Characteristic fauna of major substrate categories

The great majority of stream-dwelling macroinvertebrates live in close association with the substrate, and so they have been the main focus of organism–substrate studies. When one contrasts broad categories such as sand, stones and moss, many taxa show some degree of substrate specialization. When one examines preferences among stones of various sizes, substrate specialization is less apparent, and preference is often exhibited as statistical patterns of abundance across the particle size spectrum. However, some stream-dwelling organisms are quite restricted in the conditions they occupy, and biologists have a number of terms to describe these substrate specialists.

Lithophilous taxa are those found in association with stony substrates. Streambeds of gravel, cobble and boulders occur in a great many areas around the world, harboring a diverse fauna that Hynes (1970) remarks is broadly similar almost everywhere. Many species are equally common on stones of all sizes, some are demonstrably more likely to be found associated with a particular size class (Figure 3.9), and a few are highly restricted in their occurrence. Larvae of the water penny (Psephenidae) occur mainly on the undersides of rocks, and often under boulders in torrential flow. Pyralid moth larvae live underneath silken shelters constructed within depressions on rock surfaces. Attached and encrusting growth forms require a substrate that is not easily overturned by current, and larger stones are of course more stable. The longer the life-span, the more critical this is. Diatom populations are greatly reduced by storms that scour and flip substrates (Douglas, 1958), although populations of these single-celled organisms quickly recover. Because they grow more slowly, mosses, bryozoans and sponges are found mainly on larger stones or in locations where scouring is infrequent (Figure 4.10).

Sand is generally considered to be a poor substrate, especially for macroinvertebrates, due to its instability, and because tight packing of sand grains reduces the trapping of detritus and can limit the availability of oxygen. Nevertheless, a variety of taxa, termed psammophilous, are specialists of this habitat. The meiofauna, defined as invertebrates passing through a 0.5 mm sieve, can be very abundant, dwelling interstitially to considerable depth. In Goose Creek, a sandy bottom stream in Virginia, Palmer (1990) reported meiofaunal densities (rotifers, oligochaetes, early instar chironomids, nematodes and copepods) that averaged over 2000 per 10 cm^2 and at times reached nearly 6000 per 10 cm^2. Soluk (1985) recorded densities of very small midge larvae (Chironomidae) as high as 85 000 m^{-2} from shifting-sand regions within a large river of northern Alberta. The psammophilous fauna includes some macroinvertebrates as well, and they can exhibit distinctive adaptations, often associated with respiration. The dragonfly nymph *Lestinogomphus africanus*, found burrowing deep in sandy-bottom pools in India, has elongated respiratory siphons that reach above the sand surface (Hora, 1928). Several mayflies, including *Dolania* in the southeastern USA, have dense hairs that apparently serve to keep their bodies free of sand (Hynes, 1970). The larva of a South American species of *Macronema*, a caddisfly in the family Hydrophychidae, builds a chimney-like intake structure into its feeding chamber in order to exclude sand grains from its food-capturing net (Sattler, 1963).

Burrowing taxa can be quite specific in the particle size of substrate they inhabit. The mayflies *Ephemera danica* and *E. simulans* burrow effectively in gravel. *Hexagenia limbata* cannot, but does well in fine sediments. Substrates composed of finer sediments generally are low in oxygen, and *H. limbata* meets this challenge by beating its gills to create a current through their U-shaped burrows (Eriksen, 1964).

Xylophilous or wood-dwelling taxa illustrate that woody debris constitutes yet another substrate category of lotic environments. Dudley and Anderson (1982) considered 52 taxa in the northwestern USA to be closely associated with wood, and another 129 as facultatively associated. Wood appears to be substrate more often than it is food, although some taxa, such as the beetle *Lara avara*, feed mainly on wood and many taxa obtain some nourishment from a mix of algae, microbes and decomposing wood fiber found on wood surfaces. Woody material is an important substrate in the headwater streams of forested areas, where 25–50% of the streambed is wood and wood-created habitat (Anderson and Sedell, 1979). It is also very important in lowland rivers where 70% or more of the bed is composed of sand, and wood provides the only stable substrate. In lowland streams that flood nearby forests, wood is a significant component of habitat available seasonally.

The invertebrate taxa that live in association with aquatic plants are referred to as phytophilous. Many species utilize moss, and a few are found primarily in moss. Examples include the free-living caddis larva *Rhyacophila verrula*, and a number of mayflies with backward-directed dorsal spines, evidently to prevent entanglement (Hynes, 1970). Many taxa that utilize moss do so opportunistically, however, as in Minkley's (1963) observations described earlier. A substantial number of invertebrates are also found on the surface of submersed macrophytes. Tokeshi and Pinder (1985) examined the aquatic insect assemblage occurring on two architecturally distinct aquatic plants, *Potamogeton × zizii*, which has oblong crenulate leaves, and *P. pectinatus*, which has narrow, thread-like leaves. Living on these plants were 18 species of the Chironomidae, along with two members of the Simuliidae, three trichopterans, and one ephemeropteran. Aquatic insect species richness did not differ between the two species of *Potamogeton*. However, many of the insect species had distinctive micro-distributions on the plant surface, and these somewhat differed between the two structurally distinct aquatic macrophytes.

A number of species of fishes and other vertebrates of rivers tend to occur on or near particular substrates, although most are not so constrained to life on the riverbed as are the invertebrates. Some fishes such as sculpins (*Cottus*) occur within the substrate down to depths of many centimetres in coarse gravel, but most species live on or above the streambed and the majority occur over a wide variety of substrates (Hynes, 1970). However, some fishes are quite specialized in their substrate affinities. The darters live an epibenthic existence, and a number of them are quite restricted in occurrence. So, for example, the mud darter (*Etheostoma asprigene*) is restricted primarily to the backwaters of larger rivers of the Mississippi River, as is the southern sand darter (*Ammocrypta meridiana*) to clean, sandy substrates of the Mobile River basin, and the blenny darter (*Etheostoma blennius*) to the gravel and rubble bottom of fast riffles in Tennessee River tributaries (Lee *et al.*, 1980).

Even when substrate associations are as clear-cut as with these darters, fish often do not 'use' the substrate as directly or extensively as a bottom-dwelling invertebrate in the same locale. At spawning, however, and for some time after hatching, many fishes of running waters require particular substrate conditions. The majority of freshwater fish select hard substrates for reproduction, from individual large stones to some mix of gravel, and it is likely that the availability of substrate for spawning affects the distribution and abundance of many fishes (Hynes, 1970). An advantage of coarse mineral substrates is that they can be sculpted into nests, where eggs and sperm can mix without being swept away by the current. In addition, because water flows into the interstices of coarse substrates, ample oxygen is transported to buried eggs. Mineral substrates also allow behavioral elaboration. A number of species move rocks or pebbles with their mouths; the North American chub *Nocomis* erects structures as high as 30 cm and 1 m in diameter, evidently for the purpose of attracting mates. A few species such as the logperch *Percina caprodes* spawn in sand, depositing sticky eggs that become camouflaged with a sandy coating. Among the species-rich darters, examples also exist of species that spawn on such specialized surfaces as rotting vegetation (*Etheostoma exile*), the macroalga *Cladophora* (*E. blennioides*), and other rooted plants (*E. lepidum, E. punctulatum*) (Hynes, 1970).

3.2.3 The influence of substrate on organism abundance and diversity

Within almost any stretch of river or stream we can observe a diversity of substrate types, and each substrate is itself complex. This naturally leads one to investigate the faunal assemblages found in particular substrates, and also to ask whether some substrates favor greater biotic diversity and abundance. These questions have been addressed by thorough field surveys and a variety of experiments. In general, diversity and abundance increase with substrate stability and the presence of organic detritus. Other factors which appear to play a role include the mean particle size of mineral substrates, the variety of

TABLE 3.4 Abundance and species diversity of aquatic insects found in five habitats (characterized mainly by their substrates) in a Quebec stream. Values are annual averages (From Mackay and Kalff, 1969)

Habitat	*Abundance (No. m^{-2})*	*No. of Species*	*Diversity*[a]
Sand	920	61	1.96
Gravel	1300	82	2.31
Cobbles and pebbles	2130	76	2.02
Leaves	3480	92	2.40
Detritus[b]	5680	66	1.73

[a] Diversity = $(S - 1)/\log_e N$
[b] Finely divided leaf material in pools and along stream margins.

sizes, and surface texture, although it is difficult to generalize about their effects.

(a) Field surveys of entire assemblages

Small woodland streams are very heterogeneous. Within at most a few tens of meters, one can usually find locations of sand, gravel and perhaps larger stones, and areas both rich and poor in organic matter such as sticks, leaves and finer particles. Mackay and Kalff (1969) demonstrated that the numbers and species diversity of aquatic insects did indeed vary among these substrates (Table 3.4). Numbers were much higher on organic than on mineral substrates, and least on sand. Some 120 species were identified overall; leaves supported the most, and sand the least.

In lowland rivers, submerged wood is often the only stable substrate present. Because 70% or more of the substrate of lowland rivers is composed of sandy material, fallen trees (snags) can be disproportionately important as a substrate. In the Satilla River, Georgia, Benke *et al.* (1984) estimated that snag, mud and sand substrates occurred in the ratio 1:1.4:14 at an upriver site, and 1:3.6:18 at a downriver site. They also made a thorough study of the numbers and biomass of aquatic invertebrates, and estimated their production (biomass produced throughout the year) at both sites. As Table 3.5 demonstrates, snags supported more taxa and a far higher biomass of invertebrates than did mud, where values were somewhat higher than in sand. Interestingly, total numbers per unit area did not differ markedly between snags and sand. However, the invertebrates in the sand substrate were mostly oligochaetes and psammophilous midges of very small size, and so their biomass was modest. Integrating among all habitats in the river channel, snag surfaces account for over half of the invertebrate biomass. Although sand was found to support a low biomass per unit surface area, it constituted 70–80% of the substrate, and as a consequence was responsible for most of the remainder of total invertebrate biomass.

Much fine-scale heterogeneity is apparent in streams with stony beds. Pools and stream margins generally have finer substrates than riffles and the stream center, and a number of studies have investigated whether macroinvertebrates vary accordingly. Often the differences are minor, and this might be expected in small, high-gradient streams where substrate conditions are not very different between riffle and pool areas, and small spatial scale allows easy access to all locations. Dudgeon (1982) made a detailed study of the distribution of 47 macroinvertebrate taxa across the width of a shaded riffle in a Hong Kong stream. About 25% of the taxa showed no obvious substrate or microhabitat affinity, although others did to varying degrees. Species diversity was maximal midway between the bank and middle of the stream,

TABLE 3.5 The number of taxa and standing crop biomass of invertebrates found in snag, sand and mud habitats in the Satilla River, Georgia. These substrates occurred in the ratio 1 (snag): 3.6 (mud): 18 (sand) at a downriver site and 1 : 1.4 : 14 at an upriver site (From Benke *et al.*, 1984)

	Wood substrates			*Sand*			*Mud*		
	No. of genera	*Biomass (mg m^{-2})*		*No. of genera*	*Biomass (mg m^{-2})*		*No. of genera*	*Biomass (mg m^{-2})*	
		Lower site	*Upper site*		*Lower site*	*Upper site*		*Lower site*	*Upper site*
Diptera	17	243	696	15	64	124	11	148	309
Trichoptera	9	4222	1581	0	–		3	24	30
Ephemeroptera	5	97	56	0	–		0	–	
Plecoptera	2	137	109	0	–		0	–	
Coleoptera	3	218	117	1	8	11	0	–	
Megaloptera	1	379	259	0	–		0	–	
Odonata	3	529	578	1	–		0	–	
Oligochaeta	0	–		3	22	22	0	420	290
Totals	40	5825	3396	20	94	157	17	592	629

corresponding to the region of greatest substrate heterogeneity. When investigators have compared clean, stony riffles to pools rich in silt (e. g. the Black River, Missouri; O'Connell and Campbell, 1953) the former are generally much richer. In small streams where riffles and pools both contain mainly gravel and larger material, differences are often less noticeable.

(b) Experimental studies of organism–substrate interactions

As just discussed, field surveys provide ample evidence of the importance of substrate to the diversity and abundance of the biota of running waters. Some but by no means all aquatic invertebrates exhibit distinctive substrate preferences. Numbers of individuals and species appear to be greatest on some substrate types (mosses and coarse mineral substrates, for example) and to vary with the availability of detritus. These observations suggest a number of experimental tests, and substrate characteristics are relatively easy to manipulate experimentally. As a consequence there is a substantial body of experimental evidence to further our understanding of the role of substrate (Minshall, 1984). The size, stability and heterogeneity of (primarily mineral) substrates have been the subject of much study, as have the modifying effects of silt and organic matter. Some subtler aspects such as texture, surface area and the amount of interstitial space have received less thorough study.

Many studies have placed baskets of mineral substrates on the streambed to determine the colonization response of individual species and entire assemblages. In general, diversity and abundance of benthic invertebrates increase with median particle size (MPS), and some evidence suggests that diversity declines with stones at or above the size of cobbles (Figure 3.10; Minshall, 1984). There is little doubt that diversity and abundance increase as one proceeds from sand to gravel substrates. Several investigators have found higher numbers and greater diversity using baskets containing pebbles compared with cobbles (Minshall and Minshall, 1977; Wise and Molles, 1979), consistent with the declining right arm of Figure 3.10. However, when stones are combined in size mixtures, more individuals and species are found on cobbles than pebbles, indicating that results in heterogeneous

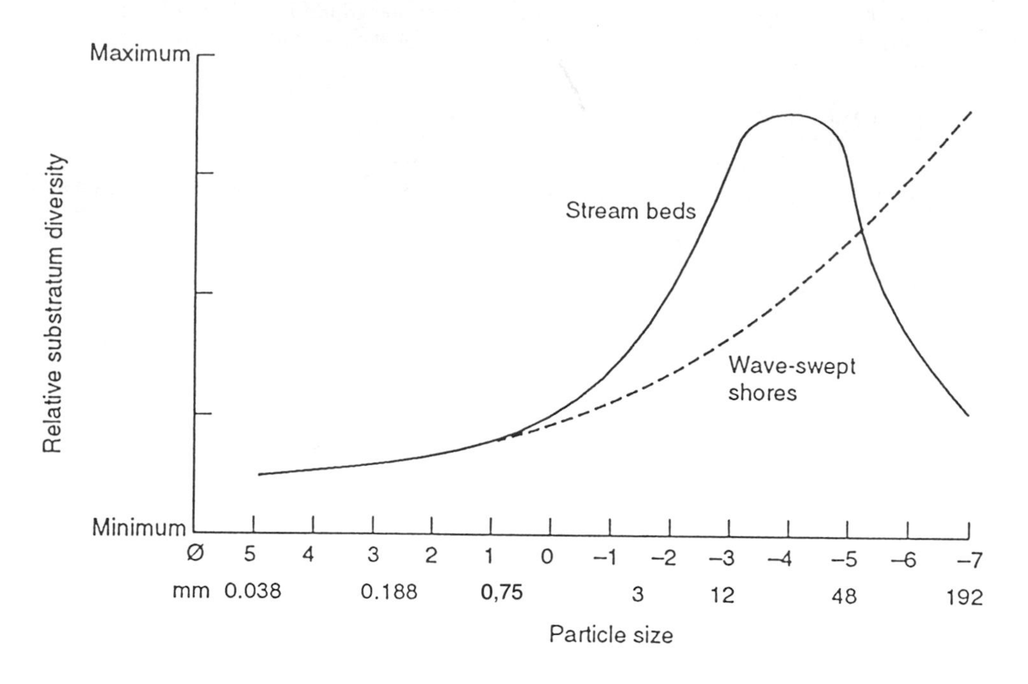

FIGURE 3.10 Hypothetical relationship between substrate diversity (variance in particle size) and median particle size for mineral substrates in streams. (From Minshall, 1984.)

substrates cannot be predicted from studies using single substrates (Minshall, 1984). Simultaneous sampling of substrate size and benthic invertebrates in a small river in southern Manitoba corroborates these points. A positive correlation was found between MPS and number of species in spring, when particles were well sorted, but the relationship broke down later in the year after spring floods altered substrate heterogeneity (de March, 1976).

From the foregoing it is apparent that the size of mineral substrates is important, but is not the sole consideration. Nor do these results explain why larger stones are more favorable. Substrate stability is thought to be part of the answer, and presumably this is one disadvantage of sand, at least to the larger invertebrates. The amount of detritus trapped within the crevices is also likely to be important, and substrates of intermediate size are superior in this regard. A variable mix of substrates ought to accommodate more taxa and individuals, and particle size variance usually increases with MPS (e.g. de March, 1976). Stone surface area and amount of interstitial space must also vary with both MPS and substrate heterogeneity, and these relationships presumably are complex. There have been some attempts to dissect these factors experimentally (Minshall, 1984), but generalizations remain elusive. For example, while one might expect greater species richness in mixed than in uniform substrates, size class mixtures alone (i. e. in the absence of a change in MPS or detritus) apparently do not augment abundance or diversity (Wise and Molles, 1979; Erman and Erman, 1984).

The importance of detritus, demonstrated so clearly in Egglishaw's (1964) early study, is strongly confirmed by subsequent experiments. Using baskets of mineral substrates with and without attached leaves in a small woodland stream, Reice (1980) observed that invertebrate

TABLE 3.6 Number of invertebrates per m^2 of available surface on patches of different sized rocks in New Hope Creek, North Carolina (From Reice, 1980)

	With leaf packs			*Without leaf packs*		
	Gravel	*Pebble*	*Cobble*	*Gravel*	*Pebble*	*Cobble*
No. of individuals	37.5	85.8	177.4	25.5	129.7	133.3
No. of species	2.8	10.1	16.6	2.1	9.2	21.2

abundance and diversity were usually augmented by the presence of leaf packs (Table 3.6). Rabeni and Minshall (1977) found greater numbers of invertebrates colonizing trays of very coarse sand (1–2 cm) than medium gravel (4.5 × 7 cm) in a cool desert stream in Idaho. However, they also found more small detritus within the coarse sand, whereas more sticks and twigs accumulated in the gravel. Indeed, when they compared colonization of sand and gravel containing similar amounts of detritus, overall abundances no longer favored the smaller substrate. Evidently the amount and type of detritus contained within the sediments is sufficiently dependent on the size and mix of the mineral substrates that it is unwise to measure substrate preference without concurrent study of trapped organic matter.

Silt, in small amounts, benefits at least some taxa (Cummins and Lauff, 1969). When silt was added to larger mineral substrates in laboratory preference tests, silt enhanced the preference for coarse substrates in the mayfly *Caenis latipennis* (Figure 3.9) and the stonefly *Perlesta placida.* Eight other species were unaffected, however. Greater amounts of silt can alter the substrate sufficiently to cause marked changes in the invertebrate assemblage, as Flecker (1993) found in a Venezuelan piedmont stream. Heavy silt deposition was uncommon on substrates accessible to abundant bottom-feeding fishes, but in areas the fish avoided, or where they were experimentally excluded, silt deposition was considerable. The outcome was a substrate mosaic, with discernibly different faunas occupying silted and silt-free stone surfaces. In large amounts silt generally is detrimental to macroinvertebrates. It causes scour during high flow, fills interstices thus reducing habitat space and the exchange of gases and water, and reduces the algal and microbial food supply. The adverse effects of siltation were apparent in a study designed to test the influence of predaceous stoneflies on macroinvertebrate colonization into mesh cages (Peckarsky, 1985). Unanticipated high discharge caused heavy silt deposition into some, but not all cages. Under low siltation the stoneflies reduced invertebrate abundance as expected, but heavily silted cages showed substantial overall reduction and obscured any effect of predator treatment.

Substrate texture refers to surface properties such as hardness, roughness, and perhaps ease of burrowing, along with other aspects. It is somewhat subjective, and has not received much experimental study (Minshall, 1984). Erman and Erman (1984) found that more invertebrates colonized granite and sandstone, which have comparatively rough surfaces, than the smoother quartzite. Using substrates constructed from ceramic clay and placed in a stream, Hart (1978) found diversity and abundance to be greater on irregular than on smooth substrates of the same overall size. Clearly there is ample opportunity for further experimental study of these and other facets of substrate.

3.3 Temperature

The temperature of running waters usually varies on seasonal and daily time scales, and among locations due to climate, elevation, extent of streamside vegetation and the relative importance of groundwater inputs. Stream temperatures can be quite constant under certain circumstances. The temperature of groundwater usually is within 1°C of mean annual air temperature, and so wherever groundwater inputs are important, as in springs and some headwater streams, seasonal temperature variation will be slight. A spring source in northern Colorado remained between 8 and 10°C year-round, despite the much greater annual variation in air temperature at this site (Ward and Dufford, 1979). In tropical locations with very constant air temperatures throughout the year, water temperatures also are constant. The Amazon at Manaus, at 29 ± 1°C, is one of the most thermally stable water masses in the world (Sioli, 1984).

Seasonal changes in water temperature in rivers closely follow seasonal trends in mean monthly air temperature, except that in winter the water temperature does not fall below 0°C, and air warms more rapidly in the spring than does water. For temperatures above freezing, Crisp and Howson (1982) found that mean weekly water temperatures (and the growth rate of brown trout) could be predicted very accurately from air temperatures using a 5–7 day lag. Some 60% of their estimates were within ± 1°C, and 80% within ± 1.5°C, of the measured stream temperatures.

The annual temperature range in temperate rivers usually is between 0 and 25°C, but desert streams can reach nearly 40°C, which is near the thermal tolerance even of fishes adapted to these extreme environments (Matthews and Zimmerman, 1990). At high latitudes and elevations, maximum temperatures rarely exceed 15°C, and they can be cooler yet, in very cold climates where ice-cover can extend for over half the year. Because long rivers originate at higher elevations with cool climates and flow into the warmer lowlands, a longitudinal temperature increase is the norm. In Central African streams that arise from ice water on mountains over 4000 m high, the temperature increases from near freezing to the high twenties over their length (Hynes, 1970).

The temperature of large rivers is unlikely to be affected much by shading, as their size conveys considerable thermal inertia and virtually ensures that they are largely exposed to the sun. In small streams, however, shading can substantially alter summer temperatures. Gray and Edington (1969) closely monitored temperature in a small stream in northern England before and after much of the catchment was deforested. The summer maximum increased from 15 to 21.5°C the year after trees were felled. Loss of streamside vegetation due to agriculture and other human activities likely has increased the summer temperatures of many streams, and is an important cause of deterioration in faunal composition (Karr and Schlosser, 1978).

Impoundment of a river in the temperate zone usually alters thermal regime, even in large rivers, and the type of dam has a great influence on downstream temperatures. If the flow through a reservoir is relatively slow, it will develop the thermal stratification typical of lakes (Wetzel, 1983). During the summer, surface waters likely will be warmer than is typical for river water, and deep water will be quite cool, often between 6 and 10°C. During the winter, surface water will be near freezing and deep water will be 4°C, which is the temperature at which water is most dense. A dam that releases surface water will usually increase the annual temperature range immediately downstream, whereas a deep release dam will lessen annual variation. The biological consequences of altered temperature regimes can be substantial (chapter 14).

Diel temperature fluctuations are common in small streams, especially if unshaded, due to day *versus* night changes in air temperature and absorption of solar radiation during the day. Diel temperature changes can be extreme in small, exposed streams at high elevations, where sub-surface flow of the melting snowpack cools the stream at night to below 5°C while daytime warming elevates temperatures to above 15°C (Allan, 1985). Large rivers are less variable than

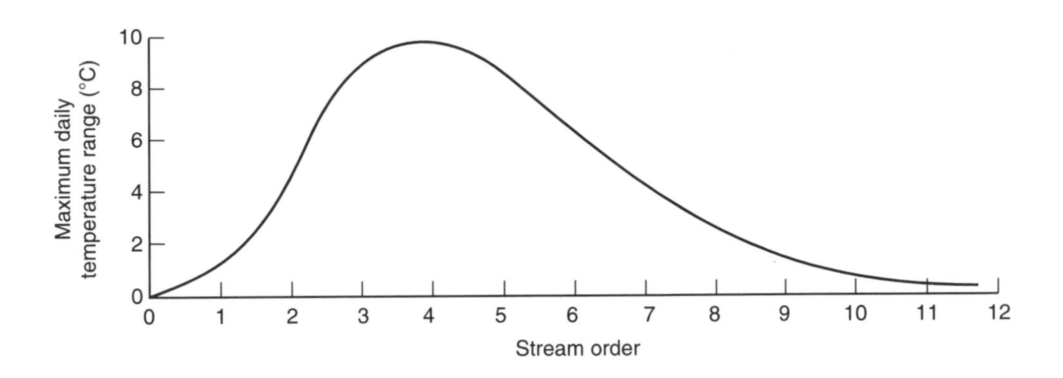

FIGURE 3.11 Maximum daily temperature range in relation to stream order in temperate streams. (From Vannote and Sweeney, 1980.)

small streams because of their greater volume, and springbrooks often are less variable because of groundwater influence and shading. As a consequence, the extent of diel temperature variation usually is greatest in streams of intermediate size (Figure 3.11).

Unlike lakes, which exhibit pronounced vertical thermal stratification in summer, river water is sufficiently well mixed that temperatures are nearly constant from surface to bottom. However, groundwater inputs can result in sub-surface water that is markedly cooler and less variable in summer than surface waters, and also warmer in winter (Shepard, Hartman and Wilson, 1986, see Figure 13.8). As a consequence, invertebrates living deep within the sediments often experience different temperatures to surface-dwelling fauna.

Organisms experience a lengthy growing season in some climates and only a brief period of warm temperatures in others. This suggests that it will be more useful to compare the cumulative temperature experienced under each circumstance than the summer maximum, which might not be very different. Degree-days are useful for this purpose, and are calculated by summing daily mean temperatures above 0°C, or above a developmental threshold if one is known to exist for the species of interest. This approach clearly demonstrates that even when warmest summer temperatures are comparable, degree-days can be very different. For example, three sites in a Rocky Mountain stream all experienced similar maximum summer temperatures near 16°C, but the annual degree-days were estimated to be 800 at 3350 m, 1000 at 3050 m and 1500 at 2740 m (Allan 1985).

Annual degree-days increase with decreasing latitude (Figure 3.12) because of the strong dependency of river water temperature on air temperature. Values in the range of 3000–5000 degree-days are typical of temperate, low elevation sites. At subtropical Silver Springs, Florida, annual degree-days were in excess of 8000 (Odum, 1957), which is an order of magnitude greater than at the highest elevation site studied by Allan. Moreover, the same number of degree-days can result from very different seasonal temperature patterns (Vannote and Sweeney, 1980). Six locations along a stream in Pennsylvania each accumulated about 4200 degree-days annually, but temperature variation throughout the year was much more pronounced at downstream sites (Figure 3.13).

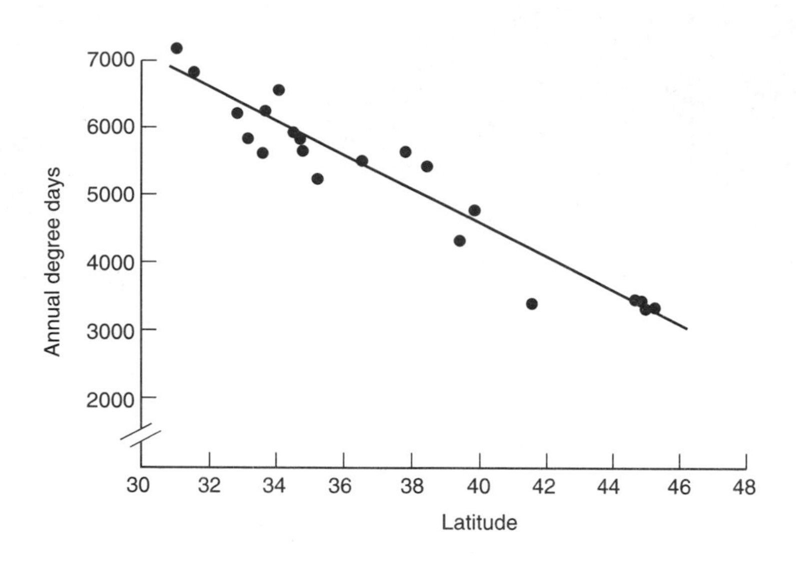

FIGURE 3.12 Total annual degree-day accumulation (>0°C) as a function of latitude for various rivers of the eastern USA. (From Vannote and Sweeney, 1980.)

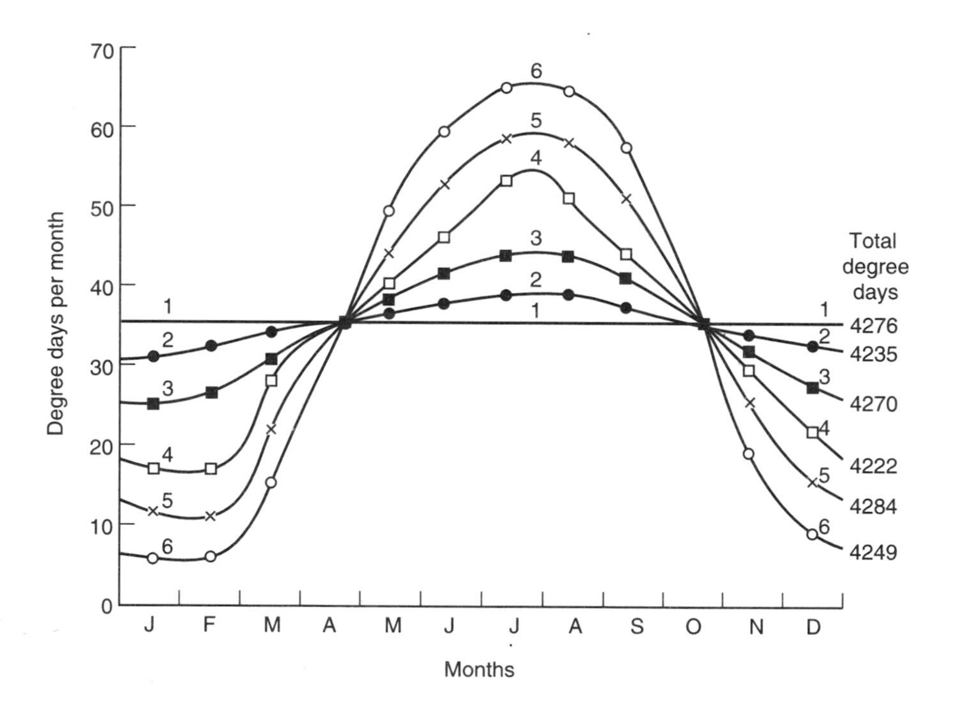

FIGURE 3.13 Degree-day accumulations and monthly totals at six sites along White Clay Creek, Pennsylvania. 1, Groundwater; 2, spring seeps, 3, first-order springbrooks; 4, second-order streams; 5, upstream segment of third-order stream; 6, downstream segment of third-order stream. (From Vannote and Sweeney, 1980.)

3.[illegible] [illegible]es of thermal regime

[illegible] in its distribution to [illegible] e and altitude, and [illegible] [illegible]perature range as well. Species found in cool climates typically are cool-adapted, and those of warm climates warm-adapted. This should not be taken as evidence that temperature preferences are the ultimate determinant (in an evolutionary sense) of where a species lives. It is equally likely that other factors set range limits, and the species has adapted to the conditions where it occurs. Nonetheless, temperature unquestionably sets limits on where a species presently can live. Strayer (1991) illustrates a useful application of such knowledge by forecasting the future distribution of zebra mussels in North America, based on their known distribution with regard to temperature in Europe.

There are a number of reasons for the specific temperature requirements of a particular species. The taxonomic lineage to which it belongs may have originated and diversified in cool waters at high latitudes (e. g. Trichoptera (Ross, 1967) and Plecoptera (Hynes, 1988)) or in warmer water at low latitudes (e. g. Odonata (Corbet, 1980)). The timing of an insect's life cycle, which often is both cued and regulated by temperature, determines the seasons when it most actively grows, consumes resources and is exposed to predators. Thus resource availability and predation risk, two topics that will re-appear in later chapters, can be important evolutionary pressures that are responsible for a particular life cycle and suite of temperature adaptations. A number of studies document how closely related species hatch, grow and emerge in such a neatly staggered sequence that their life cycle separation appears to ameliorate competition (see Figure 9.4). Although this might in fact be the case, the role of competition in shaping life cycle evolution is difficult to establish (chapter 9). By determining when an insect hatches and grows, temperature synchronizes the life cycle to changing seasonal conditions, coordinating growth with resource availability and ensuring the availability of mates.

Species that occupy a narrow temperature range are referred to as stenothermal, while those that thrive over a wide range are called eurythermal. Several species of triclads studied in Indiana by Chandler (1966) illustrate the use of these terms. *Phagocata gracilis* is found only in cool springs, hence is a cold stenotherm, whereas *Dugesia tigrina* is a warm stenotherm that becomes inactive below 11 °C. *Cura foremani* is successful under both temperatures, and so is eurythermal.

Although species have adapted to the cooler and warmer extremes of most natural waters, few taxa are able to cope with very high temperatures. Cold-water fishes cannot survive water above 25 °C for very long, and most warm-water fishes including esocids and many cyprinids have upper limits near 30 °C. Some fishes of desert streams can tolerate nearly 40 °C, a few invertebrates live at up to 50 °C, and specialized cyanobacteria of hot springs survive 75 °C (Hynes, 1970). These are all exceptional cases, however.

Whereas cool-adapted species often cannot withstand summer temperatures much above those they normally experience, warm-adapted species of the temperate zone usually must survive freezing temperatures, because they experience these seasonally. Their absence from streams that are cool in summer is because their efficiency of feeding and ability to grow are insufficient to maintain populations, or because their reproductive needs are not met. Thus trout are absent from warm-water streams partly because they cannot meet their energetic demands at temperatures much above 20 °C, but also because they simply cannot survive above 25 °C. Bass and other centrarchids that dominate warm-water streams in the USA are absent from cool-water streams because they are suboptimal for growth (Hynes, 1970).

Very cold environments can be physically harsh environments even to cool-adapted species, because of the effects of ice on the biota. Scour of the streambed during ice break-up can directly harm the biota. Ice also forms on the streambed as frazil ice, which forms a slush on the stream bottom, and as anchor ice, which forms in shallow waters, first on riffles but later spreading to cover larger areas (Hynes, 1970). A severe winter decline in the bottom fauna of a Utah stream was attributed to scouring of the substrate by frazil and anchor ice (Gaufin, 1959).

Because species differ in their temperature requirements, the flora and fauna of rivers change gradually in composition with latitude or altitude. Changes in species composition along a river's length establish that longitudinal zonation is commonly observed in both the fish (Huet, 1949) and invertebrate fauna (Illies and Botosaneanu, 1963). These authors proposed the existence of some number of distinct zones separated by relatively abrupt transitions. Huet recognized four fish zones as one proceeds downstream, characterized by their dominant species: trout, grayling, barbel and bream. Illies and Botosaneanu (1963) recognized three major zones characterized by their environmental conditions rather than particular species. The spring-fed headwaters or crenon is a restricted region of reasonably constant and cool temperatures, the rhithron refers to the cooler upland section, and the potamon is the warmer and less steep river of the lowlands.

Studies of invertebrate zonation unquestionably demonstrate both additions and losses to the faunal list as one proceeds from higher to lower elevation (Allan, 1975). Fishes typically show an increase in number of species with stream size, although some also disappear (Kuehne, 1962; Horwitz, 1978). Kuehne reported one fish species in a first-order stream, 12 in a second-order stream, 20 in a third-order stream and 27 in a fourth-order stream in Kentucky. Some species that dropped from the faunal list as stream order increased were cold-water stenotherms (brook trout and sculpins), while others such as the fathead minnow (*Pimephales promelas*) and green sunfish (*Lepomis cyanellus*) were specialists of intermittent headwater streams.

From the many studies of longitudinal zonation, it seems that additions usually exceed losses, and few investigators today report their findings in terms of zonation. Although temperature and faunal composition both change substantially as one proceeds from headwaters to river mouth, it should be emphasized that temperature is just one of many environmental variables exhibiting a longitudinal trend, and might not be causal.

In addition to species transitions, movement along an altitudinal or latitudinal gradient often results in a change in the number of generations per year. The same species can complete its life cycle in one year under a warm thermal regime, and require two or three years under a cooler regime (e. g. the caddisfly *Rhyacophila evoluta*, (Décamps, 1967)). Univoltine species are those that complete one generation annually, whereas bivoltine and multivoltine refer to two and many generations per year, respectively. Species of the mayfly genus *Baetis* are univoltine at high elevations and multivoltine at lower elevations (Allan, 1985).

Temperature has diverse effects on the activity and the life cycles of the biota of running waters. It affects voltinism and triggers development, as just described. In addition, temperature influences the rate at which eggs develop and juveniles grow. These in turn determine voltinism, rates of growth and the productivity of the biological community.

Seasonal changes in temperature often act as a cue to development, and the resulting synchronization of the life cycle to seasonal changes in the environment is critical to efficient utilization of resources and the availability of mates. The eggs of *Ephoron album* diapause during the long winter experienced by this high latitude (40–50° N) mayfly. Egg hatching and growth are rapid and synchronous, and the mayfly com-

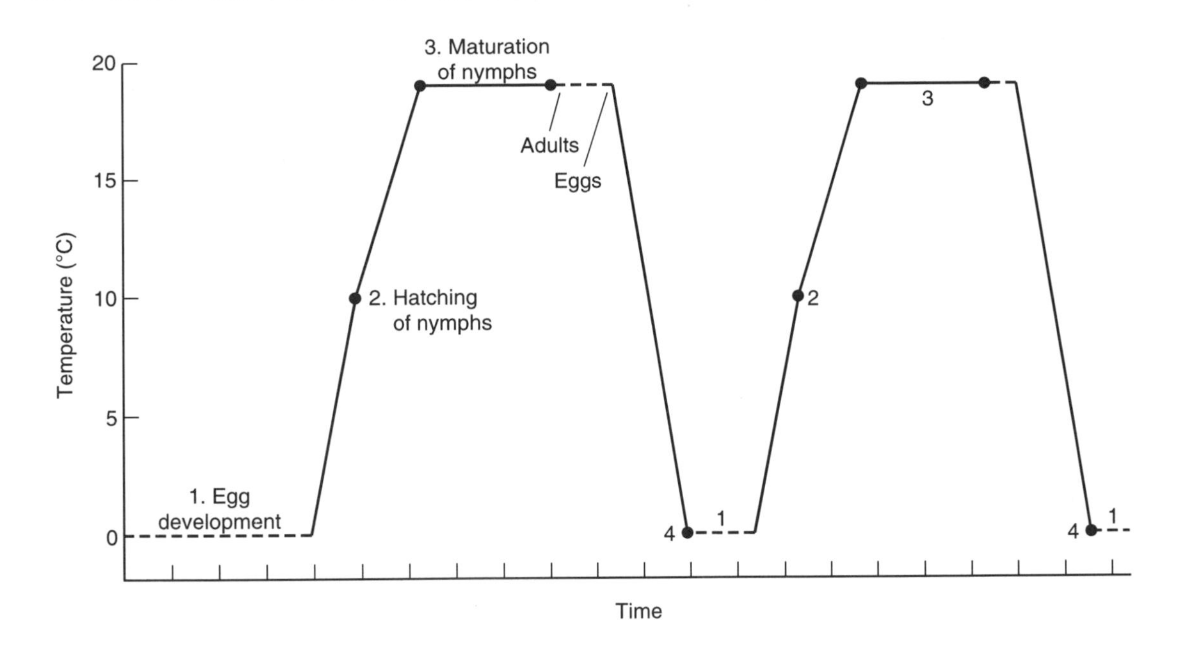

FIGURE 3.14 The sequence of changes in water temperature necessary for the mayfly *Ephoron album* to complete its life cycle. Egg diapause during the winter months at 0°C (1) is broken by rapid warming in the spring which cues egg hatching (2). Nymphs mature and reproduce during the short, warm summer in Saskatchewan, Canada (3) and produce eggs that will not develop (4) until they experience an extended period of freezing temperatures. (From Lehmkuhl, 1974.)

pletes its life cycle in only a few months (Britt, 1962). However, exposure to near-freezing temperatures followed by a rapid increase in temperature is required to break egg diapause. When a deep release dam was built on the Saskatchewan River, winter temperatures were maintained near 4°C and the fauna no longer experienced a prolonged period near freezing, followed by a rapid temperature rise, and so the cue to ending egg diapause was eliminated (Figure 3.14). Not only was *E. album* eliminated from the stretch of river in which the temperature regime was modified, but virtually all other taxa disappeared as well. A fauna previously comprising 12 orders, 30 families and 75 species was reduced to only the midge family Chironomidae (Lehmkuhl, 1974).

Studies of egg hatching clearly illustrate the strong influence of temperature on developmental rates. Some nymphs of the mayfly *Ephemerella ignita* appear in the autumn in small, cool streams of northern England, but the majority appear in the spring. The reasons for this variation were unknown, and made more puzzling by evidence that early hatching nymphs did not survive the winter (Elliot, 1978). By monitoring egg hatching of *E. ignita* at constant temperature, Elliott established that eggs hatched over a wide range of values, although the proportion hatching was reduced at both high and low temperatures (Figure 3.15). Time to 100% hatching was shortened by rearing eggs at higher temperatures, and then lengthened again at the highest temperatures used (Figure 3.16). By combining his experimental results with the annual temperature cycle in a nearby stream, Elliott showed that the duration of egg hatching in nature extended over nearly 8 months (Figure 3.17).

Growth is also highly temperature dependent,

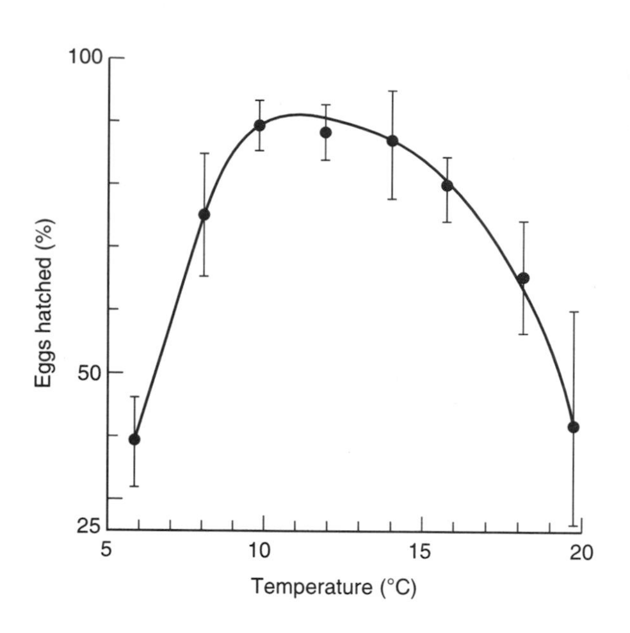

FIGURE 3.15 Percentage of eggs of *Ephemerella ignita* that hatched when maintained at different constant temperatures. Vertical lines give 95% confidence limits. (From Elliott, 1978.)

and this has been demonstrated in many studies using fish (Brett, 1971; Ricker, 1979), crustaceans (Sutcliffe, Carrick and Willoughby, 1981), and insects (Sweeney, 1984). The growth rate of brown trout increased with temperature from 5.6°C to between 12 and 15°C, and then declined drastically at 19.5°C (Elliott, 1975). Although warmer temperatures increase feeding activity and digestion rate, they also raise respiratory rates, and for brown trout the energetic costs of high metabolic rates significantly curtail growth when reared at constant temperatures much above 15°C. Growth rate probably declines in most species when they are near their upper temperature limit, but perhaps not as dramatically as in cold stenothermal species, such as trout. A midge larva studied in a warm-water river in Georgia showed some decline in daily growth at temperatures above 25°C, whereas a mayfly and black fly did not (Figure 3.18; Benke, 1993).

The extent to which cold-adapted aquatic insects remain active and grow at near-freezing temperatures is surprising. The Plecoptera are

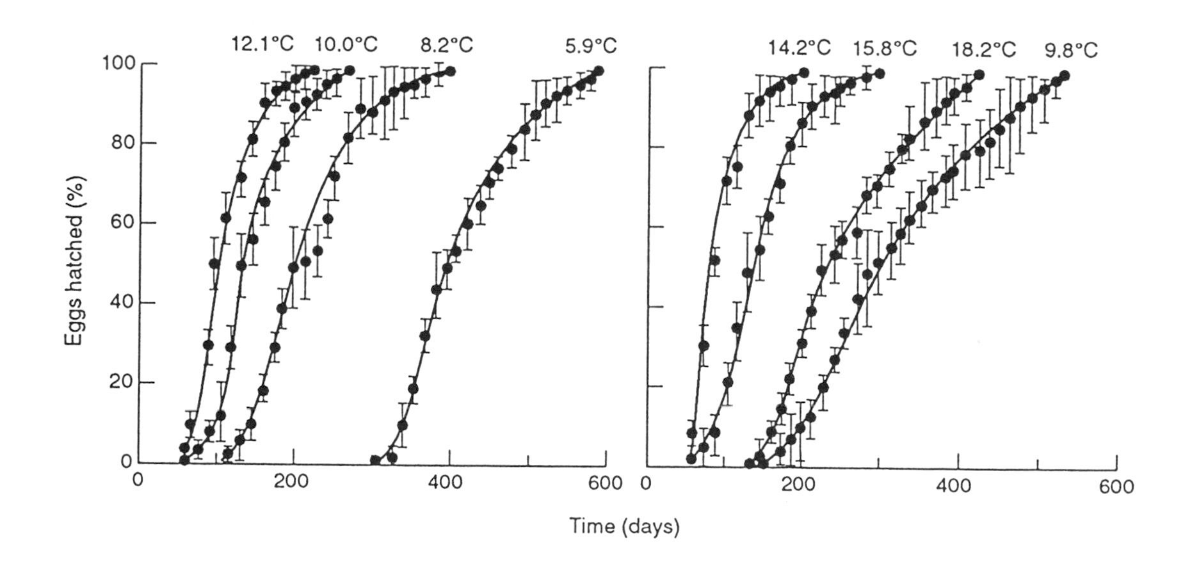

FIGURE 3.16 Cumulative percentage of *E. ignita* eggs hatching *versus* incubation duration at different temperatures. Each point is the mean with 95% confidence limits. (From Elliott, 1978.)

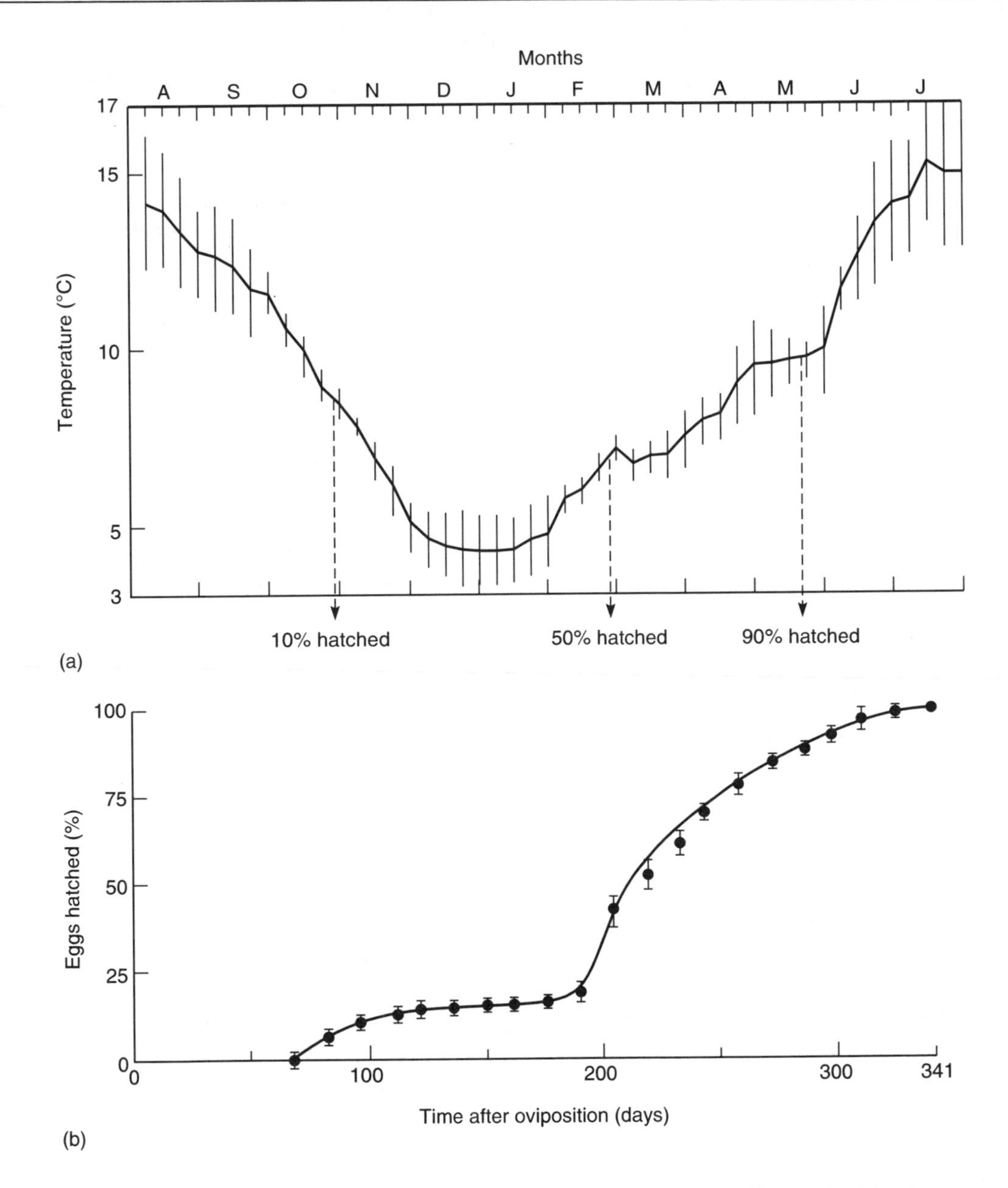

FIGURE 3.17 Predicted and observed hatching of eggs of *E. ignita* in Wilfen Beck. (a) Mean, maximum and minimum water temperatures with predicted dates on which 10%, 50% and 90% of eggs were expected to hatch. (b) Cumulative percentage of eggs hatched from egg masses placed in the stream. (From Elliott, 1978.)

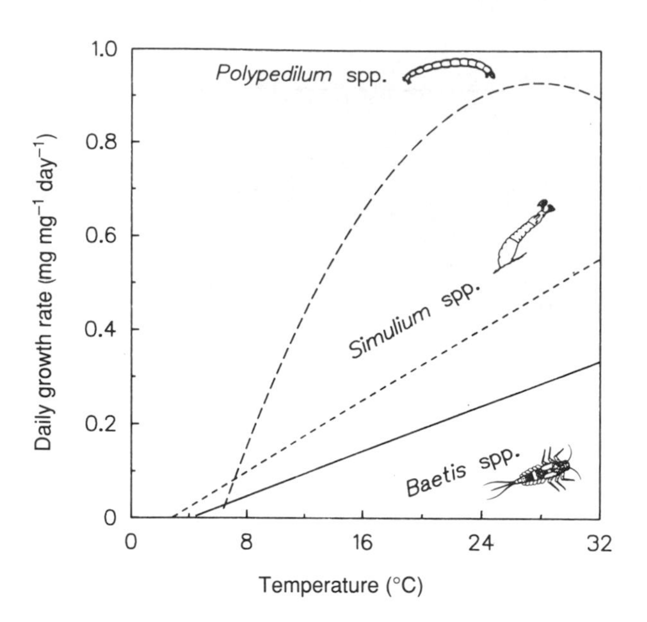

FIGURE 3.18 Daily growth rates (mg mg^{-1} day^{-1}) as a function of temperature for three aquatic insects found on snag habitat in the Ogeechee River, Georgia, and reared in streamside artificial channels. Insects include the midge *Polypedilum*, the black fly *Simulium*, and the mayfly *Baetis*. (From Benke, 1993.)

noteworthy in their ability to grow at near-freezing temperatures, surpassing most other insect groups in this regard. For example, the stonefly *Taeniopteryx nivalis* accomplishes significant growth between November and March at water temperatures that are only a few degrees above freezing (Knight, Simmons and Simmons, 1976), and other 'winter stoneflies' must do so as well.

Temperature also influences body size, which in turn affects fecundity because body size and number of eggs are positively correlated (Vannote and Sweeney, 1980). Adult body size in ectotherms is often inversely related to the temperature at which growth occurs, and a number of studies confirm this pattern in mayflies. *Ephemerella dorothea* emerging from cool tributaries of White Clay Creek were nearly twice the mass of individuals emerging from a warm tributary (Vannote and Sweeney, 1980). Final instar nymphs of *Baetis* from the over-wintering generation in high elevation streams are considerably larger than summer generation individuals, which experience much warmer temperatures during their development (Allan, 1985). There is also evidence that species have an optimal temperature range where both body size and fecundity are maximized. Rearing studies with the mayfly *Centroptilum rufostrigatum* showed adult body size to be reduced at both high and low temperatures, suggesting that intermediate values were optimal (Sweeney and Vannote, 1978).

Of the many influences of temperature on the biota of streams, its effect on the life cycles of aquatic insects, many of which are strongly seasonal in temperate environments, is especially important (Hynes, 1970). Hynes considers species that live longer than a year to be non-seasonal, such as some large stoneflies and the Megaloptera. Most other insects complete their growth in a year or less and exhibit some sort of seasonal pattern (Figure 3.19). In slow seasonal cycles, egg hatching is prolonged and growth occurs over a year, and sometimes longer. Adults emerge in early (S1) or late spring (S2), or in early summer (S3). In fast life cycles, rapid growth follows a long egg diapause, and so immatures are present for a relatively restricted time period. The distribution of timing of adult emergence is approximate. It probably is not fruitful to attempt to apply this scheme rigidly, but it does usefully sum up how critical is seasonal variation in temperature to the life cycles of aquatic organisms.

3.4 Oxygen

As has been mentioned previously, dissolved oxygen in unpolluted flowing water is usually near saturation, and under these circumstances the concentration of oxygen is of little biological significance. Oxygen exchange within surface microlayers may be important, but little is known of this subject. However, the supply of

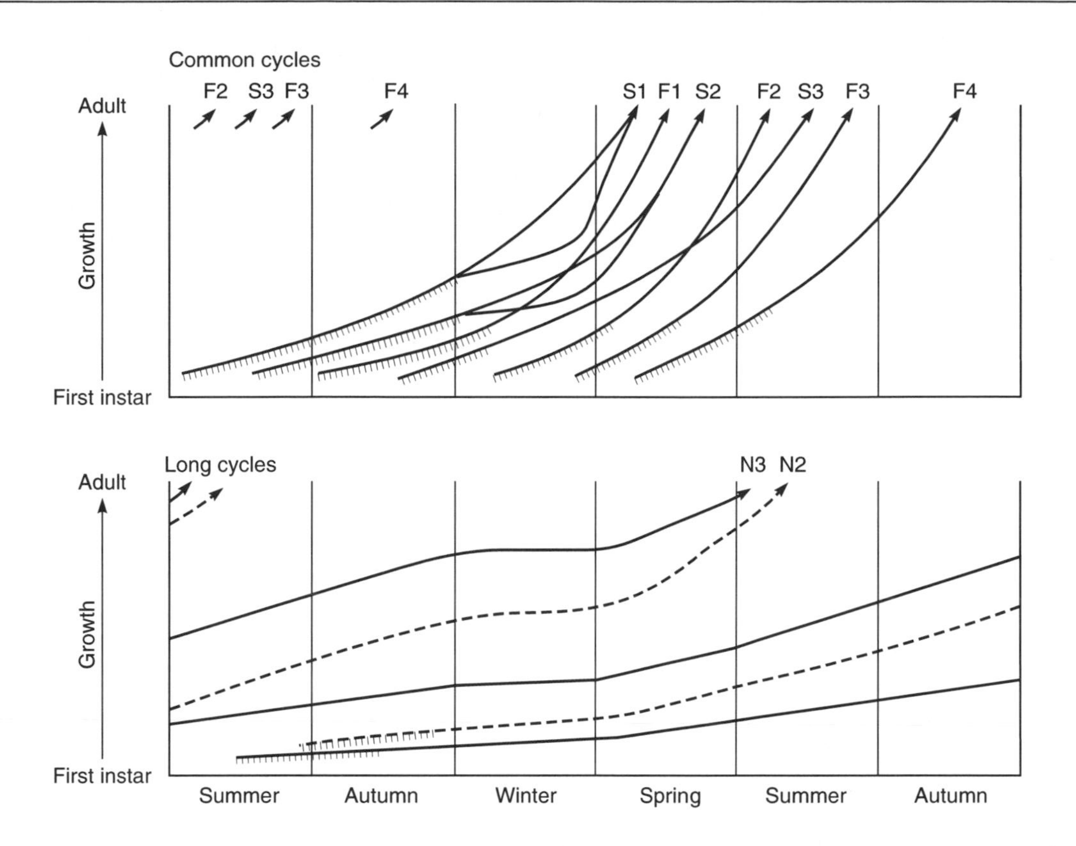

FIGURE 3.19 A conceptualization of the various life cycles exhibited by aquatic insects in temperate seasonal streams. The vertical axis represents size. Hatching of young is shown by vertical hatching. S and F are slow and fast seasonal cycles, and N refers to species whose lifespan is greater than one year. See text for further explanation. (From Hynes, 1970.)

oxygen can be limiting to the macrofauna under certain circumstances. These are always most related to current, temperature or substrate conditions, and so we return to the topic of oxygen in the context of these interactions.

Oxygen is transported across the gills and other respiratory structures of aquatic organisms by diffusion. The rate is dependent upon the concentration gradient, and oxygen in the water in contact with the organism becomes depleted by respiratory uptake. The effect of current is continually to renew the water in contact with respiratory structures. The biota of running waters is in several ways highly dependent upon this ready availability of oxygen (Hynes, 1970). Many invertebrate taxa rely on the integument rather than specialized gills for respiration and have no ability to generate water movements, relying solely on the current in this regard. Compared with the fauna of standing waters, lotic invertebrates are more likely to be respiratory conformers, meaning that their respiration rate closely follows oxygen concentrations, than to be respiratory regulators, meaning that they have some degree of metabolic independence from ambient oxygen. Lotic invertebrates also tend to have higher metabolic rates than lentic taxa.

The influence of current on the ability of aquatic insects to survive low oxygen concentrations has been demonstrated convincingly

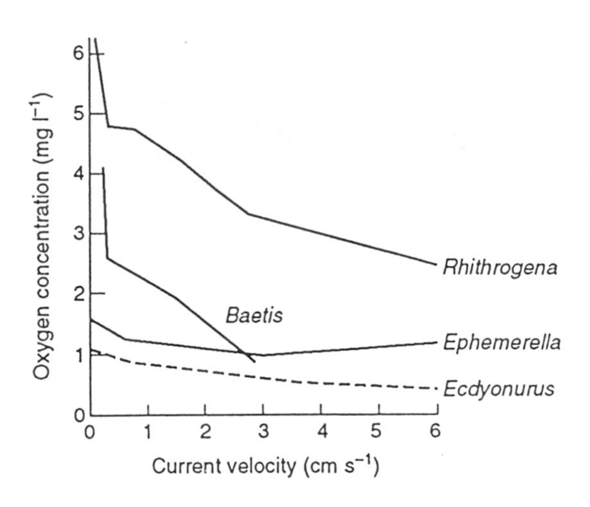

FIGURE 3.20 Oxygen concentrations that are just below the threshold for survival of four mayfly nymphs at different current speeds. (From Jaag and Ambühl, 1964.)

in a number of laboratory studies. The stonefly *Hesperoperla pacifica* experienced 100% mortality in low-oxygen water at a current velocity of 1.5 cm s^{-1}, and no mortality at 7.6 cm s^{-1} (Knight and Gaufin, 1963). The mayflies *Baetis* and *Rhithrogena*, both active, fastwater species, likewise tolerate quite low values of oxygen except in slow currents (Jaag and Ambühl, 1964). However, *Ephemerella* is able to generate water exchange by rapidly moving its gills back and forth, and as a consequence it is unaffected by a drop in current (Figure 3.20). A number of aquatic insects are able to perform ventilatory movements. Stoneflies placed in standing water soon begin 'push-ups', a number of mayflies can move their gills, and some burrowing mayflies move water through their tunnels by body undulations. The cases of caddisfly larvae have a posterior hole, and the anterior end of course is open, so undulatory movements create a current through the case. Because lotic insects differ in the extent to which they are dependent upon current, taxa will also differ in their ability to tolerate conditions of low flows.

Aquatic organisms are much more likely to experience respiratory distress in warm than in cool water. First, the solubility of oxygen in water decreases as temperature increases (Table 2.1). Second, the metabolic activities of organisms, including oxygen consumption, increase with temperature. As mentioned previously, active cold stenotherms such as trout experience a sharp rise in their respiratory rate at temperatures above 15 °C. This is the principal reason why, even when fed to excess, their growth rate declines at higher temperatures (Elliott, 1978). Indeed, as Hynes (1970) points out, the cold stenothermy of such species can be attributed as much to the effects of oxygen availability at higher temperatures, as to temperature itself.

Some aquatic insects respond behaviorally to the respiratory challenges of higher temperatures. The caddis larva *Hydropsyche pellucidula* increased the percentage of time engaged in ventilatory movements, from 25% at 10 °C to 55% at 25 °C (Philipson, 1978). Larvae of the caddis *Glossosoma nigrior* spent more time on stone surfaces at high than at low temperatures, which Kovalak (1976) interpreted in terms of exposure to current offsetting the lower solubility of oxygen in warmer water. In laboratory streams, mayfly nymphs took more or less current-exposed positions according to their tolerance of low oxygen (Wiley and Kohler, 1980).

Substrate has also been implicated as exerting some influence over insect respiration. When the burrowing mayfly *Ephemera simulans* was provided with different substrates in the laboratory, its rate of oxygen consumption was highest on the least preferred substrate and lowest on the most preferred (Eriksen, 1964). Whether this was due to stress, or ease of burrowing, is uncertain.

In sum, while it is true that oxygen is usually not limiting to the biota of running waters, under certain conditions it can be important, and these conditions depend on interactions involving other physical and biological factors. Many rheophilic organisms will be unable to obtain sufficient oxygen under conditions of low flows and high temperatures. Thus droughts can be particularly serious, and intermittent streams that frequently experience these conditions contain a restricted fauna. The decomposition of organic matter causes oxygen depletion, and this can be substantial in stagnant backwaters. The availability of oxygen within the substrate is influenced by its permeability and the amount of mixing with surface waters. Deep groundwater can be devoid of oxygen owing to microbial decomposition as it passes through the soil. Locations receiving mainly groundwater inputs, including some springs and deeper areas of the streambed, can also be oxygen poor.

3.5 Summary

The directional flow of water is what distinguishes rivers and streams from other aquatic environments. Studies of ecological distributions and of the anatomical and behavioral adaptations of organisms, along with more recent experimental studies, provide ample evidence that current is of direct importance to the biota of these environments. Because many of the animals and plants of running waters live in contact with the substrate, much attention has focused on the nearbed flow environment.

Water in contact with non-eroding substrate has zero velocity. Above the substrate, current speed is reduced due to friction with the channel bottom and sides. Thus, velocities are highest at or near the surface of running waters. The boundary layer extends from the streambed or other surface upward to the depth where friction no longer affects streamflow, which might be the surface in a shallow stream. In practice, the upper extent of the boundary layer is often defined as the point where current velocity reaches 90% of the mean channel velocity. Under certain circumstances, a laminar or viscous sublayer exists in the lower region of the boundary layer next to substrate surfaces. This region of greatly reduced flow potentially permits benthic organisms, especially those of low vertical aspect, to avoid fluid forces.

A number of hydraulic variables and equations are useful for describing streamflow and the fluid forces acting upon particles on substrate surfaces. The Reynolds number, which can be calculated for the stream channel (bulk flow), near the streambed (Roughness Re_*), and through an animal's frontal aspect (R_L), is a particularly useful measure. Although direct measurement of laminar sublayers presently is not feasible under natural stream conditions, estimates of Re_* and other information on near-bed hydrodynamics strongly suggest that stable laminar sublayers are unlikely under the usual conditions of turbulent flow and bed roughness found in streams. Microorganisms and the smallest invertebrates might experience current speeds that are only a small fraction of the mean velocity of the water column, but macroinvertebrates on substrate surfaces must usually experience high velocities and complex, three-dimensional flow. Organism size and shape remain important, not because of sheltering from the current, but because they influence the ratio of inertial to viscous forces that animals experience.

Substrate has numerous direct and indirect effects on the biota of running waters. It provides a surface to cling to or burrow in, shelter from current, material for construction of cases and tubes, and refuge from predators. The availability of interstitial space as habitat, the supply of water, oxygen and nutrients, the availability of food, and various other factors interact with substrate in complex ways that are difficult to separate. Substrate is highly heterogeneous, consisting of both inorganic and organic material that is difficult to combine into a linear scale, although inorganic substrates can be readily separated into size classes. There are many examples of taxa of algae, invertebrates, and fishes that are associated with a particular substrate category. These affinities are clearest when substrates are most distinct, such as soft *versus* stony bottoms, or organic *versus* inorganic substrates. When one compares stony substrates that differ in average particle size, preferences differ among the taxa, but are relatively weak.

Temperature affects the growth and respiration of individual organisms and the productivity of ecosystems through its many influences upon metabolic processes. Running waters exhibit a wide range of thermal regimes, including some that are extremely constant and others that vary substantially on daily and seasonal scales, and from headwaters to river mouth. Organisms generally perform best within the subset of possible temperatures that corresponds to where they are found, but some tolerate a wider temperature range than others. Because temperature influences both the availability and the demand for oxygen, it is difficult to say whether temperature directly excludes cold-water species from occupying warmer environments, or if oxygen supply is responsible. Under most conditions in rivers, oxygen is not limiting to biological activity, but under certain conditions, particularly high temperatures and low flows, oxygen can be a critical environmental variable.

Chapter four

Autotrophs

In this and the following chapter we examine the sources of energy to lotic food webs. Autotrophs are organisms that acquire their energy from sunlight and their materials from non-living sources. Green plants, as well as some bacteria and protists, are important autotrophs in running waters. Heterotrophs obtain energy and materials by consuming living or dead organic matter. All animals of course are heterotrophs, as are fungi and bacteria that gain nourishment through the processing of dead organic matter. Together, the autotrophs and microbial heterotrophs of running waters make organic energy available to consumer organisms at higher trophic levels. The major autotrophs of running waters include large plants, referred to as macrophytes, and various small autotrophs. The latter are referred to as periphyton when found on substrates, and phytoplankton when they occur in suspension in the water column.

Macrophytes include vascular plants, represented by aquatic angiosperms; non-vascular plants, principally the bryophytes (the mosses and liverworts); and some members of the periphyton when they become large (filaments of the green alga *Cladophora* have been estimated to reach 6 m). The periphyton of running waters include representatives of three kingdoms. Photosynthetic protists of running waters are the diatoms, yellow–brown algae (included with diatoms in the Chrysophyta) and Euglenophyta (called euglenoids or phytoflagellates). True members of the plant kingdom include the Chlorophyta (green algae) and the Rhodophyta (red algae). Bacterial autotrophs are the Cyanobacteria (blue–green algae). With the exception of the Rhodophyta, all of the taxa found in the periphyton also have representatives among the phytoplankton.

Angiosperms require moderate depths and slow currents, and so are most common in springs, rivers of intermediate size, and along the margins and in backwaters of larger rivers. Bryophytes are restricted in distribution but can be abundant in cool climates and in shaded headwater streams. Periphyton occur on virtually all surfaces within rivers, typically in intimate association with heterotrophic microbes and an extracellular matrix, to which the all-inclusive term *Aufwuchs* applies. Planktonic autotrophs, or phytoplankton, are unable to maintain populations in fast-flowing streams, but can develop sizeable populations in slowly moving rivers and backwaters where doubling rates exceed downstream losses due to current. Thus, according to a somewhat idealized view of the longitudinal profile of a river system, periphyton and occasional bryophytes predominate in headwater and upper stream sections, macrophytes occur mainly in mid-sized rivers and along the margins of larger rivers, and substantial phytoplankton populations develop only in large, lowland rivers (Vannote *et al.*, 1980).

This chapter focuses primarily on the role of autotrophs in the overall ecology of flowing waters, and in particular on their contribution to other trophic categories. In addition there is a bias toward streams and small rivers, where periphyton usually are the major plant representives. The reader looking for a fuller treatment of the ecology of aquatic plants, especially with regard to macrophytes and river phytoplankton, should consult volumes by Gessner (1955, 1959), Round (1981), Haslam (1988), and appropriate chapters in Hynes (1970) and Whitton (1975b).

4.1 Periphyton

Virtually all surfaces receiving light, from small streams to large rivers, sustain a periphyton community. Benthic autotrophs can be further categorized as macroalgae (benthic forms having a mature thallus visible to the naked eye, bearing in mind that many are not truly algae) *versus* microalgae; and by their occurrence on stones (epilithon), soft sediments (epipelon) and other plants (epiphyton). Lists of epipelic, epilithic and epiphytic species indicate considerable habitat specialization, although some certainly occur in multiple settings (Round, 1964). Epipelic taxa form films or mats on silt and mud bottoms, and typically are motile and easily swept away by increased current. Because of their motility, glass slides placed in Petri dishes on the sediment surface are readily colonized, which serves as one sampling technique (Round, 1964). Epiphytic taxa occur on macrophytes, particularly angiosperms, where epiphytic loading can be detrimental to the host plant. Unlike epipelic species, epiphytic and epilithic taxa are usually firmly attached by mucilaginous secretions or via a basal cell and stalk; thus, they are much less likely to be carried away by currents unless flow is substantial.

Some periphyton species are in contact with the substrate (or host epidermis in the case of epiphytes) along the entire cell wall, colony or filamentous system. This growth form is termed adpressed, and contrasts with erect (pedunculate) forms in which only a basal cell or basal mucilage contacts the substrate. As a consequence of this variety in growth form and lifestyle, a close look at a periphyton community reveals much structural diversity (Figure 4.1).

Diatoms typically comprise the majority of species within the periphyton, although green algae and cyanobacteria are well represented and can dominate the biomass of benthic autotrophs under some circumstances. The prevalence of diatoms is apparent from Patrick's (1961) cell counts from glass slides placed in three US rivers, Moore's (1972) study of the epipelon of a southern Ontario stream, and Chudyba's (1965) study of epiphytes of *Cladophora glomerata* in the Skawa River, Poland (Table 4.1). In an extensive survey of the macroalgae of 1000 20 m long stream segments in North America, Sheath and Cole (1992) recorded 259 taxa, of which 35% were Chlorophyta, 24% were Cyanobacteria, 21% were diatoms and other Chrysophyta, and 20% were Rhodophyta. Many diatoms do not form mats, gelatinous colonies or filaments, and so would not be included in a survey of visible macroalgae.

The species composition of the periphyton assemblage varies seasonally, and as this occurs even in constant temperature springs, changing light conditions must be partly responsible (Gessner 1955; Hynes, 1970). Seasonal abundance data from the tropics are few, but numerous studies of stony streams from Europe, Japan and North America suggest a fairly regular seasonal pattern in temperate streams. Diatoms dominate during winter and continue to be a major component of the flora in spring and early summer, although the species composition changes. Total abundance generally is greatest in the spring, and a secondary peak can occur in autumn. Other groups can become abundant during summer, particularly green algae and cyanobacteria. Benthic autotrophs often decrease during summer and increase again in the fall, due to reduced shading or changes in other environmental conditions. Moore's (1972) study of epipelic taxa in a stream in southern Ontario is a good example of this general scenario.

Within a particular region and type of stream, certain taxa tend to occur together and include

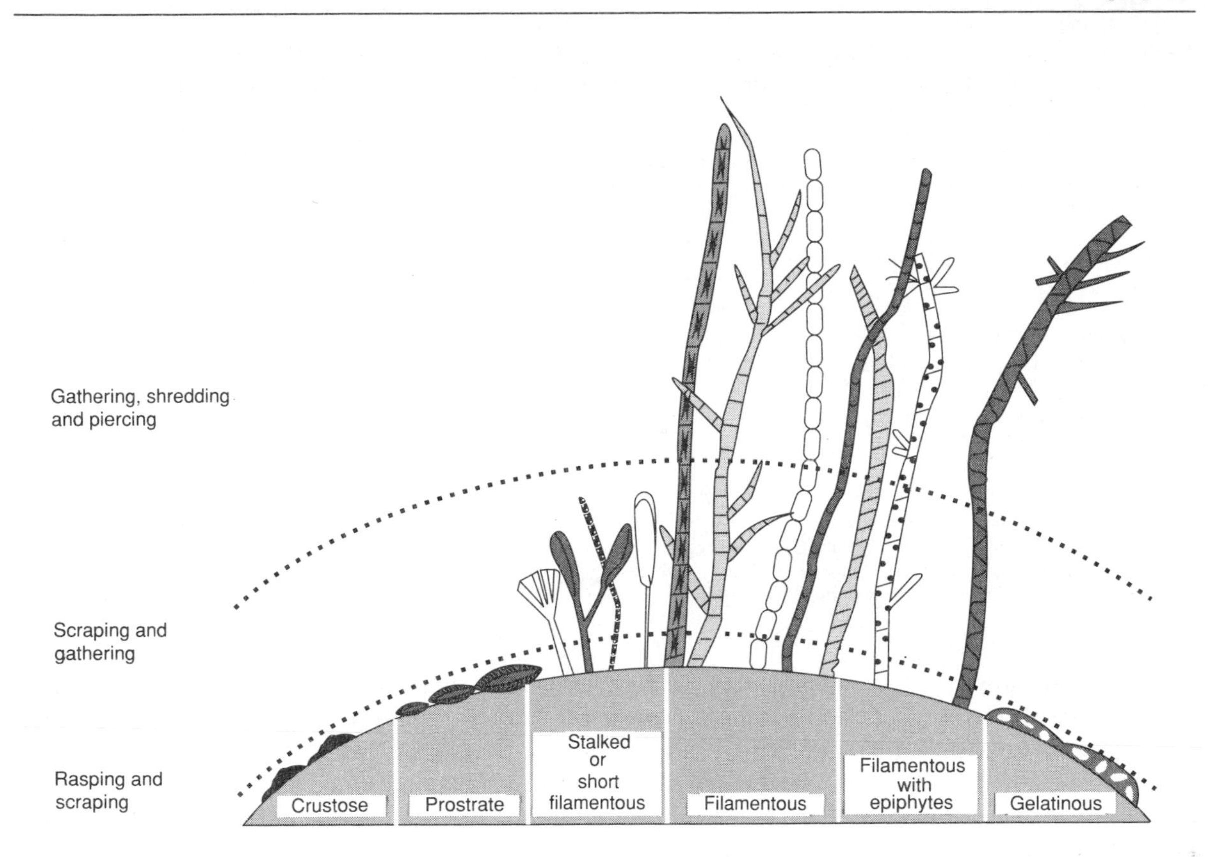

FIGURE 4.1 The major growth forms of periphyton illustrating the considerable variation in shape and vertical layering. Their vulnerability to consumers, discussed further in Chapter 6, is expected to be influenced by these structural attributes. (From Steinman, 1994 after Gregory, 1980.)

one to several dominant species, which often are used as a convenient name for the larger assemblage. Margalef (1960) recognized three major associations in European rivers: (1) the upper, fast-flowing region (*Hydrurus/Ceratoneis*); (2) middle reaches (*Diatoma/Meridion*); and (3) the downstream region (*Melosira*). These are all pedunculate growth forms; Margalef also recognized an upstream *versus* downstream association of adpressed taxa, and quite a number of local variants. Obviously, such classifications can become so complicated that they differ little from the site-specific description they are meant to replace. Moreover, the implication that associations are communities of interdependent taxa lacks supporting evidence (Hynes, 1970). Nonetheless, these associations appear to have sufficient regularity to be recognizable to workers in other geographic regions. For instance, Margalef's *Melosira* association occurs in lowland English streams, where it extends almost to the headwaters (Round, 1981).

Patchiness in the distribution of benthic autotrophs is also apparent at a small spatial scale, even within apparently uniform stream sections (Marker, 1976). Often this fine-scale pattern is associated with obviously distinct microhabitats (Saunders and Eaton, 1976; Tett *et al.*, 1978;

TABLE 4.1 Representation of major periphyton taxa in collections where all habitats were sampled, and from studies emphasizing epipelic and epiphytic assemblages. Patrick's (1961) data are from one time of year and include only those species represented by a minimum of six specimens in a very large sample (a count of 8000 individuals). Inclusion of rarer species would at least double the species list. The studies of Moore (1972) and Chudyba (1965) probably represent closely the entire flora for the site

	No. of taxa				
	All habitats			*Epipelon*	*Epiphyton*
Diatoms	81[a]	80[b]	59[c]	321[d]	176[e]
Chlorophyta (green algae)	12	12	7	32	27
Cyanobacteria (blue-green algae)	9	9	6	14	19
Euglenophyta (phytoflagellates)	17	15	7	20	–[f]
Chrysophyta (yellow-brown algae)	0	1	1	1	2
Rhodophyta (red algae)	1	3	0	0	1
Total	120	120	80	388	225

[a] Potomac River, Maryland, [b] Savannah River, Georgia, [c] White Clay Creek, Pennsylvania (Patrick, 1961); [d] clay and detritus bottom stream, southern Ontario, (Moore, 1972); [e] epiphytes on *Cladophora glomerata* in the Skawa River, Poland (Chudyba, 1965); [f] flagellates present but not identified to species.

Keithan and Lowe, 1985). Using glass slides and an artificial substrate composed of sand and agar that released nutrients, Pringle (1990) demonstrated how an intricate micro-scale distribution of periphyton taxa resulted from differences in type of substrate, and the availability of nutrients in the water column and within the substrate itself (Figure 4.2).

4.1.1 Factors controlling the distribution and abundance of the periphyton

Factors that potentially influence periphyton populations include light, temperature, current, substrate, the scouring effects of floods, water chemistry and grazing (Hynes, 1970; Whitton, 1975b). By making appropriate comparisons across locations or seasons, it is often possible to demonstrate the primary influence of one or another of these variables. Because more than one factor can vary at the same time, experimental studies are especially useful in sorting out the effects of multiple covarying factors.

The simplest framework for discussing limiting factors is to take them one at a time, although in reality these factors interact in complex ways. This is because the importance of any one factor in limiting periphyton growth depends on whether some other factor is in even shorter supply. Thus light might be limiting if nutrients are plentiful, and their importance can alternate seasonally or from place to place. In addition, it is possible that two or more factors simultaneously limit growth. For these reasons it is difficult to address any one factor without specifying the context of other environmental variables.

Before considering singly and in detail the factors that potentially limit populations of periphyton, it may help to present a brief overview of the role played by each. This will admittedly be general and simplified, but it should serve as a framework against which to judge the evidence.

The hydrologic regime exerts important control over the biota of rivers. High currents restrict the establishment of macrophytes and influence the distribution of periphyton, in terms of both taxa and growth forms. Where discharge is variable, flood events and scour by suspended sediments can cause major reductions in periphyton standing crops. As a consequence, in very rainy climates the growing season is

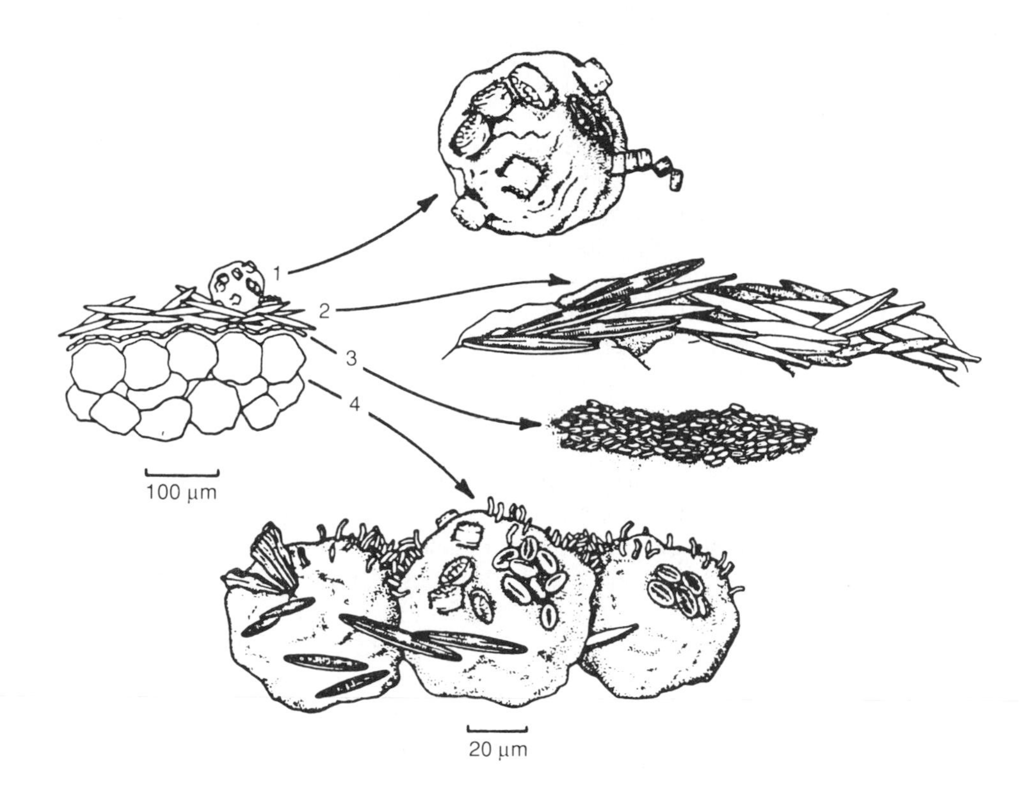

FIGURE 4.2 Schematic cross-section of periphyton growing on sand. Canopy layers in the mat include: (1) bedload sand grains dominated by stalked cells of *Opephora martyi*, along with small numbers of adnate *Achnanthes lanceolata*, and a stalked, chain-forming diatom *Fragilaria pinnata*; (2) thick upperstory mats of motile diatoms including *Navicula tripunctata*, *Nitzschia amphibia* and *N. filiformis*; (3) mucilaginous layers of *Navicula pelliculosa*; (4) understory layer dominated by sessile and relatively sessile taxa (*Achnanthes minutissima*, *Cocconeis placentula*, *Gomphonema* spp. and others). (From Pringle, 1990.)

restricted to the time between the last spring flood and the first autumn flood, and summer standing crops depend on the number of consecutive flood-free days. However, in running waters where extremes of discharge are minimal, direct effects of current are probably less important to the periphyton.

Light can be a limiting factor in small streams under dense forest cover, where periphyton populations tend to reach maxima just prior to canopy development and then decline through the summer. However, other factors often override light. Present indications are that benthic autotrophs do quite well under moderate light levels, yet are not inhibited by high incident light. Nutrients, particularly major ones like phosphorus and nitrogen, might be expected to exert a critical influence over autotrophs in rivers just as they do in standing freshwaters, but this turns out to be rather complicated. Flow provides continual delivery of water, thus minimizing the depletion of nutrients, and thermal stratification does not occur to restrict mixing of nutrients throughout the water column, so nutrient exhaustion should be less likely to occur in rivers than in lakes, where its importance is well established. In some studies, enrichment of streamwater with nitrogen and phosphorus had no effect on periphyton development, and in others it did. Too little is known at present to attribute this solely to differences in the nutrient status of particular streams. However, where nutrient enrichment stimulates growth of autotrophs, phosphorus has proved more effective than nitrogen, as long as N:P ratios are high enough (greater than 25). Other nutrients, trace metals and bicarbonate are probably of some importance, temperature increases rates of metabolism, and grazing limits periphyton populations in some circumstances. Overall, however, it appears that the effects of scour, light and nutrients are of greatest importance and each can be strongly limiting at a particular location or time. Now, what is the evidence for this scenario?

(a) Light

The influence of light on periphyton assemblages can be substantial. Green algae are usually associated with high light levels, and Hynes (1970) reports a red alga (*Batrachospermum*) to be found exclusively in shaded areas. In contrast, many diatoms seem to be unaffected by seasonal changes in light. The Metolius River in Oregon is essentially constant throughout the year in temperature (7–10°C), current velocity, dissolved substances and turbidity, leading Sherman and Phinney (1971) to infer that only variation in light intensity and photoperiod could account for any seasonal changes in periphyton in this system. Abundances were indeed reduced during winter, presumably an effect of reduced light, and some diatoms such as *Diatoma hiemale* were restricted to the spring–summer time period. However, of the 60 species sufficiently common for detailed study, only nine were seasonal while the rest occurred throughout the year, suggesting only minimal influence of changing light regime.

A series of studies conducted in laboratory streams by McIntire and colleagues have examined how light and current velocity influence the growth and species composition of periphyton assemblages. Water was supplied from a natural stream after passage through a coarse sand filter. Thus water chemistry and temperature were representative of local streams in coastal Oregon, and the flora was established by waterborne propagules. Natural, sun-dried substrate and glass slides were placed in the streams for later sampling, and *in situ* rates of photosynthesis and respiration were determined using a recirculating chamber that provided current and allowed monitoring of dissolved oxygen concentrations. An ambitious study by McIntire and Phinney (1965) examined the production and metabolism of periphyton communities that developed under two light regimes. One received approximately 6000 lux and was considered light-adapted, the other received 2500 lux and was considered shade-adapted. Photosynthetic rate, measured over a range of light levels, increased steadily with increasing light level up to about 10 000 lux, whereupon the rate of increase slowed and eventually leveled off (Figure 4.3). In nearby streams, maximum light intensity occurs in March, before the canopy develops, and is about 22 500 lux; minimum values occur in late summer and are as low as 1100 lux (McIntire, 1973). This makes it likely that, during the period that a forest canopy shades the stream, light levels are low enough to restrict the photosynthetic rates of benthic autotrophs. Figure 4.3 also indicates that the shade-adapted assemblage was able to maintain higher photosynthetic rates at low light levels, although whether the mechanism involved physiological adjustment or changes in species composition is uncertain. The light-adapted assemblage consisted of 46% diatoms, 42% cyanobacteria and 12% green algae; corresponding figures under low light levels were 67% diatoms, 26% cyanobacteria and 7% green algae. In another study, McIntire (1968) again found that diatoms predominated at low light levels whereas filamentous green algae,

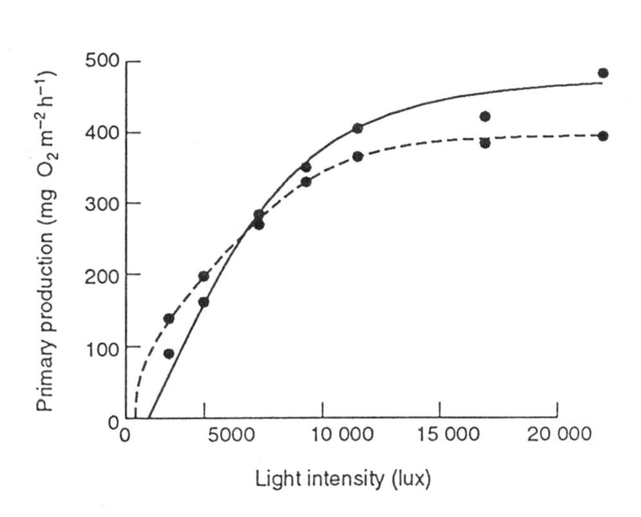

FIGURE 4.3 Primary production as a function of light intensity in two laboratory stream periphyton communities, measured using a flow-through respirometer. The shade-adapted community (----) developed at about 2500 lux, the light-adapted community (——) at about 6000 lux. (After McIntire and Phinney, 1965.)

cyanobacteria and yellow–brown algae were abundant at higher light levels. He also noted some instances where different diatom species were associated with either high or low light regimes.

A number of field studies have also documented light limitation of benthic autotrophs. In the Fort River, Massachusetts, periphyton abundance varied seasonally in response to shading by surrounding trees (Figure 4.4). Benthic chlorophyll *a* peaked in the spring just prior to leaf-out, and showed a minor autumn peak just after leaf fall. A number of studies comparing clear-cut with forested small streams have demonstrated enhanced periphyton growth under greater light (Murphy and Hall, 1981; Keithan and Lowe 1985; Lowe, Golladay and Webster, 1986). In contrast, primary production was only marginally enhanced by greater light penetration after clear-cutting in nutrient-poor streams of coastal British Columbia (Stockner and Shortreed, 1978). Only if streams experienced nutrient enrichment or if light declined to very low levels did these authors find evidence that available light limited periphyton growth.

Strong herbivory might also account for a limited response of periphyton biomass to increased light levels. When Steinman (1992) substantially increased irradiance in a heavily shaded stream using halide lamps, the biomass-specific carbon fixation rates of periphyton increased significantly. However, an increase in periphyton biomass was observed only when snail populations were drastically reduced. At least in this system, intense herbivory prevents the positive influence of light on periphyton production from being translated into a greater biomass of autotrophs.

(b) Nutrients

Phosphate, nitrate and silica are generally considered the most critical nutrients for autotrophic production, although other chemical constituents also can limit growth under some circumstances (Hutchinson, 1967). In nutrient-poor freshwaters, inorganic phosphate is often the principal factor limiting the growth of plants, algae and other primary producers. Nitrate–nitrogen tends to become limiting when phosphate is plentiful and when the atomic ratio of N:P falls below some level. In theory, a ratio of N:P below 16:1 leads to nitrogen limitation (Redfield, 1958), while in practice the shift from phosphorus to nitrogen limitation occurs over a wider range, perhaps 10–30:1. Because of their ability to fix atmospheric nitrogen, certain cyanobacteria typically increase under nitrogen limitation, sometimes to nuisance levels. The supply of silica also might be expected to become limiting, because the frustules of diatoms are composed of silicious material and diatoms are a major component of the periphyton of cool, shaded streams. The dynamics of lake diatom populations have been shown to depend on silica concentrations (Wetzel, 1983). However, in

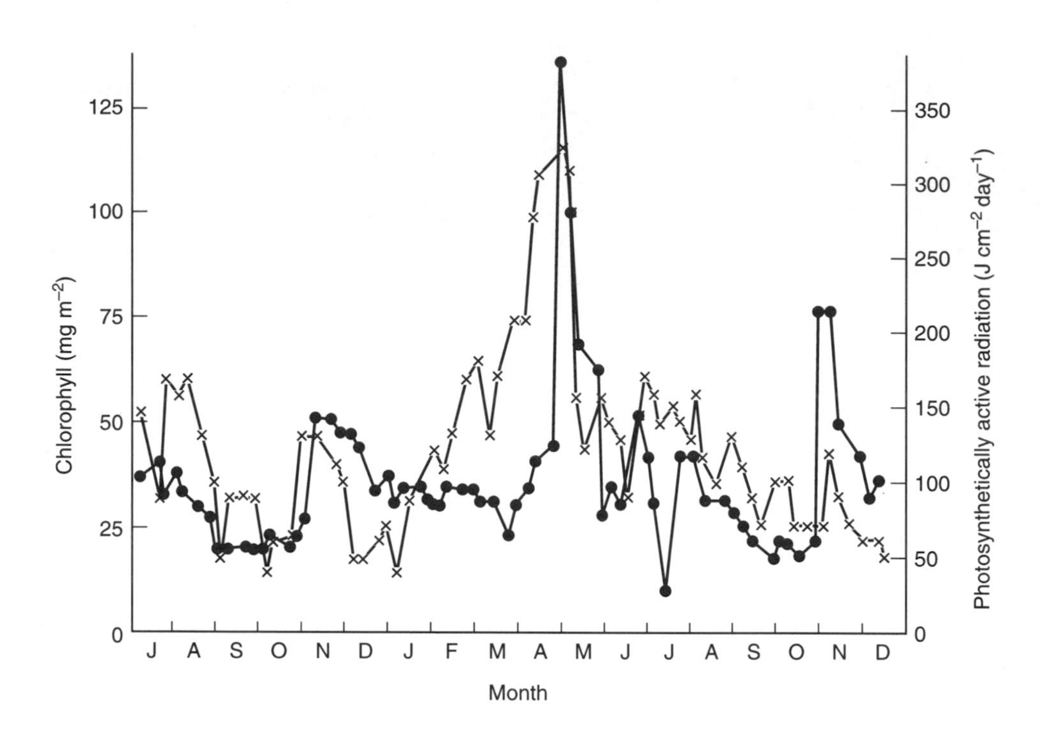

FIGURE 4.4 Seasonal change in mean periphyton abundance, measured as chlorophyll *a*, in a small Massachusetts river flowing through mostly agricultural land but with riparian shading. The shaded period extended from 10 May until 20 October. Note the major peak in chlorophyll (3.9 × mean summer values) just prior to leaf-out, and the minor peak (1.7 × mean summer values) just following leaf fall. Water temperatures were highest throughout the summer. ● = Chlorophyll *a*; × = photosynthetically active radiation. (After Sumner and Fisher, 1979.)

running waters silica is rarely in short supply (chapter 2) and consequently it may seldom limit diatom growth.

Nutrient limitation of phytoplankton populations is a well-established aspect of the ecology of lakes, for which the above description generally holds true. Its applicability to flowing waters has long been questioned, however, based on the reasoning that current continually brought a fresh supply of nutrients to the vicinity of an individual cell. Thus, no matter how low the concentration of nutrients in the water, the cell's environment is 'physiologically enriched' (Ruttner, 1926) by the flow of water carrying dissolved materials over its surface.

There can be little doubt that current indeed provides for the continual renewal of nutrients, gases, and other water-borne materials. Whitford and Schumacher (1964) measured the uptake of ^{32}P in two algae under currents ranging from 0 to 40 cm s^{-1}, and found uptake rates to increase linearly over these conditions. The degree of physiological enrichment is likely to be affected by nearbed flow conditions and also by the thickness of periphyton mats. Saturation of cellular uptake of phosphate in thin film periphyton communities occurs at very low levels (near 1 $\mu g\, l^{-1}$ PO_4), but a thick matrix impedes diffusion so that saturation of uptake by the community might require concentrations ten

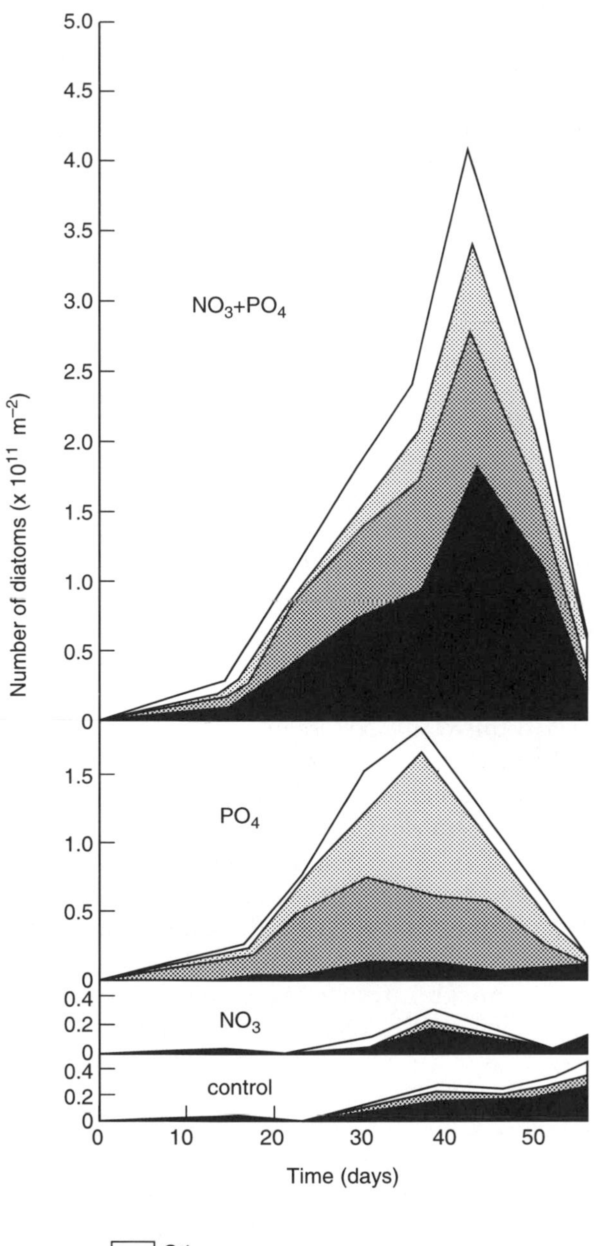

FIGURE 4.5 Changes in the numbers of the dominant diatom species in troughs enriched with NO_3-N, PO_4-P, or both in combination. Troughs were placed in Carnation Creek, Vancouver Island, allowed 4 weeks to colonize, and then fertilized for 52 days. Note that periphyton populations peaked after 30–40 days, and then declined sharply, prior to termination of the fertilization experiment. (After Stockner and Shortreed, 1978.)

times greater (Bothwell, 1989). This helps to explain why accrual of periphyton biomass continues in response to phosphorus enrichment well after saturation conditions for individual cells have been surpassed (Perrin, Bothwell and Slaney, 1987).

Field studies now provide ample evidence that nutrient supply can limit periphyton growth in nature. Experimental confirmation of the stimulatory effect of phosphorus enrichment has been obtained by several approaches. One technique has been to build simple troughs in or beside the stream and add the desired nutrients to the streamwater entering some of the troughs. A series of studies of periphyton dynamics in small streams of the west coast of Vancouver Island, a temperate rainforest setting, has demonstrated this system to be strongly P-limited (Stockner and Shortreed, 1976, 1978; Shortreed and Stockner, 1983). Plexiglas roughened with steel wool was attached to 50 kg concrete blocks (to withstand frequent floods) and sampled for chlorophyll *a*, ash-free dry mass, and cell counts. Phosphate concentrations were found to be very low, one to two orders of magnitude less than the Oregon streams studied by McIntire (1973), and periphyton standing crops were among the lowest on record. Stockner and Shortreed then constructed four wooden troughs at a shaded site, allowed 4 weeks for colonization, and then enriched streamwater with nitrate, phosphate or both N and P. Nitrate enrichment produced little response, but periphyton accumulated rapidly in the two troughs receiving phosphate, exhibiting roughly a fivefold increase in response to phosphate alone, and a 7- to 8-fold increase where both nutrients were added (Figure 4.5).

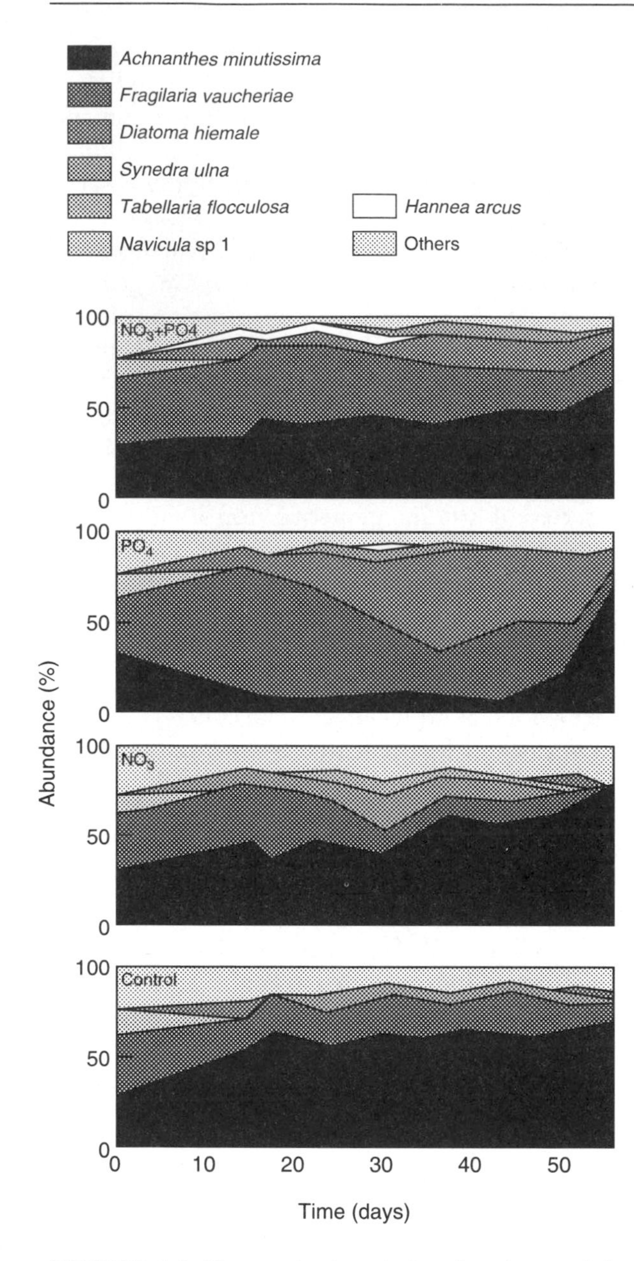

FIGURE 4.6 Changes in the relative abundance of the major diatoms in response to nutrient manipulation. Note the decline in *Achnanthes minutissima* in the trough receiving only phosphorus. (After Stockner and Shortreed, 1978.)

Filamentous green algae responded dramatically to phosphate enrichment. Cyanobacteria did not flourish under the nitrate treatment, despite the fact that the N:P ratio fell to 11 and below. However, the frequency of the diatom *Achnanthes minutissima* decreased in the nitrate-only treatment while it increased in the other troughs (Figure 4.6), and this may be an indication that reduction in the N:P ratio can have adverse effects. Subsequently, this watershed was logged, allowing a direct test of the effect of increased light on periphyton biomass. Essentially no change occurred, and the case seems clear that low phosphorus concentrations are the primary limiting factor in this environment. Comparison of the flora between natural substrates and Plexiglas samplers indicated a close correspondence, enhancing our confidence in these experimental results.

Peterson, Hobbie and Corliss (1983) built a tethered, floating apparatus (Figure 4.7) to bioassay nutrient effects in an Alaskan tundra river, apparently because floods had destroyed the troughs they had first constructed. Addition of PO_4-P significantly enhanced chlorophyll *a* and $^{14}CO_2$ uptake, and subsequent enrichment of a large section of stream with phosphate confirmed this result (Peterson *et al.*, 1985).

Another novel approach to the topic of nutrient limitation of benthic autotrophs is the use of nutrient-releasing substrates. In one version clay flowerpots are filled with an agar solution containing the desired mix of nutrients and capped with a plastic Petri dish (Fairchild and Lowe, 1984). Nutrients leach through the pot, although rates can be somewhat variable depending on thickness and consistency of the clay walls of the pots. Pringle and Bowers (1984) used sterilized sand from a streambed consolidated with agar and nutrients in Petri dishes, thus achieving very consistent release rates. Studies in Carp Creek, a small, nutrient-poor stream in northern Michigan, demonstrated that phosphate was limiting to periphyton, and additions into the water column and from substrates could each enhance periphyton abundance (Figure 4.8). This method appears to have considerable potential as a bioassay, and likely

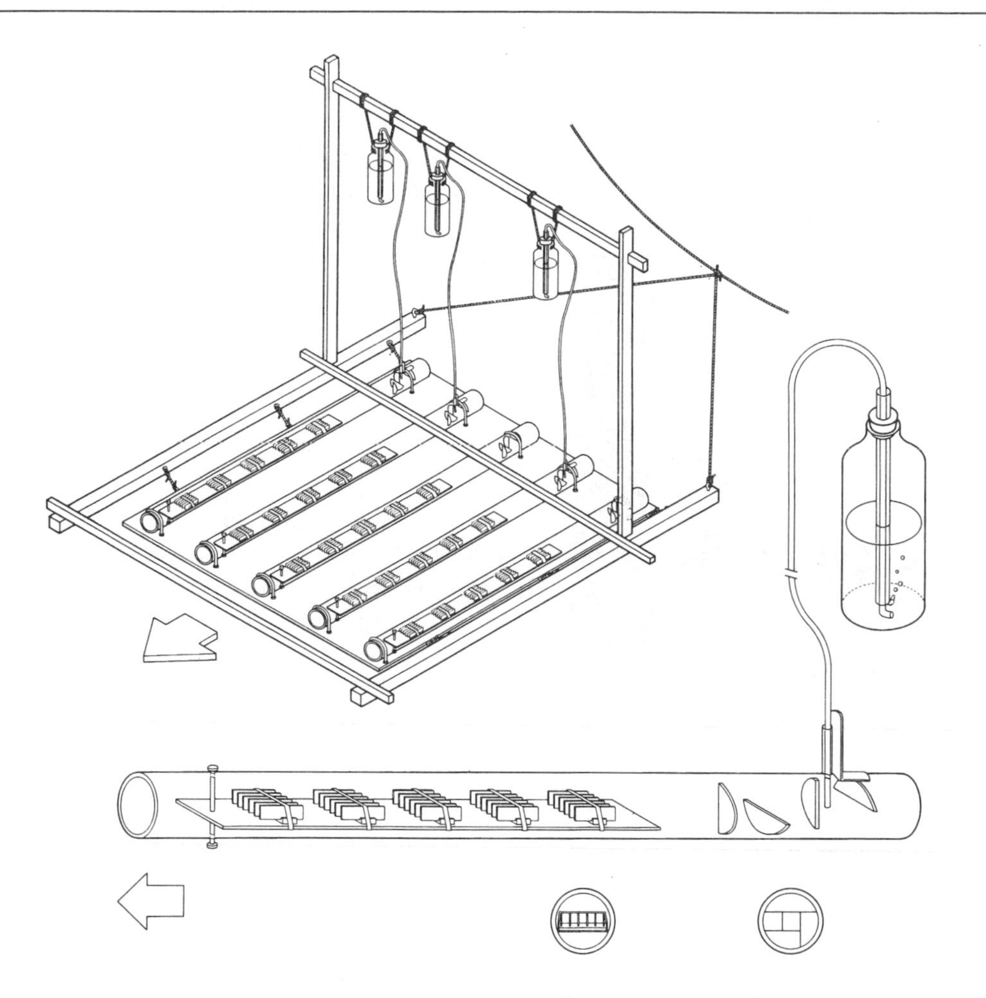

FIGURE 4.7 Diagram of continuous-flow periphyton bioassay system which is designed to float at the water surface, tethered by a rope across the stream, and thus withstand floods. The individual tubes of clear plastic (4 in or 9.2 cm i.d.) are attached to a Plexiglas sheet and contain sets of microscope slides which are later removed to provide samples. Nutrients are added by siphoning from a Mariotte bottle to achieve desired concentrations. (After Peterson, Hobbie and Corliss, 1983.)

is more suitable for epiphytic and epipelic taxa, where the substrate does exude nutrients, than for periphyton growing on stones and other substrates that release few or no nutrients.

In contrast to the substantial evidence of phosphorus limitation of benthic autotrophs, nitrogen is often shown not to limit growth of periphyton in streams. A consistent result in the nutrient enrichment experiments discussed above (Stockner and Shortreed, 1978; Peterson, Hobbie and Corliss, 1983; Pringle and Bowers, 1984) is the absence of a response when only NO_3-N was enhanced. However, in each instance simultaneous addition of nitrate and phosphate resulted in greater periphyton growth than did phosphate alone. Fertilization with phosphate alone inevitably reduces the ratio of N:P, and in lakes this can have the undesirable effect of stimulating cyanobacteria because of their capacity to fix atmospheric nitrogen.

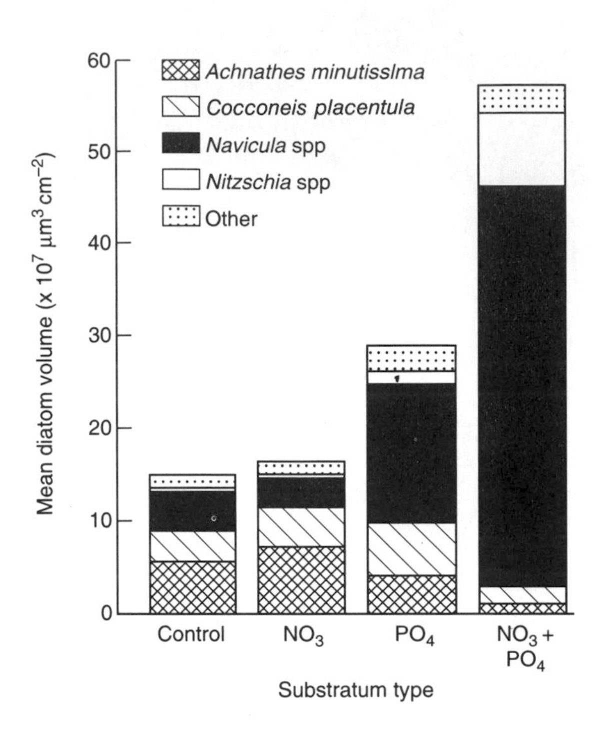

FIGURE 4.8 The relative biovolume abundance of diatoms colonizing nutrient-releasing substrates made of a sand–agar mixture and placed in a nutrient-poor stream in northern Michigan. The response was due to two motile pennate diatoms, *Navicula* and *Nitzschia*, and occurred only with phosphate enrichment. (After Pringle and Bowers, 1984.)

Such an effect is uncommon in stream studies, however.

A strong case for nitrogen limitation might be expected from streams where nitrogen is normally in very short supply. Most streams of the eastern USA have quite high (about 68:1) molar ratios of N:P, but nitrate levels are low in some mountainous regions, especially in the Pacific northwest (Omernik, 1977). Low N:P ratios have been reported from a desert stream in Arizona (Fisher *et al.*, 1982) and from a number of sites from northern California through Washington state (Gregory, 1980; Triska *et al.*, 1983). Gregory (1980) obtained a response to whole stream nitrogen enrichment in Mack Creek, Oregon, but the response was not great and did not occur under normal, highly shaded conditions. Filamentous cyanobacteria increased, apparently in response to experimental phosphate enrichment, in the study of Elwood *et al.* (1981), and as nitrogen naturally became in short supply following a flood event in the study of Fisher *et al.* (1982). Triska *et al.* (1983) conducted a nutrient enrichment experiment using troughs in a northern California stream where N:P ratios near 4 led them to suspect nitrate limitation. However, the amount of shade had a stronger effect than nutrient manipulations. In fact, this study, Gregory's (1980) experiments in an Oregon stream, and a study using nutrient-releasing flowerpots in southern Appalachian streams (Lowe, Golladay and Webster, 1986) all document situations where light limitation overrides any effect of nutrient shortages.

As more such studies are completed, we may be able to generalize concerning the circumstances where nutrients are limiting, and which nutrients. Factors that might prevent any nutrient effect, in addition to heavy shade, include high ambient nutrient levels (e. g. Moore, 1977), grazing, and the possibility that micronutrients and trace metals exert a greater limiting effect than the more commonly studied major nutrients. This last consideration might explain why periphyton biomass did not respond to enrichment of groundwater with nitrate and phosphate in large (75 m) outdoor channels at a Swiss facility for water pollution study, whereas addition of small amounts of sedimented sewage from the city of Zurich caused a significant biomass increase and a shift from diatoms to filamentous green algae (Wuhrmann and Eichenberger, 1975). However, the causal factor or factors have yet to be identified.

(c) Current

The flow of water influences the distribution and abundance of plants and other autotrophs of rivers in a number of ways. Taxa presumably differ in their ability to remain attached at high velocities. Current influences substrate characteristics, which in turn will affect site suitability for attachment and growth. In addition, flow results in the continual renewal of gases and nutrients, and so current speed affects diffusion rates of needed materials into cells.

Many species have been found to occupy specific flow regimes (Blum, 1960; Whitford, 1960; Hynes, 1970). In Carnation Creek, British Columbia, *Achnanthes minutissima* was abundant everywhere except in riffles, where *Hannaea arcus* was dominant (Stockner and Shortreed, 1976). Keithan and Lowe (1985) found different periphyton taxa associated with particular regions in small Tennessee streams, and their description of individual growth forms is consistent with a direct effect of current speed. In slower currents they found diatoms to be more densely packed, with a higher proportion growing in an erect position and a greater abundance of large colonial forms (e. g. *Meridion*). Many of the same species were found at faster currents, but in the prostrate position. At the highest velocities most of the diatoms were prostrate, many were in crevices, and tightly adherent species were prevalent. The differential distribution of periphyton between upstream and downstream faces of stones (Gessner, 1955; Blum, 1960) probably has a similar explanation. There also are examples where macrophyte distribution is clearly limited by ability to attach firmly and withstand breakage. Growth form in *Cladophora glomerata* is reported to vary with current velocity. Thick plumose growths occur in slow water, whereas long, tough, rope-like strands are found in faster water (Whitton, 1975a).

Several laboratory studies have shown an effect of current, and physiological enrichment most likely plays a part. Whitford and Schumacher (1964) showed that rates of ^{32}P uptake and CO_2 liberation in *Spirogyra* and *Oedogonium* increased with current up to $40\,cm\,s^{-1}$, the highest velocity tested. When periphyton was allowed to colonize laboratory streams which were similar except for current velocities (9 *versus* 38 $cm\,s^{-1}$) and received the same propagules from common source water, different assemblages developed (McIntire, 1968). At the higher current, initial colonization was slower but biomass eventually exceeded that of the slower channel, chlorophyll *a* content was higher, and more biomass was exported. Filamentous chlorophytes and chrysophytes became abundant in the slow stream, whereas diatoms, especially *Synedra*, dominated under faster current. A subsequent experiment using two light levels (1650 and 7650 lux) and three velocities (0, 14 and $35\,cm\,s^{-1}$) indicated little effect of current (as long as it was over zero) at low light, whereas biomass, export and species composition were clearly influenced by current at the higher light level (McIntire, 1968). Physiological enrichment rather than withstanding dislodgement seems most likely to be the cause of these differences, but unfortunately that is only speculation.

Physiological enrichment of periphyton might occur in natural streams in response to moderate storms. Stevenson (1990) attributed stimulation of diatom growth following a 24 hour storm in a gravel-bed stream to an increased nutrient supply, caused either by greater flow through mats or increased delivery of nutrients from the catchment.

Extreme discharge can have a strongly negative impact on lotic periphyton populations. Scouring of cells from surfaces can result simply from increases in current velocity, from overturning substrates (Robinson and Rushforth, 1987), and from abrasion due to tumbling (Power and Stewart, 1987) and perhaps suspended sediments as well. In streams on the west coast of Vancouver Island, winters are very rainy and limit the growing season to the period between the last spring flood and the first autumn flood. A flow index was devised by Shortreed and Stockner (1983) to give greater

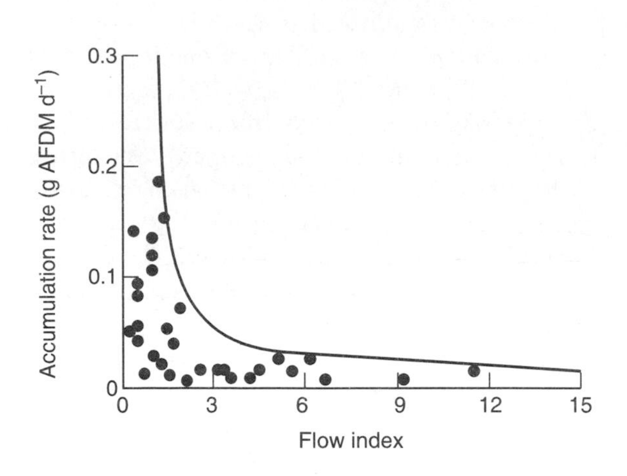

FIGURE 4.9 The relationship between periphyton accumulation rate and a flow index in a small stream (Carnation Creek) in the high rainfall environment of the west coast of Vancouver Island. See text for definition of flow index. (From Shortreed and Stockner, 1983.)

weight to more recent as well as more severe events:

$$flow\ index = \sum_{i=1}^{d} \frac{F_i}{i} \tag{4.1}$$

where F_i is the maximum daily flow i days prior to the sampling date and d is the number of days in the sampling interval. As Figure 4.9 illustrates, periphyton accumulation demonstrated a clear inverse relationship with flood events. Tett *et al.* (1978) reached similar conclusions from their study of the Mechums River in Virginia. During a study that extended from 15 May until the following 31 January there were 11 flood-free periods of seven days or longer, the median duration of low stable flows was 12 days, and the longest was 33 days. The mean density of chlorophyll *a* increased during low flows and decreased abruptly following floods, indicating that chance variation in rainfall was the main factor controlling periphyton abundance.

The importance of freshets to the control of periphyton populations clearly is dependent upon the combination of hydrological and watershed characteristics that determine the extremes of discharge (chapter 1), and probably on the availability of suspended sediments as well. In agricultural watersheds human modification will enhance runoff and transport of sediments, but undisturbed watersheds also can experience discharge extremes. Fisher *et al.* (1982) provide a particularly dramatic example from a desert stream in Arizona. Intense flash-floods virtually eliminated periphyton, which then recovered and increased steadily (biomass and chlorophyll *a*) for approximately 60 days, when another flood occurred.

Thick periphyton mats are particularly vulnerable to dislodgement due to senescence of the bottom-most layers, which weakens their attachment to the substrate and renders the entire mat vulnerable to sloughing. Shading, the buildup of metabolites, and reduced rates of exchange of gases and nutrients all can contribute to the lower-most layer being unable to support the weight of the overlying mat. In experimental channels with constant discharge, periphyton biomass commonly goes through a rapid increase followed by a precipitous decline over a time span of 3–6 weeks (e. g. Stockner and Shortreed, 1978; Triska *et al*, 1983), although given more time, channels might develop a more stable flora (e. g. Wuhrmann and Eichenberger, 1975). The tendency of periphyton mats to break loose in such experimental studies derives at least partly from their artificiality.

Storm events, in addition to dislodging and burying periphyton, can shift the substrate to which they are attached. This will of course depend on the surface occupied and how increases in current are experienced at that location. For example, Tett *et al.* (1978) found that mud habitats were relatively undisturbed over a wide range of flows, and Douglas (1958) reported that only very severe floods affected epilithic populations of *Achnanthes* in a stony stream. In a Montana mountain stream where

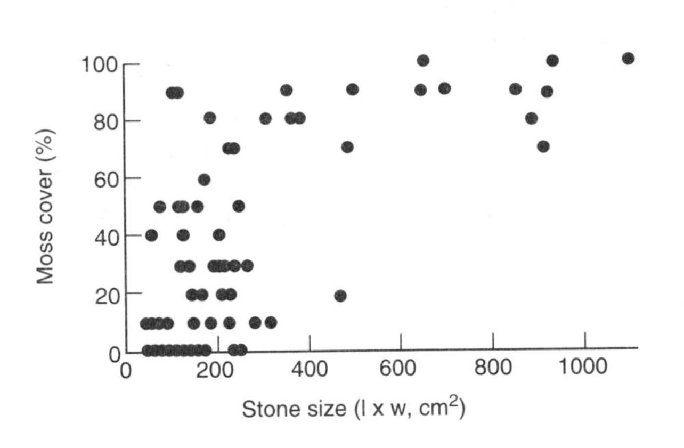

FIGURE 4.10 Amount of stone surface covered by the moss *Hygrohypnum* as a function of stone size in a mountain stream. (From McAuliffe, 1983.)

winter anchor ice and spring floods are the principal scouring agents, a moss was found to cover a high percentage of surface only for stones larger than about 400 cm^2 (Figure 4.10). Evidently, smaller stones in this system are too unstable for extensive moss cover to accumulate (McAuliffe, 1983). Bryophyte abundance was also associated with stable substrates in a Tennessee woodland stream (Steinman and Boston, 1993).

(d) Substrate

Substrate influences periphyton populations in addition to providing stability against high flows. Nutrients and other aspects of the chemical environment can vary with substrate conditions, particularly for epipelic and epiphytic populations, but perhaps also for epilithic taxa on occasion. In a Montana stream, *Monostroma quaternarium* was largely confined to iron-rich rocks, while *Hydrurus* occurred mainly on limestone and sandstone, and *Batrachospermum* showed no substrate specificity (Parker, Samsel and Prescott, 1973). Physical aspects of the substrate's surface might also affect its suitability for periphyton growth, as when the availability of crevices apparently permits some taxa to persist in very fast waters (Keithan and Lowe, 1985).

(e) Temperature

The influence of water temperature on periphyton assemblages can be inferred from seasonal and distributional reports of species found mostly at either low or high temperatures (Hynes, 1970), but apparently there has been little experimental verification. Seasonal changes in the taxonomic composition of periphyton assemblages of temperate rivers are observed, as described previously, and warmer temperatures probably are partly responsible for the greater representation of green algae and cyanobacteria during summer months (Whitton, 1975a). However, total biomass often does not vary greatly over the year, especially where climatic fluctuations are not extreme. Douglas (1958), who studied *Achnanthes* spp. by cell counts in a stony stream in the north of England, and Marker (1976), who monitored chlorophyll concentrations on the gravel-bed of a chalk stream in the south of England, both found seasonal variation to be modest. Apparently a mid-winter minimum and a spring or summer maximum is the most seasonality often seen for periphyton in streams where climactic extremes are not great.

(f) Grazing animals

Until recently one could only speculate on the potential importance of grazing as a limiting factor for stream periphyton, based on a few examples of inverse correlations between abundances at these two trophic levels (Hynes, 1970). In most of the studies discussed in this chapter, grazing was thought to play a minor role in limiting periphyton populations. However, total elimination of grazers by an insecticide has been followed by spectacular increases in periphyton biomass, in channels in Switzerland (Eichenberger and Schlatter, 1978), and in a mountain stream in Japan (Yasuno *et al.*, 1982).

A growing body of experimental evidence now argues convincingly for strong interactions between herbivores and small autotrophs in running waters, and these will be described in greater detail in chapter 8. In addition to reducing the biomass of periphyton, grazers can alter the species composition by consuming more of particular taxa, while primary production is expected to peak at an intermediate grazing pressure and periphyton standing crop (Lamberti and Moore, 1984).

4.1.2 Primary production by periphyton

Primary production is the formation of new energy by photosynthesis. The autotrophs of running waters, which include higher plants, algae, some protists and some bacteria, use some of this energy for their own metabolism. Thus it is useful to distinguish between gross primary production, which is the total amount of energy fixed per unit of time, and net primary production, which is the amount remaining after energy expenditures by the autotroph in its own metabolic activities.

In contrast to lakes and oceans, where productivity patterns are well known, measurements of primary production in running waters are few. The challenge of measuring primary production in running waters is a major reason for this scarcity of data. There are essentially three approaches, each with particular strengths and weaknesses (Wetzel, 1975).

Biomass accrual over time is a widely used measure of net primary production, and for macrophytes it is usually the preferred method. However, for small autotrophs this method is less satisfactory due to the challenge of removing accurate samples from the substratum or, if artificial substrates are employed, the necessity of extrapolating to natural substrates. Moreover, population turnover rates can be rapid relative to the sampling interval, rendering the method less accurate. Measurement of open stream gas exchange (usually O_2 but CO_2 also can be used) treats the entire stream as a unit and so has the advantage of estimating rates of photosynthesis and respiration for the whole ecosystem. This method can be very useful where circumstances are appropriate (Figure 2.1). However, it requires an estimate of diffusion between air and water, and hence is difficult to apply to streams of low productivity and high turbulence. In addition, the gas exchange method assumes that night fluxes accurately reflect respiration over the diel cycle, and that the measured respiration includes only the autotrophs. Neither of these assumptions is easily checked, and the latter one might result in substantial over-estimation of gross primary production.

The third method is based on efforts to apply the light and dark bottle approach to periphyton and use gas exchange (oxygen) or radiotracer uptake (^{14}C) to measure rates. This is a well-developed methodology for autotrophs in suspension, and therefore easily applicable to river phytoplankton. However, its application to streambed periphyton is challenging (e. g. McIntire, 1966; Hansmann, Lane and Hall, 1971; Bott *et al.*, 1978; Hornick, Webster and Benfield, 1981). Rather than a simple bottle, the apparatus is a chamber that is buried flush with the streambed, filled with substrate, and left for days to weeks to recover from the disturbance of its installation. Care must be taken to simulate current and avoid unnatural changes in gas or nutrient concentrations. The substrate bearing the periphyton has to be introduced into the chamber, limiting which substrates can be examined and raising questions of disturbance during the transfer. Eventually, use of light and dark chambers and measurement of either O_2 changes or ^{14}C uptake likely will become the standard approach for the periphyton. Unfortunately, due to the pronounced spatial heterogeneity of benthic autotrophs (Marker, 1976), a thorough characterization of periphyton primary production appears to be well in the future.

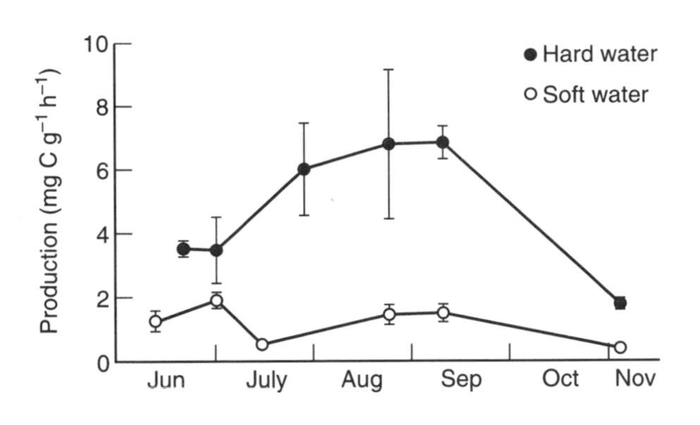

FIGURE 4.11 Primary production of periphyton measured by ^{14}C uptake using substrate placed in recirculating chambers, at two sites in the New River, Virginia. The river has a wide, shallow, bedrock channel and swift flow; the soft-water site has about 15 mg l^{-1} $CaCO_3$, compared with about 45 mg l^{-1} at the hard-water site. (From Hill and Webster, 1982a.)

(a) The magnitude of primary production by periphyton

At present, a score or more estimates of lotic primary production are available, but they vary widely, and detection of general trends must await a larger data base obtained using well validated methods. Tentatively, it appears that maximal daily net primary production (P_N) in deciduous biome studies ranges from below 0.01 to about 0.1 g C m^{-2} d^{-1} in shaded areas (Minshall, 1967; Elwood and Nelson, 1972; Hornick, Webster and Benfield, 1981), whereas values from 0.25 to about 2 g C m^{-2} d^{-1} are common at sites where the canopy is open (King and Ball, 1966; Berrie, 1972; McDiffett, Carr and Young, 1972; Sumner and Fisher, 1979; Hill and Webster, 1982a; Bott, 1983). Estimates from streams flowing through grasslands (Prophet and Ransom, 1974), in desert regions (Minshall, 1978; Fisher *et al.*, 1982), and open coniferous biome streams (Thomas and O'Connell, 1966; Wright and Mills, 1967) appear to give maximal values between 1 and perhaps 6 g C m^{-2} d^{-1}. Thus the magnitude of daily primary production does appear to be greater in unshaded compared with shaded sites, although in large rivers, turbidity again results in lowered primary production (e. g. the Danube: Ertl and Tomajka, 1973). Laboratory estimates of maximal production are comparable: McIntire and Phinney (1965) estimated gross primary production (P_G) between 2.4 and 6 g C m^{-2} d^{-1} in laboratory streams provided with adequate light, and P_N typically is slightly more than one-half of P_G. Thus it appears that the maximal values just cited might prove to typify many streams of average productivity, although some very productive flowing water systems substantially exceed this range, achieving daily P_G approaching 20 g C m^{-2} d^{-1} (e. g. Odum, 1956: Duffer and Dorris, 1966).

Most of these estimates of primary production are efforts to quantify its magnitude in the field, perhaps with contrasts among sites and seasons (e. g. Minshall *et al.*, 1983). Any information concerning limiting factors is therefore inferential, but results appear at least generally consistent with what is known from studies of autotroph biomass. For instance, periphyton production varies seasonally, and is greater in hard compared with soft waters (Figure 4.11), which Hill and Webster (1982a) attribute to the availability of free CO_2.

(b) The fate of benthic primary production

Primary production in flowing waters has a limited number of fates. Consumption by herbivores appears to be an important loss term for periphyton, and probably also for phytoplankton but not for macrophytes. The reasons for this, and a more detailed consideration of the interactions between herbivores and autotrophs, are discussed in chapter 8. Other fates of primary production include sloughing and downstream export, and burial, both of which become inputs to the particulate detrital pool. In addition, healthy cells at times exude organic compounds, and these constitute inputs to the pool of dissolved organic carbon. A complete reckoning of these quantities is not possible at present, but their inclusion should serve to emphasize that grazing is not the only fate of periphyton production.

Directional flow in streams inevitably leads to downstream transport of phytoplankton, dead or dislodged macrophytes, and sloughed periphyton. Sloughing of benthic periphyton is commonly observed in laboratory and nutrient enrichment studies, field studies report depletion of benthic populations following freshets, and downstream export apparently can be considerable. In their one-year study of periphyton dynamics in laboratory streams, McIntire and Phinney (1965) recorded substantial export, particularly of filamentous green algae under higher light conditions and during periods of high turbidity of the inflowing water. Sycamore Creek, Arizona, was found to be a net exporter of organic matter during a 63 day recovery between floods (Fisher *et al.*, 1982). Of the 56% of P_N not utilized at the site of production, the major amount was exported downstream and a smaller portion was stranded laterally by drying.

Flood events cause erosion and re-deposition of sediments, and this must result in burial of some material such as sloughed phytoplankton mats. Even in the absence of extreme discharge, cells die and become lodged in the sediments. This can be documented by epifluorescent microscopy, which distinguishes living from dead cells, and also by the accumulation of chlorophyll degradation products (Marker *et al.*, 1980).

Lastly, autotrophs can exude substantial quantities of dissolved and colloidal matter. Energy budgets constructed for periphyton populations in laboratory streams (P_G should equal respiration plus biomass accumulation plus biomass export) revealed a discrepancy of 17–28% (McIntire and Phinney, 1965). This indicates a substantial export term, which the authors attributed to dissolved matter. Extracellular release by periphyton in White Clay Creek, Pennsylvania, was found to dominate dissolved organic carbon (DOC) dynamics and result in a pronounced diel cycle in DOC concentrations at certain times of the year (Kaplan and Bott, 1982). Particularly in the spring, when periphyton biomass was maximal, allochthonous inputs were few and discharge was low, extracellular release by autotrophs was estimated to account for as much as 20% of daily DOC export. Clearly, a substantial fraction of P_G can be exported as DOC under circumstances of high primary production.

It should be apparent that we really know very little about the utilization of net primary production in flowing waters. It might be expected that energy fixed by benthic autotrophs enters grazing food chains, from where it can reach higher consumers or enter the detritus pool via the feces of herbivores. However, some amount of plant production enters the detritus pool directly, via sloughing, burial, and organic matter excretion. In the absence of quantitative information on these pathways, the relative magnitude of each simply is unknown.

4.2 Macrophytes

Flowering plants, mosses and liverworts, a few species of encrusting lichens, the Charales and other large algal species constitute the macrophytes of flowing waters (Hynes, 1970). Most can also be found in standing water, but as one proceeds to faster flows the flora becomes restricted to the small number of species able to withstand current. Those taxa almost entirely restricted to very fast currents include two flowering plant families of the tropics, the Podostemaceae and Hydrostachyaceae, and a number of bryophytes (Westlake, 1975a).

Macrophytes exhibit few adaptations to life in flowing water and are most successful in slow current areas such as deltas and backwaters. Certain characteristics permit establishment and maintenance of populations in appreciable current. Tough, flexible stems and leaves, firm attachment by adventitious roots, rhizomes or stolons, and vegetative reproduction typify most macrophytic species (Hynes, 1970; Westlake, 1975a). However, the Podostemaceae and Hydrostachyaceae of tropical torrential rivers possess aerial flowers and sticky seeds, and so are able to reproduce sexually.

Macrophytes can be classified according to their growth form, their manner of attachment, and even more specifically by the range of environmental conditions that a species inhabits. Four major growth forms are recognized by Westlake (1975a). Emergents occur on river banks and shoals. They are rooted in soil that is close to or below water level much of the year, and their leaves and reproductive organs are aerial. Floating-leaved taxa occupy margins of slow rivers, where they are rooted in submerged soils. Their leaves and reproductive organs are floating or aerial. Free-floating plants are usually not attached to the substrate and can form large mats, often entangled with other species and debris, in slow tropical rivers. Submerged taxa are attached to the substrate, their leaves are entirely submerged, and they typically occur in midstream unless the water is too deep. The categories preferred by Hynes (1970) are fairly similar, but focus on attachment. He also recognizes the free-floating life form. The attached plants include the bryophytes, certain lichens and the aforementioned tropical angiosperms, all of which inhabit fast-flowing water and attach firmly to fixed substrate. Hynes' remaining category includes all plants that grow in substrate loose enough to penetrate with roots (the majority of the flowering plants) or rhizoids (Charales).

Particular species occur in specific habitats based on growth form and attachment characteristics, as well as other environmental factors. Bryophytes occur world-wide in relatively cool streams and characteristically are found in headwater regions, associated with high currents and low light. In a woodland stream in Tennessee, a liverwort and two mosses were the most abundant bryophytes, achieving photosynthetic rates similar to periphyton in late summer and fall (Steinman and Boston, 1993). Along with some angiosperms, bryophytes are unable to use bicarbonate and require free CO_2, which is most available in turbulent soft waters. A combination of harder water and slower current allows rooted plants to become established, with a corresponding decline of bryophytes (Westlake, 1975a). In longitudinal view, therefore, one expects the macrophytic composition to exhibit a downstream succession from bryophytes to fastwater angiosperms such as *Ranunculus*, to flowering plants such as *Potamogeton* and *Elodea* which are typical of slower and more fertile waters, to emergent and floating-leaved plants in the slowest and deepest sections. However, local conditions can profoundly modify such a general scheme.

4.2.1 Limiting factors

In contrast to land, where coverage of the soil surface by vegetation often is near 100%, only a small percentage of stream bottom supports growth of higher plants. As Hynes (1970) points out, aquatic botanists have an understandable bias toward studying areas where plants are fairly abundant. Even so, studies of macrophyte production in two mid-size Appalachian rivers estimated percent cover between 27 and 42% (Hill and Webster, 1983; Rodgers *et al.*, 1983), and in Bavarian streams 37% of the area studied was found to have less than 10% cover (Gessner, 1955). Clearly, the amount of macrophyte cover varies enormously with locale, and in many stream habitats macrophytes are of little importance.

Where macrophytes do occur, the growing season can be quite long if water temperatures stay above freezing throughout winter. In British rivers many species simply grow slowly or cease growth during winter, although others, emergent plants in particular, shed leaves and die back to rhizomes and stolons (Hynes, 1970). In tropical waters there likely is little seasonality to growth, unless due to changes in river flow.

Major plant nutrients, particularly phosphorus but possibly nitrogen and potassium, can be limiting in nutrient poor waters such as mountain streams. In eutrophic lowland waters, there is presumed to be an excess of nutrients, based on the relationship between annual throughput versus demand (Westlake, 1975a). Wong and Clark (1976) found the daily photosynthetic rate of *Cladophora* in small rivers in southern Ontario to be linearly related to total phosphorus concentrations up to levels that would be considered eutrophic for lake waters. However, there was no correlation with nitrogen, or with phosphorus beyond 60–80 $\mu g\,l^{-1}$. Of course, rooted plants obtain nutrients via their roots as well as their shoots (Chambers *et al.*, 1989), and so water column measurements of nutrient availability can be misleading.

Hardness of water, or its correlates including calcium, alkalinity and pH, influences the distribution of particular macrophyte species and also limits the occurrence of bryophytes, probably by affecting the availability of free CO_2 as mentioned above. Indeed, the flora of British rivers can be reasonably categorized on the basis of hardness, current and substrate with the result that certain taxa are regularly associated with particular environmental conditions (Butcher, 1933; Hynes, 1970). Interestingly, even within an area favorable to a certain species, its distribution and abundance can be highly variable. Although some of this must be due to heterogeneous environmental conditions, the growth of a rooted plant can modify local current so as to enhance both deposition and erosion (Gessner, 1955), or favor a second species that replaces the first (Ladle and Casey, 1971). As a consequence, the mapping of vegetated areas over several years reveals them to be continually shifting mosaics.

Along with current, Westlake (1975a) considers light to be among the most important factors limiting macrophytes. Shading by forest canopy commonly reduces surface irradiance by 35–95%, and heavy shade results in complete exclusion of angiosperms. Light is also attenuated with depth. This inhibits the establishment of macrophytes in deeper rivers, although the precise depth at which photosynthesis can no longer balance respiration varies with turbidity and species-specific light requirements. Some very turbid tropical rivers lack higher plants for this reason.

Herbivory on freshwater macrophytes generally has been viewed as unimportant in limiting their growth and abundance (Hutchinson, 1981; Wetzel, 1983). Indeed, few representatives of the major groups of aquatic invertebrates are able to graze on macrophytes until after death and decomposition of the plant. Interestingly, some insects are effective herbivores, but these are species phylogenetically allied with families and orders of insect groups that are primarily terrestrial (Newman, 1991). Some of these herbivorous insects are successful agents of biological control, as is the manatee, against waterweeds that in many areas pose a serious nuisance. Most macrophytes do not become so common, however, and the bulk of their biomass is consumed only after it enters the detritus pool following plant senescence.

4.2.2 Production and its fate

Estimates of net primary production (P_N) indicate a maximum of roughly 3 $g\,C\,m^{-2}\,d^{-1}$ for submerged macrophytes compared with about 10–20 $g\,C\,m^{-2}\,d^{-1}$ for emergents, and with somewhat higher values in tropical compared with temperate environments (Westlake, 1975b). Macrophyte production naturally varies with environmental conditions, being highest in rivers of medium size where light is ample, current is moderate, turbidity is low, and strong fluctuations in depth and discharge are minimal. Even in circumstances that favor their growth, macrophytes generally contribute only a small fraction of the total energy base in natural streams (Fisher and Carpenter, 1976; chapter 12). Exceptions include ditches and canals in fertile regions, as well as marshes and river mouths where macrophytes might occupy virtually all available habitat.

The possible fates of macrophyte production include consumption as living tissue, secretion of dissolved organic matter and decomposition. Very little is consumed as living plant biomass, although terrestrial herbivores feed on emergent vegetation and some vertebrates, including waterfowl, the manatee, and grass carp (see chapter 8), consume submerged aquatic macrophytes while others such as the muskrat harvest plant material for construction of lodges (Westlake, 1975b). Little is known concerning the release of dissolved organic compounds by living plants, but this secretion can be substantial. Epiphytic autotrophs and heterotrophs undoubtedly gain immediate benefit, and in addition an unknown amount enters the downstream pool of dissolved organic carbon. Wetzel and Manny (1971) suggest as a conservative estimate that 4% of P_G of macrophytes in lakes is released as exudate. The majority consists of labile compounds of low molecular weight.

The principal fate of macrophytic primary production is to enter the detritus food chain. In addition to seasonal die-back of plants, up to 50% of new biomass is lost during the period of growth (Westlake, 1975a). These inputs to the detritus pool, discussed more fully in chapter 5, undergo rapid breakdown during periods when freshets are few, and so local utilization is favored (Fisher and Carpenter, 1976). Because macrophytic detritus is of relatively high quality, and appears when summer periphyton might be waning and prior to autumnal leaf fall, Hill and Webster (1983) argue that these inputs are of greater importance to consumers than would be anticipated solely on the quantity produced.

4.3 Phytoplankton

Small autotrophs suspended in the water column and transported by currents, including algae, protists and cyanobacteria, comprise the phytoplankton. Whether a river phytoplankton could be self-sustaining was in doubt for some time, because downstream flow would seem to prevent the persistence of their populations. It was suggested that any river plankton was the result of displacement of cells from the benthos, backwaters and lakes or impoundments along the river's course, and reflected wash-out and export rather than a true 'potamoplankton'. These are indeed major sources of phytoplankton in river water. In small, fast-flowing streams, sloughing of attached autotrophs likely is the primary source, and any cells in the water column are simply sloughed material in transit (e. g. Swanson and Bachmann, 1976). However, in sluggish, lowland streams, in side channels and within macrophyte beds, and in rivers of considerable length, the residence time of a water mass can be sufficient for true plankton to colonize and reproduce. Under these conditions phytoplankton and zooplankton almost always are present, and at times they can develop substantial populations (Hynes, 1970).

It is doubtful that any planktonic organisms are restricted only to flowing water, and so the truly planktonic species found in rivers are drawn from the same pool of species found in standing water. Thus the presence of lakes, ponds and backwaters, and more recently the creation of impoundments, can be of great importance in seeding the river with plankton. As Cushing (1964) showed in the Montreal River, Saskatchewan, the occurrence of lakes along the river's length results in marked increases in phytoplankton abundance. These lacustrine inputs usually undergo rapid attrition in flowing waters due to settling, consumption and the filtering effects of macrophytes when present (Reif, 1939). Reif found that diatoms, cyanobacteria and rotifers persisted farther downstream than did Chlorophyceae, desmids and crustacean zooplankton, which suggests that the species composition of river plankton is partly influenced by differential settling and trapping. However, it is the circumstances that permit reproductive increase of plankton populations that determine the overall abundance.

Diatoms, particularly centric diatoms, have been found to dominate the composition of river phytoplankton in a number of studies (the Nile, (Talling and Rzóska, 1967); the Thames, (Lack, 1971); the Mississippi, (Baker and Baker, 1979)). Abundances are greatest in spring and summer, when additional taxa are likely to be found. A 300 mile section of the Sacramento River included 15 genera of Cyanobacteria, 38 greens and 13 flagellates, as well as 29 diatom genera (Greenberg, 1964). Diatoms dominated numerically, however, averaging 75% and occasionally comprising 99% of cell counts. Cyanobacteria have been reported to form dense blooms in the Nile and during summer in temperate rivers (Talling and Rzóska, 1967; Bennett, Woodward and Schultz, 1986), apparently in response to nitrate depletion. These and other studies clearly establish what was once a contested issue, namely that phytoplankton and zooplankton can be abundant in virtually all major rivers examined. They are present throughout the year, even when not apparent, as Sze (1981) demonstrated by collecting samples of Potomac River water from which he successfully cultured several major phytoplankton groups. Interestingly, the order of appearance in culture: centric diatoms, then chlorococcolean green algae, then pennate diatoms and finally the Cyanobacteria, paralleled their seasonal appearance in the river.

4.3.1 Limiting factors

Factors affecting the growth of phytoplankton in running waters include all the same variables, such as light, temperature and nutrients, that limit growth in lakes. However, discharge regime has a profound influence over river phytoplankton, and the influence of light and nutrients differs in some ways from what is seen in standing waters. In addition, adjacent stagnant waters are critical to the establishment of river phytoplankton, and so can affect the size of the inoculum. This can be of considerable importance, especially when the residence time of the water mass is short enough to limit the buildup of populations.

Of all the factors that influence the plankton populations of rivers, those associated with current and discharge clearly are of overriding importance. An inverse relationship between river discharge and phytoplankton abundance is perhaps the most common finding of detailed investigations of river phytoplankton (Rzóska, Brooks and Prowse, 1952; Dorris, Copeland and Lauer, 1963; Greenberg, 1964; Williams, 1964; Lack, 1971; Décamps, Capblanq and Tourenq, 1984). As a mass of water moves downstream and the entrained plankton multiply, one expects maximal abundances to be associated with a water mass that is traveling slowly and is uninterrupted over a long distance. Talling and Rzóska (1967) estimated that a water mass traversed the 357 km section of the Blue Nile between Sennar Reservoir and Khartoum in 40 days at low flows, but required only 2 days at high flood. Since phytoplankton are capable of a maximum of about 1–2 doublings per day, the consequences for eventual population size are enormous.

Using a data set that included 345 sites on large rivers and 812 lakes and impoundments within the continental USA, Søballe and Kimmel (1987) concluded that rivers and lakes occupy two ends of a continuum, and impoundments fall in-between. Along the gradient from rivers to impoundments to lakes, residence time increased (mean values in days of 18.4, 528.5 and 1073.5, respectively), transparency increased, total phosphorus declined, and phytoplankton counts increased several-fold. Interestingly, water residence time appeared to act as a threshold factor,

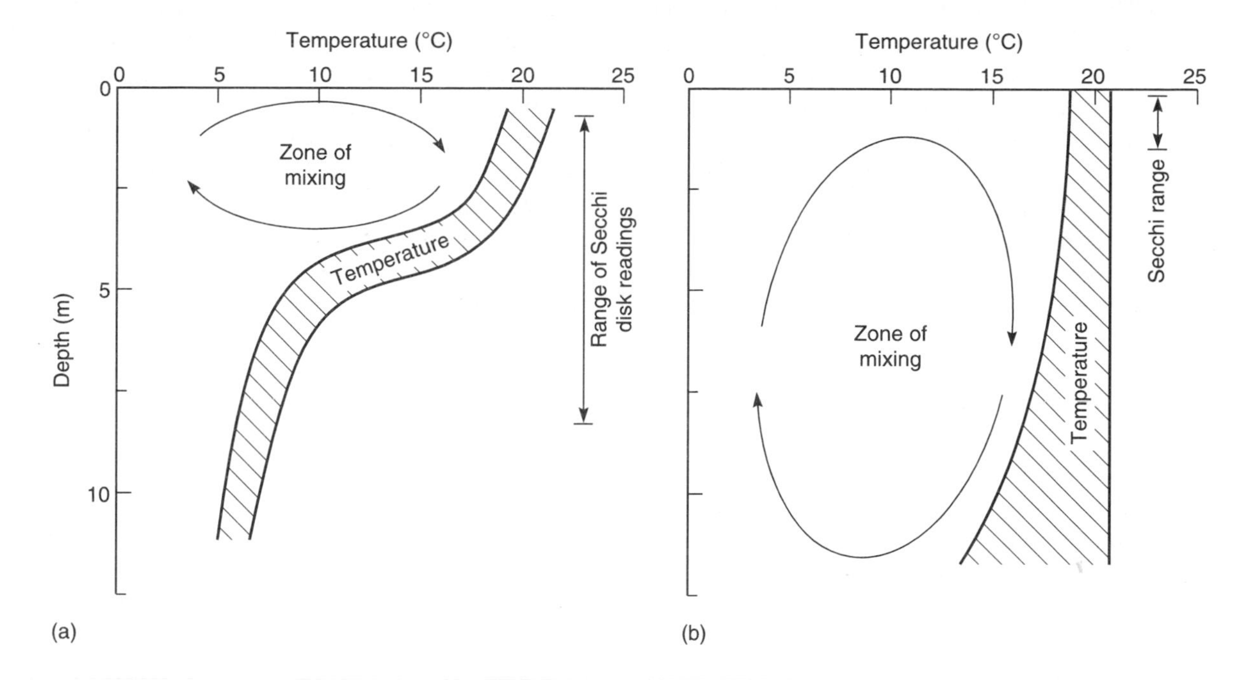

FIGURE 4.12 Schematic diagram comparing effect of depth of mixing on primary production in phytoplankton of a lake *versus* a river. In a lake (a), establishment of a temperature barrier between surface and deep waters restricts mixing to the upper few meters. In a river (b), temperature stratification is impeded by turbulence of flow, and the water column typically mixes from top to bottom. Secchi disk transparency, a rough measure of depth of light penetration for photosynthesis, ranges from about 1 to 8 m in mesotrophic and eutrophic lakes, and can be much deeper in clear, unproductive lakes. Large rivers commonly are 5–20 m or more in depth, but due to their sediment loads, light penetration is usually restricted to at best the upper 1–2 m.

being of great importance at values less than 75–100 days, and of little importance when residence time was longer.

Although the usual effect of flooding is dilution and downstream transport, under certain circumstances floods might augment river plankton by washing in populations from stagnant areas. The lower Orinoco and several of its tributaries comprise a large tropical river system with fringing floodplain regions that are in contact with the river for up to 180 days in wet years (Lewis, 1988). Primary production per unit volume was greatest during the period of low water, but nonetheless was quite low, due to a combination of light limitation and the short residence time of river water. Total phytoplankton transport exhibited a minimum just as discharge began its seasonal increase and was maximal at high water. Lewis concluded that flushing of backwaters in or adjacent to the river channel accounted for increased transport during the rise of floodwaters, while the floodplain contributed phytoplankton during the flushing and draining periods associated with peak and declining flood stages.

In a large river with ample nutrients and a transit time long enough to permit multiplication of phytoplankton, it is likely that autotrophs are limited by light via an interaction among turbidity, depth and turbulence. If the water column mixes to a depth greater than the photic zone, an individual cell will spend part of the day at light levels too low to support photosynthesis (Figure 4.12). In the Hudson River, for a 12 h day, Cole, Caraco and Peierls, (1991) estimate that the average phytoplankton cell would spend from 18 to 22 h below the 1% light level. Rather than growing, the cell would be expected to lose biomass. This is a real puzzle, because phytoplankton biomass does increase during the spring and summer. One possible explanation is that phytoplankton blooms originate only in river sections less than 4 m in depth (Cole, Caraco and Peierls, 1992). Similarly, Lewis (1988) reasoned that the bulk of phytoplankton biomass transported in the lower Orinoco system originates from stagnant waters in or adjacent to the channel, because phytoplankton within the channel spent too little time at light levels sufficient for growth.

Turbidity and depth of mixing will of course vary from place to place and with seasons, greatly affecting the opportunities for growth of phytoplankton populations. Lewis (1988) determined that phytoplankton productivity was suppressed to a greater degree in the whitewater Apure River than in the blackwater Caura River, corresponding to differences in light penetration. Depth of the euphotic zone varies seasonally with the sediment load, and also due to self-shading when phytoplankton are abundant. In the Lot River of south-central France, the maximum depth at which photosynthesis can occur ranges between 0.7 and 5.3 m (Décamps, Capblanq and Tourenq, 1984), depending on season. In the Blue Nile, passage of the flood crest reduces Secchi disk readings to zero (Rzóska, Brooks and Prowse, 1952). Reservoirs usually enhance the conditions for phytoplankton growth because greater water clarity results from settling of sediments. Under these conditions, self-shading by the plankton replaces sediments in suspension as the principal limitation to light penetration.

Nutrient limitation of river phytoplankton does not appear to be usual. Downstream transport and light are typically overriding variables, and nutrient concentrations in rivers are often considerably higher than found in lakes (chapter 13). In their survey of over 100 sites on lakes, impoundments and rivers, Søballe and Kimmel (1987) observed a positive relationship between phosphorus concentrations and phytoplankton biomass, a well established trend of lakes. However, river phytoplankton abundance was several times lower than would be expected based on phosphorus availability, and the relationship also was more variable, suggesting that factors other than nutrients exerted primary control.

4.3.2 Primary production by river phytoplankton

The magnitude and fate of primary production by river phytoplankton have received some study (e. g. Prowse and Talling, 1958; Décamps, Capblanq and Tourenq, 1984; Filardo and Dunstan, 1985; Bennett, Woodward and Schultz, 1986). Most data are from locations where rivers approach reservoir or estuarine conditions, and more studies along unaltered large river sections would be helpful. Søballe and Kimmel's (1987) analysis showed that phytoplankton cell counts in large rivers are several times lower than would be found in lakes of comparable phosphorus concentration, and Lewis' (1988) study shows that phytoplankton productivity can be very low in large rivers where environmental conditions are unfavorable. Annual estimates of net primary production are scarce, but likely will be in the range of unproductive lakes. Production is greater under impoundment conditions, although light limitation with depth again can be great (Prowse and Talling, 1958). Nuisance levels of phytoplankton have been reported under conditions of high nutrient loading (Bennett, Woodward and Schultz, 1986), and low flows (Décamps, Capblanq and Tourenq, 1984).

Zooplankton grazing does not appear to constitute a major loss term for phytoplankton under most conditions. Compared with lakes, rivers typically support a zooplankton biomass less than would be expected based on amount of phytoplankton (Pace, Findlay and Lints, 1992). In the Danube (Bothar, 1987), a whitewater tributary of the Orinoco (Saunders and Lewis, 1988), and the Hudson (Pace, Findlay and Lints, 1992), zooplankton grazing was believed to have little impact on phytoplankton.

The largest species of microcrustaceans, which are capable of the highest filtering rates, rarely are abundant in rivers, because their slow rates of population growth do not compensate for downstream transit. Rotifers and smaller crustaceans usually predominate because of their shorter generation times, and even these taxa can build up their numbers only during low-flow periods. For example, transit times at moderate flows are two- to threefold longer in the White Nile compared with the Blue Nile. Crustacean zooplankton predominate in the former, while the Blue Nile contains mainly rotifers (Rzóska, Brooks and Prowse, 1952). Thus discharge conditions determine the species and size composition of zooplankton, and strongly constrain their ability to exert strong grazing pressure.

Nonetheless, grazing does consume significant amounts of phytoplankton production under some circumstances. In the fresh, tidal Potomac River in Maryland–Virginia, phytoplankton abundance was reduced by 40–60% on passage through a section of high densities of the Asiatic clam *Corbicula fluminea*. The filtering rates of these clams could reduce chlorophyll *a* from a river water sample by 30% in 2 h (Cohen *et al.*, 1984). Because it is based on an exotic species this example should be viewed with caution, but it does make sense that an abundant sessile grazer could produce a substantial abundance 'sag' during phytoplankton passage.

Because of low grazing rates, under most circumstances the majority of the primary production of rivers is exported downstream. At the post-flood phytoplankton maximum in the White Nile, the magnitude of this export was estimated at 40×10^6 g dry mass per day (Talling and Rzóska, 1967). What happens when this material reaches a lake or reservoir is uncertain, with catastrophic mortality as one possibility (e. g. Søballe and Bachman, 1984) and a gradual blending with lentic populations as another possibility. However, on reaching an estuary the likely result is mass mortality due to increased salinity. High oxygen demand has been reported at low salinity boundaries in estuaries (Morris, Bake and Howland, 1982), indicating that transported river plankton provides a substantial input to estuarine detrital dynamics.

4.4 Summary

Periphyton, macrophytes and river phytoplankton constitute three very different groups of autotrophs occurring in streams and rivers. The attached periphyton is found on surfaces and substrates virtually everywhere, and is especially common on the beds of streams and small rivers. Macrophytes are most abundant in mid-sized rivers, and in backwaters and along the margins of larger rivers. Phytoplankton populations are likely to develop in large lowland rivers during intervals of modest flow. The relative importance of these very different groups of primary producers changes greatly with river size and conditions. In each case we can ask about limiting factors, the amount of production, and its fate. Relatively more attention is given to the periphyton in this treatment, partly because they have been more extensively studied, and partly out of a bias toward smaller rivers where attached autotrophs predominate.

The periphyton, comprising an abundance of diatoms, the Chlorophyta (green algae), Cyanobacteria (blue–green algae), and a few other groups, occur on virtually every surface in running waters including stones, soft sediments and macrophytes. Their abundance varies seasonally and spatially due to multiple factors, any one of which can be of overriding importance in certain circumstances. High discharge, by dislodging cells, flipping stones and scouring surfaces, often restricts periphyton growth to quieter periods. Light can be limiting, particularly under dense forest canopies. There is growing evidence that nutrient limitation of periphyton is widespread, most often due to a short supply of phosphorus. Small autotrophs are vulnerable to a wide variety of herbivores, and grazing can be a significant loss term.

Macrophytes, including some large algae, bryophytes and vascular plants, are found in flowing water mainly where neither the depth nor current is great. Rivers of intermediate size, canals, and river margins usually support the greatest biomass of these groups. Four major categories based on growth form include the emergents found on banks and shoals, floating-leaved taxa on the margins of slow rivers, free-floating plants such as occur in great mats on large tropical rivers, and submerged taxa that attach to the substrate. The length of the growing season, current and light appear to be major limiting factors for aquatic macrophytes. Grazing on living plants is in most instances a minor factor, and the bulk of plant production enters the detritus pool after senesence.

The phytoplankton, comprising cells and colonies of small autotrophs in the water column, develop self-sustaining populations only under certain conditions typically found in large lowland rivers. The first condition that must be met is a sufficient residence time to allow biomass to increase faster than it is transported downstream. Nutrients are usually not a critical limiting factor to river phytoplankton. Instead, light often becomes the limiting factor wherever discharge is low enough to permit phytoplankton blooms to develop. The depth of light penetration usually is only a small fraction of depth to the bed of large turbid rivers, and because the water column typically is well-mixed, the phytoplankton experience little or no light much of the time. Indeed, the opportunities for photosynthesis might be so brief that phytoplankton populations cannot be maintained without inputs from shallow regions, embayments, and floodplain lakes. In comparison to standing waters of comparable nutrient status, riverine phytoplankton biomass is substantially lower. Moreover, although our knowledge of grazing pressure on river phytoplankton is scant, this also does not appear to be a strongly limiting factor. Thus, in contrast to standing waters where phytoplankton frequently are limited by some combination of nutrient supply and grazing, in rivers these factors usually are considerably less important. Current evidence suggests that downstream export rather than *in situ* energy processing is the dominant fate of large river phytoplankton production.

Chapter five

Heterotrophic energy sources

Particulate and dissolved non-living organic matter are important energy inputs to most food webs, and this is especially true in running water ecosystems. While primary production by the autotrophs of running waters can be substantial (chapter 4), much of the energy support of lotic food webs derives from non-living sources of organic matter. These energy pathways are referred to as heterotrophic and the immediate consumers of this material are decomposers and detritivores, in contrast to autotrophic pathways linked to higher trophic levels by herbivores (chapter 6).

Heterotrophic production requires a source of non-living organic matter, and the presence of microorganisms (bacteria, fungi) to break down the organic matter and release its stored energy. Plant litter and other coarse debris that falls or blows into stream channels, fine particulates that originate from many sources including the breakdown of larger particles, and dissolved organic matter constitute the three main categories of non-living organic matter in most situations (Table 5.1). Some of this material originates within the stream (such as dying macrophytes, animal feces, extracellular release of dissolved compounds) and some is transported into the stream from outside (such as leaf fall, soil particulates and compounds dissolved in soil water). Collectively these sources can substantially exceed the energy transformed within streams by photosynthesis. Although these sources of organic matter vary widely in nutritional quality, they nonetheless constitute important energy pathways.

Heterotrophic pathways are of greatest importance where the opportunities for photosynthesis are least (Vannote *et al.*, 1980). Small streams in forested regions are one example: their heavily wooded stream banks provide abundant inputs of plant litter and other detritus, while at the same time algal growth is reduced by the shade of the forest canopy (Figure 4.4). Large rivers also are likely to be dominated by decomposition processes, because turbidity and depth limit the availability of light (Figure 4.11), while downstream transport and the floodplain are potential sources of organic matter inputs. Autotrophy is likely to dominate only when conditions favor high primary production. This may occur seasonally, such as prior to spring leaf-out; in unshaded streams and small rivers, especially in dry regions including deserts and grasslands; and in large rivers when conditions favor phytoplankton blooms. Even under such conditions, however, non-living organic matter usually is a substantial energy source.

The division of non-living organic energy into size classes is useful in studying detrital dynamics

TABLE 5.1 Sources of organic matter to running waters. Many originate from outside the section where they are found, with certain exceptions (marked with an asterisk) where energy fixed by photosynthesis within the stream enters heterotrophic pathways

Sources of Input	*Comments*
Coarse particulate organic matter (CPOM)	
Leaves and needles	Major input in woodland streams, typically pulsed seasonally
Macrophytes during die-back*	Locally important
Woody debris	May be major biomass component, very slowly utilized
Other plant parts (flowers, fruit, pollen)	Little information available
Other animal inputs (feces and carcases)	Little information available
Fine particulate organic matter (FPOM)	
Breakdown of CPOM	Major input where leaf fall or macrophytes provide CPOM
Feces of small consumers	Important transformation of CPOM
From DOM by microbial uptake	Organic microlayers on stones and other surfaces
From DOM by physical–chemical processes	Flocculation and adsorption, probably less important than microbial uptake route
Sloughing of algae*	Of local importance, may show temporal pulses
Sloughing of organic layers	Little infomation available
Forest floor litter and soil	Influenced by storms causing increased channel width and inundation of floodplain, affected by overland *versus* sub-surface flow
Stream bank and channel	Little known, likely related to storm events
Dissolved organic matter (DOM)	
Groundwater	Major input, relatively constant over time, often highly refractory
Sub-surface or interflow	Less known, perhaps important during storms
Surface flow	Less known, perhaps important during storms
Leachate from detritus of terrestrial origin	Major input, pulsed depending upon leaf fall
Throughfall	Small input, dependent on contact of precipitation with canopy
Extracellular release and leachate from algae*	Of local importance, may show seasonal and diel pulses
Extracellular release and leachate from macrophytes*	Of local importance, may show seasonal and diel pulses

in streams. The usual categories are coarse particulate organic matter (CPOM, greater than 1 mm), fine particulate organic matter (FPOM, less than 1 mm and more than 0.5 μm) and dissolved organic matter (DOM, less than 0.5 μm). Each category can be divided further (e. g. Cummins, 1974), but the dividing lines are arbitrary. In some instances, particularly the breakdown of forest leaves that enter streams, we have a detailed understanding of the processes involved. In other instances, such as the pathways involving fine particulates, we know considerably less. However, it is clear that the dynamics of organic matter in streams are complex, microorganisms are critical mediators of these complex interactions, and the surrounding landscape critically influences what goes on within the stream (Hynes, 1975).

5.1 Decomposition of coarse particulate organic matter

The fate of CPOM is best known for autumn-shed leaves, which are major energy inputs to many small, forested streams. The literature on this topic is reviewed by Webster and Benfield (1986), and methods of study by Boulton and Boon (1991). The breakdown of macrophytes is similar to that of leaves of terrestrial origin, although some minor differences are noted below. The breakdown of woody material is, not surprisingly, much slower than leaves and of lesser importance to higher trophic levels. Other sources of CPOM that enter heterotrophic pathways in running waters and may be locally or seasonally important, such as flower parts, hippopotamus feces and carcasses of large animals, have received little study.

Studies of organic matter breakdown start with the source material, often using leaves picked from riparian woody plants just prior to abscission, and follow its disappearance over time. The breakdown of CPOM culminates in a limited number of possible fates: mineralization, storage and export. However, much of the CPOM becomes fine particulates, which are difficult to follow, and so the dynamics of CPOM once it becomes FPOM are not well understood.

The loss of leaf mass over time is approximately log-linear, although some data have been interpreted as linear or as consisting of two or more distinct stages. Webster and Benfield (1986) convincingly argue that a simple exponential model,

$$W_t = W_i e^{-kt} \qquad (5.1)$$

where W_t = dry mass at time t, W_i = initial dry mass, and t is measured in days, provides a general and utilitarian description of the breakdown process. The statistic k (in units days^{-1}), which is the slope of the plot of $\log_e$ of leaf mass *versus* time, provides a single measure of breakdown rate.

The rate of leaf breakdown is determined by intrinsic differences among leaves, a number of environmental variables, and the feeding activity of detritivores. Petersen and Cummins (1974) suggested a continuum of decomposition rates from slow to fast, based on the breakdown of leaves from six deciduous tree species in a small Michigan stream (Table 5.2). They also recognized that this variation in leaf breakdown rates, which they termed a 'processing continuum', had important consequences for invertebrate consumers by extending the time interval over which microbially colonized leaf litter was available. The wide variation in the breakdown rate of the leaves of different plant species has now been amply documented (Figure 5.1). Non-woody plant leaves decompose much more quickly, on average, than do leaves of woody plants (mean half-lives from data in Figure 5.1 are 65 days and 200 days, respectively). Submerged and floating macrophytes are among the fastest to decay, presumably because they contain the least amount of support tissue.

A number of environmental factors also influence k. Although leaf breakdown can occur quite rapidly at near-zero temperatures (e.g. Short, Canton and Ward, 1980), breakdown rates generally are faster at warmer temperatures. Higher breakdown rates also occur in more nutrient-rich systems, apparently due to the greater availability of nitrogen. Laboratory studies typically show acceleration of leaf breakdown in response to nitrogen addition (Meyer and Johnson, 1983). Enrichment with phosphate alone does not appear to affect breakdown rates, and other possible nutrients have received little study (Webster and Benfield, 1986). Low pH retards breakdown through

TABLE 5.2 Leaf breakdown rates for leaves of six deciduous trees in a Michigan stream (From Petersen and Cummins, 1974)

	Slow	*Medium*	*Fast*
k (days^{-1})	0.005	0.005–0.10	0.10–0.15
Time to 50% disappearance (months)	4.6	2.3–4.6	1.5–2.3
Time to 90% disappearance (months)	⩾15	8–15	⩽8
Example	*Quercus alba* *Populus tremuloides*	*Salix lucida* *Carya glabra*	*Cornus amomum* *Fraxinus americana*

inhibition of microorganisms and invertebrates, and hydrologic regime can have complicated effects including abrasion, burial and influencing the spatial distribution of litter, thus indirectly affecting the chemical environment and invertebrate densities.

5.1.1 Stages in the breakdown and decay of leaves

Kaushik and Hynes (1971) proposed the first detailed scenario describing the fate of autumn-shed leaves. While it has since been amplified considerably (see reviews by Bärlocher, 1985; Webster and Benfield, 1986), the basic model has proved to be a good description of events (Figure 5.2). Leaves fall directly or are wind-blown into streams, become wetted, and commence to leach soluble organic and inorganic constituents. Most of the leaching occurs within a few days and is followed by the second stage, a period of microbial colonization and growth, causing numerous changes in leaf condition. The third stage, fragmentation by mechanical means and invertebrate activity, usually requires some prior microbial conditioning of the leaf and is complete when no large particles remain. Up to 25% of the initial dry mass of leaves is lost due to leaching in the first 24 h (Webster and Benfield, 1986). The amount of soluble constituents lost into the stream is influenced by prior leaching by rain while leaves are still on trees and especially in the litter. Experimental results therefore will be influenced by whether leaves were picked from trees or from the ground. Constituents lost during leaching are primarily soluble carbohydrates and polyphenols (Suberkropp, Godshalk and Klug, 1976). Leaves of different plants show species-specific leaching rates: alder (*Alnus rugosa*) lost only about 4% of dry mass, whereas maple (*Acer saccharum*) and elm (*Ulnus americana*) lost 16% (Figure 5.3). Animals and microbes have little influence on leaching rates during the first week or so, but probably influence subsequent losses due to leaching during later stages of decay (Meyer and O'Hop, 1983).

During the second stage of leaf processing, microbial populations colonize and proliferate on the leaf substrate. Leaf weight-loss continues, but usually at a detectably slower rate than initial leaching. Differences in leaf chemistry and structure result in a wide variation in decay rates (Webster and Benfield, 1986). Leaves with a high initial nutrient concentration breakdown more rapidly than leaves of lower nutrient content. For example, Kaushik and Hynes (1971) established a positive relationship between initial nitrogen concentration and rapidity of breakdown. Conversely, a high lignin content slows

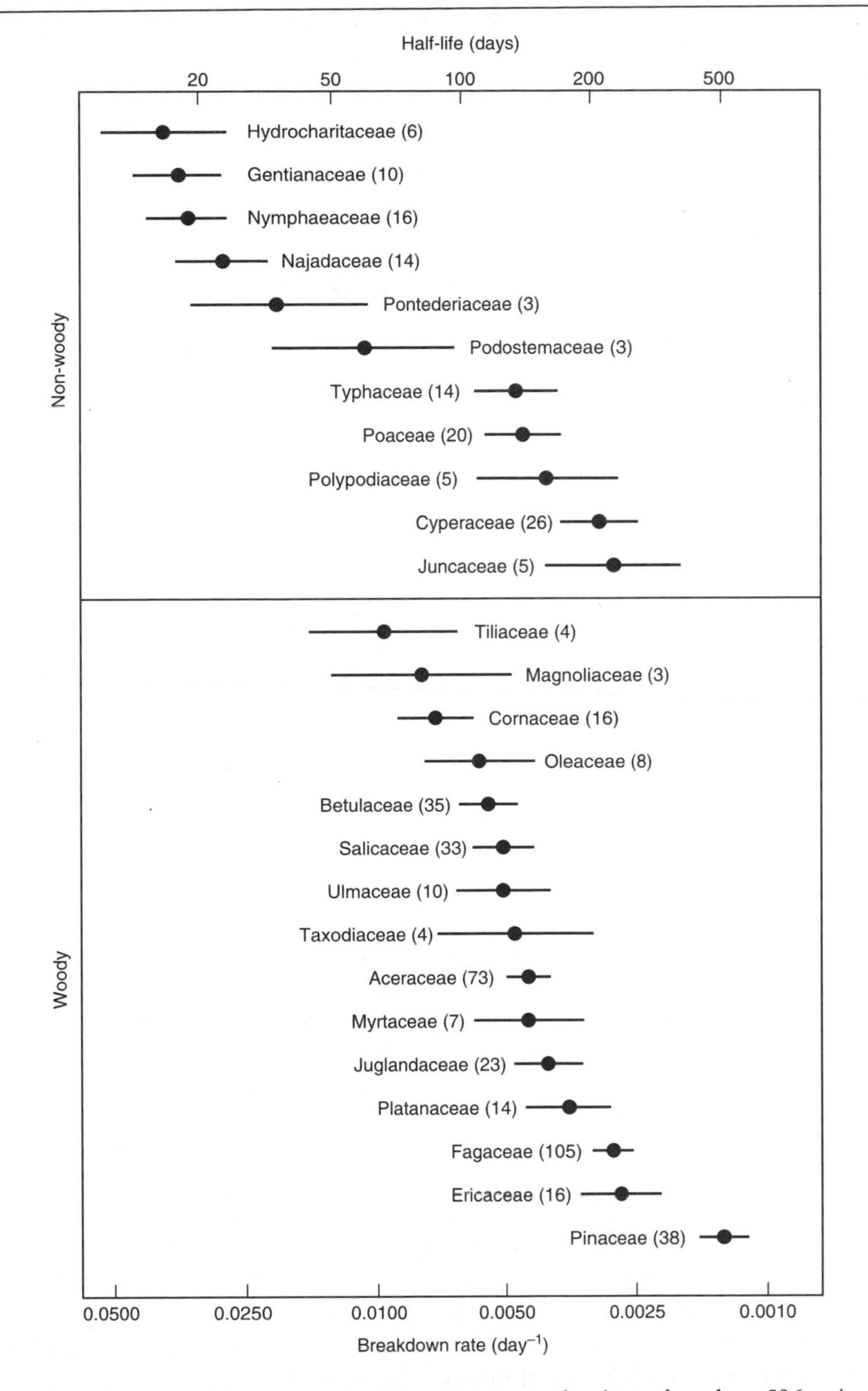

FIGURE 5.1 The breakdown rates for various woody and non-woody plants, based on 596 estimates compiled from field studies in all types of freshwater ecosystems. Means ± 1 standard error are shown, and the variation is due to (at least) effects of site, technique, and numerous environmental variables. The number of individual rate estimates is shown in parentheses. (After Webster and Benfield, 1986.)

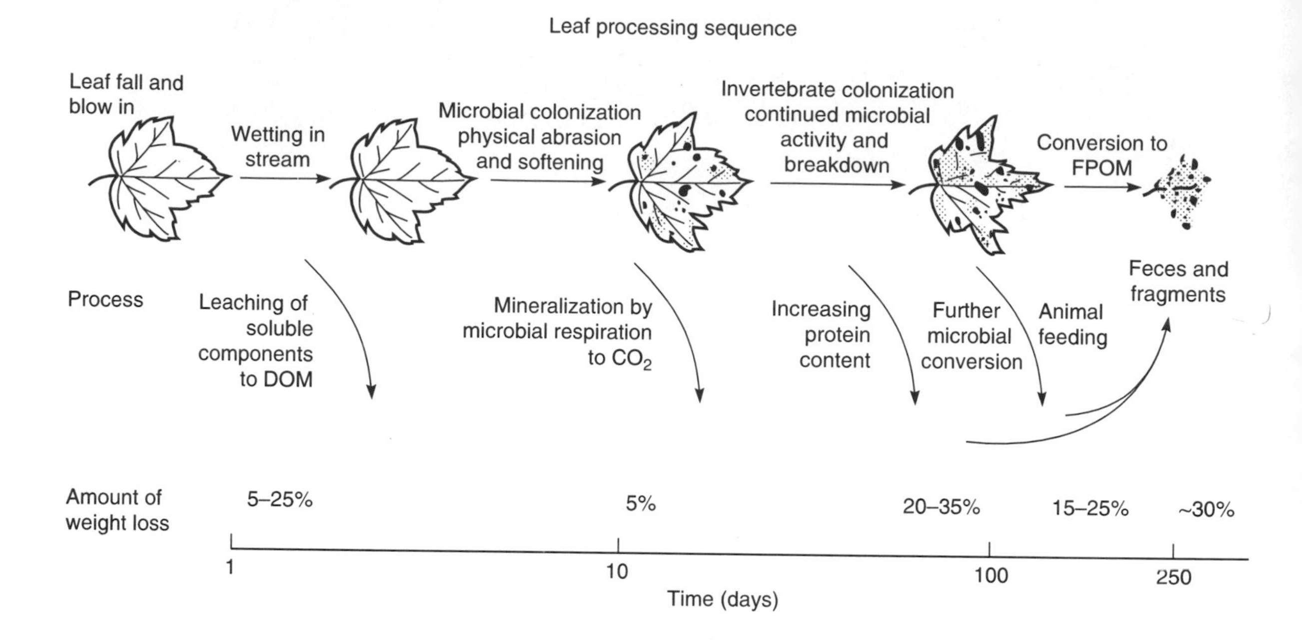

FIGURE 5.2 The processing or 'conditioning' sequence for a medium-fast deciduous tree leaf in a temperate stream. Details of the fate of material converted to fine particulate organic matter (FPOM) are unknown. Leached dissolved organic matter (DOM) is thought to be rapidly transferred into the sediment layer, primarily by microbial uptake.

breakdown. A combination of initial nitrogen and lignin proved to be an effective predictor of the breakdown rate of six species of leaves in the terrestrial litter of a New England forest (Melillo, Aber and Muratore, 1982) and of wood chips from five species of trees in streams in eastern Canada (Melillo *et al.*, 1983). Lastly, chemical inhibitors impede leaf decay in several ways. Tough outer coatings such as the cuticle of conifer needles slow fungal invasion (Bärlocher, Kendrick and Michaelides, 1978) and complexing of protein to tannins is a principal cause of slow breakdown in many broadleafed woody plants. Toxic effects of chemical constituents also may influence breakdown rates (Webster and Benfield, 1986), although evidence is scant. Stout (1989) suggested that phytochemical differences among leaves in tannins, secondary plant compounds that defend against terrestrial herbivores, might contribute to species differences in breakdown rates. Somewhat surprisingly, chemical measures of tannins (total phenolics and condensed tannins) of 48 deciduous trees were unrelated to published breakdown rates (Ostrofsky, 1993). This appears to rule out one of the most likely ways for leaf chemistry to affect leaf processing rate.

Leaves undergo a number of chemical changes as they decay. Nitrogen typically increases as a percentage of remaining dry mass and sometimes increases in absolute terms as well. Because protein complexed to lignin and cellulose is very resistant to breakdown, nitrogen compounds remain while other leaf consitutents are lost,

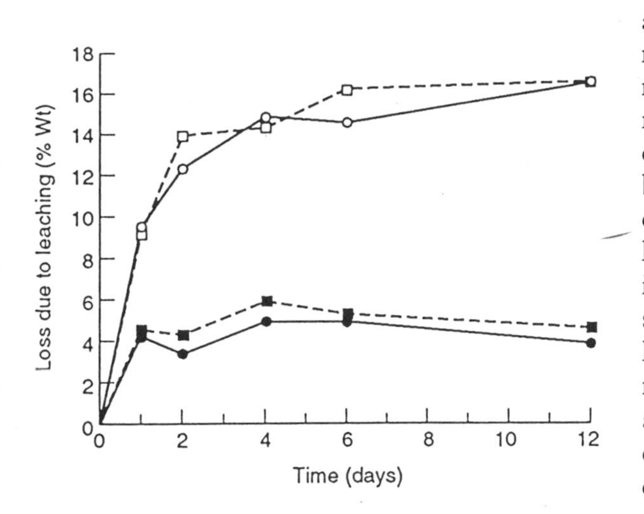

FIGURE 5.3 The time course of leaching of soluble organics from elm (*Ulmus americana*) leaves at 10°C (○) and at 20–22°C (□) and from alder (*Alnus rugosa*) leaves at 10°C (●) and at 20–22°C (■), as evidenced by weight loss. Note that most leaching occurs in 1–3 days. (After Kaushik and Hynes, 1971.)

resulting in a relative increase. When increases in total N are recorded, this immobilization of nitrogen is usually attributed to an increase in microbial biomass and incorporation of nitrogen from the surrounding water into new protein. However, it is also possible that a variety of refractive, non-protein nitrogen compounds may accumulate, with the result that available nitrogen might be less than that estimated by the commonly used micro-Kjehldahl technique (Odum, Kirk and Zieman, 1979). Rice (1982) suggested that nitrogen enrichment could be due to non-labile humic nitrogen rather than living microbial protein, with a correspondingly much lower biological availability.

The time course of loss of non-soluble leaf constituents is well illustrated by a study of oak and hickory leaves incubated in a Michigan stream over the winter (Suberkropp, Godshalk and Klug, 1976). Cellulose and hemicellulose declined at about the same rate as total leaf mass, while lignin was processed more slowly and increased as a percentage of remaining mass. Lipids were lost more rapidly than total mass, and thus were a declining fraction of remaining dry mass of leaf. Differences between oak and hickory were observed, and attributed by Suberkropp, Godshalk and Klug to greater complexing of proteins by polyphenols or lignin-like compounds in oak leaves. Thus, after the initial rapid loss of soluble compounds, non-soluble constituents were lost more slowly, and lignin and complexed compounds were lost even more slowly. A buildup of microbial biomass associated with the leaf not only is a major cause of these changes, but is also reflected in the observation of net nitrogen immobilization.

The importance of microbes to the decomposition process can be demonstrated by comparing weight loss with and without the presence of antibiotics (Table 5.3). Suppression of microbial activity prevented essentially any loss in mass or increase in protein content (Kaushik and Hynes, 1971). When fungal growth was inhibited the effect was great; when only bacterial growth was inhibited the effect was small and, for percentage protein, not significant. Suberkropp and Klug (1980) tested the efficacy of five species of fungi at macerating disks of hickory (*Carya glabra*) leaves in the absence of macroinvertebrate consumers. At 10°C, all five species of aquatic hyphomycetes metabolized and converted 66–75% of the initial leaf mass to FPOM in just 6 weeks.

While a good deal is known about the importance of microbes as a food source, described below and in chapter 6, considerably less is known about the ecology of the bacteria and fungi themselves (Bärlocher, 1982). Suberkropp and Klug (1976) followed in detail the succession of dominant mycoflora and bacterial flora on oak and hickory leaves in a Michigan stream, from November until June. Fungi, primarily aquatic hyphomycetes, dominated during the first half (12–18 weeks) of the processing period. Bacteria, whose numbers gradually increased throughout, dominated the terminal processing

TABLE 5.3 Influence of microbes on the breakdown of elm leaves[a]

	No antibiotics	*Anti-fungal and anti-bacterial antibiotics*	*Anti-fungal antibiotics only*	*Anti-bacterial antibiotics only*
Loss in mass (%)	21.9	1	9.3	17.5
Protein content (% final mass)	12.5	4.3–5.9	7.3	11.3

[a] Kaushik and Hynes (1971) placed 15, 1-cm leaf discs in 300 ml of streamwater, enriched with nitrogen and phosphorus, at 10°C. Antibiotics were added as shown and renewed twice weekly. Results after 4 weeks demonstrated that inhibition of microbial activity prevented almost all loss in mass, and protein content remained close to its initial value of 4.3%. Fungi evidently contributed more than bacteria to changes in mass and percentage protein.

stage and perhaps were benefited by fungal-induced changes in leaf surface area or by the release of labile compounds. Propagules of soil fungi, although commonly carried into the stream on falling leaves, appeared to contribute little to decomposition (Suberkropp and Klug, 1976). Soil fungi could be more important during summer, however, but not at colder temperatures which favor aquatic fungi (Bärlocher, 1985). Bärlocher (1982) reported that typically 4–8 species of aquatic fungi dominate throughout the decomposition of leaves, while a similar or larger number of rare species appears erratically. Apparently no particular succession occurs on a single leaf; whichever fungal species arrives first as a waterborne spore establishes numerical dominance.

Because of variation in timing of leaf fall, variation in species-specific rates of leaf conditioning by fungi, and the diversity of fungi present (Suberkropp and Klug, 1976; Bärlocher, 1982; Shearer and Lane, 1983), leaves on a stream bottom are a mosaic of patches of microbial populations. The extent of this variation is shown by Bärlocher's (1983) study of a Swiss stream flowing through an alder–willow–maple forest. The standing crop of CPOM was maximal in October–November, and by April only veins and petioles remained for an 85% loss of leaf mass. Soluble protein also decreased after November, indicating that nutritional value declined from that time on. However, on any one date, the quality of individual leaves was so variable that the amount of soluble protein in the richest 10% of leaves in mid-April exceeded the median value of leaves sampled in mid-November. Detritivores capable enough or fortunate enough in patch choice could enjoy food quality well above average.

5.1.2 Influence of detritivores on litter decomposition

The fragmentation of leaves by invertebrate feeding and abrasion constitutes the third stage of leaf breakdown. During the breakdown of autumn-shed leaves in temperate woodland streams, microbial populations play a central role not only in decomposing the leaf substrate, but also in rendering it more palatable and nutritious to consumers (chapter 6). In turn, the feeding activities of detritivores significantly accelerate the decomposition process. Their contribution to the fragmentation of coarse particles through feeding activities and production of feces significantly accelerates breakdown rates and influences subsequent biological processing of the original CPOM inputs.

Several lines of evidence indicate that consumer organisms accelerate the breakdown of leaves in streams (Webster and Benfield, 1986). The finding that leaf packs in mesh bags decomposed more slowly than those tethered to bricks with fishing line indicated that the former method underestimated k. Exclusion of detritivores is the likely cause of this difference. Subsequent studies revealed more rapid leaf breakdown when bags with larger mesh size were used (e. g. Benfield, Paul and Webster, 1979), presumably because invertebrate access was greater. Furthermore, breakdown rates are higher where invertebrates are more abundant (Benfield and Webster, 1985), and there is a positive relationship between invertebrate preferences and decay rates (Webster and Benfield, 1986). Each line of evidence is indirect and inconclusive, however. Litter bags probably affect the circulation of water and thereby influence chemical conditions; variation in physical or chemical regime might independently influence breakdown rate and invertebrate densities; and the preference of consumers for rapidly decomposing leaves does not demonstrate that consumers influence those rates. However, other evidence more directly establishes a causal role for detritivores.

Comparison of breakdown rates in experiments with and without insect detritivores establishes that as much as 25% of leaf degradation is attributable to the presence of animals. Processing rates in two experimental streams, one lacking invertebrates and another stocked with detritivores (*Tipula*, *Pycnopsyche* and *Pteronarcys*) at densities believed to represent natural maxima, indicated that 21–24% of the loss of hickory leaves was due to the influence of detritivores (Petersen and Cummins, 1974). The contribution of macroinvertebrates to the breakdown of *Phragmites*, a macrophyte, was comparable (Polunin, 1982). Other studies of similar design found weight loss to be several times greater in the presence of invertebrate consumers than in their absence (Herbst, 1982; Mulholland *et al.*, 1985a). In addition to direct consumption, the possible influences of detritivore feeding include release of nutrients and dissolved organic matter, comminution of litter, and modification of water circulation (Polunin, 1984).

The experimental removal of detritivorous insects from a small mountain stream in North Carolina provides a particularly convincing demonstration that animal consumers regulate rates of litter decomposition. Wallace, Webster and Cuffney (1982) added the insecticide methoxychlor to one small stream in February, with supplemental treatments in May, August and November. Massive downstream drift of invertebrates occurred and densities subsequently were reduced to less than 10% of an adjacent, untreated reference stream, while oligochaetes increased roughly threefold. Leaf breakdown rates were significantly retarded in the treated stream, presumably due to the great reduction in insect density, and the magnitude of the effect was greatest for the most refractory leaf species (Table 5.4). Export of suspended fine particulates was also reduced in the treated stream, consistent with the finding of reduced leaf processing. It bears mention, however, that this was a study of very small streams. The importance of detritivory may not be as great in larger systems, possibly because detritus inputs may not be sufficient to support a leaf-consuming assemblage (Benfield, Jones and Patterson, 1977).

5.1.3 Processing of other CPOM

Macrophytes are an important source of detritus where they are abundant, typically in larger rivers and in floodplains (chapter 4). Polunin (1984) reviewed studies of the decomposition and fate of this material, which is similar to that of terrestrial leaves. Breakdown rates are relatively fast, although less so in emergent macrophytes that contain more support tissue. Bacteria appear to play a greater role in macrophyte decomposition than is true for leaves of terrestrial origin (Webster and Benfield, 1986).

Woody debris ranging from small branches to large tree trunks can be an abundant source of CPOM, especially in small streams flowing through mature forests (Anderson and Sedell, 1979). Not surprisingly, the breakdown and decay of wood occurs very slowly. Woody debris influences channel structure and stream habitat in a number of ways (chapter 12), and also contributes to the nutrition of some consumers (Anderson *et al.*, 1978; Anderson and Sedell, 1979). The high lignin and cellulose content of wood fiber, coupled with relatively small surface area and low penetrability, results in very slow breakdown with microbial activity confined to surface layers. Even small wood chips (0.75–1.5 cm size range) placed in coarse mesh bags into Quebec streams showed very slow loss rates (Melillo *et al.*, 1983). Alder chips (*Alnus rugosa*) had a half-life of about 7 months, while spruce (*Picea mariana*) chips would require roughly 17 years to experience a 50% reduction in mass. Anderson *et al.* (1978) retrieved wooden sticks (2.5 × 2.5 × 92 cm) of alder, hemlock (*Tsuga*) and Douglas-fir (*Pseudotsuga*) after 15 months in an Oregon stream, with similar results. At present, decay rates for woody debris are little more than guesses, but it seems reasonable to think in terms of 0.5–1 decade for twigs less than 1 cm in diameter, possibly five decades for wood 5–10 cm in diameter, and 10–25 decades for larger trees. Microbial and invertebrate processing contributes to this loss, perhaps significantly over the extended time period (Anderson *et al.*, 1978). However, chemical and physical processes also contribute to the breakdown of large woody debris, and downstream transport and burial also are likely to be important (Harmon *et al.*, 1986).

5.2 Fine particulate organic matter

Less is known about the energy pathways involving fine particulate organic matter (FPOM) than those involving CPOM. One source of FPOM obviously is the breakdown of leaf litter (Figure 5.2). Roughly one-third of the original leaf mass is unaccounted for at the point where leaf fragmentation prevents further tracing of its fate. When one includes the production of feces and fragments by leaf-shredding insects and the eventual contribution of leached DOC to formation of fine particles (discussed below), it is apparent that a large fraction of leaf litter eventually becomes fine particulate matter. FPOM is also delivered to streams from the terrestrial landscape and formed from dissolved organic matter in several ways. These sources are difficult to trace and potentially of greater magnitude than FPOM derived from leaf fragmentation.

The presence and activity of microbial populations on fine particulates is measured in a number of ways. Cell counts and ATP concentration provide estimates of biomass, while respiration rates and uptake of labeled substrate provide estimates of metabolic activity. Because

TABLE 5.4 Estimated half-lives (days) based on exponential decay in ash-free dry mass of four leaf species in a stream treated with an insecticide, compared with a reference stream (From Wallace, Ross and Meyer, 1982)

	Reference stream	*Treated stream*	*Change in half-life*
Dogwood (*Cornus florida*)	41.0	65.4	(1.6×)
Red maple (*Acer rubrum*)	50.2	135.9	(2.7×)
White oak (*Quercus alba*)	64.2	173.3	(2.7×)
Rhododendron (*R. maxima*)	128.4	577.6	(4.5×)

bacteria become more abundant in the later stages of CPOM processing, and the small size of FPOM suggests a reduced role for fungal hyphae, bacteria are likely to dominate microbial populations on fines. However, compared with the extensive data on leaf breakdown, little is known about the eventual fate of FPOM (Ward and Woods, 1986; Ward *et al.*, 1990). Rates of processing, resemblance between FPOM and its source, nutritive value and eventual fate all are poorly understood.

5.2.1 FPOM originating in leaf decomposition

Feeding by leaf-shredding insects produces FPOM not only by fragmenting larger into smaller particles but also through the production of feces. With assimilation efficiencies of about 10–20% and ingestion rates in the range of the animal's mass per day (chapter 6), many consumers of CPOM produce copious amounts of feces. Coprophagy has been reported to be a dietary mainstay in some instances (Hynes, 1970) and an important supplement in others (Wotton, 1980). Fecal pellets and their associated microflora undoubtedly are important in the trophic economy of certain systems (Hargrave, 1976). Some stream-dwelling invertebrates such as amphipods and isopods produce pellets enclosed in a peritrophic membrane, while most insects apparently produce feces that are less discrete and more variable in size (Shepard and Minshall, 1981). Ladle and Griffiths (1980) provide a pictorial description and commentary on size, shape, texture, cohesiveness and so on. Most such particles appear to vary between 100 and 1000 μm in longest dimension, but since a correlation exists between size of particles and the organisms that produced them, the smallest invertebrates probably produce even smaller fecal particles.

Maceration by aquatic hyphomycete fungi can accomplish substantial fragmentation in the absence of either physical abrasion or animal consumption. Of five fungi tested for their effect on discs of hickory leaves, Suberkropp and Klug (1980) found that all were able to soften and skeletonize leaf tissue. In 6 weeks at 10°C between 41 and 51% of the original dry mass was converted to FPOM, 27–35% remained as CPOM, and the rest was lost to respiration or converted to DOM. FPOM produced by fungal maceration is lower in absolute amounts of cellulose and hemicellulose and in percent lignin, compared with the whole hickory leaves from which they came (Suberkropp and Klug, 1980). In addition, consumers differ in how they attack leaf material, which can affect the appearance of resultant FPOM. *Tipula* and many limnephilid caddis larvae eat all parts of the leaf, both mesophyll and venation, while peltoperlid stonefly nymphs avoid venation and concentrate mainly on mesophyll, cuticle and epidermal cells. Fecal pellets from *Tipula* visually resembled macerated leaf fragments and were similar to source material in lignin, hemicellulose and cellulose content. If one included the non-ingested fragments, resultant FPOM even more faithfully resembled its source. In contrast, *Talloperla cornelia* produced a macerated FPOM in which lignin content was substantially reduced, especially from leaves with highest initial concentrations. Cellulose also was reduced, while hemicellulose remained similar or increased (Ward and Woods, 1986).

Because the most readily assimilated material is likely to be processed in the steps prior to FPOM production (Figure 5.2), much of what remains is likely to be quite refractory. This is borne out by the finding that respiration rates associated with native detritus were much lower than conditioned and mechanically ground oak and hickory leaves (Ward and Cummins, 1979). As FPOM is decomposed and reduced in size, one might expect particles to become more refractory to microbial action and lower in nutritional value. A study of the chemical composition and microbial activity of FPOM from a southern Appalachian headwater stream, in relation to particle size, supports this expectation (Peters, Benfield and Webster, 1989). As particle size decreased from 500 to 10 μm, the organic content declined while lignin and cellulose content increased.

The influence of leaf-shredding insects on the availability of fine particulates probably is small overall, yet of ecological significance under certain circumstances. The previously described elimination of shredding insects from a small Appalachian stream reduced downstream transport of FPOM by about 14%. An ecosystem model constructed by Webster (1983) estimated that, even when shredder ingestion is low (13% of CPOM inputs), they can be significant in making FPOM available to consumers under conditions of low streamflow. Further studies of CPOM processing and FPOM export in methoxychlor-treated streams indicated an even greater effect than in Webster's model (Wallace *et al.*, 1991).

5.2.2 Other sources of FPOM

However tentative may be our understanding of the dynamics of FPOM originating in leaf litter, even less is known about other FPOM sources, which could be of equal or greater magnitude (Table 5.1). Dissolved organic carbon enters the particulate pool via a number of pathways, and probably in substantial amounts. Input of fine particulates from the forest floor, soil water and banks and channel of the streambed all can be substantial. Algal cells sloughed from periphyton mats and washed out of lakes and beaver ponds also contribute to total FPOM.

Not surprisingly, the origin of fine particulates is difficult to infer simply by examining FPOM (Lush and Hynes, 1973). However, studies using a variety of techniques give us some insight into the importance of the multiple sources of FPOM just described (Ward *et al.*, 1990). Inspection of fine particles by scanning electron microscopy sometimes allows identification of source material. Another approach makes use of the fact that neither algae nor bacteria contain lignin, and so its presence establishes that the material originated as vascular plant tissue. Lignin oxidation products are indicative of the type, concentration and degree of preservation of the plant source. Stable carbon isotope ratios distinguish grasses from other plants, and the atomic C:N ratio is highest (about 20) in relatively unaltered plant detritus, whereas lower values (about 10–12) are characteristic of well-decomposed soil organic matter.

Employing such techniques, Hedges *et al.* (1986) analyzed FPOM from the Amazon River by comparing the 'signatures' of river particles with various potential organic sources. They concluded that the majority originated as soil humic material, and at least for large rivers, this conclusion appears to be general. Based on the ratio of C:N atoms in river seston world-wide (Meybeck, 1982), the majority of riverine FPOM most closely resembles soil organic matter. Even in headwater streams, our detailed knowledge of the leaf breakdown process notwithstanding, it seems likely that most FPOM originates from sources other than comminution of leaves. Based on rough calculations of the magnitudes of inputs attributable to soil organic matter and the breakdown of wood, Ward and Aumen (1986) concluded that leaves and needles were minor sources of FPOM. Sollins, Glassman and Dahm (1985) used flotation to separate organo-mineral particles from fragmented plant material collected from small forested streams in Oregon. The majority of detrital carbon and nitrogen was present as organic material adsorbed on mineral surfaces, rather than as plant fragments.

5.2.3 Transport of FPOM

The small size of FPOM makes it more susceptible than CPOM to downstream transport. As a consequence, FPOM availability to consumers is strongly influenced by flow regime and instream obstructions that help to retain it in place. Transport distance depends on many variables, including size of material, bed and channel features (Webster *et al.*, 1987), presence of physical obstructions that trap debris and reduce flow (Bilby, 1981), stream gradient and discharge (Jones and Smock, 1991), magnitude and timing of high flow events (Golladay, Webster and Benfield, 1987), and uptake by suspension feeders (Georgian and Thorp, 1992). With so many variables, it is difficult to generalize about transport distances, except that distances are shorter for wood than leaves, while fine particles travel the farthest. Leaves and small pieces of wood typically travel less than 10 m before being retained or buried, except under flood conditions when transport distance may exceed 100 m. By comparison, fine particulates have mean transport distances in the 5–10 m range only under very low flows in small channels (Webster *et al.*, 1987; Jones and Smock, 1991). Mean transport distance was estimated at 100–200 m in a second-order stream in New York, using corn pollen grains (Miller and Georgian, 1992). Using natural FPOM labeled with ^{14}C, Cushing, Minshall and Newbold (1993) obtained estimates of 800 and 580 m in a small Idaho stream at a water velocity of $0.27\ m\ s^{-1}$ and 630 m in the Salmon River headwaters at a velocity of $0.29\ m\ s^{-1}$. Using an average transport distance per event of 0.5 km and particle resuspension estimated to occur every 1.5–3 h, Cushing, Minshall and Newbold calculate an average downstream transport of $4–8\ km\ d^{-1}$. Moreover, these estimates were obtained for small streams at baseflow. Transport distances for large rivers and during stormflows must be larger yet.

Although biological processes are important to the generation of suspended FPOM, they apparently are insignificant to its removal. Suspension feeders such as hydropsychid caddis larvae are the obvious candidates for capture and depletion of organic seston in transport, yet estimates show them to be incapable of removing more than a small fraction of organic seston (Benke and Wallace, 1980). This is hardly surprising for large rivers with little available substrate for organism attachment, but it seems to be true for smaller streams as well (Georgian and Thorp, 1992).

5.3 Dissolved organic matter

Dissolved organic matter (DOM) typically is the largest pool of organic carbon in running waters (Fisher and Likens, 1973; Hobbie and Likens, 1973; McDowell and Fisher, 1976; Moeller *et al.*, 1979). All DOM originates as natural biological products from soil, plant or aquatic organic matter. Some derives from instream processes described previously, including leachate from leaves and other particulate organic matter (POM), and by extracellular release from plants. In addition, soil and groundwater are major sources of DOM in river water. While the size of the DOM pool indicates its potential importance to heterotrophic energy pathways, the bulk of this material is highly refractory and probably of limited biological significance. However, river water also contains a smaller fraction of labile DOM, and this material constitutes a potentially important heterotrophic energy pathway.

DOM was for some time measured by chemical oxidation, and therefore reported as DOM. Many studies now rely on automated carbon analysis by combustion of water samples, and so report DOC. For all practical purposes these terms can be used interchangeably, and interconverted by assuming that DOM is 45–50% organic carbon by mass. The division between FPOM and DOM is one of convenience, dictated by what passes a 0.45 μm filter. In reality the dissolved fraction is likely to include some smaller bacteria, viruses and some colloidal organic matter. Lock, Wallis and Hynes (1977) used ultracentrifugation to examine the colloidal fraction, which was defined by a sedimentation coefficient and estimated to correspond to a spherical diameter between 0.021 and 0.45 μm (perhaps 0.01–0.5 μm should be considered the general size range for colloidal organic matter).

In water from a variety of sources in Canada, the colloidal fraction comprised between 29 and 53% of total DOM. According to Thurman (1985), however, the colloidal fraction typically is less than 10%.

Approximately 10–25% of DOM consists of identifiable molecules of known structure: carbohydrates, and fatty, amino and hydroxy acids. The remainder (50–75%, up to 90% in colored waters) can be placed in general categories such as humic and fulvic acids and hydrophilic acids. Humic acids separate from fulvic acids by precipitating at a pH less than 2, while fulvic acids remain in solution. Fulvic acids also are smaller (less than 2000 molecular weight) than humic acids, which often form colloidal aggregates (over 5000 molecular weight) and may be associated with clays or oxides of iron and aluminum. Fulvic acids generally are the majority of humic substances (Thurman, 1985). In the Amazon, for example, fulvic acids were approximately 50% and humic acids 10% of riverine DOM (Ertel *et al.*, 1986).

Soil organic matter is a quantitatively important organic matter source to river ecosystems, originating in above-ground and below-ground terrestrial production. Grasslands contain the highest soil organic matter, deserts the least, and forests are intermediate. The interstitial water of soils usually contains high DOC concentrations, in the range 2–30 $mg\,l^{-1}$, due to solubilization of organic litter (Thurman, 1985). As DOC penetrates to deeper soil horizons, soil microorganisms metabolize this energy source and concentrations decline markedly. In small streams in North Carolina, Meyer and Tate (1983) recorded DOC concentrations of 2–12 $mg\,l^{-1}$ in soil water in contact with the active root zone, compared with 0.2–0.7 $mg\,l^{-1}$ in sub-surface seeps. Similarly, in the watershed of an Alberta stream, the median DOC concentration in soil interstitial waters was 7 $mg\,l^{-1}$, (range 3–35) whereas shallow groundwater in the saturated zone contained 3 $mg\,l^{-1}$ DOC (Wallis, Hynes and Telang, 1981).

Groundwater contains low concentrations of DOC as a consequence of biological and chemical degradation of organic matter during passage from the soil surface through deeper soil horizons, and because groundwater residence time may be in the 100s to 1000s of years. Median values are usually less than 1–2 $mg\,l^{-1}$. However, where organic-rich surface waters recharge groundwaters, values up to 15 $mg\,l^{-1}$ have been found (Thurman, 1985). Precipitation also is a source of DOC. The DOC content of precipitation is highly variable, depending on contact with plant vegetation. Concentrations above the canopy generally are about 1 $mg\,l^{-1}$, while values below the canopy typically are 2–3 $mg\,l^{-1}$. When rain-water is intercepted by leaves, leaching removes significant amounts of organic matter and canopy drip can reach 25 $mg\,l^{-1}$ DOC (Thurman, 1985). Fisher and Likens (1973) estimated an average value of 17.8 $mg\,l^{-1}$ for canopy drip in a hardwood forest in New England.

It is evident that DOC concentrations in streamwater depend greatly on how the water reaches the stream channel. Inputs from deep groundwater are likely to be very low in DOC and have a diluting effect on streamwater. Plant organic matter on the soil surface or in soil interstitial water has a much shorter residence time, resulting in relatively high DOC concentrations. This water usually enters the stream channel during storms and wet season flushing, after a residence time that at most is tens of years (Wallis, Hynes and Telang, 1981; Thurman, 1985) and perhaps is within the current growing season. As a consequence, DOC concentrations can vary depending on the magnitude of these various flow pathways.

5.3.1 Instream transformations affecting DOC

Dissolved organic carbon is removed from streamwater by both abiotic and biotic processes. The principal biotic processes are uptake by microorganisms, especially bacteria, assimilation of the organic carbon into microbial biomass, consumption of this heterotrophic production, and its eventual re-mineralization to CO_2 by community respiration. DOC is also removed from the water column by abiotic uptake, and can be converted into particulate form.

The episodic input of leaf litter and the speed with which leaching occurs imply a strong pulse in leaf leachate DOC during autumn in woodland streams. McDowell and Fisher (1976) estimated that 42% of the autumnal DOC inputs to a small New England stream were due to this source. However, despite its strongly pulsed nature, stream concentrations of DOC are rarely observed to change during leaf fall. This implies that leaf leachate DOC is removed rapidly and within a short distance of its generation.

Studies of DOC removal from streamwater establish that highly labile sources such as leaf leachate and simple sugars are taken up rapidly, usually within 48–72 h (Lock and Hynes, 1976; Lush and Hynes, 1978; Dahm, 1981). Processes within the streambed rather than the water column account for most of the removal, at least in smaller streams. Using a leachate made from maple (*Acer saccharum*) leaves, Lock and Hynes (1976) found that disappearance of DOC from the water column of laboratory chambers was minor (less than 25%) over 4 days. When river sediments were present, however, concentrations of leachate were reduced to 15% of initial values within 9 h. In experiments with alder leachate (Figure 5.4), Dahm (1981) estimated that abiotic uptake by adsoption onto clays and chemical complexing with oxides of aluminum and iron accounted for up to one-third of the removal of DOC from the water column. This step occurred very rapidly. Over a period of several days, however, microbial uptake was responsible for the majority of DOC disappearance from the water column into the sediment layer. Dahm pointed out that reported rates of uptake of leaf leachate vary over almost two orders of magnitude. Characteristics of the sediments and microbial populations, rate of supply of available leachate, and various chemical conditions are among the possible causal factors in need of further study. Clearly, however, biotic and abiotic uptake are able to immobilize labile leachate very efficiently.

A number of abiotic processes remove DOC from solution, including adsorption, flocculation, precipitation, and photochemical destruction. The adsorption and chemical complexing just described is one example of abiotic immobilization of DOC. The importance of flocculation is suggested by studies in marine systems showing that DOC can adsorb onto the surface of bubbles, leading to particle formation when bubbles collapse (Riley, 1970). Noting that fast-flowing streams entrain air bubbles and develop foam, Lush and Hynes (1973) investigated whether fine particulates might similarly develop from DOC in streamwater. Using hand-picked leaves leached in a chemically defined water to approximate natural concentrations of leachate, they compared sterilized water, in which particle formation should be mainly by abiotic processes, with water where biotic processes were permitted. Particles were formed in the autoclaved water, some as large as a few mm, but generally less than 60 μm. In appearance they looked like structureless clumps and plate-like sheets. Subsequent microbial colonization of this material led to further changes and eventual breakdown. Lush and Hynes suggest that the species of leaf determines the amount of leachate released, while the ionic composition of the water, pH, and turbulence all influence particle formation. Once formed, particles settle, become invaded by microbes, and are a potential energy source for consumers.

Growing recognition of the importance of microbial uptake of DOC, and of the association of microbial populations with the sediments, has spurred efforts to investigate these attached communities in greater detail (Geesey, Mutch and Costerton, 1978; Rounick and Winterbourn, 1983; Lock *et al.*, 1984). Organic

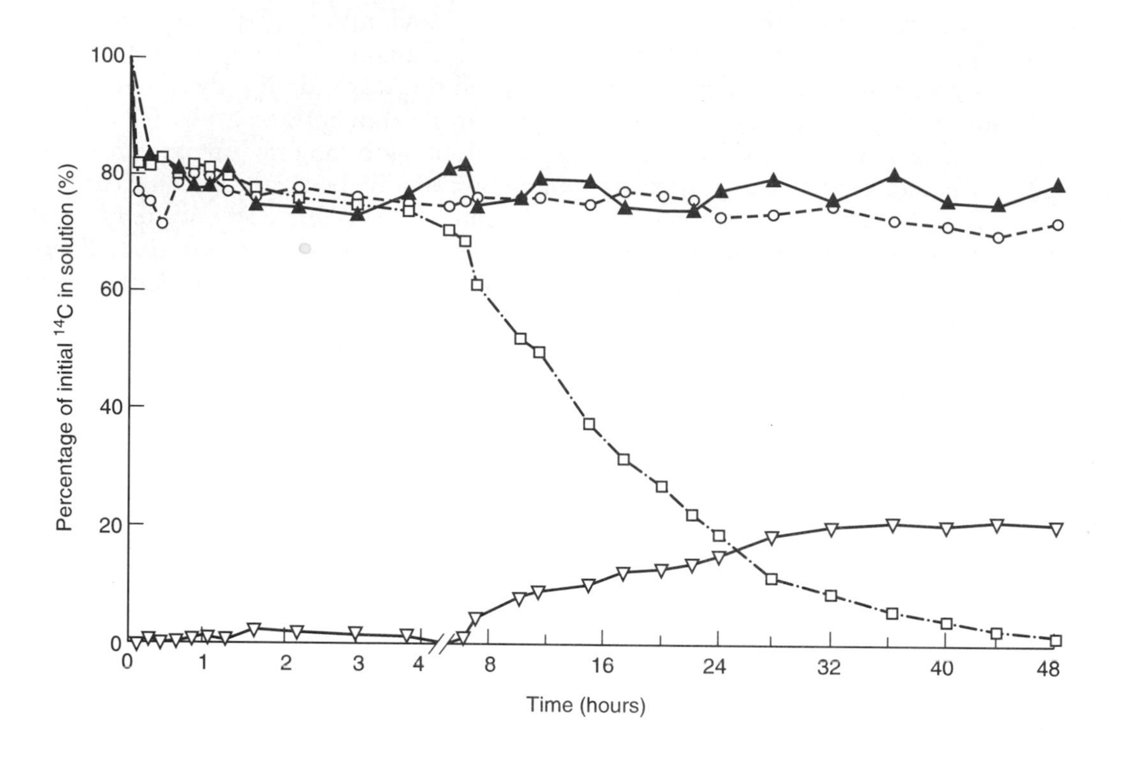

FIGURE 5.4 Pathways of removal of DOC (^{14}C-labeled leachate of alder leaves) from the water column. Three experimental treatments each contained sediments, FPOM and microbial populations. Treatment 1 (□) was the control, treatment 2 (○) had mercuric chloride added as a poison, and treatment 3 (▲) received the poison and H_2O_2 to oxidize the particulate organic matter. An initial rapid decline in labeled DOC in solution under each treatment was due to physical-chemical absorption onto sediments or suspended particles. Thereafter, DOC concentrations in both poisoned chambers evidenced little change, indicating that microbial utilization associated with the sediments was responsible for most of the decline of DOC in solution in the control treatment. After 48 hrs, 57% of labeled DOC from the control chamber had been incorporated into microbial biomass, and 22% of this initial DOC had been re-mineralized to CO_2 (△). Any increase in DOC uptake in the poisoned chamber also treated with an oxidizing agent would have been attributable to particulate organic-dissolved organic interaction. (After Dahm, 1981.)

microlayers on stone surfaces have received the most study, but biofilms also form on surfaces of plants, decaying wood and leaves, and on suspended particles in larger rivers. A gelatinous polysaccharide matrix secreted by microorganisms binds together algae, bacteria, fungi, detrital particles, various exudates, exoenzymes and metabolic products in an organic microlayer (Figure 5.5). Key features of this model include tight internal nutrient cycling between heterotrophs and autotrophs and the accumulation of exoenzymes.

Lock (1981) discussed several ways in which the polysaccharide matrix could influence energy and mineral transfers. By reducing diffusion rates, the organic microlayer tends to retain and concentrate compounds, particularly those of higher molecular weights. Retention of nutrients and organic molecules presumably enhances microbial transformations, although diffusion of oxygen, DOC and other substances into the layer could potentially become limiting. Exoenzymes and enzymes derived from cell lysis might similarly be retained and remain active,

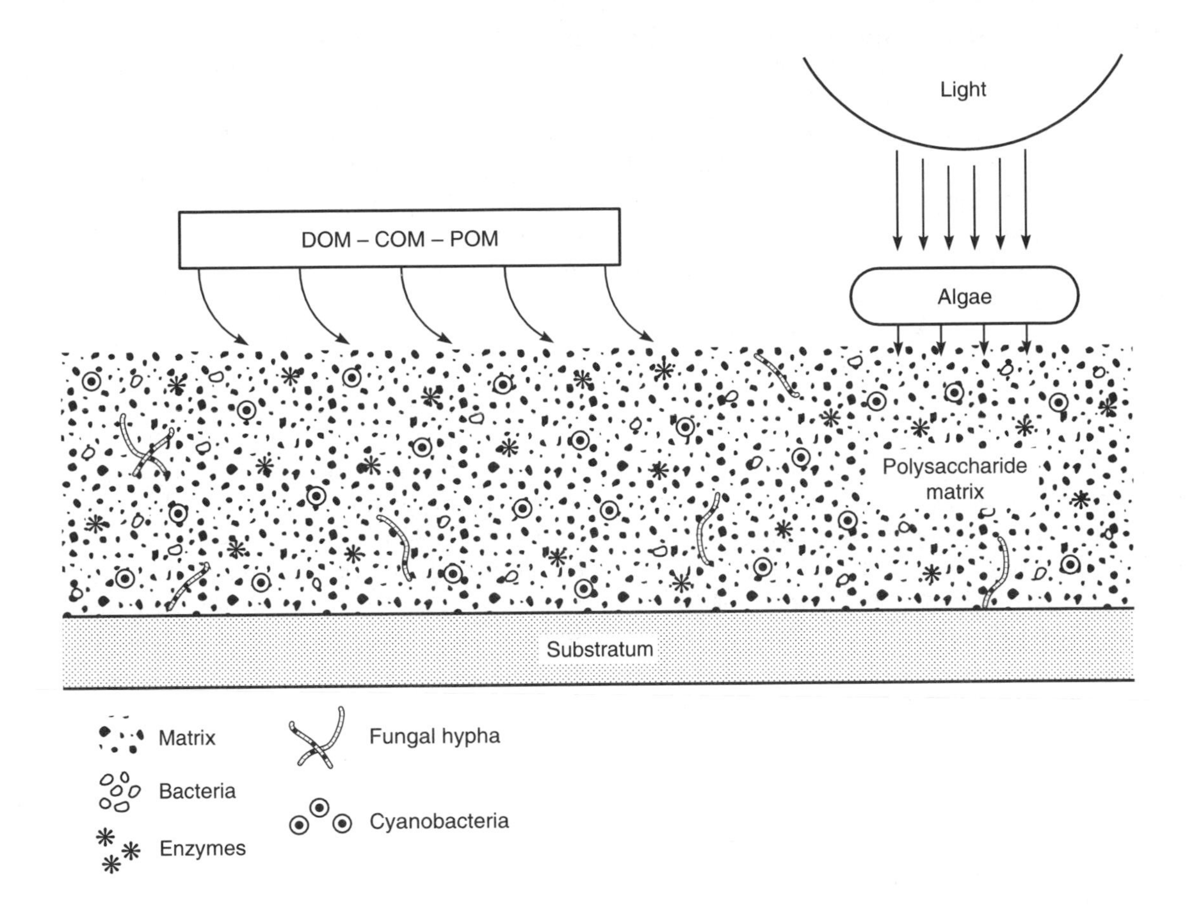

FIGURE 5.5 A structural and functional model of the organic microlayer–microbial community found as a surface 'slime' on stones and other submerged objects in streams. The matrix of polysaccharide fibrils produced by the microbial community binds together bacteria, algae and fungi, and is inhabited by protozoans and micrometazoans which graze on this material. Heterotrophic inputs include dissolved, colloidal and fine particulate organic matter, while light energy is trapped by algal photosynthesis. Within the matrix, extracellular release and cell death result in enzymes and other molecular products that are retained due to reduced diffusion rates, and available for utilization by other microorganisms. (Modified from Lock, 1981.)

facilitating the release of molecular products and reducing the energy demands on microorganisms for enzyme synthesis. In addition, the polysaccharide matrix can act as an ion-exchange system, attracting and binding charged organic molecules, anions and cations.

Energy transformations within this layer include the conversion of light to chemical energy by algal photosynthesis, adsorption and microbial uptake of heterotrophic carbon, and internal transfers due to extracellular release and cell lysis. Both autotrophs and heterotrophs are likely to benefit from these internal fluxes. Members of the periphyton can make substantial amounts of dissolved organics available for bacterial uptake (Kaplan and Bott, 1982) and probably are responsible for secreting most of the 'slime' structure (Geesey, Mutch and Costerton, 1978), while the products of bacterial metabolism would in turn be available for algal uptake. Geesey, Mutch and Costerton, (1978) found that measures of bacterial biomass (ATP) and periphyton biomass (chlorophyll *a*) in the epilithon fluctuated together, and suggested that bacterial populations were dependent on the periphyton for their establishment and maintenance.

The advantage of accumulation of exoenzymes within the microlayer is in permitting surface film bacteria to divert resources from enzyme synthesis to microbial growth. Collection of biofilm samples from the surfaces of stones (Sinsabaugh *et al.*, 1991) and organic substrates (Golladay and Sinsabaugh, 1991) documented that exoenzyme accumulation occurs as suggested under the Lock model. This has the beneficial effect that enzyme activity can be spatially distant from the microorganisms that produced the original enzymes. However, the same properties of slow diffusion that act to retain enzymes might also retard the entry of DOC, making the organic microlayer slow to respond to alterations in DOC supply.

Although few in number, studies of the development of biofilms in nature are beginning to identify the sequence of events and important variables. Golladay and Sinsabaugh (1991) followed the establishment of organic layers on sugar maple leaves and white birch ice-cream sticks in a boreal river in upper New York. Biofilms developed rapidly on both surfaces. Fungi provided an important structuring element to surface biofilms, especially for leaves, which are susceptible to fragmentation. The extent of development of organic microlayers is likely to be affected by flow regime, disturbance history, and bioturbation by the meiofauna and macrofauna. Golladay and Sinsabaugh reported higher ATP and chlorophyll *a* concentrations on wood at fast relative to slow flows, but no differences were observed on leaves. Exoenzyme activity was unaffected by current velocity. In general, microbial biomass and exoenzyme accumulation were greater on wood than leaves, suggesting that wood may be an important site of biofilm development in streams.

By placing granite discs (150 mm diameter, 10 mm thick) on the bed of an Alberta boreal stream for 15–23 weeks, Lock *et al.* (1984) examined the establishment of epilithic biofilms. Although the tops of discs showed more extensive community development, bottom surfaces also were well colonized, including chlorophyll-containing microorganisms. Cyanobacteria were about equally abundant on top and bottom surfaces, whereas diatoms occurred primarily in the upper layer. Chironomid tunnels in the thick epilithon implied grazer utilization of this organic layer. Light is not necessary for the development of an epilithon, as Rounick and Winterbourn (1983) established by studying the development of organic layers in a darkened stream section. Using radioisotopes, they also showed that this epilithon was ingested and assimilated by a number of invertebrate taxa.

5.3.2 Microbial production

Estimation of bacterial abundance and production in running waters is challenging. However, cell density and biomass can be estimated by various techniques including epifluorescence microscopy, ATP analyses and scanning electron microscopy. Metabolic activity can be assessed by measuring the incorporation of ^{3}H-thymidine, which measures DNA replication and thus cell division rate. Respiration rates of bacteria in suspension can be accomplished by oxygen depletion methods after size fractionation to remove phytoplankton. Although estimates are few and tentative, indications are that bacterial production can be substantial. In the Breitenbach, a small stream in Germany, Marksen (1988) estimated an annual production of roughly 750 $g\,C\,m^{-2}\,yr^{-1}$. This value exceeds the expected maximum based on available carbon and presumably is too high, surpassing any others in the literature for freshwater. However, other estimates of daily bacterial production (Findlay, Meyer and Risley, 1986a; Crocker and

Meyer, 1987; Kaplan and Bott, 1989) include quite high values, indicating that the flux of carbon through bacteria can indeed be an important term in stream energy flow.

Microbial production is likely to be influenced by environmental conditions, especially the amount and quality of organic substrate available. A survey of published estimates of bacterial production demonstrated a significant positive relationship with amount of organic matter in the sediments (Cole, Findlay and Pace, 1988). Findlay, Meyer and Risley (1986a) reported daily bacterial production to be an order of magnitude higher in sediments of backwater areas than in sandy regions of two blackwater rivers of the southeastern USA, corresponding to differences in organic content between sites. Similarly, bacterial production was higher in an Appalachian mountain spring (Crocker and Meyer, 1987) and in a Pennsylvania stream (Kaplan and Bott, 1989) than in the two blackwater rivers, and sediment organic matter content also was higher at the former sites. Studies to date have concentrated on sites with sandy sediments, where bacterial densities are an order of magnitude or more higher than found on stony substrates (Kaplan and Bott, 1989).

Bacterial production in the water column has received limited study. Bacterial densities are known to be one to several orders of magnitude lower in the water column than in the sediments. Production by bacteria suspended in the water column is often thought to be minor, originating from sloughing of cells from the surface of sediments and epiphytes, and subject to continual washout. This likely is true for small streams. In larger rivers, however, production by suspended bacteria can be substantial, as Edwards, Meyer and Findlay (1990) document for the Ogeechee River, a sixth-order blackwater river in Georgia. Even so, bacterial populations apparently cannot be accounted for without inputs from floodplain and tributary swamps, and much more community respiration takes place in the sediments than in the water column.

The quality of the organic matter source is expected to be important to bacterial production, perhaps more so than total amount. DOC of low molecular weight, monosaccharides and certain 'high quality' sources are taken up most rapidly by bacteria. Preferential removal of low molecular weight DOC has been reported in several studies (Kaplan and Bott, 1982; Meyer, Edwards and Risley, 1987; Kaplan and Bott, 1989). Meyer, Edwards and Risley separated DOC from the Ogeechee River into three size fractions (less than 1000, 1000–10 000 and over 10 000 nominal molecular weights (nMW)) and concentrated these fractions for subsequent enrichment of river water. Bacterial growth after a 3 day incubation showed no difference between unenriched river water and the middle size fraction. Enrichment with the largest and smallest size fractions significantly stimulated growth, and the greatest effect was due to the smallest size fraction. Reduction in DOC by bacterial activity also was of the order:

low MW DOC > high MW DOC > middle MW DOC = unenriched river water.

Hydrolysis of the high molecular weight DOC established that the biologically available portion consisted of low MW DOC complexed to otherwise refractory high MW DOC. These results are consistent with observed dynamics of water column DOC in the Ogeechee River and its tributary, Black Creek (Meyer, 1986). Intermediate weight DOC (1000–10 000 nMW) comprised over half of the total, correlated most closely with river discharge, and was least altered by instream biological processing.

During springtime periphyton blooms, stream DOC concentrations have been noted to increase as much as 37% from a pre-dawn minimum to a late afternoon maximum, apparently due to extracellular release by autotrophs (Kaplan and Bott, 1989). A substantial daily increase in carbohydrates was caused mainly by changes in the amount of polysaccharides, while monosaccharides stayed fairly constant. Because DOC release exceeded heterotrophic uptake, resulting in net export under these bloom conditions, the likely explanation is that microbial activity selectively removed simple sugars.

The evidence indicates that some constituents of the heterogeneous mix of molecules that comprises DOC are more available than others. Rapid uptake of high quality sources suggests that the most labile material is removed within relatively short distances. Thus, DOC in transport presumably is dominated by refractory material. While it is not yet known what portion of riverine DOC is utilizable, two examples illustrate that water column DOC indeed supports bacterial growth. Bott, Kaplan and Kuserk (1984) exposed stream sediments to water from a seep and a location 30 m downstream where DOC concentrations were more than twice as great. Bacterial populations were 55% greater in sediments exposed to water from the downstream site, indicating that DOC in transport contributed significantly to the support of bacterial biomass in surface sediments. However, as only a very small fraction of the DOC in transport was converted into bacterial biomass, this energy source was of low quality, despite its quantitative significance. In the blackwater river studied by Meyer, Edwards and Risley (1987), only low MW DOC supported bacterial growth. However, low MW DOC amounted to 2–4 mg l^{-1}, and hence was a sizeable amount. Furthermore, growth efficiencies as high as 30% were obtained on unenriched river water, suggesting that bacterial production can be considerable.

(a) The fate of microbial production

Consumption and export are likely fates of bacterial production. In lotic ecosystems there have been few studies of either process, and it has often been assumed that wash-out accounted for what little bacterial production occurred. However, the amount of carbon present as DOC and POC in river ecosystems is large, bacterial production now appears to be substantial, and there is growing evidence that carbon flux through microbial food webs might be much greater than previously recognized (Pomeroy and Wiebe, 1988). At present we do not know whether microbial production eventually reaches the 'usual' metazoan food web and thus constitutes an energy source for larger animals, or whether bacterial production is exhausted within a series of microbial transformations. Microbial food webs will be examined more closely in chapter 6, and their role in carbon and mineral dynamics returned to in chapters 12 and 13. Here we briefly describe the two main fates that bacterial production is likely to meet.

Export of benthic bacteria presumably results from sloughing of cells from the surfaces of sediments and epiphytes. Perhaps this will be most common when biofilms exceed some thickness and under scouring flows, as is observed for periphyton mats, but this is only speculation. Suspended bacteria will be transported downstream in the water column, but in larger rivers a parcel of water can last days to weeks, while bacterial doubling times require only hours to days. At least under modest flows, microbial growth can exceed export, indicating the potential for accumulation of bacterial biomass and support of consumers by this microbial production.

Protozoans, mainly ciliates and flagellates, are the most likely consumers of bacterial production. Based on the first rough estimates, it appears that protozoan grazing can account for virtually all of bacterial production under some circumstances. For the sixth-order blackwater Ogeegee river, Carlough and Meyer (1990) estimated that protozoa grazing was sufficient to clear almost half the water column of suspended bacterial cells each day. Rates of bacterial removal from streambed sediments by ciliates and flagellates indicated that protozoan feeding was able to keep pace with bacterial production in the much smaller White Clay Creek (Bott and Kaplan, 1990). Thus, though the evidence at present is scant, it appears that microbial transformations of bacterial production are internal mechanisms of great importance to the flux of organic carbon in running waters.

5.4 Summary

A great deal of the energy supply to rivers and streams derives from non-living organic matter. CPOM, FPOM and DOM comprise a diverse array of potential food sources for consumers in lotic food webs. In many instances, heterotrophic production constitutes a more significant energy supply than does primary production. The sources are numerous, and most POM and DOM originates as external inputs from outside the stream channel.

The breakdown of autumn-shed leaves is an important source of CPOM to small woodland streams. An initial leaching of some chemical constituents, microbial colonization primarily by fungi, invertebrate feeding, and physical abrasion all contribute to leaf breakdown into FPOM. Environmental conditions and leaf characteristics are well-studied variables that determine the timespan of the breakdown process. A substantial fraction of leaf mass is consumed and mineralized by detritivores and microbial populations, and the remainder enters the pools of FPOM and DOM.

FPOM derives not only from the breakdown of leaves, but from a number of sources, and perhaps mainly from the soil and forest floor. Fine particles often are the end product of extensive processing, and so should be of low nutritional value. The FPOM of most use to consumers likely includes sloughed algal cells, recently produced leaf fragments, animal feces, and perhaps particulates produced by immobilization of labile DOM. Unlike CPOM, which is likely to be processed near to its point of entry into stream channels, FPOM is believed to be transported considerable distances from its point of origin until final transformation into mineralized carbon.

DOM also derives from many sources, mostly outside the stream channel, and often is the largest single pool of organic carbon in lotic ecosystems. However, much DOM is a poor source of energy to the bacteria that are its primary entry point to the biota. Simple sugars, leachate from freshly shed leaves, exudate from periphyton blooms, and low molecular weight compounds are DOC sources that are taken up most rapidly. The sparse data presently available indicate that microbial production based on both DOM and POM can be large, but its contribution to higher trophic levels is uncertain.

Microbes associated with organic microlayers on substrates are the primary site of uptake in small streams, although in larger rivers bacteria in suspension also can be important. Organic microlayers are composed of bacteria, algae, fungi, protozoa and micrometazoa, along with exoenzymes and detrital particles enmeshed in a gelatinous polysaccharide matrix. The high metabolic activity rates of surface film microbes may be partly due to the accumulation of exoenzymes within the microlayer, permitting bacteria to divert resources from exoenzyme synthesis to microbial growth.

Lotic ecosystems obtain their energy from instream primary production by plants (chapter 4), and both instream and externally produced non-living organic matter (this chapter). This diverse mix of autotrophic and heterotrophic energy inputs provides the support for higher trophic levels. We now examine the many ways that consumers make use of these diverse energy sources.

Chapter six

Trophic relationships

Trophic organization in river ecosystems can be both complex and indistinct. Many consumers are polyphagous rather than monophagous, and exhibit considerable overlap with one another in their diets. The gut contents of invertebrates usually are difficult to distinguish, so these consumers are often characterized by the unspecific term of herbivore–detritivore. At least in temperate waters, the vast majority of fishes eat invertebrates. As a consequence, while a particular species may be classified solely on the basis of what it eats (herbivore, carnivore, detritivore and so on) the resulting categories are of limited usefulness because they offer too few distinctions among feeding roles. Some improvement may occur as advances are made in the characterization of food sources. However, it has proved more useful to distinguish among feeding roles on the basis of how the food is obtained, rather than solely in terms of what food is eaten. When several species consume a common resource and acquire it in a similar fashion, they are considered members of the same guild. Thus, a fish species that captures invertebrate prey directly from the bottom would occupy a different guild from another species that consumes the same prey, but captures it from the water column.

The guild concept is useful because it provides a reasonable degree of subdivision in feeding roles for both invertebrate and vertebrate consumers in streams, where the high degree of polyphagy frustrates adequate subdivision using food type alone. The particular species in a guild may change seasonally or geographically with, one presumes, little effect on trophic function. Moreover, even if taxonomic knowledge is incomplete, it may still be feasible to characterize fishes into guilds.

Invertebrates also can be divided into feeding guilds on the basis of what is eaten and how the resource is obtained. It is customary to refer to these categories as functional groups (Cummins, 1973), but the meaning is the same. It is important to note that members of different invertebrate functional groups may consume more or less the same resource. For example, fine particulate organic matter can be captured from the water column or collected from depositional locales. The main difference is not the resource, but the organism's method of capturing it.

A complete characterization of food webs and trophic relationships in running waters should include a description of the microbial food web or microbial loop, mentioned briefly at the end of the last chapter. There is considerable evidence that fungi contribute significantly to the nutritional value of shed leaves of temperate woodland streams. The energy contained in fungal biomass apparently is transferred directly to macro-consumers, and is reasonably well studied. Until recently, however, there was scant evidence that bacteria played an important role in lotic ecosystems. This view is changing, and even though there is relatively little one can say about the trophic pathways followed by microbial production, the growing importance of this topic requires that it receive whatever mention is possible.

6.1 The microbial loop

Mounting evidence from studies of lakes and marine systems suggests that carbon flux through microbial food webs may be much greater than previously recognized (Pomeroy and Wiebe, 1988). In these ecosystems, uptake of DOC by bacteria is thought to be a significant energy pathway. The microbial biomass that results from DOC uptake is consumed by protozoans and micro-metazoans, and perhaps subsequently by macro-metazoans. There is much uncertainty regarding the amount of energy that reaches the 'usual' metazoan food web by this route, for several reasons. Our knowledge of bacterial production is scant. It is suspected that several trophic transfers take place within the microbial web, with energy loss at each step. The extent of consumption of protozoans and micro-metazoans by larger metazoans is unknown. Thus it is possible that little of the biomass produced by bacterial growth ever reaches larger consumers. Instead, energy may be dissipated and minerals recycled within a largely distinct and separate 'microbial loop'. Even if no energy is transferred to larger consumers, however, the loop is still critically important in re-mineralizing organic matter (Edwards, Meyer and Findlay, 1990).

In all likelihood the microbial loop is of considerable significance in streams and rivers. The amount of carbon present as DOC and in FPOM in running waters is substantial (chapter 12). Some portion of this energy source is processed by microbial populations, constituting a carbon flux about which little is known. In the open water of lakes and oceans, bacteria are the significant microbial producers. As discussed in chapter 5, the importance of hyphomycete fungi in the decay of leaves and other organic debris in temperate streams is well established. However, our present understanding is that this microbial production is directly ingested by larger consumers, and so these fungi will be considered together with the consumption of CPOM.

The microbial food web of running waters (Figure 6.1) can be described only in broad outline. Bacteria, and to some degree fungi and algae, transform carbon obtained from dissolved and fine particulate sources into microbial biomass. Sites of potentially significant microbial production include the water column of larger rivers, floodplains and backwaters, depositional areas where FPOM may accumulate, and surface microlayers as depicted in Figure 5.5. Thus we must consider the bacteria associated with sediments, inorganic and organic surfaces, and also those suspended in the water column. Current evidence indicates that benthic bacteria are far more active and abundant than suspended bacteria (Edwards, Meyer and Findlay, 1990). In small streams, water column bacteria presumably originate by sloughing from populations associated with sediments and epiphytes, and are subject to wash-out once in suspension. Their role is likely to be minor, relative to bacteria associated with surfaces and depositional areas. In larger rivers, however, especially those with floodplains and back channels, bacterial populations may become much more important (Meyer, 1990). In the Ogeechee River of Georgia, bacteria originating in the extensive floodplain and tributary swamps resulted in substantial production in the water column of the main channel. Even under the most favorable assumptions for suspended bacteria, however, benthic bacteria dominated community respiration (Edwards, Meyer and Findlay, 1990).

Likely primary consumers of microbial production include protists and micro-metazoans. Among the protists, it appears that flagellates and ciliates are able to exert significant grazing pressure on bacteria in streambed sediments (Bott and Kaplan, 1990) and in the water column (Carlough and Meyer, 1990). The next trophic level may be occupied by various micro-

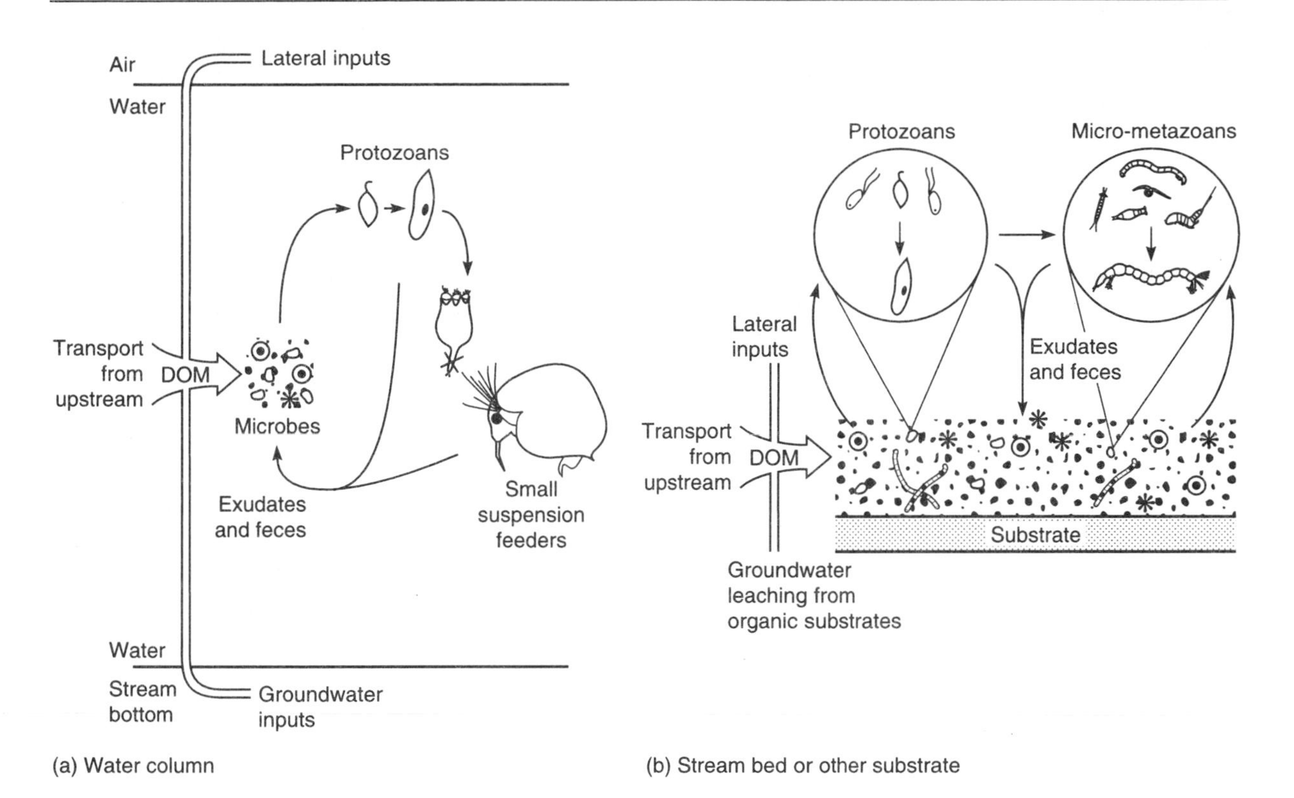

FIGURE 6.1 Microbial food webs within the water column of a large river (left) and on the streambed or other substrate of a small stream or larger river (right). Microbes within the water column, primarily bacteria, are consumed by flagellates and ciliates, which in turn are grazed by zooplankton such as rotifers and micro-crustaceans. Exudates, waste products and decomposing consumers are likely to be utilized by microbes, completing the "microbial loop". The amount of microbial production that is re-mineralized to CO_2 within the microbial loop, *versus* transferred to the macroinvertebrate food web by various pathways including suspension feeding and predation, is uncertain. Microbial production in organic microlayers (see Figure 5.5) on stones, wood, sand grains, and other surfaces is due to a complex organic matrix of bacteria, fungi, diatoms and algae. Micro-consumers again include flagellates and ciliates, as well as rotifers, nematodes, early instars of macroinvertebrates such as midge larvae, and harpacticoid and cyclopoid copepods. It is likely that both the organic layer and these micro-consumers are ingested by larger invertebrates, such as predaceous midge larvae and a variety of grazers, but linkages to macroinvertebrate consumers are poorly understood.

metazoans, particularly in benthic habitats. This includes copepods, oligochaetes, rotifers, nematodes and early instar insect larvae, which are found interstitially as well as on the substrate surface. Although they are of very small size, their numbers and rapid growth argue that their potential as consumers should not be underestimated. Note also that these micro-metazoans are two trophic transfers removed from bacterial production, and three from the presumed DOC or FPOC energy source. If only 10% of the energy transfers from one level to the next, considerable energy dissipation will occur. Should these micro-metazoans be consumed by larger invertebrates, and then by invertivorous and piscivorous fish, the energy loss is several-fold greater.

The number of trophic links in the microbial

web is strongly influenced by the small size of its members. The average size of a bacterial cell is about 0.5 μm. Except for black fly larvae and the Asiatic clam *Corbicula*, few suspension feeders are able to capture particles of this size (Wotton, 1994). Thus the ingestion of bacteria by flagellates (about 5 μm) and ciliates (about 25 μm) may be necessary for this energy to reach larger consumers. On the other hand, bacteria associated with periphyton in a surface microlayer may be ingested directly by benthic grazers of attached material and by deposit feeders that pass organic matter and associated microbes directly through their guts (e. g. Rounick and Winterbourn, 1983). Clearly, the number of trophic transfers between bacterial production and macroinvertebrate consumer can be one or many, with significant consequences for energy dissipation, and this may vary between water column and benthic habitats. It is conceivable that in some habitats the microbial and metazoan webs are linked, and in others the microbial loop is an energy sink, internally dissipating whatever energy is obtained from dissolved and fine particulate carbon sources. Elucidation of the magnitude of the 'link *versus* sink' role of the microbial loop is an important current challenge in lotic ecology.

6.2 Invertebrate consumers

Food resources for invertebrate consumers include periphyton and other surface layer complexes, macrophytes, detritus, and other animals (Table 6.1). The distribution and abundance of these various resources are influenced by size of stream or river, shading, substrate and many other variables discussed in chapters 4 and 5. It has been suggested that the relative availability of food resources changes predictably from headwaters to river mouth, causing food webs also to vary in a predictable fashion (Vannote *et al.*, 1980), and we consider this idea in chapter 12. The periphyton are an important food source to some invertebrates, particularly in shallow streams with minimal shading. In addition, organic microlayers occurring on stones and other substrates have been shown to be a food source for aquatic insects (Rounick and Winterbourn, 1983) and to be sites of active microbial uptake of DOC (Dahm, 1981). As yet, the contribution of this energy source is poorly understood. Although the frustules of diatoms are easily detected in gut analyses, other small autotrophs are usually unrecognizable, and the contribution of the microbes can only be studied using special methodologies such as radiotracers.

Macrophytes are relatively restricted in their occurence and are rarely attacked by aquatic herbivores in freshwater (Hutchinson, 1981), due to their low digestibility and high levels of cellulose and lignin (Cummins and Klug, 1979). They typically are consumed only after entering the detrital pool, although some fishes are capable of direct consumption. Interestingly, Newman (1991) reports that herbivory on living macrophytes can be significant, but is usually by specialized oligophagous herbivores derived from primarily terrestrial insect groups.

Detritus includes all non-living, particulate organic matter greater than 0.5 μm, and associated microorganisms (fungi, bacteria, protists and microinvertebrates). The latter are included because they are impractical to analyze separately, and also because detritus and microorganisms are normally found in close association (Anderson and Sedell, 1979). The nutritional quality of detritus depends largely on its microbial colonizers, and thus on the temporal sequence of detrital processing (chapter 5).

Animal prey represents a high quality resource for predators and parasites capable of locating and capturing other animals. Because of the comparative ease with which laboratory studies and gut analyses can be conducted, predator–prey linkages are often the best documented features in lotic food webs.

TABLE 6.1 The feeding roles of invertebrate consumers in running waters. (Based on Cummins, 1973; Cummins and Klug, 1979; Anderson and Sedell, 1979; Wallace and Merritt, 1980)

Feeding role	*Food resource*	*Feeding mechanism*	*Examples*
Shredder	Non-woody CPOM, primarily leaves; and associated microbiota, especially fungi	Chewing and mining	Several families of Trichoptera, Plecoptera, and Crustacea; some Diptera, snails
Shredder/gouger	Woody CPOM and microbiota, especially fungi; primarily surficial layers are utilized	As above	Occasional taxa among Diptera, Coleoptera, Trichoptera
Suspension feeder/ filterer-collector	FPOM and microbiota, especially bacteria and sloughed periphyton in water column	Collect particles using setae, specialized filtering apparatus or nets and secretions	Net-spinning Trichoptera, Simuliidae and other Diptera; some Ephemeroptera
Deposit feeder/ collector-gatherer	FPOM and microbiota, especially bacteria, and organic microlayer	Collect surface deposits, browse on amorphous material, burrow in soft sediments	Many Ephemeroptera, Chironomidae and Ceratopogonidae
Grazer	Periphyton, especially diatoms; and organic microlayer	Scraping, rasping and browsing adaptions	Several families of Ephemeroptera and Trichoptera; some Diptera, Lepidoptera and Coleoptera
	Macrophytes	Piercing	Hydroptilid caddis larvae
Predator	Animal prey	Biting and piercing	Odonata, Megaloptera, some Plecoptera, Trichoptera, Diptera and Coleoptera

6.2.1 Consumers of CPOM

Figure 6.2 depicts the links between shredders, CPOM, fungi and bacteria typical of a small stream in the temperate zone. The series of events culminating in the disappearance of leaf CPOM has already been discussed (chapter 5), and the critical role that microorganisms play both directly and indirectly in influencing the nutritional quality of this resource is elaborated below. CPOM that becomes available in wetlands and larger rivers following macrophyte die-back enters the decomposer trophic web in a very comparable fashion (Polunin, 1984). Woody debris, the coarsest of the CPOM, has been viewed in similar terms (Anderson *et al.*, 1978), although fewer taxa are able to consume this material and its rate of utilization is very slow.

The consumption of autumn-shed leaves in woodland streams by various invertebrates is the most extensively investigated trophic pathway involving CPOM (Cummins, 1973; Anderson and Sedell, 1979; Cummins and Klug, 1979) and will serve as our model here. Invertebrates that feed on decaying leaves include crustaceans (especially amphipods, isopods, crayfish and freshwater shrimp), snails and several groups of insect larvae (Cummins *et al.*, 1989). The latter includes crane fly larvae (Tipulidae), and several families of trichopterans (Limnephilidae,

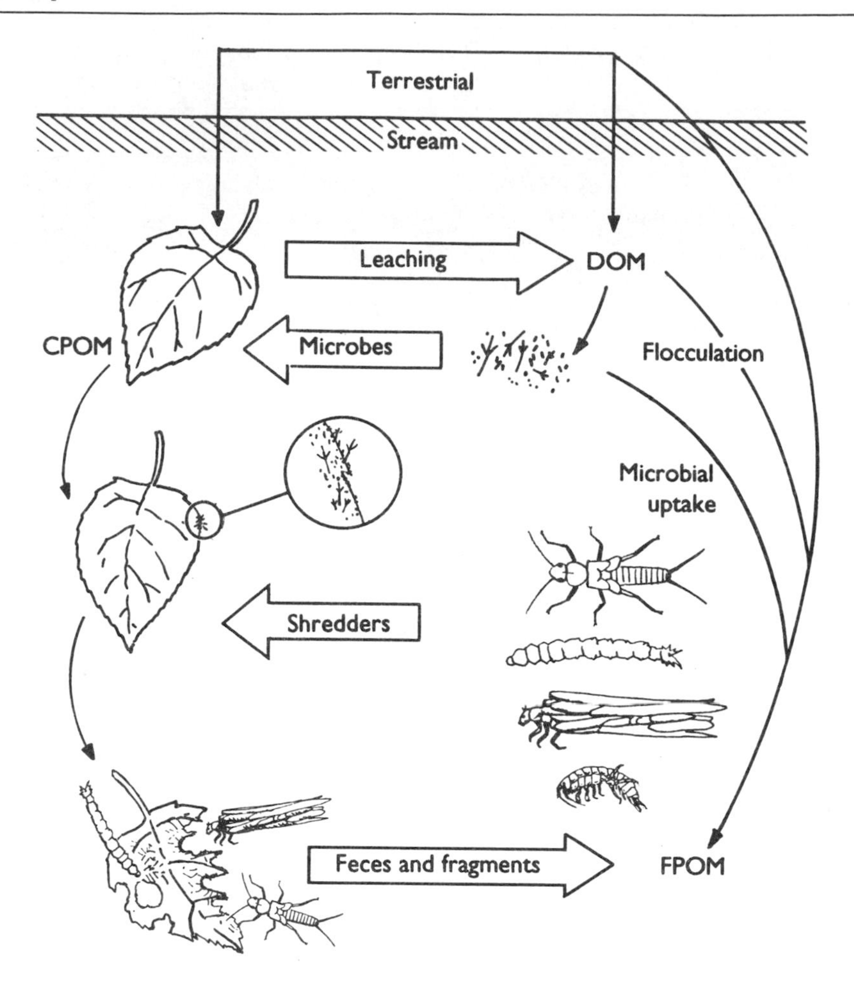

FIGURE 6.2 The links between shredders and CPOM, fungi and bacteria modeled for a small stream within a temperate deciduous forest. Physical abrasion, microbial activity (especially by fungi) and invertebrate shredders reduce much of the CPOM to smaller particles. Chemical leaching and microbial excretion and respiration release DOM and CO_2, but much of the original carbon enters other detrital pools as feces and fragmented material. (After Cummins and Klug, 1979.)

Lepidostomatidae, Sericostomatidae, Oeconesidae) and plecopterans (Peltoperlidae, Pteronarcidae, Nemouridae). The leaf-shredding activities of insect larvae and gammarid amphipods are particularly well studied. *Tipula* and many limnephilid caddis larvae eat all parts of the leaf, both mesophyll and venation, whereas peltoperlid stonefly nymphs avoid venation and concentrate mainly on mesophyll, cuticle and epidermal cells (Ward and Woods, 1986; Wallace, Woodall and Sherberger, 1970). The radula of snails and mouthparts of *Gammarus* are most effective at scraping softer tissues, and the bigger crustaceans are able to tear and engulf larger leaf fragments (Anderson and Sedell, 1979). These differences caution us that not all consumers of CPOM are alike in feeding mechanism, and a greater understanding of the functional morphology of these organisms would be of considerable value.

The nutritional quality of leaves is intimately linked with the microorganisms that contribute

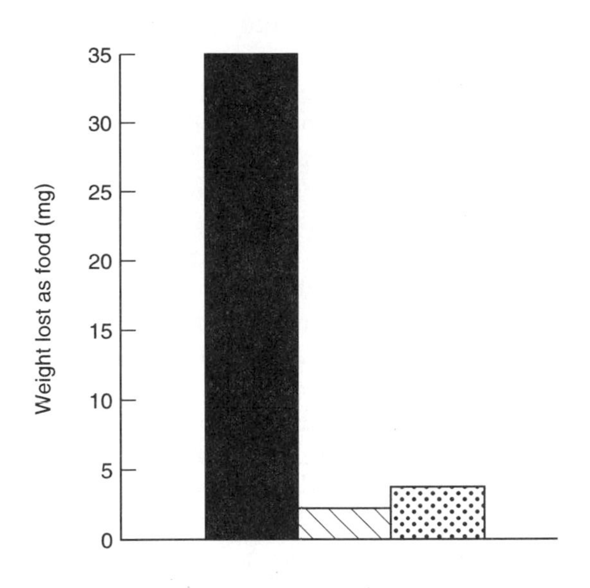

FIGURE 6.3 Feeding preference of *Gammarus*, measured by quantity of elm leaves consumed when microbial growth was permitted (■), compared with leaves consumed where microbial growth was inhibited with antibiotics (▧) or by autoclaving (▦). (After Kaushik and Hynes, 1971.)

so greatly to leaf breakdown. Indeed, it has been questioned whether the leaf substrate contributes anything to detritivore nutrition. Cummins (1974) suggested the interesting analogy that microorganisms on a leaf are like peanut butter on a cracker, with most of the nourishment provided by the peanut butter. Much effort has been directed at determining how microorganisms directly (as food) and indirectly (by modifying the substrate) contribute to the nourishment of consumer organisms, and what capabilities the detritivores possess to digest the various components of their diet.

Invertebrate detritus feeders unquestionably prefer leaves that have been 'conditioned' by microbial colonization in comparison to uncolonized leaves. When presented with elm leaves that were either autoclaved or cultured with antibiotics (see Table 5.3) to inhibit microbial growth, and normal colonized leaves, *Gammarus* consumed far more of the latter (Figure 6.3). Subsequent work confirms that preference is greatest for leaves at a 'peak' stage of conditioning that roughly corresponds to the period of greatest microbial growth (Arsuffi and Suberkropp, 1984; Suberkropp and Arsuffi, 1984). The benefits to the consumer include greater efficiency of converting ingested leaf biomass into consumer biomass and a higher individual growth rate (Lawson, Klug and Merritt, 1984; Figure 6.4).

Detailed studies of leaf decomposition document the dominant role played by fungi, particularly aquatic hyphomycetes, and the changes in nitrogen and protein content over time in detrital material. Studies of assimilation efficiency, growth, and survival of detritivores (reviewed in Barlöcher, 1985) confirm the nutritional superiority of conditioned leaves and fungal diets. Ward and Cummins (1979) compared the growth of the midge larva *Paratendipes albimanus* on diets of conditioned hickory and oak leaves, *Tipula* feces, and native detritus collected from a Michigan stream. Growth rate decreased in the order just given and correlated well with measures of microbial biomass and respiration, but not with total nitrogen or C:N ratio. A similar conclusion was reached in studies of young *Gammarus pulex*, which grew fastest on well-colonized elm leaves. Leached leaves lacking a flora permitted growth at a markedly lower rate (Sutcliffe, Carrick and Willoughby, 1981).

Further investigations using a variety of fungal species in pure cultures and inoculated onto various leaves demonstrate that the nutritional quality of fungi varies among species. When presented with leaf disks supporting negligible microflora, the preference rank of *G. pseudolimneaus* was ash > maple > oak. However, this order could be reversed depending on the fungal species allowed to colonize discs (Barlöcher and Kendrick, 1973a). Growth rate and survival of caddis larvae on some fungi, both terrestrial and aquatic, was quite high whereas other fungal species clearly provided inadequate diets. The palatability of various fungal species, assessed by

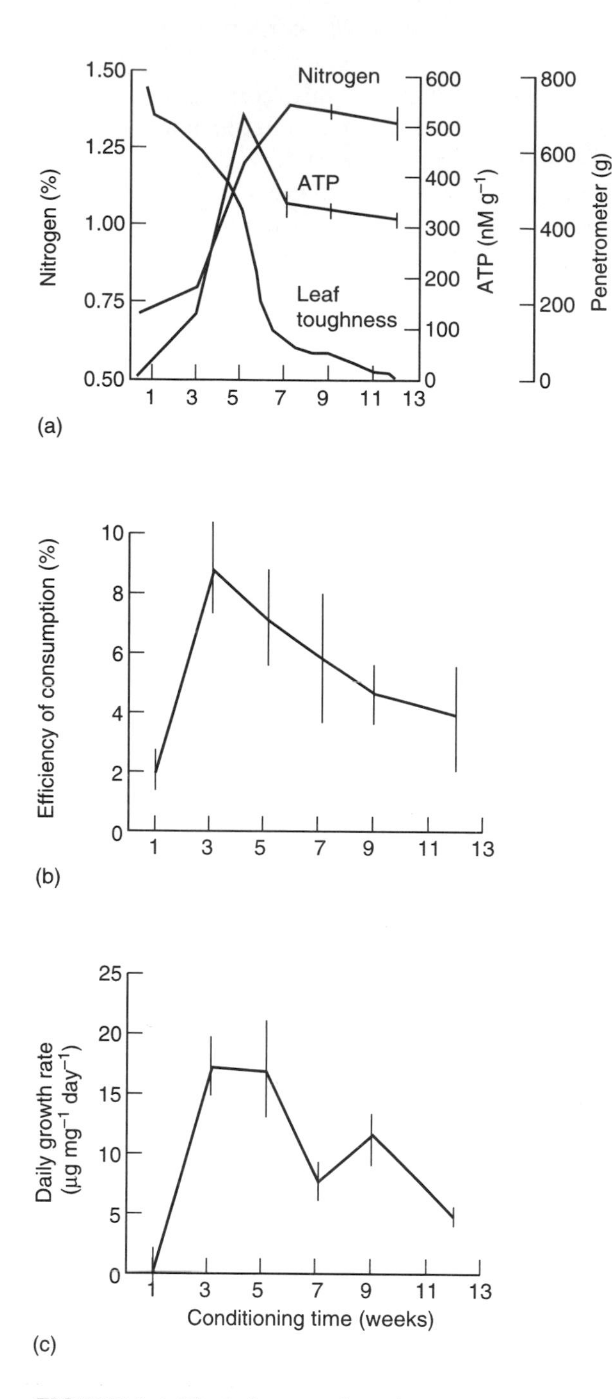

FIGURE 6.4 The influence of conditioning time of discs of hickory leaves on utilization by *Tipula abdominalis*. The food appears to reach peak condition in 3–5 weeks, as indicated by buildup of ATP (an index of microbial biomass), relative nitrogen content, and the progressive softening of leaf discs (a). Changes in efficiency of conversion of ingested material into consumer biomass (b) and daily growth rate (c) correspond to the time course of conditioning. The greater nutritional value of leaf discs at peak conditioning time apparently was due to peak digestibility of the leaf itself, rather than ingestion of microbial biomass. (After Lawson, Klug and Merritt, 1984).

feeding preference trials, and their nutritional value, assessed by assimilation or growth rate, are correlated but do not correspond perfectly (e. g. Arsuffi and Suberkropp, 1986).

Microorganisms may enhance the palatability and nutritional quality of leaves in at least two distinct ways (Barlöcher, 1982, 1985). One, termed microbial production, refers to the addition of microbial tissue, substances, or excretions to the substrate; in essence the role originally proposed by Kaushik and Hynes (1971). Because assimilation efficiencies on fungal mycelia and mixed microflora have been shown to exceed 60%, while values for conditioned and unconditioned leaves average about 20% (Martin and Kukor, 1984; Barlöcher, 1985), indications are that the nutrient content per unit mass in microorganisms exceeds that of the leaf substrate several-fold.

The second potential role for microorganisms is microbial catalysis, and encompasses all of the changes that render the leaf more digestible. This includes partial digestion of the substrate into sub-units that detritivores are capable of assimilating, and production of exoenzymes that remain active after ingestion. As support for this proposition, Barlöcher (1985) pointed out that structural carbohydrates (cellulose, hemicellulose, pectin) may be partially digested by microorganisms into intermediate products which the gut fluids of invertebrates are then able to degrade. Indeed, leaves subjected to partial hydrolysis with hot HCL were preferred by *G. pseudolimnaeus* over untreated leaves (Barlöcher and Kendrick, 1975). Barlöcher (1982) also showed that fungal exoenzymes extracted from decomposing leaves remained active in the presence of gut enzymes of *G. fossarum* for up to 4 h at the foregut's pH, indicating that ingested exoenzymes can aid in the digestion of polysaccharides.

Although the ability to synthesize cellulase and thereby derive nutrition from plant cell wall polysaccharides occurs in some detritus feeders, including representatives of the mollusks, crustaceans and annelids (Yokoe and Yasumasu, 1964; Bjarnov, 1972; Monk, 1976) aquatic insects generally show negligible enzymatic activity toward cellulose and other plant structural polysaccharides, and this has been a principal reason for arguing the importance of microorganisms as an energy source. A series of studies of leaf-shredding insect larvae revealed both similarities and differences in gut biochemistry among dipterans (*Tipula abdominalis* (Martin *et al.*, 1980; Sharma, Martin and Shafer, 1984); *Prosimulium fuscum* (Martin *et al.*, 1985); plecopterans (*Pteronarcys pictetii* and *P. californica*; Martin *et al.*, 1981a); and trichopterans (*Pycnopsyche guttifer*, *Phryganea* sp., *Agrypnia vestita*; Martin *et al.*, 1981b). All showed high proteolytic activity indicating efficient digestion of bacteria and microinvertebrates, limited capacity for polysaccharide digestion which probably would allow utilization of fungal and periphyton tissue, and essentially no activity against vascular plant detritus. The groups differed in the alkalinity of gut fluids. Aquatic Diptera have gut pH values in excess of 10–11 (Martin *et al.*, 1985), while existing data for Plecoptera and Trichoptera indicate a pH range of 7–7.5. In terrestrial herbivorous insects, highly alkaline gut fluids are believed to aid in the digestion of protein that is complexed with tannins, lignin and other polyphenols. Whether the aquatic Diptera are able to digest proteins unavailable to other aquatic insects remains unknown at present.

The picture that emerges, then, is one in which the nutritional value of the leaf–microorganism unit depends upon the species of leaf and fungus, the time course of microbial production and catalysis, and the ability of the detritivore to assimilate the various leaf constituents and breakdown products. As a result, different shredder species may feed on CPOM in quite different ways.

Barlöcher (1983) makes the point that *Gammarus* can possess a digestive machinery adapted to leaves at peak stage of conditioning because its high mobility allows it to search widely, whereas the low mobility of *Tipula* places a higher premium on being able to digest whatever it encounters (Table 6.2).

Several lines of evidence suggest that a substantial fraction of the energy needs of detritivorous invertebrates must be furnished by the decomposing vegetation itself. Microbial biomass usually is less than 10% of the ingested detritus (Iverson, 1973; Barlöcher, 1985) and probably cannot meet the carbon and nitrogen needs of the consumer, despite the high assimilation efficiency of a microbial diet (Cummins and Klug, 1979; Martin and Kukor, 1984). In addition, while insect larvae may lack the ability to synthesize cellulytic enzymes, Sinsabaugh, Linkins and Benfield (1985) demonstrated, using radiolabeled cellulose substrate, that leaf-shredding insects were able to digest and assimilate plant cell wall polysaccharides. Tissue-level synthesis of cellulose degrading enzymes seems unlikely to be the primary mechanism in any of the species studied (the stonefly *Pteronarcys proteus*, crane fly *Tipula abdominalis* and caddisfly *Pycnopsyche luculenta*). A high fraction (40%) of the digested material passed the cell wall in *Pteronarcys* as neutral sugar, and its simple gut tract was not thought to harbour endosymbionts (Cummins and Klug, 1979). For these reasons, Sinsabaugh, Linkins and Benfield inferred that ingested exoenzymes accounted for digestion in this species. In *Tipula*, in contrast, the distinctive rectal lobe of the hind gut and dense bacterial flora were indicative of an endosymbiont community, and this was further supported by the finding that over 90% of the labeled digestion products were organic or amino acids, consistent

TABLE 6.2 The contrasting feeding strategies of two CPOM detritivores (Based on Barlöcher, 1983)

	Gammarus fossarum	*Tipula abdominalis*
Feeding mechanism	Scrapes at leaf surface	Chews entire leaf
Gut pH and digestive biochemistry	Anterior gut slightly acid	Foregut and midgut highly alkaline (up to 11.6)
	Its own enzymes and fungal exoenzymes attack leaf carbohydrates	Result is high proteolic activity but inactivation of fungal exoenzymes, thus little activity toward leaf carbohydrates
	Posterior gut is alkaline, would digest microbial proteins and some leaf proteins	
Efficiency	Highly efficient at processing conditioned leaves at low metabolic cost	Less dependent upon stage of conditioning, probably good at extracting protein, but at high metabolic cost
Other attributes of feeding ecology	Highly mobile	Low mobility
	Polyphagous	Obligate detritivore

with microbial metabolism of the substrate. Lastly, experiments using radiotracers to separate the microbial and substrate contributions to individual growth show the bulk of the energy coming from the leaf. Using radiolabeled food sources and inhibitors of DNA synthesis, Findlay, Meyer and Smith (1984, 1986b) demonstrated that only 15% of the respired carbon in the freshwater isopod *Lirceus* and 25% in the stonefly *Peltoperla* was met by consumption of microbes. Virtually all of the assimilated microbial carbon was derived from fungi, as the bacterial contribution was less than 1%. Similarly, Lawson, Klug and Merrit, (1984) concluded that only 11–27% of growth in *Tipula* was derived from microbial sources, thus 73–89% of its gain in biomass must be attributable to the leaf matrix.

In summary, while the relative contributions of microorganisms and leaf substrate to detritivore nutrition are not yet fully elucidated, the importance of leaf substrate is being revised upwards, aided by an improved understanding of microbial processes. Furthermore, the answer is likely to vary depending on the identity of the fungus, leaf and detritivore (Barlöcher, 1985). Overall, recent estimates of the contribution of the microbial portion of the diet to consumer energy needs show that it may be surprisingly low, indicating that the 'cracker', mediated by microbial processing, indeed contributes a substantial fraction of the assimilated energy.

Woody debris has been largely ignored as an energy source, because few invertebrates feed on it directly, and wood appears to be a poor food. Nevertheless, wood may contribute 15–50% of total litter fall in small, deciduous forest streams, and even more in coniferous regions (Anderson and Sedell, 1979). Although its importance diminishes downstream and it is utilized only very slowly (a residence time of at least years to decades, in comparison to weeks to months for leaves), wood provides food and habitat for a substantial number of species. Anderson *et al.* (1978) found some 40 taxa associated with this resource in wood-rich Oregon streams. Prominent aquatic xylophages included a midge (*Brilla*) which was an early colonizer of phloem on newly fallen branches, two species that gouged the microbially conditioned surface of waterlogged wood (the elmid, *Lara*, and the caddis, *Heteroplectron*), and a cranefly (*Lipsothrix*) that consumed nearly decomposed woody material.

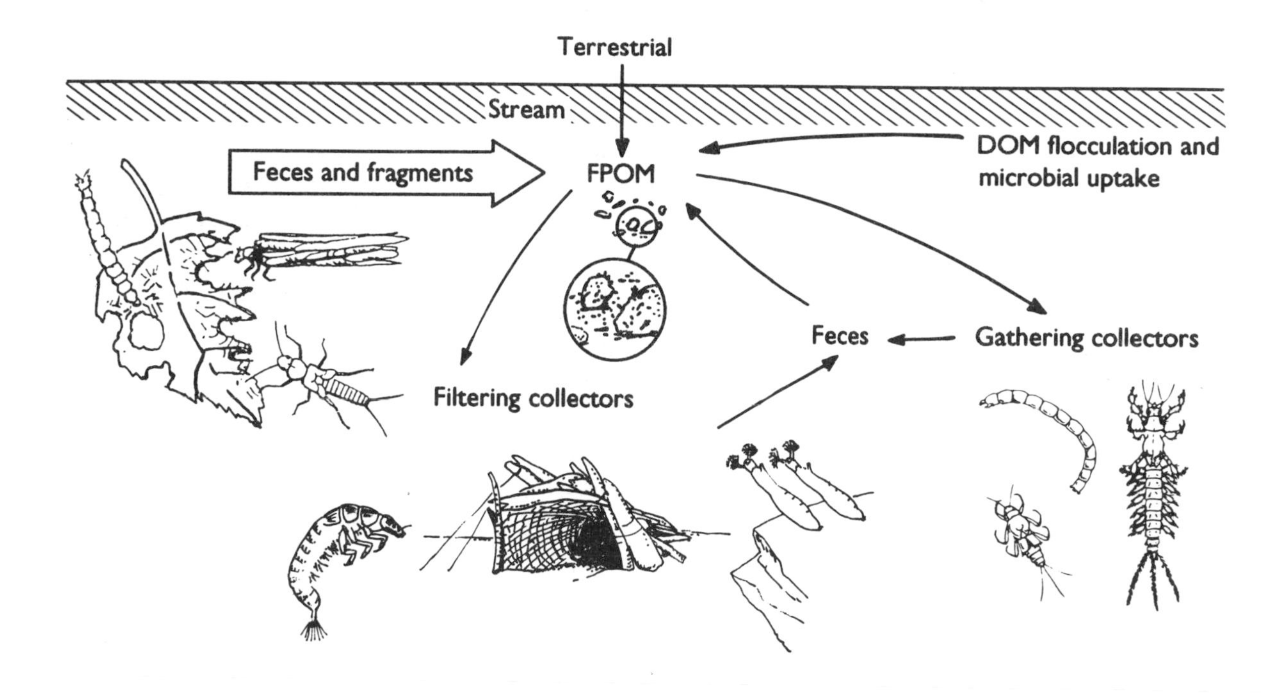

FIGURE 6.5 The collector–FPOM–bacterial linkage modeled for a small stream within a temperate deciduous forest. Detrital particles less than 1 mm produced by fragmentation of larger particles and from terrestrial inputs are surface-colonized by microorganisms. Additional carbon may accrue via flocculation and microbial uptake. Fecal matter, small animals and cells from the periphyton also contribute to FPOM. Suspension and deposit feeding are primary modes of FPOM acquisition, and further distinctions are made on the basis of particle size. (After Cummins and Klug, 1979.)

In comparison to leaves, invertebrate standing crop biomass on wood was about two orders of magnitude lower per kilogram of substrate.

Lotic consumers are relatively unspecialized for xylophagy. The beetle *Lara avara* possesses robust mandibles capable of slicing away thin strips of wood, but apparently lacks digestive enzymes or gut symbionts to aid digestion. Microscopic inspection of material progressing through the gut indicated no change to the wood (Steedman and Anderson, 1985); presumably the larva is nourished by microbiota and their exudates occurring on the wood surface. Not surprisingly, *L. avara* grows very slowly and requires 4–6 years to attain maturity.

6.2.2 Consumers of FPOM

The link between collectors, FPOM and bacteria (Figure 6.5) depends on FPOM captured from suspension or from the substratum. As discussed in chapter 5, FPOM is as yet a poorly characterized food source, and it originates in a number of ways. Categories considered to be among the richest in quality include sloughed periphyton, organic microlayers and particles produced in the breakdown of CPOM. Morphological and behavioral specializations for suspension feeding are diverse and well studied (Wallace and Merritt, 1980), while deposit feeders are less well known (Berg, 1994; Wotton, 1994).

The importance of suspended particulates as a food resource was first recognized from observations of enormous concentrations of suspension-feeding insect larvae at lake outlets (e. g. Müller, 1954a). Densities decrease sharply within a kilometer or so, as the rich seston declines downstream from lake outflows due to removal and dilution (Carlsson *et al.*, 1977). Black fly larvae in Swedish Lapland were roughly 15-fold more abundant at a lake outflow compared with merely 2 km downstream.

Caddisflies in the superfamily Hydropsychoidea (which includes the Philopotamidae, Psychomyiidae, Polycentropodidae and Hydropsychidae) spin silken capture nets in a variety of elegant and intricate designs. Most are passive filter feeders, constructing nets in exposed locations, but some nets act as snares (*Plectrocnemia*) or as depositional traps where undulations by the larvae create current (*Phylocentropus*, (Wallace and Malas, 1976); an Amazonian *Macronema*, (Sattler, 1963)). Filter-feeding hydropsychids vary considerably in mesh size and microhabitat placement of their nets (Wallace, Webster and Woodall, 1977; Wallace and Merritt, 1980). As a rule, larger mesh catch-nets tend to be found at higher velocities, and there is reason to believe that the expected decrease in filtering efficiency is offset by higher volumes filtered. At the other extreme, fine-mesh nets should be more efficient but also more resistant to flow, which favors their location in microhabitat of low velocity, as shown by Edington (1968). The longitudinal distribution of Nearctic Hydropsychidae apparently reflects these different suspension-feeding tactics (Wiggins and Mackay, 1978). Members of the Arctopsychinae spin coarse nets, capture a good deal of animal prey and larger detritus, and tend to occur in headwaters. The Macronematinae occur in larger rivers, spin fine nets and capture small particles. The Hydropsychinae are intermediate in net mesh size, more widely distributed, and perhaps because of the broad range of resources utilized, also are richer in genera.

The amount and kind of suspended particles available must influence the distribution of suspension feeders. This in turn will depend not only on environmental features, such as lake outlets, but also on the food processing activities of other consumers. Two suspension feeders, a hydropsychid and a simuliid, acquired more label from ^{32}P-alder leaves when a shredder (*Pteronarcys*) was present, compared with its absence, demonstrating that feeding on CPOM made more food available to consumers of FPOM (Short and Maslin, 1977).

Impressive as the nets of caddis larvae are, they are but one of the many specialized adaptations for capturing particles from suspension that have arisen frequently and repeatedly among aquatic invertebrates (Wallace and Merritt, 1980). These various devices are collectively referred to as filtering or sieving adaptations, but the actual details of particle capture and retention may be more complex (Rubenstein and Koehl, 1977). Although many use suspension and filter feeding interchangeably, the former term is preferred as more general and less likely to promote misconceptions of the filtering mechanism based on childhood experiences in sand boxes.

The larvae of black flies are highly specialized suspension feeders. They have been studied extensively because the adults include important disease vectors as well as nuisance pests. Larvae attach to the substrate in rapid, often shallow water (Figure 6.6). The paired cephalic fans of suspension-feeding larvae each consist of primary, secondary and medial fans (Chance, 1970; Currie and Craig, 1987). Particles apparently are snared by sticky material on the primary fans, which are the main suspension-feeding organs, while secondary and medial fans act to slow and deflect the passage of particles. Food items are removed by the combing action of mandibular brushes and labral bristles, further adaptations to a filtering existence and lacking in some black fly species that scrape substrates instead. Fans are opened for feeding and closed

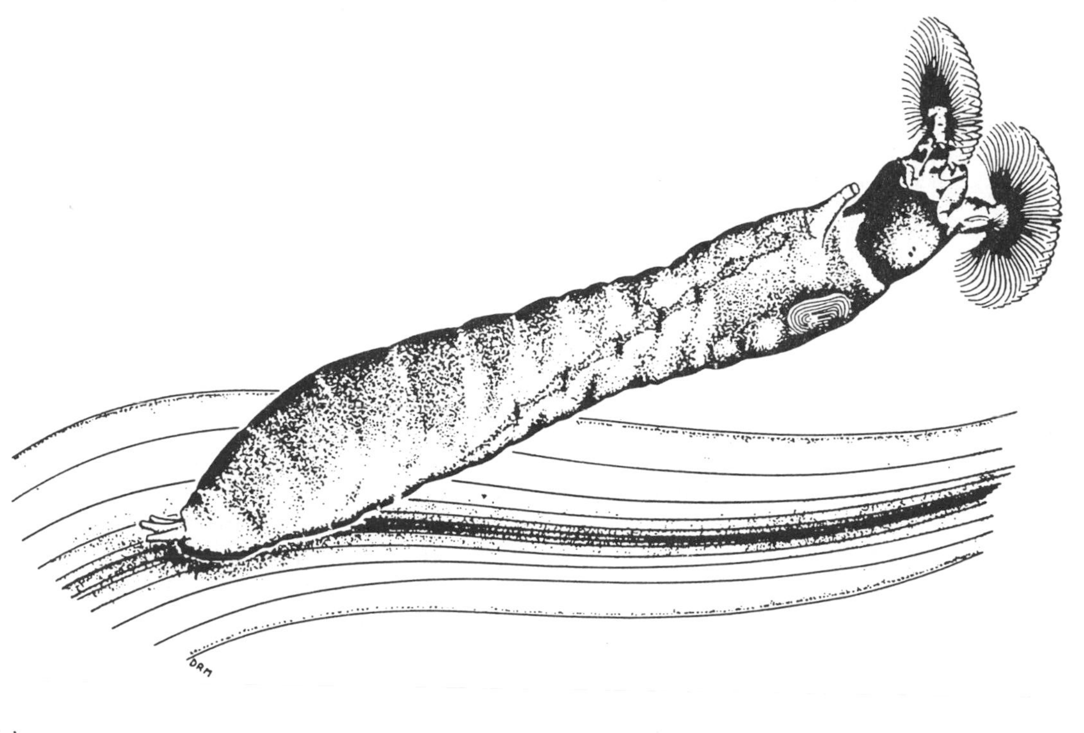

(a)

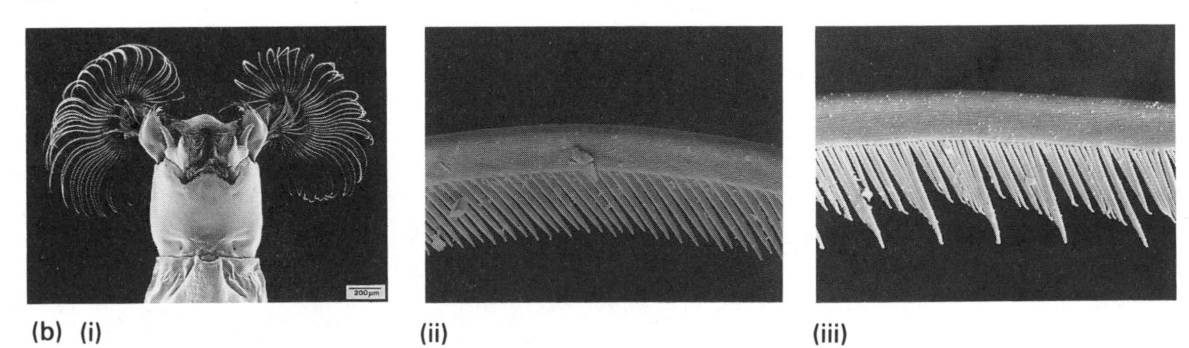

(b) (i) (ii) (iii)

FIGURE 6.6 (a) The typical filtering stance of a black fly larva (*Simulium vittatum* complex). The larval body extends downstream at progressively greater deflection from vertical with increasing current velocity, and is rotated 90–180° longitudinally as can be seen by following the line of the ventral nerve cord. The position of the paired cephalic fans is upper and lower, rather than side by side. The boundary layer (depth where U falls below 90% of mainstream flow) begins at roughly the height of the upper fan (Chance and Craig, 1986). (b) Details of cephalic fans: (i) head of a normal larva seen from beneath, with cephalic fans fully open; (ii) *Simulium tahitiense* with uniform fringe of microtrichia; (iii) *S.rorotaense* with long and short microtrichia. (From Crosskey, 1990; SEM photographs courtesy of D.A. Craig.)

at other times (Crosskey, 1990). The four species studied by Chance (1970) ingested particles from less than 1 μm to over 350 μm. Field studies generally report the majority of ingested particles to be less than 10 μm in diameter (e. g. Merritt, Ross and Larson, 1982). Some 80% of the particles ingested by *Metacnephra tredecimatum*, a lake outlet species, were less than 2 μm (Wotton, 1978), and Fredeen (1964) successfully reared black fly larvae on a bacterial suspension.

A detailed investigation of the flow regime surrounding individual simuliid larvae demonstrates that this hydrodynamic environment is exceedingly complex (chapter 3), and larvae may be able to manipulate flow vortices to enhance feeding (Chance and Craig, 1986; Lacoursiere and Craig, 1993). Of further interest, despite the evident elegance of the adaptations of larval simuliids for suspension feeding, this is by no means the only feeding mode employed. Currie and Craig (1987) state that scraping the substrate using mandibles and labrum is the second most important method of larval feeding, not including species that lack cephalic fans and are obligate scrapers. There is at best a blurry line between scraping cells of the periphyton and deposit feeding on fine particulate organic matter. In addition, black fly larvae occasionally ingest animal prey. This diversity is a useful reminder that even those taxa displaying great specialization for a particular trophic role also may be capable of great versatility.

Other dipteran families with representatives adapted to a suspension-feeding existence in running waters include the Culicidae, Dixidae and Chironomidae (Wallace and Merritt, 1980; Berg, 1994). Some Chironominae construct tubes or burrows with catchnets and create current by body undulations; others such as *Rheotanytarsus* passively suspension feed by means of a sticky secretion supported by rib-like structures on the anterior end of the case. Figure 6.7 from the excellent review by Wallace and Merritt (1980) depicts some of the marvelous variety of suspension feeding lifestyles found among lotic invertebrates.

Mechanisms of deposit feeding on FPOM are either less diverse in comparison to the suspension-feeding mode, or less is known about the subject. Also called collector–gatherers in Cummins' (1973) functional group classification, this feeding role is well-represented in most running water environments in terms of both species and total abundances. Among the macroinvertebrates in swifter streams, representatives of the mayflies, caddisflies, midges, crustaceans and gastropod mollusks are prominent deposit feeders. In slower currents and finer sediments one would expect in addition oligochaetes, nematodes and other members of the meiofauna (invertebrates passing through a 0.5 mm sieve, often interstitial in lifestyle, and important in the microbial loop within the benthos). It would be surprising if these animals all fed in the same way and consumed the same food. In addition to their particular food-gathering morphologies, these taxa differ in their ability to produce mucus, in mobility and body size, in their digestive capabilities, and in whether they are surface-dwellers or live within the sediments.

Our limited understanding of FPOM dynamics places further restrictions on our ability to assess the trophic role of lotic collector–gatherers (Ward and Woods, 1986). Some insight can be gained by comparison with the considerable literature on marine deposit feeders (Lopez and Levinton, 1987), yet there as well, progress has been limited by similar difficulties in characterizing the mix of microorganisms, organic molecules, micro-metazoans, and plant and animal debris that provides nourishment. Microbes rather than the substrate have traditionally been considered the prime source of energy. As was true for leaf CPOM in streams, this may be too one-sided a view, although the efficiency with which food is absorbed unquestionably varies from microbes and small autotrophs at one extreme to lignin-rich, refractory detritus at the other (Lopez and Levinton,

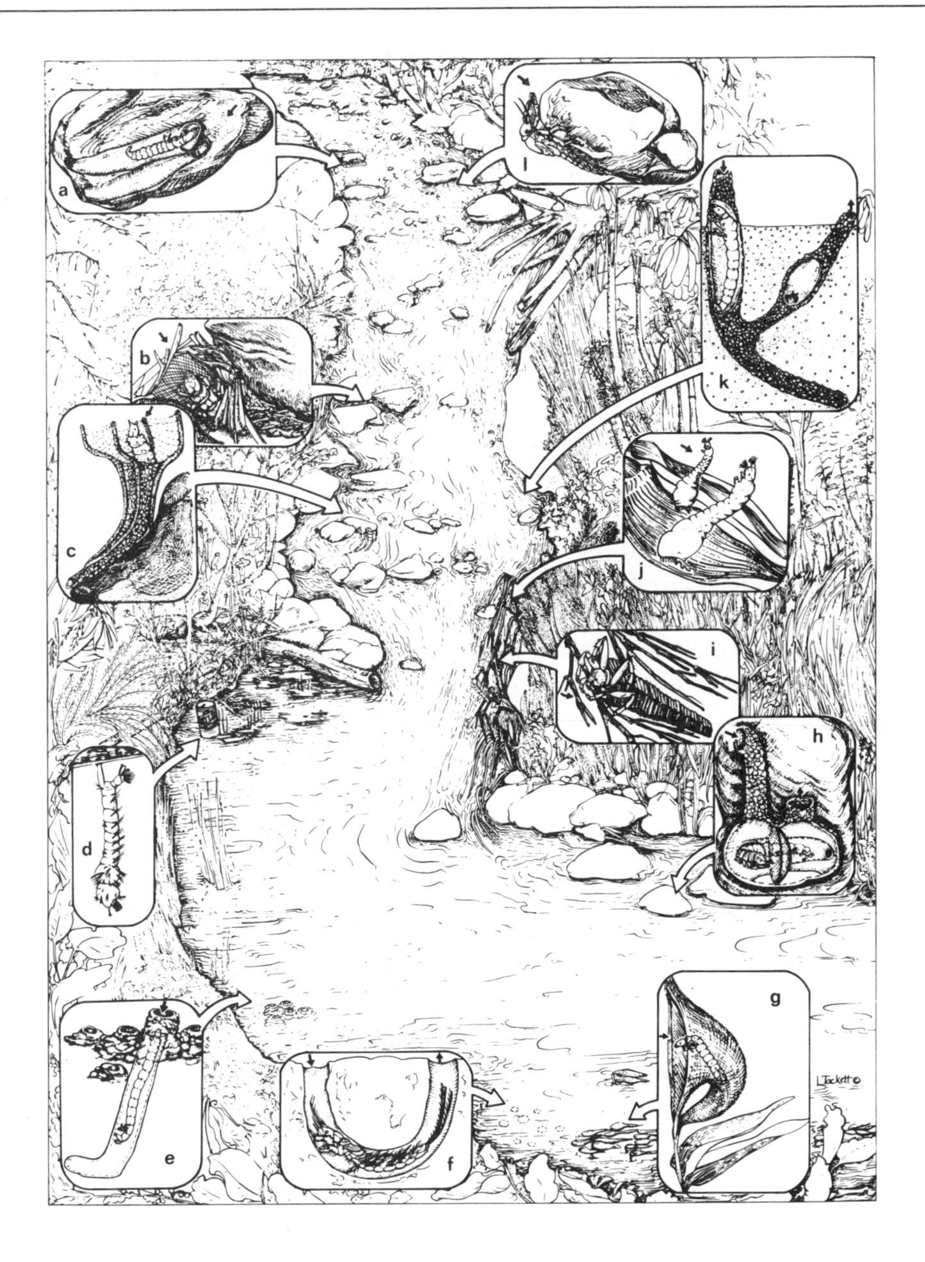

FIGURE 6.7 Diversity of suspension-feeding modes in running water. (a) Philopotamid caddis larva with tube-like net on the lower surface of a stone; (b) a hydropsychid caddis larva feeding on materials trapped on its capture net; (c) a chironomid midge larva, *Rheotanytarsus*, and its tube case; (d) a culicine mosquito larva in a discarded container; (e) a *Chironomus* larva in its J-shaped tube, better suited to deposit feeding. This larva also suspension feeds when U-shaped tubes are constructed; (f) the mayfly nymph *Hexagenia* in its U-shaped burrow; (g) the polycentopodid caddis *Neureclipsis* and its cornucopia-shaped net; (h) the larval dwelling and filtering apparatus of the hydropsychid *Macronema*, with its water intake opening projecting above the surface; (i) a *Brachycentrus* caddisfly larva in filtering position; (j) a black fly larva with extended cephalic fans; (k) the polycentopodid caddis larva *Phylocentropus* in its branched dwelling tube located in regions of finer sediments; (l) the mayfly *Isonychia* filtering with setae of its forelegs. (From Wallace and Merritt, 1980.)

1987). A variety of methods are available for the study of FPOM dynamics. These include direct examination by light and scanning electron microscopy; measurement of lignin oxidation products; assessment of microbial activity by epiflourescent microscopy, ATP content and respiration rates; stable isotope analysis; use of various tracers; and use of consumer assimilation and growth as a bioassay of nutritional quality. At present, however, we have limited knowledge of the average nutritional value of FPOM, and do not know the extent of natural variation in its quality (Ward *et al.*, 1994).

The feeding activity of shredders on CPOM can be a significant source of FPOM for deposit feeders. The mayfly *Stenonema* provided with leaf litter grew faster when leaf-shredding invertebrates were present (Cummins *et al.*, 1973), although exactly how this mayfly benefitted is uncertain. Production of feces is one possibility, and these feces together with the attached microflora can provide an adequate diet for fine particle feeders. The growth rate of *Paratendipes albimanus* on feces of *Tipula* was superior to that obtained on native detritus, although inferior to conditioned oak or hickory leaves (Ward and Cummins, 1979). Because leaf-shredding invertebrates vary in how they consume the leaf, there is also variation in the feces and fragments they release (Ladle and Griffiths, 1980; Ward and Woods, 1986).

While it is unquestionably true that fragmentation of CPOM produces fine particulates and these may best be harvested by a collector–gatherer feeding mode, other sources of FPOM may be of equal or greater importance. Any highly labile DOM present in the water column is rapidly incorporated into the sediments by microbial uptake, entering the organic layer and microbial tissue. As already discussed, how much of this material remains within a closed microbial loop, and how much reaches larger deposit feeders, is an open question. The presence of chironomid tunnels within organic layers (Lock *et al.*, 1984) shows that smaller invertebrates use it as food. Tracer techniques confirm that macroinvertebrates do so as well. Rounick and Winterbourn (1983) used radiotracers to label organic mats that had developed in darkness and so presumably lacked algae and other autotrophs. In 24 h feeding trials with several common stream invertebrates, all but the predaceous taxa ingested this food source, with assimilation efficiencies ranging from 18 to 74%.

Stable carbon isotope analysis is a new and useful way to identify certain energy pathways (Petersen and Fry, 1987). The ratio $^{13}C/^{12}C$ differs sufficiently among terrestrial leaves, grasses and algae to allow identification of each. Furthermore, the ratio in an animal's tissue is a record of its recent feeding history reflecting assimilation, not just ingestion. By combining visual inspection of gut contents with stable carbon isotope ratios, Winterbourn, Cowie and Rounick (1984) established that most of the fauna fed on fine particulates, the ultimate origin of which was terrestrial. DOC exuded from the periphyton and captured within organic microlayers was evidently an important energy source at one site. Rau and Anderson (1981) demonstrated that leachate (from alder leaves in their study) was assimilated by macroconsumers, presumably after the DOM is incorporated into microbial biomass.

Browsing on easily assimilated organic layers may allow consumers to meet their energy needs without having to ingest large quantities of material. This is not the case for animals that ingest refractory POM mixed with sediments. Many deposit feeders 'bulk feed', processing each day from one to many times their body mass of sediments and assimilating a low fraction of what they ingest (Lopez and Levinton, 1987). The burrowing mayfly *Hexagenia limbata* ingests 100% of its dry mass daily (Zimmerman and Wassing, 1978). Data for FPOM deposit feeders in streams are scant, but numerous studies of leaf-shredding insects document assimilation efficiencies in the range 10–20% (McDiffett,

1970; Golladay, Webster and Benfield, 1983) and daily ingestion rates in excess of one body mass per day.

Under the reasonable assumption that detritus varies widely in food value, one may ask whether deposit feeders adjust to different feeding opportunities. Taghon and Jumars (1984) argue that selection can be accomplished by either differential ingestion, which usually involves some method of particle rejection in the buccal region, or differential digestion, based on digestive physiology and gut retention time (GRT).

GRT varies inversely with feeding rate; thus high quality foods that can be absorbed rapidly should favor high feeding rates and short GRT, whereas feeding should slow to allow longer GRT on poor foods (Lopez and Levinton, 1987). Calow (1975a) demonstrated an inverse relation between ingestion rate and absorption efficiency in two freshwater gastropods. When starved, snails increased GRT by slowing the rate of passage of food through the hepatopancreas, the main site of absorption and digestion. The effect of changing food quality on GRT apparently varies with the quality of the food, however. Calow (1975b) found that the herbivorous limpet *Ancylus fluviatilis* increased GRT on poor quality food (the expected result), but the detritivorous snail *Planorbis contortus* did the opposite. It may be that whenever the food carrier is highly refractory, as in the case of lignin, it pays to process material rapidly for easily removed microbes rather than attempt to extract energy from nearly indigestible substrate.

In summary, deposit feeders are among the least well understood of functional feeding groups in running waters, partly for lack of analysis of feeding mechanisms, and partly because of current shortcomings in understanding the sources and pathways of FPOM. Some taxa may shift opportunistically between this role and shredding (e. g. *Gammarus*), and others between the collecting of FPOM and grazing of easily removed periphyton (the 'brusher' category discussed below). There is evidence to suggest that deposit feeding is common in early instars that will occupy other, more specialized guilds as they grow larger. Clearly there is room for further research into this feeding role.

6.2.3 Consumers of autotrophs

The links between grazers and periphyton and between piercers and macrophytes (Figure 6.8) comprise the principal pathways for ingestion of living primary producers. The latter refers primarily to the micro-caddisflies (Hydroptilidae), which pierce individual cells of algal filaments and imbibe cell fluids (Cummins and Klug, 1979). Descriptions of the grazing pathway often emphasize the periphyton mat and scraping mouthparts. Scraping of surfaces is an important feeding role, complete with specialized structures including the rasping radula of snails and the specialized mandibles of caddis larvae such as *Neophylax* and *Glossosoma*, the beetle larva *Psephenus*, and some mayflies. However, the issue of grazing or herbivory is more complex, partly because the resource category includes organic layers, loose FPOM, and diatoms differing in the degree to which they adhere to substrates (Figure 4.1); and partly because the functional morphology of this consumer group is under-studied.

The mayfly *Stenonema* is usually classified as a scraper, although it also is considered a gatherer (Merritt and Cummins, 1984). McShaffrey and McCafferty (1986) undertook a detailed study of the functional morphology and feeding behavior of *S. interpunctatum*, using a flow-through chamber, video-recording, scanning electron microscopy, and even constructing a moveable plastic model of its mouthparts. Feeding movements were highly stereotyped, but several feeding modes were employed depending on feeding conditions. For attached material such as algae and diatoms, a series of movements of the labium and maxillae comprised a brushing cycle. A collecting cycle occurred in the presence of loose detritus and involved similar feeding

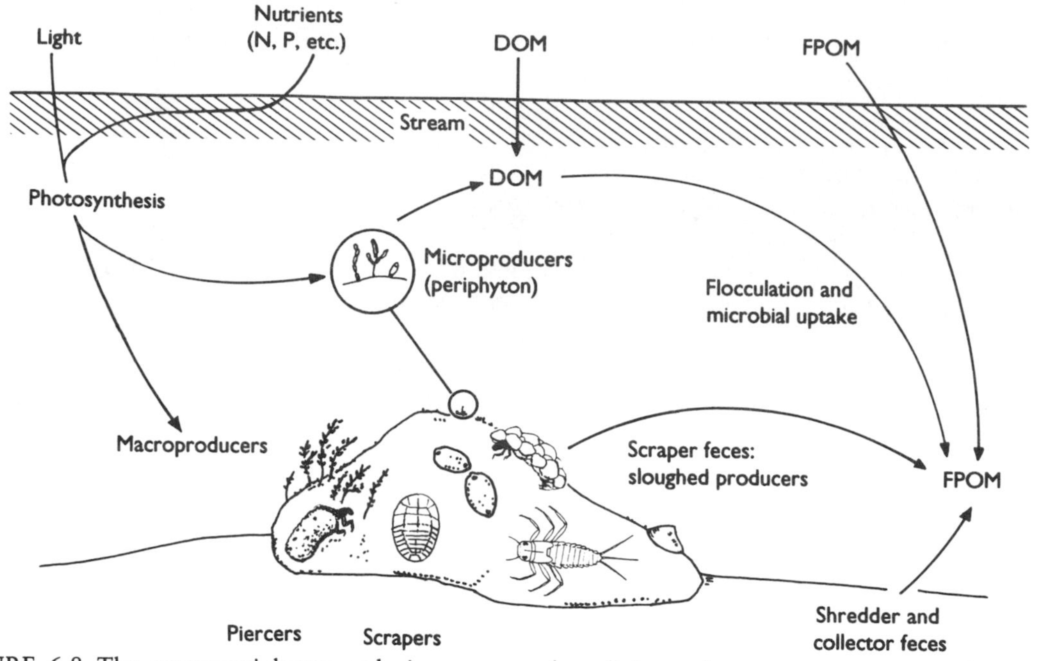

FIGURE 6.8 The grazer:periphyton and piercer:macrophyte linkages for a temperate stream. By a variety of mechanisms, the periphyton–bacteria–organic microlayer on substrate surfaces is scraped or browsed. Diatoms are a prominent constituent of this matrix. The contributions of the surface microlayer and associated bacteria, detritus and occasional very small invertebrates are difficult to quantify. Small caddis larvae (Hydroptilidae) pierce the cell walls of macroalgae and imbibe cell fluids. (After Cummins and Klug, 1979.)

movements, but mouthparts were not pressed as tightly against the substratum. Moreover, in the presence of abundant seston, *S. interpunctatum* radically altered its behavior, using the apical setae of extended maxillary palps for passive suspension feeding. In fact, this heptageniid was incapable of removing tightly adherent diatoms, and probably feeds primarily on detritus (Lamp and Britt, 1981), so scraping is the one feeding mode that does not apply. Only the labial palps and tips of maxillary palps of *S. interpunctatum* can reach the substrate, and these are setose rather than sclerotized. McShaffrey and McCafferty (1986) raise the possibility that scraping has been over-emphasized for other heptageniids as well. They introduce the term 'brusher' for taxa that remove material from the substrate using setae, and suggest that 'scraper' be used only for taxa with hardened structures that can remove adherent material.

This distinction is relevant to the interspecific interactions between two grazer insects studied by Hill and Knight (1988) in a California stream (see Chapter 8). The caddis *Neophylax* primarily depleted an adnate diatom, while the mayfly *Ameletus* substantially diminished the loose, upper layer of periphyton. Future efforts to couple functional morphology with field investigations of feeding role should be very fruitful. At present, it is uncertain whether such terms as grazing, scraping, brushing and browsing should be treated as equivalents, or constitute distinct feeding modes. Clarification is also needed on the effectiveness of these several feeding modes against various taxa and growth forms of the periphyton, FPOM and organic microlayers.

6.2.4 Consumers of other animals

The predator: prey linkage (Figure 6.9) is treated in detail in chapter 7. Most predators engulf their prey entire or in pieces, but some hemipterans

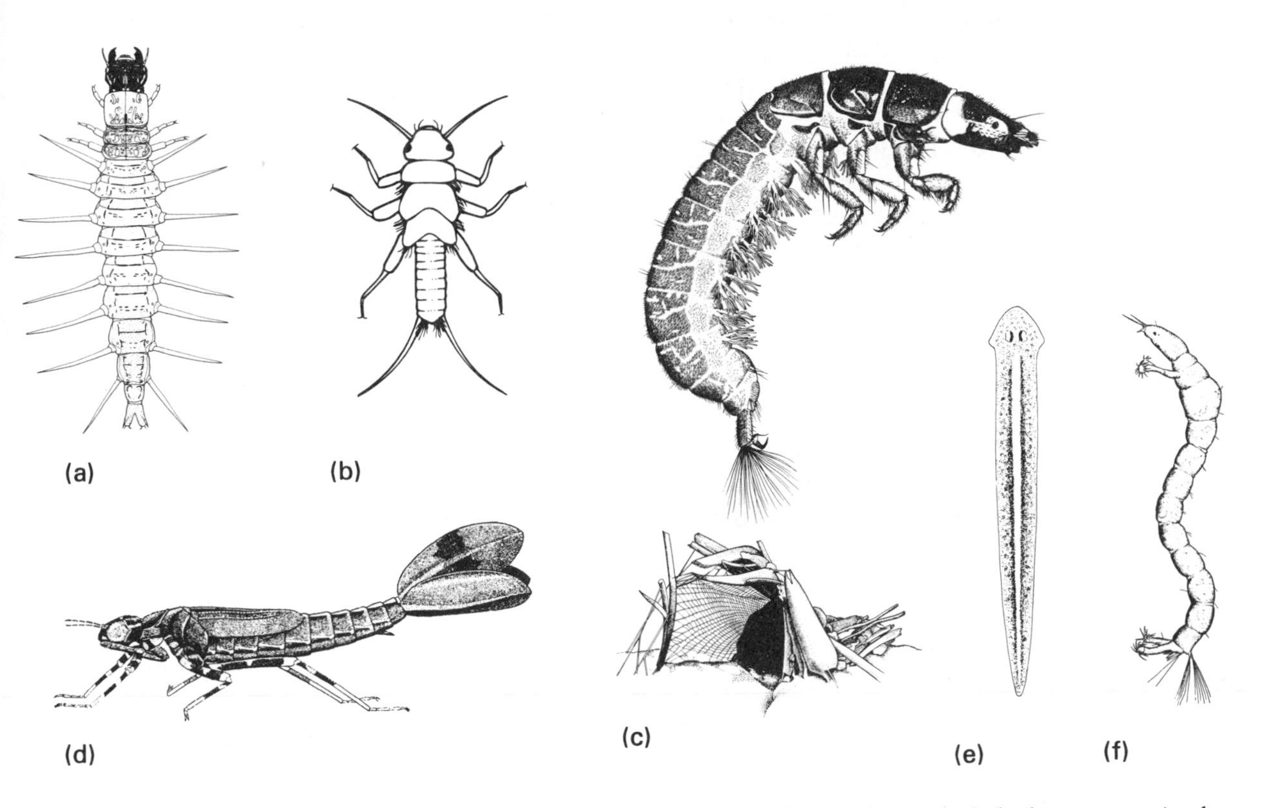

FIGURE 6.9 The predator:prey linkage for a temperate stream. Predaceous insects include those consuming large prey, illustrated by nymphs of (a) Megaloptera (Corydalidae) and (b) Plecoptera (Perlidae); those consuming prey of intermediate size, illustrated by (c) Trichoptera (Hydropsychidae) and (d) Odonata (Zygoptera); and those consuming small prey, illustrated by (e) Turbellaria (Tricladida) and (f) Chironomidae (Tanypodinae).

and rhagionid dipterans have piercing mouthparts (Cummins, 1973). Other distinctions can be made between hunting by ambush *versus* searching (Peckarsky, 1984), and whether prey are obtained from suspension, as in large hydropsychids, or strictly from the substratum, as in flatworms. Occasional predation probably is widespread, particularly the ingestion of micrometazoans, protists and early stages of macroinvertebrates. Such unpremeditated carnivory may provide high quality protein and may also form an important link between microbial and macro-consumer food webs.

6.2.5 Limitations of functional group classification for invertebrates

Changes in diet with age are well-known in fish, less so in invertebrates. However, this may simply indicate the lack of detailed study of the trophic ecology of tiny young invertebrate consumers. Predaceous stoneflies ingest more periphyton and detritus when small, and more animal prey when large (Allan, 1982a). An increased utilization of periphyton relative to detritus has been noted in grazing caddisflies and mayflies as they grow. Young nymphs of *Baetis* and *Cinygmula* in streams of Alberta, Canada, fed as collectors in late summer, and subsequently increased their consumption of diatoms

(Hamilton and Clifford, 1983). At present it is difficult to evaluate whether such diet shifts are due to changes in food availability or to changes in age and size. However, a collector–gatherer mode apparently requires little morphological specialization and in addition appears to be consistent with a relatively protected life within the sediments. As a result it may be widespread in very young invertebrates. Hynes (1941) opined that most Plecoptera feed on fine detritus in their early stages.

Changes in food availability obviously play a potentially large role in determining the seasonal and spatial abundance of the various functional groups. High growth rates of shredders in autumn demonstrate the close linkage of life cycle timing to seasonal changes in food availability. Longitudinal changes in resource abundance are the basis of the river continuum concept, discussed in detail in chapter 12. Wiggins and Mackay (1978) convincingly argue that Nearctic trichopteran species richness is tied to resource availability. Proportionately more genera of shredders are found in headwaters compared with downstream and in eastern deciduous forest regions compared with the coniferous West, in accord with expectations of greatest availability of leaves (Table 6.3).

In summary, the classification of the invertebrate consumers of streams into feeding guilds has demonstrated great utility for description and analysis. Enormous progress has been made over the last two decades toward understanding the various inputs of organic energy and how they are utilized. Characteristics of a particular river, including its size, hydrology and the vegetation of the surrounding landscape significantly influence which pathways predominate. As long as one does not lose sight of the fact that these guilds or functional groups are working conveniences that become progressively more fuzzy the closer one focuses, they serve as extremely useful building blocks towards a broader synthesis in stream ecology.

6.3 Feeding ecology of riverine fishes

Although all vertebrate classes have representatives in running waters, fishes are the principal vertebrate component of most riverine food webs. A number of attempts have been made to construct feeding categories for North American stream fishes. Allen (1969) offered a simple classification for North America: most stream fishes are invertivores, some become piscivorous for much of their life cycle, and a few are herbivores. Just as was true for early efforts to categorize invertebrates (most were herbivore–detritivores, some were predators), so simple a scheme is of limited value. Additional detail on where and how feeding occurs has allowed further subdivision of fish trophic categories (Table 6.4). Some of these categories could be subdivided further, and authors have done so where they deemed it appropriate. Herbivorous fish with scraping mouthparts such as the stoneroller *Campostoma anomalum* clearly have little in common with ooze feeders such as the bluntnose minnow *Pimephales notatus*. Some benthic invertebrate feeders utilize prey primarily from soft bottoms (the suckermouth minnow *Phenacobius mirabilis*), others from stony bottoms (the greenside darter *Etheostoma blennioides*). Particular studies may require more or fewer categories than are listed in Table 6.4. Slightly different schemes can be found in each of Horwitz (1978), Moyle and Li (1979), Grossman, Moyle and Whitaker (1982) and Schlosser (1982). While some fish can be placed in a trophic guild without difficulty, others cannot, due to flexibility in feeding habits and changes that occur over an individual's life cycle. Horwitz (1978) based his scheme on the principal food and feeding modes of adults, and placed each species in a single category. Grossman, Moyle and Whitaker recognized as many as three trophic roles for a single species, using the criterion that 70% or more of the diet fell in a given category. From invertivore to omnivore to herbivore–detritivore is essentially a continuum;

TABLE 6.3 Ratios of number of genera of Nearctic Trichoptera in three feeding categories in two North American forest biomes (From Wiggins and Mackay, 1978)

		Western montane	*Eastern deciduous*
Upstream sections	Shredders:grazers	25:26	21:15
	Shredders:collectors	25:9	21:12
Downstream sections	Shredders:grazers	6:13	9:13
	Shredders:collectors	6:6	9:8

Schlosser (1982) set the boundary between omnivore and invertivore at over 25% plant and algal matter.

In temperate warm-water streams, benthic invertebrate feeders are often numerous but, due to their small size, not a large component of biomass. Surface and water column feeders likewise are numerous, while planktivores and large predators are relatively sparse. Omnivores and herbivore–detritivores may be common, depending upon the habitat (Moyle and Li, 1979). Table 6.4 gives the percentage of fish species in each category from Horwitz's (1978) study of 15 river systems. All but one (the Powder River, Wyoming) were mid-western rivers, and the cumulative species list for each river ranged between 53 and 101 in eleven of the 15 rivers, so his results should provide a satisfactory characterization for that region. Guild proportions were very similar across basins, except for the Powder River where detritivores were many and piscivores absent. Planktivores were absent from headwaters and piscivores increased downstream; otherwise the downstream increase in species richness was unrelated to changes in trophic representation. Combining the relevant categories in Table 6.4, it would appear that fewer than 20% of the species in Horwitz's data base subsist on a diet of plant and detrital material. For all of North America, about 55 out of about 700 species are primarily herbivorous.

Extension of the guild structure for temperate zone fishes to the tropics is at best very tentative (Table 6.4). Large tropical rivers that have not been regulated by reservoirs have extensive lateral flood zones where much fish production occurs in seasonally inundated habitat (Welcomme, 1979). Allochthonous inputs are of great importance in these systems; consequently there is a greater role for mud and detritus feeding (which often supports the greatest biomass of fish) and for predation (which often dominates species richness).

The extensive flooded forests of the Amazon (Figure 6.10) make available a wide variety of plant products including seeds, fruits, flowers and leaves as well as monkey feces, numerous terrestrial invertebrates and the occasional vertebrate. Diversity of resources obviously contributes to diet diversity, and special feeding adaptations further increase the variety of feeding roles. Large characins such as the tambaqui (*Colossoma*) have evolved broad, multicusped molariform and incisive teeth in order to crush hard nuts (Goulding, 1980). A number of unrelated taxa eat particular parts of other living fishes, including scales, skin, fins, gill filaments and eyes, as well as whole chunks of the body (Roberts, 1972; Sazima, 1983). Morphological specialization is evidenced by sharp, forward-directed cutting teeth, and behavioral specialization includes aggressive mimicry. Detritivory usually is a minor feeding role for fish in temperate rivers but in tropical rivers, especially the great rivers of South America, the ingestion of dead organic matter can support the

TABLE 6.4 Trophic guilds of stream fishes, for temperate North America (modified from Horwitz, 1978) and tropical South America (modified from Welcomme, 1979). Percentages are based on the number of species, rather than the number of individuals, from the central USA only

Guild	*Description for temperate streams*	*Occurence by species*[a] *(%)*	*Comments for tropical streams*
Piscivore	Consumes primarily fish and/or large invertebrates, but includes smaller invertebrates	16	Piscivores may consume entire fish or specialize on parts of fish
Benthic invertebrate feeder	Feeds on benthic invertebrates, primarily immature insects	33	Most common in small to mid-order streams
Surface and water column feeder	Consumes surface prey (mainly terrestrial and emerging insects) and drift (zooplankton and invertebrates of benthic origin)	11	Diverse surface foods occurring in forested headwaters and during seasonal flooding
Generalized invertebrate feeder	Feeds at all depths	11	Similar category
Planktivore	Midwater specialist on phyto- and zooplankton	3	Seasonally important in large rivers
Herbivore–detritivore	Bottom feeder ingesting periphyton and detritus; includes mud feeders with long intestinal tracts	7	Herbivory may be subdivided into micro- and macrophytes, and detritus feeders separated from mud feeders
Omnivore	Ingests a wide range of animal and plant foods, and detritus	6	Similar category
Parasite	Ectoparasite (e.g. lampreys)	3	Ectoparasite (e.g. candirú catfishes)

[a] Percentage by species from Horwitz's (1978) study of 15 US river systems. An additional 9% of species could not be categorized, and one species (*Lepomis microlophus*) fed on snails.

bulk of fish biomass (Bowen, 1983). Special adaptations include a muscular stomach to grind food and an intestine with greatly increased absorptive surface due to elongation (up to 20 times body length) or elaborate mucosal folding. Bowen's definition of detritus excludes material not yet altered from its original form, such as fallen fruit, flowers and leaves; inclusion of this unaltered plant material would broaden considerably the role of detritivory.

Despite considerable specialization of dentition, jaw shape, body form and alimentary tract, many tropical fishes nonetheless display considerable flexibility in their diet. Goulding (1980) found that piranhas ingested mostly seeds and fruit during the flooded period. He concluded that seasonal fluctuation in water level was the single most important factor influencing the feeding behavior of the fishes studied, through effects on habitat and the availability of food. Seasonal comparisons in a small Panamanian stream (Zaret and Rand, 1971) revealed greater

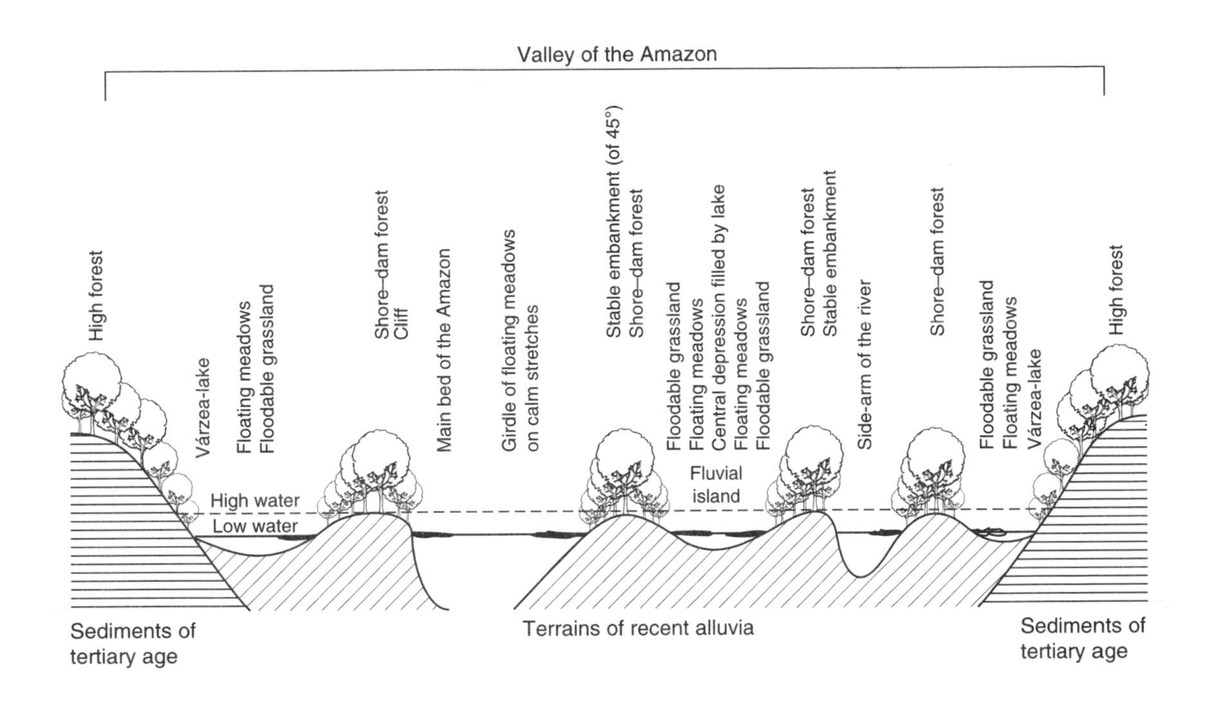

FIGURE 6.10 Profile of an Amazonian floodplain river, showing main channel, side arms, and extent of flooded forest. The amplitude from annual low to annual high water is approximatly 10 m. (Modified from Sioli (1964) and Goulding (1980).)

diet overlap in the wet compared with the dry season. Reduced food levels in the dry season evidently led to greater habitat and food specialization in that study, although it seems likely that severe food shortages could also cause diets to broaden (Lowe-McConnell, 1964).

To a greater degree than is seen in temperate zone studies, the proportional representation of guild categories in tropical rivers varies among sites. In the Rio Machado, a large, nutrient-poor clearwater tributary of the southern Amazon with extensive inundation forest, autochthonous plant food was sparse or absent and thus grazing on periphyton, plankton feeding and consumption of aquatic herbaceous vegetation were unimportant (Goulding, 1980). Quite a different result was obtained in a dry season study of feeding guild structure in nine small (1–6 m wide) Panamanian streams lying mostly in mature forest, but including some disturbed land (Angermeier and Karr, 1983). Seven feeding guilds incorporating 26 fish species were identified by cluster analysis of computed diet overlap. Mud feeders and planktivores were absent and algivores were well represented, in marked contrast to the Rio Machado and quite reasonable for small streams with primarily coarse substrate and well-defined channels. The distribution of guild biomass varied with habitat and stream characteristics, including a trend toward fewer

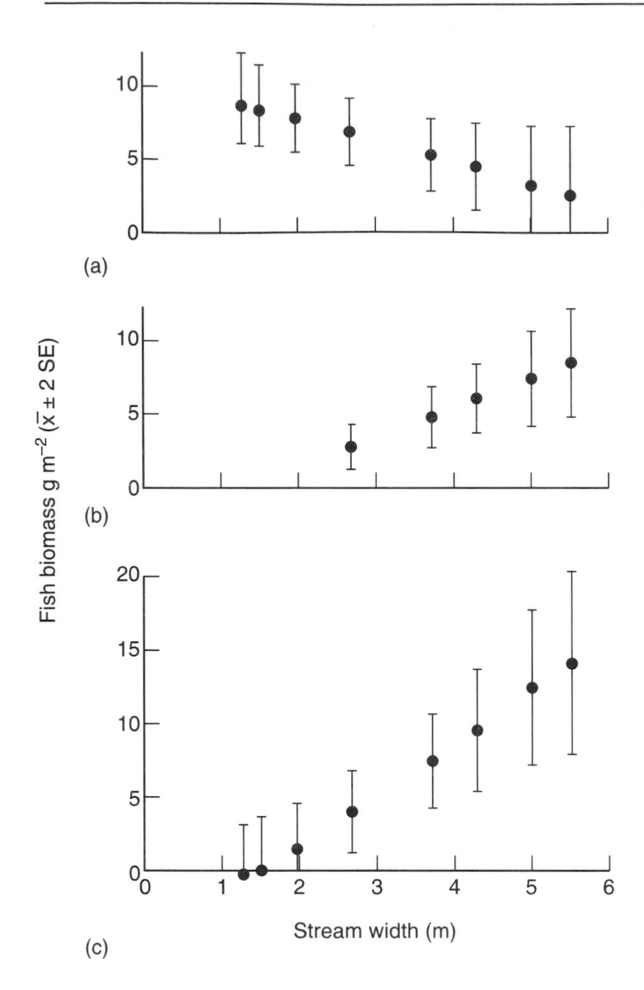

FIGURE 6.11 The abundance of three fish feeding guilds as a function of stream size in some small forested streams in Panama. (a) The generalized invertivores are represented by *Cichlasoma* (Cichlidae) and *Pimelodus* (Pimelodidae). (b) The detritivore *Brycon* (Characidae) is an invertivore when less than 80 mm, an omnivore at 80–130 mm, consumes terrestrial plant matter in pools when over 130 mm in length. (c) Catfish such as *Ancistrus* (Loricariidae) feed on periphyton attached to hard substrates. (From Angermeier and Karr, 1983.)

aquatic invertivores and more algivores with increasing stream size (Figure 6.11).

Differences in river size and flooding obviously contribute greatly to the differences among these tropical studies, and some generalizations can be made about longitudinal changes in fish trophic composition (Lowe-McConnell, 1987). In small, forested headwater streams, allochthonous inputs of aerial insects and plant matter may be of considerable importance, along with grazing of periphyton in sunlit areas and consumption of aquatic insects (e. g. Angermeier and Karr, 1983; Power, 1983). As one proceeds downstream, there may be a greater role for consumption of aquatic invertebrates, and still further downstream one finds the floodplain fauna with its preponderance of mud eaters. Inundation forests support a diverse assemblage of fruit and seed feeders but these are rare outside South America, perhaps because of past human modification of inundation forests in other parts of the world.

Although fishes of tropical streams exhibit a wide range of adult body sizes, there is a preponderance of small species relative to temperate streams (Figure 6.12), and this may affect the recorded variety of feeding roles. In the study of Angermeier and Karr (1983), most fish were less than 10 cm in total length, and small fish were disproportionately represented in a particular habitat (pools of smaller steams) and one guild (generalized invertebrate feeder). Goulding (1980) limited his efforts to fish over 10 cm, because of their importance to sustenance and commercial fishing, and this difference between the two studies must also have some effect on the outcome.

6.3.1 Specialized adaptations for particular feeding roles

Because guild classification relies as much on how food is acquired as on what is consumed, we might expect the morphological and physiological attributes of individual species to correspond to the specialization and diversity found in feeding roles. Indeed, the visual pigments, body shape and gut morphology of fishes provide evidence of such specialization.

River water varies in clarity, perhaps nowhere more evidently than in the Amazon basin. Whitewater streams are heavily colored by their alluvial loads, while blackwaters carry little silt but are darkly stained with dissolved material.

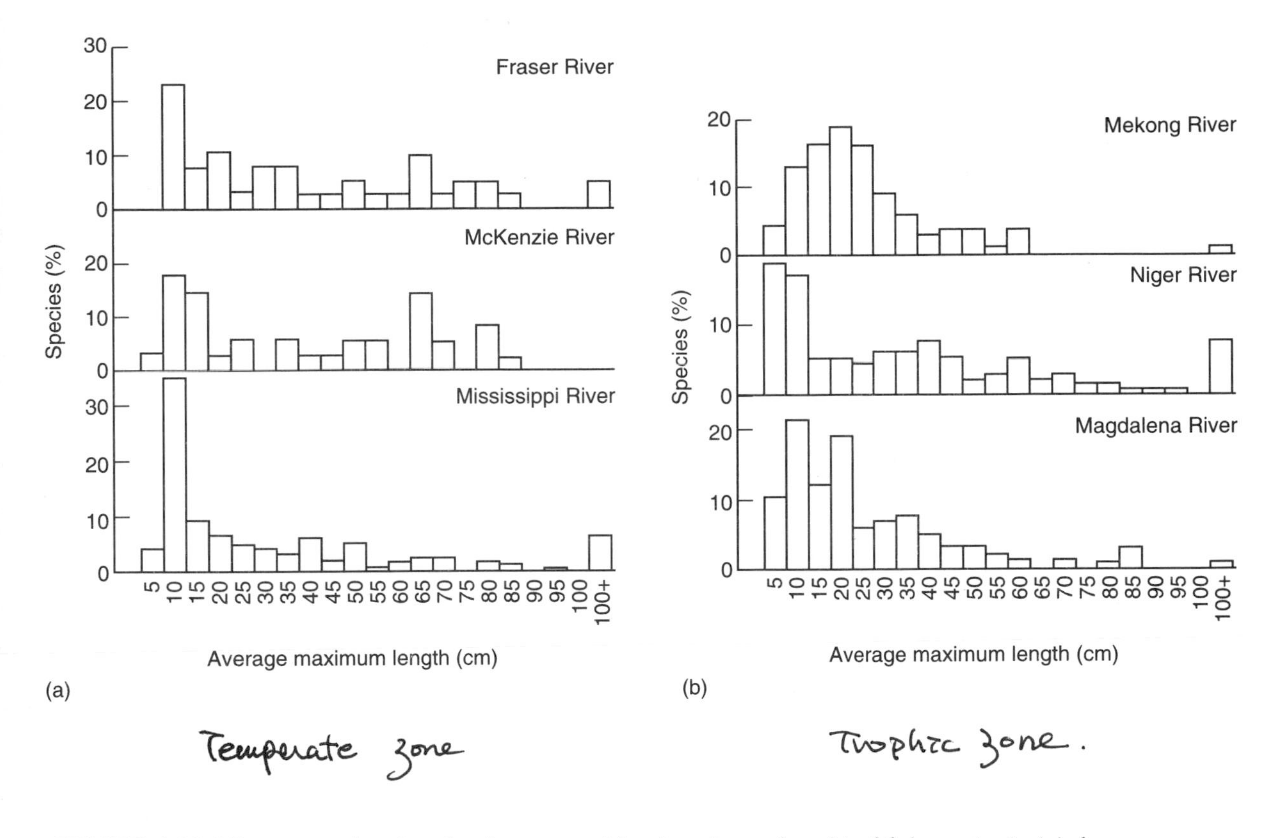

FIGURE 6.12 Histograms showing the size composition (maximum length) of fish species in (a) three temperate and (b) three tropical river systems. Tropical data from Welcomme (1979), temperate data compiled from Lee *et al.* (1980). The presence of large-bodied fishes in temperate rivers likely would appear more pronounced if European rivers were included.

Typical Secchi disk readings are less than 0.2 m in the former, 1–1.5 m in the latter (Muntz, 1982). Clearwater rivers carry comparatively little silt or dissolved organics and light penetration often equals or exceeds 4 m. These are markedly different visual environments. Absorbance of short-wavelength light is relatively great in freshwater, and more so as light penetration is reduced. Levine and MacNichol (1979) examined 43 species of (mostly tropical) freshwater fishes, and divided them into four groups on the basis of visual pigments. At one extreme fell species with strongly 'short-wave-shifted' visual pigments. These were primarily diurnal, and fed from the surface or in shallow waters. Several species exhibiting the typical behavioral and morphological characteristics of catfishes lay at the other extreme. Their visual pigments were the most long-wave sensitive; in addition they were primarily benthic and probably foraged either nocturnally or in very turbid waters.

Considerable effort has gone into attempting to interrelate the ecological role of fishes with their anatomical features, particularly for North American stream fishes. Gatz (1979) examined 56 morphological features of 44 species seined from North Carolina piedmont streams, calculated pairwise correlation coefficients among

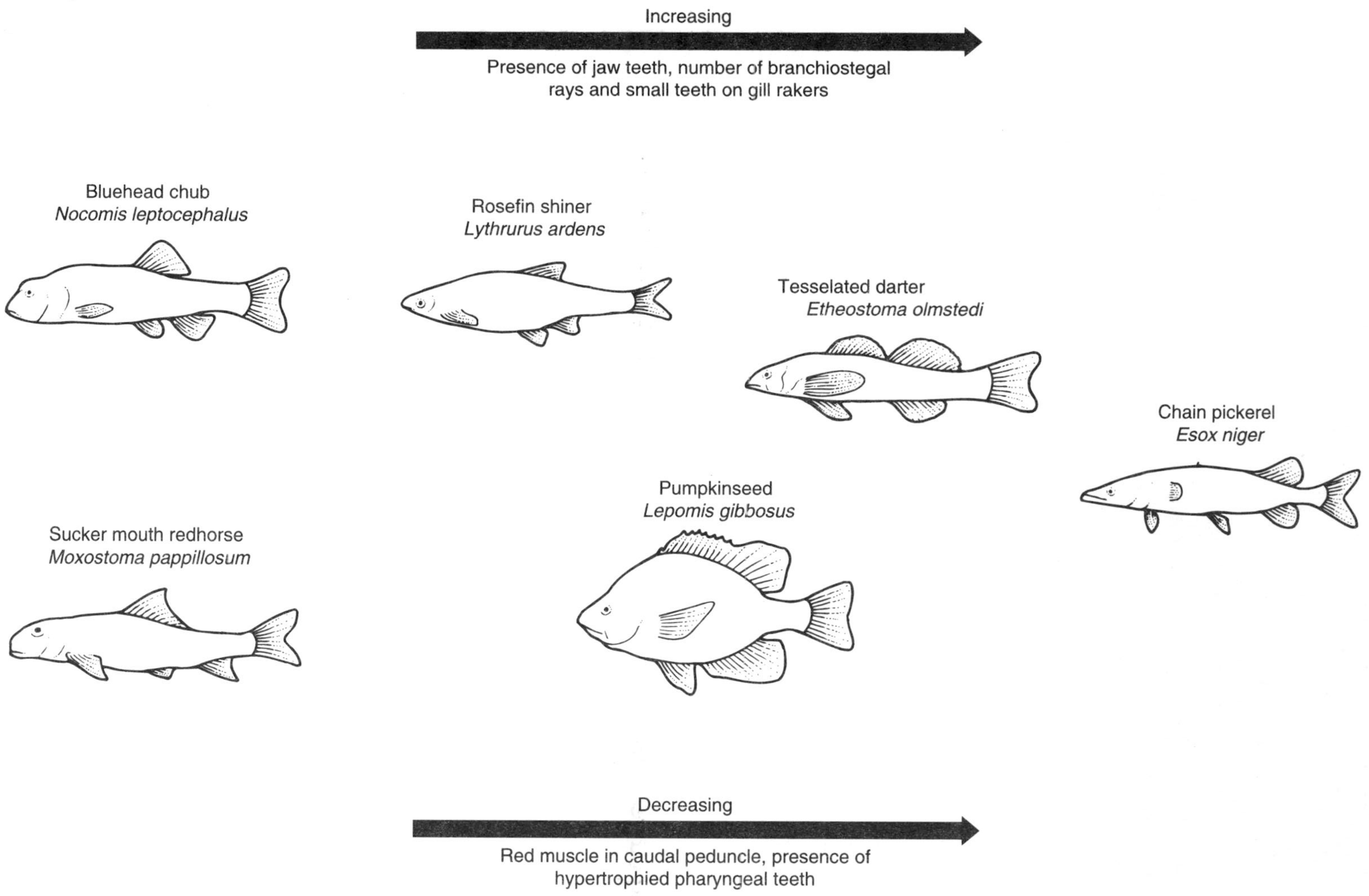

FIGURE 6.13 The ecomorphological gradation from 'lie-in-wait' ambush predators to cruising suction feeders. Pikes and pickerals lie at one extreme, benthic-feeding minnows and suckers at the other. This factor accounted for 31% of the variance in the morphological characters analyzed by Gatz (1979).

characters (3080 in total), and then used factor analysis to look for associations among characters. The first four factors together accounted for 60% of the variance in the correlation matrix. Factor 1 (Figure 6.13) separated 'lie-in-wait' biting predators from cruising suction feeders, Factor 2 reflected the differences in body shape and proportions associated with habitat use, Factor 3 separated a benthic from mid-water lifestyle, and Factor 4 separated small insectivores with short guts from other fishes.

Ecological correlates of these factors indicate that morphology does indeed influence or reflect diet and habitat preferences. Fish species with flat, deep bodies were associated with slow-water habitats. Fishes with ventral mouths obtained relatively more food from the bottom; those with terminal or anterior mouths did not. Fishes that dwell on or near the bottom in fastwater regions had reduced swim bladder volume, and relative gut length was greatest in mud feeders, to list some principal findings.

Other studies have also established an interplay between ecology and morphology. The degree of buoyancy control exhibited by fishes correlates well with lifestyle and current speed occupied (Gee, 1983; Matthews, 1985), and ecomorphological analyses of tropical fish assemblages (Moyle and Senanayake, 1984; Watson and Balon, 1984) have successfully related food and habitat partitioning to differences in body plan. It is important, however, not to overlook the limitations of an ecomorphological analysis. Only about 2% of the 3080 character correlations examined by Gatz had coefficients of correlation that explained over 50% of the variance.

In sum, attempts to strictly categorize fishes into feeding guilds inevitably bring to light the limits of any such approach. Diet changes with age because of growth in size, and season because of shifts in availability of habitat and food types. Even a single individual at a particular time and place may fall into more than one feeding category. In practice, guilds are a useful way to summarize the broad similarities in the trophic ecology of taxa that have similar feeding roles, as long as their use does not obscure the individual differences and great flexibility of which fish are capable. Because of differences in resource availability between temperate and tropical regions their fish guilds also differ, particularly in the greater roles of detritivory and ooze feeding in the tropics. Moreover, guild structure appears less uniform from place to place among river-dwelling fishes of tropical rivers, compared with temperate zone studies.

6.3.2 The trophic role of other vertebrates in running waters

Amphibians, reptiles, birds and mammals all are represented in lotic food webs. Unfortunately, as Hynes (1970) pointed out, our knowledge of the feeding role of these groups is scant and consists mostly of natural history reports. The taxa most likely to have a significant impact on riverine food webs probably are the fish-eating snakes, birds, and crocodiles, but perhaps this reflects lack of knowledge concerning other groups.

Salamanders can attain large size and high population biomass. *Megalobatrachus* of the Far East, *Cryptobranchus* (the hellbender) and *Necturus* (the mud puppy) of eastern North America, and *Dicamptodon ensatus* of the Pacific Northwest of North America are large-bodied taxa (Hynes, 1970; Nickerson and Mays, 1973). They are carnivores of invertebrates, other amphibians and fish. Small salamanders may be the principal vertebrate predators in headwater streams. Petranka (1984) concluded that larval two-lined salamanders (*Eurycea bislineata*) were opportunistic generalists, consuming a variety of insect larvae and crustaceans.

Frogs and toads do not appear to be very important in stream food webs. However, the tadpoles of some species are grazers in small streams, and a few possess powerful suckers that provide attachment and allow leech-like maneuverability (Hynes, 1970).

Reptiles that feed in rivers include the Crocodylia, many families of snakes but especially the Colubridae (water snakes), and the Chelonia (terrapins). The latter are omnivores of sluggish streams and rivers, and consume substantial amounts of invertebrate and fish prey. The former two groups are predators of fish and invertebrates; indeed, even young crocodiles feed primarily on invertebrates until a length of about 2 m (Corbet, 1959, 1960). Hynes (1970) remarks that considering the abundance of crocodiles along tropical rivers where exploitation has not reduced their numbers, they are likely to have a major impact on lower trophic levels.

At least 11 orders of birds make use of rivers and streams as feeding habitat (Hynes, 1970). Many are fish predators but some feed directly on invertebrates (e. g. the Cinclidae or dippers; Ormerod, 1985). Diving ducks also consume significant amounts of invertebrates, especially mollusks, although submerged aquatic vegetation is their primary food (Perry and Uhler, (1982) and ducks can be important consumers of macrophytes (Lodge, 1991). In some circumstances, birds may be as important as fish in their impact on prey assemblages. Brown trout in the Ausable River, Michigan, were consumed by a variety of predators including larger conspecifics, two species of mammals and three species of birds. The great blue heron and large trout accounted for the greatest amount of natural predation, while mink, otter, merganser and kingfisher each consumed an appreciable fraction of small trout (Alexander, 1979). Armored catfish in Panamanian streams experience significant predation risk from fishing birds (Power, 1984b). This causes larger individuals to avoid shallow waters, and because these fish are effective scrapers, the depth-distribution of periphyton inversely mirrors the distribution of fish.

A diversity of mammals feed within running water (Hynes, 1970). Taxa ranging from shrews to racoons to bears occasionally or frequently consume invertebrates and fish. Others such as the river otter (*Lutra canadensis*) are fully aquatic and feed almost entirely on aquatic resources. The duckbill platypus (*Ornithorhynchus anatinus*), a nocturnal hunter in Australian rivers, possesses electroreceptors capable of detecting the muscle activity of invertebrate prey (Scheich *et al.*, 1986). Very large river-dwelling mammals include the plant-eating manatees of South America and west Africa (Campbell and Irvine, 1977), and dolphins which feed on invertebrates and fish. The latter usually are temporary visitors, as in the Amazon, but one species is resident in large rivers of India (Hynes, 1970).

Although different vertebrate predators are capable of a variety of hunting tactics, most are morphologically constrained to hunt primarily by wading, diving or swimming. In addition, they are most effective within certain parts of the stream channel (Power, 1987). Wading birds typically fish in water no deeper than 20–30 cm. Leg length and striking distance must limit their success at greater depths. Diving and skimming predators such as kingfishers and bats usually fish very close to the surface, although kingfisher dives to depths of 40 cm are not unknown (Power, 1984b) and mergansers fish at depths of meters or more. Swimming predators typically fish at greater depth, either to minimize their own risk of predation or, especially if they are of large body size, to have more room to maneuver. The need to capture and swallow prey generally results in a rough correspondence between prey size and predator size, even in species able to extend their gapes or rend prey into pieces. The combination of a predator's depth range and size range may significantly affect the size and depth distribution of fishes in streams, and perhaps affect other members of the biota as well (Power, 1987). Indeed, many vertebrate predators may have their impact on riverine communities by influencing the foraging location of their prey. As we shall see in subsequent chapters, the consequences can ramify widely through the food web.

6.4 Lotic food webs

No complete food web exists for any stream, although a number of partial webs exist (Jones, 1949, 1950; Fryer, 1959; Cummins, Coffman and Roff, 1966; Minshall, 1967; Mann *et al.*, 1972; Koslucher and Minshall, 1973; Kuusela, 1979; Hildrew, Townsend and Hasham, 1985). All are based on ingestion rather than assimilation, and all illustrate the complexity of the undertaking. These food webs generally are based on species rather than guilds, and so the potential simplifying power of that approach is under-utilized. Figure 6.14a presents a somewhat abstract version of a food web in a woodland stream, based on extensive studies of Linesville Creek, Pennsylvania (Cummins, Coffman and Roff, 1966; Coffman, Cummins and Wuycheck, 1971).

Obviously it is difficult to include every species in a web, but usually the important ones are included and so the taxa omitted may not matter greatly in terms of energy flow. The food webs of Deep Creek, Idaho–Utah (Koslucher and Minshall, 1973) and Broadstone Stream, southern England (Hildrew, Townsend and Hasham, 1985, Figure 6.14b) are among the most detailed on record. The Broadstone food web is unusually complete because the acid and iron-rich conditions at this locale result in a simple community. Nonetheless, two difficulties are well illustrated. Although the Chironomidae are separated into species, oligochaetes and other microinvertebrates are not, and so further taxonomic resolution is needed before a truly complete picture is obtained. Second, the base of the web is very difficult to characterize. The farther up the food web one goes, the easier it becomes to describe in detail.

Recent developments in food web theory (Briand and Cohen, 1987) and compilation of food webs from many aquatic and terrestrial ecosystems permit us to ask how riverine webs compare with those from other environments. Briand and Cohen (1987) report shorter food chains from 'two-dimensional' environments (e. g. lake bottom, tundra or stream bed) compared with those that are 'three-dimensional' (e. g. lakewater column or forest canopy). However, such analyses tend to be based on webs that are simplified by lumping together species, and a good deal of detail may be ignored, especially near the base of the web. It is difficult to know the extent to which these conclusions are biased by differences in the degree of simplification. In the Broadstone web, unusual for its detail, Hildrew and colleagues find that both interconnectance and omnivory are greater than expected on theoretical grounds.

6.5 Summary

A great deal now is known of the feeding ecology and trophic linkages of the biota of running waters. Consumers can be assigned to one of a number of guilds or functional groups, defined as those species that eat the same food items and obtain their food in the same way (such as suspension-feeding fine particle collector, or benthic invertivore). The guild approach lends organization to the diverse and overlapping resource consumption that characterizes lotic food webs. It should also be kept in mind that guilds are only useful abstractions. This is partly because many consumers are much too versatile to fit neatly into one feeding role, and partly because energy resources of streams and rivers also can be difficult to sort into distinct categories. Microbes capable of incorporating DOC exist in intimate association with cells of small autotrophs capturing the energy of sunlight, and decomposing terrestrial leaves can become intermixed with detrital material from other sources. To understand the trophic ecology of running waters fully, we need both the detailed investigations of energy transfer that show us how complex and diffuse trophic relationships truly are, and the simplifying power of guild and web constructs to provide the conceptual framework so necessary for comparison and analysis.

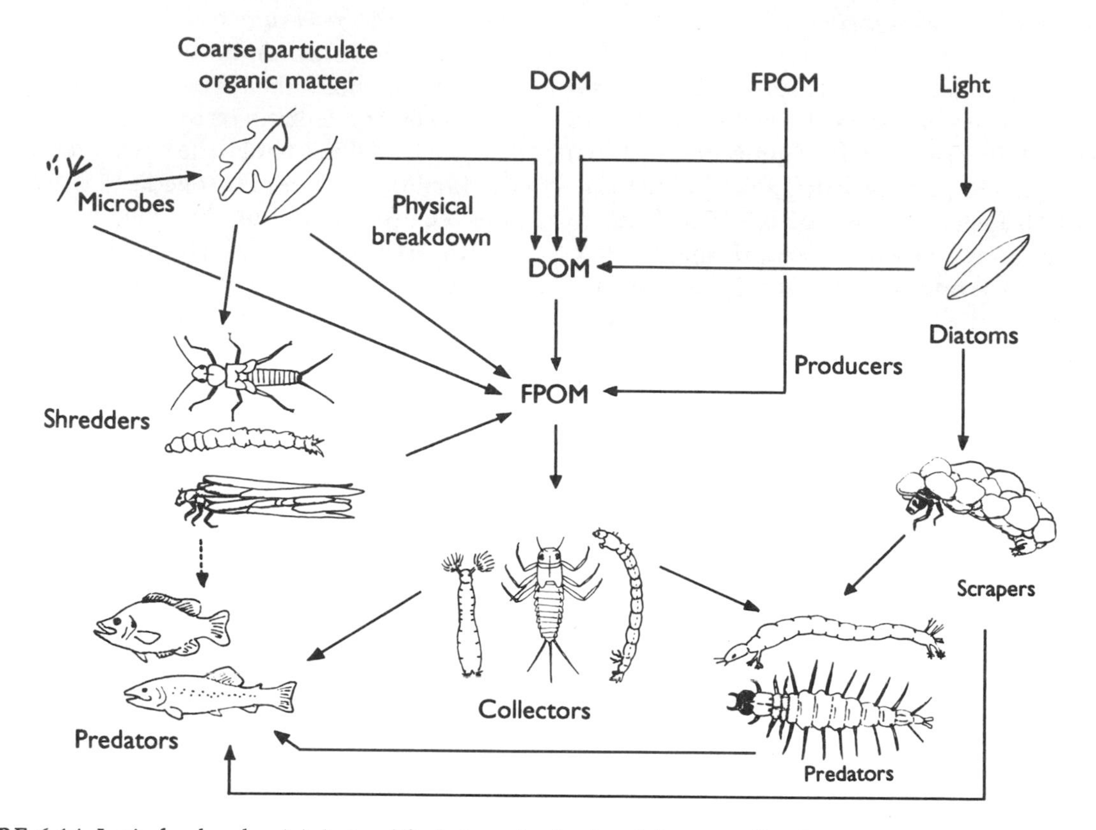

FIGURE 6.14 Lotic food webs. (a) A simplified view of a food web in a woodland stream. Energy inputs include fallen leaves, subsequently colonized by microbes; small autotrophs, primarily diatoms; and DOM and FPOM, originating from external sources and upstream. Feeding categories are based on divisions of Table 6.1: shredders include *Pteronarcys*, *Tipula* and *Pycnopsyche*; *Stenonema* is a deposit feeder, *Simulium* is a filter feeder and *Glossosoma* is a grazer. Examples of predators include *Nigronia* (Megaloptera) and two fish (*Cottus* and *Salmo*). (Modified from Cummins, 1973.) (b) Food web for a species-poor small stream in southern England. Primary consumers include: (e) *Psidium* sp., (f) Simuliidae, (g) *Niphargus aquilex*, (h) microcrustacea, (i) other microinvertebrates, (j) *Heterotrissocladius marcidus*, (k) *Micropsectra bidentata*, (l) *Prodiamesa olivacea*, (m) Oligochaeta, (n) *Leuctra nigra*, (o) *Nemurella picteti*, (p) *Brilla modesta*, (q) *Polypedilum albicornis*, (r) Tipulidae, (s) *Potamophylax cingulatus*. Predators include: (t) *Macropelopia goetghebueri*; (u) *Trissopelopia longimana*, (v) *Zavrelimyia barbatipes*, (w) *Plectrocinemia conspersa*, (x) *Sialis fulginosa*. Note that the predator *Sialis* can be four energy transfers removed from the base of the food web. (Modified from Hildrew *et al.*, 1987.)

Most of this chapter has emphasized the role of larger consumers, mainly the macroinvertebrates and fish, whose feeding ecology is reasonably well understood. These guild categories are described in Tables 6.1 and 6.4, respectively. In recent years it has become apparent that microbial transformations are of great importance to the flux of organic carbon in running waters. The contribution of fungi to the breakdown and the nutritional value of CPOM is well established, and attention now is shifting to the role of bacterial production and microbial food webs. Much remains to be learned of the processes involved, and whether this energy contributes to metazoan food chains. Regardless, the microbial loop clearly is an important internal mechanism governing energy transformations and the recycling of nutrients.

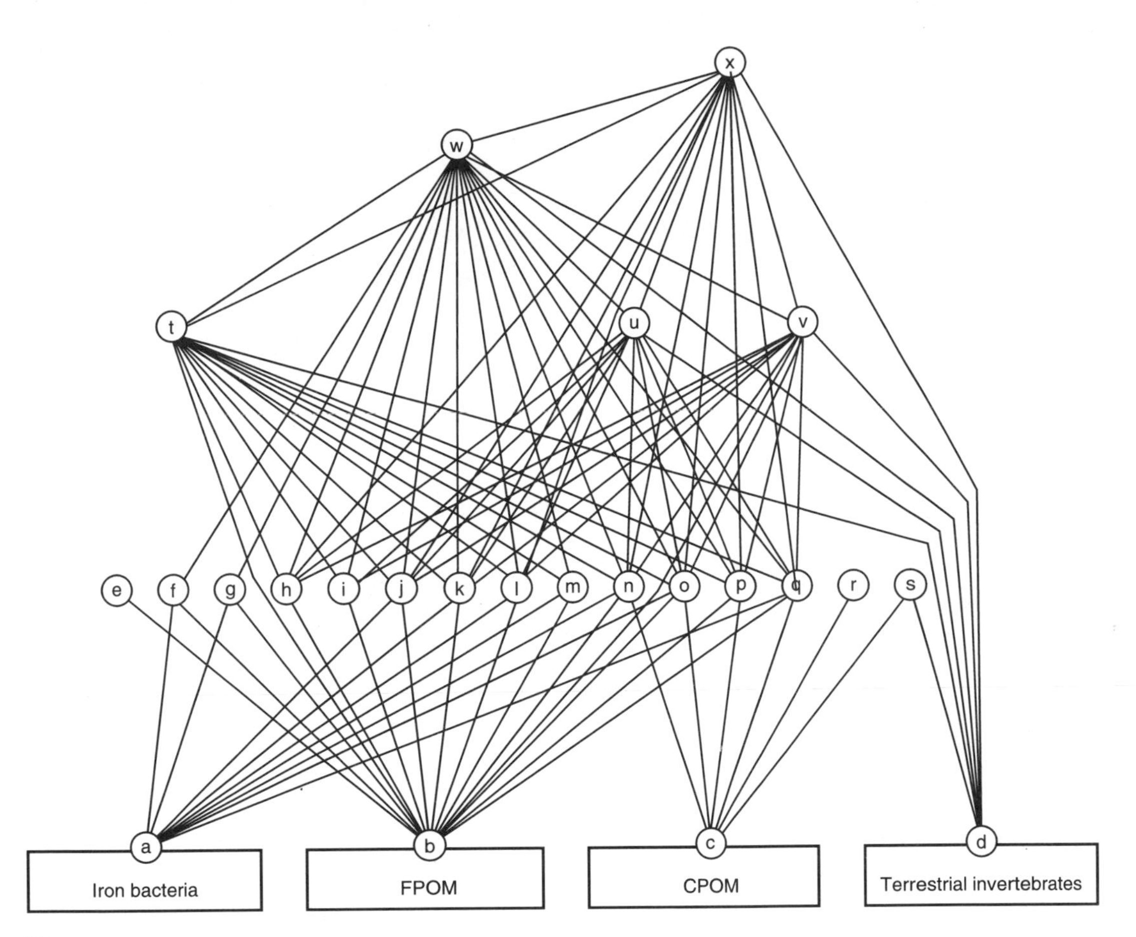

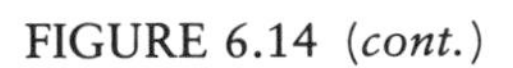

FIGURE 6.14 *(cont.)*

Chapter seven

Predation and its consequences

Predation is ubiquitous. All organisms potentially are prey for others at some stage of the life cycle, and many species encounter predation risk throughout their lives. Furthermore, the potential effects of predation are numerous. They include reduction in abundance or even the elimination of a species from a region, indirect effects on habitat use and foraging efficiency, a potential cascade of interactions through the trophic web, and adaptation via natural selection to persistent predation risk. The latter refers to predation past, while the first three items primarily refer to predation present. They will be discussed separately but with an effort to avoid any false sense of their independence. Prerequisite to evaluating each of these categories of predator effects is some consideration of why predation falls more intensely on some individuals or species relative to others. Thus we begin with the two components of what often is called preference or selectivity, choice for some prey over others on the part of the predator, and differences among prey that influence their relative vulnerability.

7.1 Choice and vulnerability

At a broad level of categorization all predators show preference, feeding mainly on fish, invertebrates or plant matter. However, as finer and finer distinctions are made among prey types, predators appear less discriminating in their diets and omnivory more widespread. Aspects of the predator that bias it towards consuming more of some prey than others include sensory capabilities, foraging mode and behavioral mechanism of prey capture. For prey, many aspects of lifestyle and body plan influence their vulnerability. These traits of predator and prey are not easily separated. We will begin by discussing the predatory behavior of vertebrate and invertebrate predators, and then examine how characteristics of the prey influence their vulnerability to each.

7.1.1 Vertebrate predators

All major groups of vertebrates include some that are predators in flowing water (chapter 6). Typically, however, fish are the major vertebrate predators. They exhibit immense diversity in body plan and feeding mode, especially the tropical fish fauna (Goulding, 1980; Lowe-McConnell, 1987b). Most feeding studies have been conducted in the temperate zone where the fish are mainly invertivores (Allen, 1969). The commercially important salmon and trout have received the most detailed study, a bias reflected herein.

Fish commonly utilize vision, smell, taste and touch in the detection of prey (Hyatt, 1979;

Ringler, 1979b). Visually dependent predation is widespread and typifies the salmonids. Although trout are able to detect prey at quite low light levels, even capturing surface-drifting ants against a starry sky (Jenkins, 1969), the effectiveness of such predators is reduced by dusk conditions (Ginitz and Larkin, 1976). Other sensory modes may be of greater use under conditions where vision is less effective. Olfaction enables the bluntnose minnow, *Pimephales notatus*, to discriminate among some aquatic insects (Hasler, 1957). Thousands of external taste buds distributed over the body and especially on the tactile barbels of the bullhead, *Ictalurus*, allow it to locate prey at up to at least 25 fish lengths, the maximum distance studied. Individuals performed true gradient search and, if barbels and taste buds were made inoperative on only one side, fish circled toward the intact side, showing that they compared concentration gradients (Bardach, Todd and Crickmer, 1967). In addition, the lateral line may aid in the detection of active prey (Alexander, 1970), and electroreceptors are known in the electric eels, knifefishes and elephantfishes, some catfishes and the duckbill platypus, all of which feed in murky waters.

Accompanying this diversity in sensory modality, fish show considerable variety in morphological modifications of jaws, teeth, gill arches and fins that influence capture and handling of prey (Ringler, 1979b). Ability to create suction with the buccal cavity and to extend the mouth into crevices are further adaptations that influence diet. Fish inhabiting rapid water tend to be specialized for either a midwater or bottom-dwelling life (Allen, 1969). The former are well streamlined and active swimmers such as the salmonids and some cyprinids. The high percentages of red muscle in their axial musculature (Gatz, 1973) bears witness to their ability for sustained cruising. Bottom dwellers such as sculpins and darters are usually depressed vertically, may have a flattened ventral surface, and possess enlarged pectoral and sometimes pelvic fins for maneuvering. Their low amount of red muscle is indicative of short, anaerobic bursts of swimming alternating with a relatively sedentary style. These differences in morphology influence where and how different fish species feed. For instance, the minnows (*Notropis*) of a small, midwestern US stream included midwater species that fed primarily from the drift, and bottom-feeding species that utilized the benthos directly (Mendelson, 1975). Some, such as the rainbow trout *Oncorhynchus mykiss*, are capable of feeding from either the benthos or the drift (Tippets and Moyle, 1978).

Comparison of what is in the stomach to what is (apparently) available in the environment has been a traditional and useful approach to the investigation of prey choice in fish. Apart from some refinements in the quantification of gut fullness (Hyslop, 1980) and in the choice of indices that compare available to consumed prey (Chesson, 1983), this method has changed little over at least 50 years. Allen (1941) constructed an 'availability factor' to assess preference in the diet of young *Salmo salar* in two Scottish rivers studied between 1935 and 1938. This index expressed the ratio between the average percentage that a prey type contributed to the stomach contents of young salmon, and the percentage value of that prey in collections of the fauna. Activity, degree of exposure, size, and presence or absence of a hard outer covering affected prey availability. Diet composition largely resembled the composition of the fauna, except that common items were over-represented (Figure 7.1). As Allen noted, however, it is difficult using field studies to distinguish true selection from availability. Laboratory experiments are better for dissecting mechanisms, and much has been learned about predatory behavior from studies using salmon and trout.

(a) Factors affecting predatory behavior

The number of prey eaten increases with prey abundance, but at a decelerating rate due to the time limitation imposed by the handling and

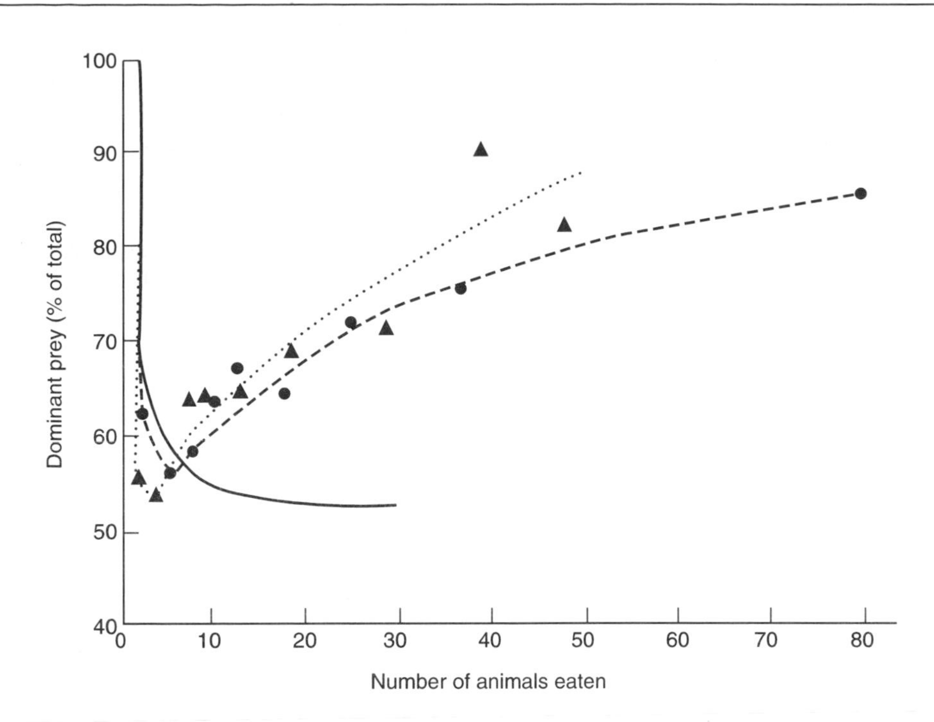

FIGURE 7.1 The percentage of total stomach contents that the dominant prey item constituted in the diet of young Atlantic salmon in two Scottish rivers. Dotted line, River Thurso; dashed line, River Eden; solid line, theoretical expectation from random feeding. (From Allen, 1941.)

ingestion of individual prey. This relationship is known as a functional response curve (Figure 7.2). Whenever more than one type of prey is present, choice is strongly influenced by contrast, motion and size, all of which serve to make certain prey more conspicuous. Ware (1973) tested against independent field data a mechanistic model of preference that included prey activity, exposure, density and size. This model was able to account for 47% of seasonal variation in prey preference in a four-species system consisting of amphipods, caddis and odonate larvae and planorbid gastropods. After exclusion of the gastropods, because of their poor fit to the model, Ware was able to explain 70% of seasonal variation in the diet of rainbow trout in the field. Prey motion was the most important single influence on predation risk, but that result may depend on whether feeding is epibenthic as in Ware's study, or from the water column.

Numerous studies have established that predation intensity increases with prey size (Figure 7.3; Ware, 1972; Metz, 1974; Allan, 1978; Ringler, 1979a; Newman and Waters, 1984). Visual predators including trout will orient toward a prey that they have sighted, and this reaction distance increases with prey size (Ware, 1972). The lower limit of prey size to which young Atlantic salmon react is very close to the mean gill raker spacing, which likely is a good approximation of the lower size limit for retaining prey (Wankowski, 1979). The maximum gape of the mouth is about 1.5 times the internal breadth, but during prey capture young salmon open their mouths at a 60° angle, at which the two major dimensions are about equal and so mouth breadth adequately characterizes gape. These minimum and maximum dimensions set broad limits on the size range of prey ingested. The fact that both increase linearly with prey

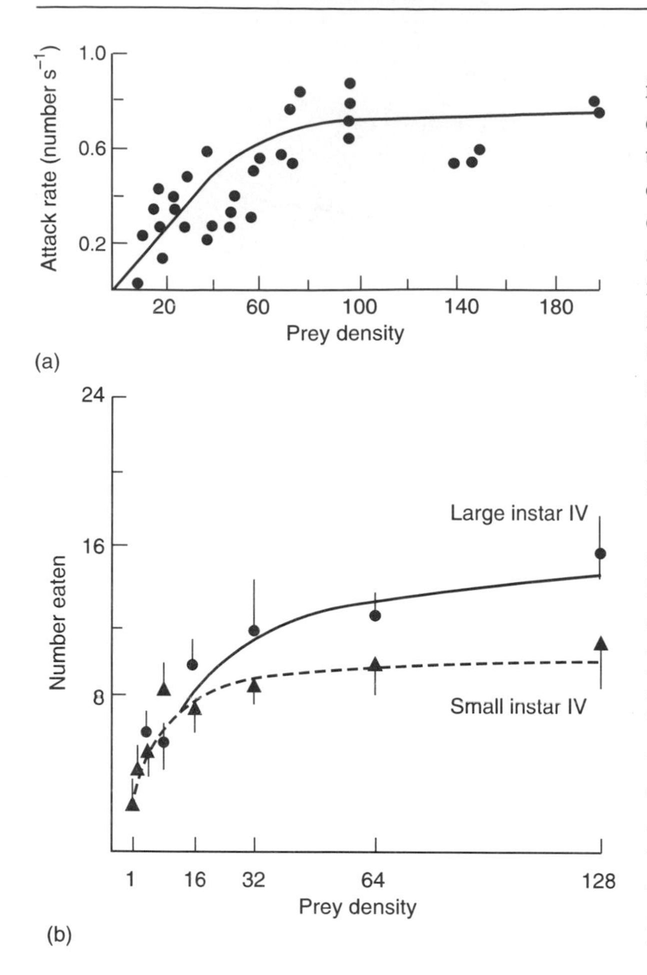

FIGURE 7.2 Functional response curves, or number of prey consumed per predator within a restricted time period, of two predator species over a range of prey densities. (a) Rainbow trout feeding on the amphipod *Hyallela* in aquaria, from Ware (1972). (b) Number of mosquito larvae eaten in 3 h by a single IVth instar *Notonecta hoffmanni* (means ± 2 S.E.). Solid curve depicts large instar IV predators, broken line depicts small instar IV predators. (From Fox and Murdoch, 1978.)

length (Figure 7.3b) may account for the frequent observation that prey size and diet breadth increase with fish size (e.g. Allen 1941; Cadwallader, 1975). This result is most apparent with fish of small size because larger invertivores rarely encounter prey that are larger than their gape.

With experience, predatory behavior changes in a number of ways. Ware (1971) found that on average 4 days elapsed before naive rainbow trout would approach a novel prey (blanched cylinders of chicken liver); thereafter the distance of attack increased and after 12 days it had about doubled. In addition, the time required for prey consumption declined. Ringler (1979a) found that brown trout preferred large prey (mealworm *Tenebrio molitor*) over small (brine shrimp *Artemia salina*); however, this preference developed gradually over 4–6 days and the least preferred prey never was completely excluded from the diet (Figure 7.3c).

Changes in fish predatory behavior due to experience result in higher rates of predation. Searching often increases via greater reactive distances, higher swimming speeds and greater path efficiency, while attack latency may decrease and capture success may increase (Dill, 1983). The result is a tendency to specialize on the prey which the predator has consumed most frequently in its recent feeding history, with an accompanying increase in foraging efficiency. In the laboratory this is termed a 'training bias' but there is evidence that it occurs naturally. The frequency of the dominant prey item consumed by young Atlantic salmon increased as the

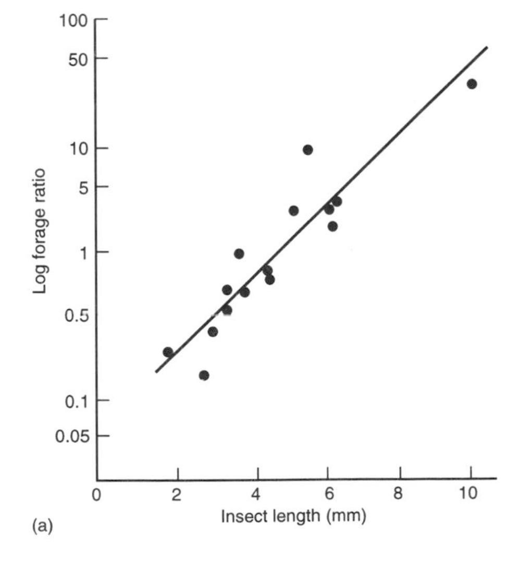

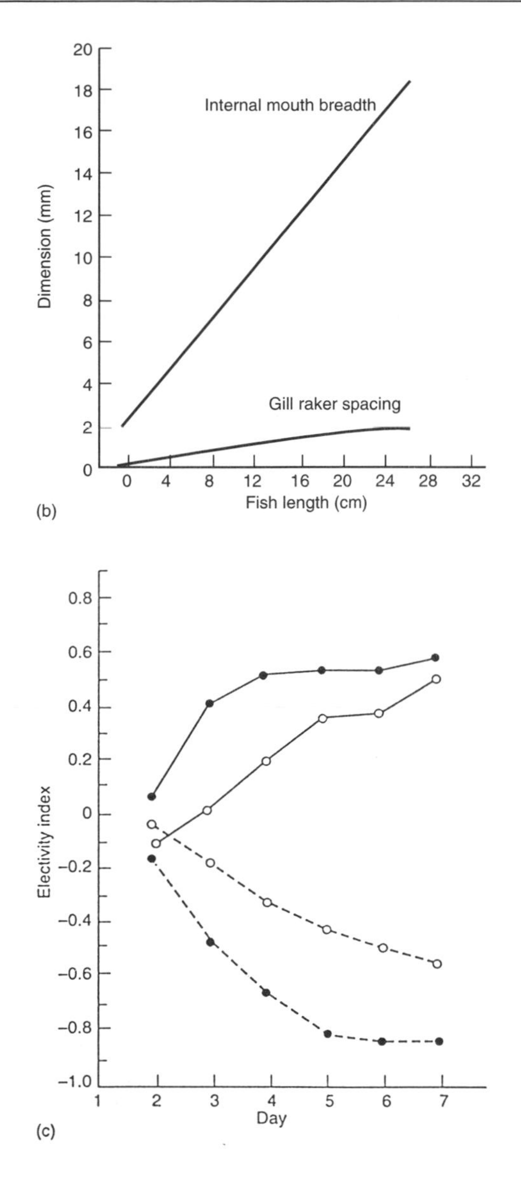

FIGURE 7.3 The size preference of trout for large prey. (a) Rainbow trout consuming surface drift in a German stream, mainly emerging insects and adults of Ephemeroptera and Chironomidae. (From Metz, 1974.) (b) Internal mouth breadth and mean gill raker spacing as a function of fish length in young Atlantic salmon. (From Wankowski, 1979.) (c) Electivity (Ivlev, 1961) of drift-feeding brown trout in a laboratory stream. Wild trout were maintained on a diet of brine shrimp. In this experiment, brine shrimp (dashed lines) only were provided on Day 1, and the larger mealworms (solid lines) were added on Day 2. Drift rates were 5 (○, low) and 10 (●, high) per minute. (From Ringler, 1979b.)

number of prey items per stomach increased (Figure 7.1), suggestive of learning (Allen, 1941). Individually marked wild trout that were sampled over a 6 month period by flushing their stomachs and returning them unharmed to the stream exhibited consistent individual specialization (Bryan and Larkin, 1972), also indicative of a learned specialization. Variation between individual trout in their feeding response has been noted frequently (e. g. Allan, 1981; Ringler, 1983), and while environmental heterogeneity undoubtedly is part of the cause, the effect of experience on subsequent behavior also appears to be important (Dill, 1983).

Hunger can influence predation rate by modifying any of several aspects of predatory behavior (Beukema, 1968; Dill, 1983). As hunger decreases, apparently sensed through stretch receptors in the gut, searching decreases owing to changes in movement speed and reactive distance. In addition, the probability that an attack will follow an encounter declines and handling time increases (Ware, 1972). Capture rate consequently varies with hunger level.

(b) Environmental variables

Although visually dependent predators can feed under quite dim light and some trout detect prey by starlight (approximately 0.001 lux; Jenkins, 1969; Robinson and Tash, 1979), prey capture success declines with falling light levels. A light intensity of 0.1 lux, corresponding to late dusk, often is the lower threshold for effective visual location of prey (Hyatt, 1979). Even within the range we consider daylight, however, gradation in light level can be influential. Wilzbach, Cummins and Hall (1986) compared the feeding of cutthroat trout in pools from forested sections of streams with pools from open (logged) sections. Prey were captured at higher rates in open pools, and artificial shading lowered capture rate to that of shaded pools (Figure 7.4).

Predation intensity increases with temperature, within normal bounds, because warmer temperatures allow higher metabolic rates and

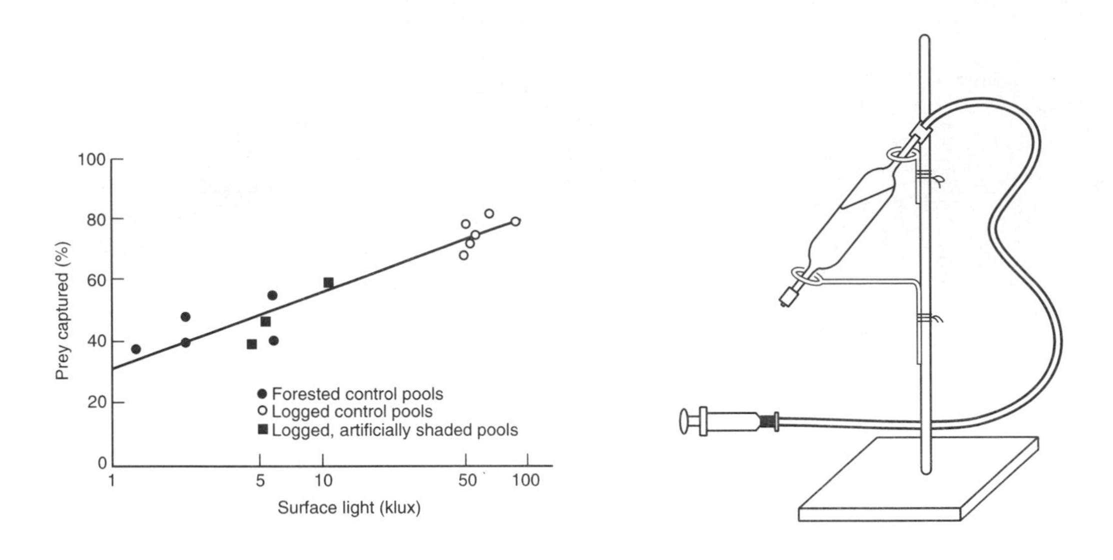

FIGURE 7.4 The percentage of prey captured by cutthroat trout (*Oncorhynchus clarki*) as a function of light intensity at the water surface. Prey were mosquito larvae introduced with a syringe apparatus and captures were recorded by underwater observation. (From Wilzbach, Cummins and Hall, 1986.)

cause fish to require larger rations for maintenance and growth (Elliott, 1976). As a consequence, diel and seasonal changes in water temperature can both result in considerable variation in predatory activity. Young Atlantic salmon show a pronounced seasonal (August–September) reduction in size of the area around the head of the fish within which it reacts to the presence of prey. Apparently this is an adaptation to the reduced energy demands associated with normal autumn temperatures (Wankowski, 1981).

Habitat characteristics can influence predation intensity in a number of ways. The type of prey available is habitat dependent, and the availability of cover for predators that themselves face risk of predation will influence the area searched. Especially for epibenthic predators, however, capture rate declines with increasing complexity of the substrate (Ware, 1972). For example, sculpins feeding in the laboratory were able to capture prey more readily from a sandy bottom than from a more heterogeneous cobble habitat (Brusven and Rose, 1981).

(c) Foraging behavior and energy maximization

Any demonstration of non-random feeding behavior leads to the question of whether the predator's rate of return is greater than would be anticipated under random feeding. Brown trout offered a mix of brine shrimp and mealworms gradually improved their performance over 7 days, by which time their energy intake was over three times the prediction from random feeding (Ringler, 1979a; Figure 7.3c). They achieved between 54 and 91% of a hypothetical maximum diet, falling short because brine shrimp were never completely excluded from the diet.

The main reason for non-random feeding being adaptive is the difference in energy reward among prey, commonly expressed as:

$$\text{Energy gain} = \frac{\text{Dry mass or calories obtained}}{\text{Energy expended in prey acquisition}} \quad (7.1)$$

where energy expended is usually quantified by handling time, or handling time plus search time. As prey size increases, the energy content of the prey at first rises more rapidly than does the cost of acquisition, but larger prey eventually become less profitable because of the increased capture effort. Prey profitability thus should increase with prey size up to some optimum, and then decline. This has been confirmed experimentally by Wankowski and Thorpe (1979), who reared young Atlantic salmon on pellets of various sizes. Expressed as the ratio between particle diameter and fish fork length (PFR), maximum growth was obtained in the range 0.022–0.026 PFR. The optimum ratio was constant for fish from 4.2 to 20.3 cm, therefore optimum prey size increased with fish size. Additionally, the reaction distance of fish to pellets, and the striking distance (maximum distance traveled), both showed a curvilinear response to increasing values of PFR (Figure 7.5). Wankowski (1979) concluded that the reduction in striking distance at PFR below the optimum could at first be explained on visual mechanics alone, but that with even smaller prey (less than 0.018 PFR), negative selection was operating. Fish feeding on larger than optimum prey retained maximal reaction and striking distances up to a PFR of 0.051. However, rejection of prey increased sharply over this size range.

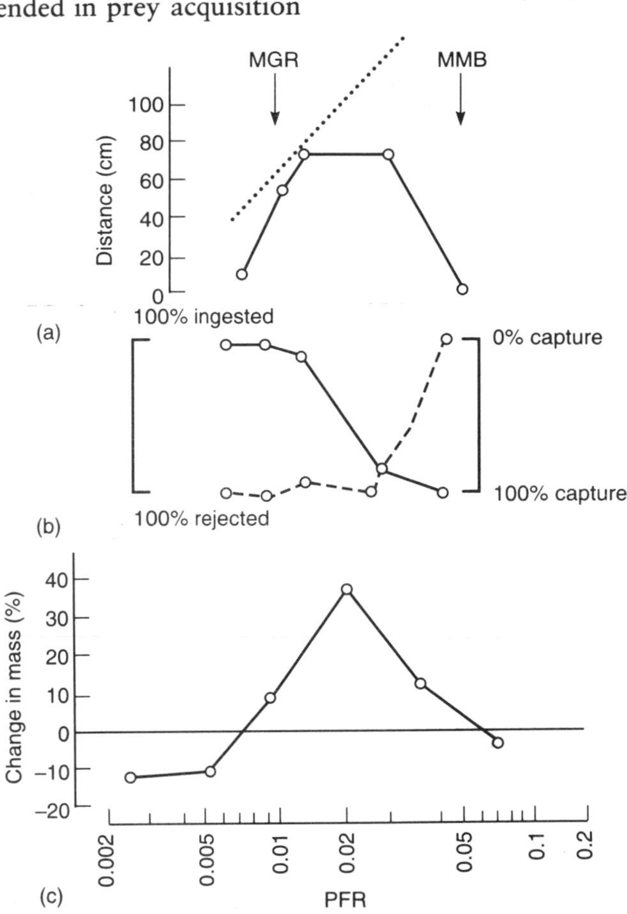

FIGURE 7.5 Feeding behavior, selectivity and growth in Atlantic salmon of mean length 8.6 cm in June, as a function of the ratio of prey diameter to fish fork length (PFR). (a) MGR, mean gill raker spacing; MMB, mean mouth breadth; expected response distance (dotted line); striking distance (solid line). (b) Capture success and percentage of prey rejected after striking. (c) Growth, measured as percentage change in mass. (From Wankowski, 1979.)

This model cannot be applied to nature very exactly because real prey are not spherical. Nonetheless, Wankowski's re-analysis of Allen's field data for 4–13 cm Atlantic salmon indicated good correspondence. However, it seems likely that for stream-dwelling salmonids over 12–15 cm in length, the optimal prey size is larger than what normally is available. Prey shaped like cigars but with the same area as 0.026 times fish fork length will exhibit a prey length:fish length ratio of about 0.05–0.06. For an 8 cm salmon, the optimal prey would be 4–5 mm long. Allen's field data suggested that the prey length:fish length ratio declined in salmon larger than 7.5 cm (Wankowski and Thorpe, 1979), which might be due to the unavailability of larger prey. Bannon and Ringler (1986) measured the length distribution of drift in the early morning in a New England stream, for comparison to brown trout diet. Few drifting prey were longer than 4 mm, and most were smaller. The size categories were: less than 1 mm, 1–2 mm, 2–3 mm, and over 4 mm and the respective frequencies were 2–20%, 30–60%, 20–30% and 5–18%. Wilzbach, Cummins and Hall (1986) present similar data for Oregon streams. Since the optimal prey would be roughly 10–12 mm long for a 20 cm trout or salmon, such a diet would almost certainly be unobtainable. Presumably this explains why no upper limit to prey size preference is apparent from studies of large trout and natural invertebrate prey.

Prey size is the most obvious cause for variation in equation 7.1 and it has received the most study. However, the denominator also can change in response to a variety of factors that affect ease of detection and capture. Diet specialization may be energetically beneficial even if two prey items offer identical rewards to naive fish, simply because the time required per prey capture decreases with experience. This is a likely explanation for the pattern in Figure 7.1. Lastly, prey of course vary in their vulnerability because of differences in availability and ease of capture, and are heterogeneously distributed in space and time. The common finding that what is eaten closely corresponds to what is available (Allen, 1941; Allan, 1981) attests to the considerable importance of prey availability. Flexibility in the predatory behavior of individual fish is an adaptation to this heterogeneity (Dill, 1983), but environmental variation and evolutionary responses of the prey must set limits to what any predator can attain.

7.1.2 Invertebrate predators

Peckarsky (1982, 1984) provides thorough surveys of freshwater invertebrate predation. Mechanical detection is the most widespread and varied modality for sensing prey. In many instances this means actual contact, for instance with antennae and setal fringes of limbs as in the stonefly *Dinocras cephalotes* (Sjöström, 1985). Vibrations in the water also serve as signals, as in the hemipteran *Notonecta* (Lang, 1980), which feeds on the water surface, and net-spinning caddis larvae that detect vibrations of prey in their nets (Tachet, 1977). Possibly this will prove to be more widespread than we presently realize. Visual cues are not usually important to invertebrate predators, because eyes are not well developed and many species dwell in crevices or are not active by day (Peckarsky, 1982). Odonates and hemipterans rely more on vision, however, and prey motion may enhance this mode of detection (Pritchard, 1966; Corbet, 1980). Chemical detection also seems to be rare, although this may reflect inadequate study. Lake-dwelling triclads exhibit a chemosensory response to their isopod prey (Bellamy and Reynoldson, 1974), and presumably stream-dwelling triclads do so also.

Cummins' (1973) functional group analysis separates invertebrate predators into engulfers and piercers, and an additional separation can be made between sit-and-wait predators and searchers. In addition to simply remaining motionless until the prey approaches within striking range, sit-and-wait predators may use nets (e. g. caddis larvae) or lay mucus traps (e. g. flatworms; Adams, 1980). Odonates that usually ambush also will stalk prey (Johnson and Crowley, 1980), perhaps influenced by hunger level (Corbet, 1980). The caddis larva *Rhyacophila nubila* captured agile mayfly nymphs from sheltered positions, whereas sedentary black fly larvae were captured on random search (Otto, 1993). Sjöström (1985) reported that *D. cephalotes* searched in darkness, but was primarily a sit-and-wait predator in very low light. Risk from its own predators is the most likely explanation, although ability of prey to escape may be an additional factor.

Attack behavior also varies considerably, from the eversible proboscis of a glossiphoniid leech to the rapidly extended labial structure of an odonate. Cooper, Smith and Bence (1985) speculate that ambush predators often have a more rapidly accelerating attack than do searching predators.

(a) Field studies of prey preference

Investigation of food habits by gut analysis has been carried out with a number of predaceous invertebrates of running water (e. g. Chironomidae (Roback, 1968); Ephemeroptera (Tsui and Hubbard, 1979); Odonata (Koslucher and Minshall, 1973); Megaloptera (Devonport and Winterbourn, 1976; Townsend and Hildrew, 1979); Plecoptera (Sheldon, 1972; Siegfried and Knight, 1976; Allan, 1982a); Trichoptera (Wallace, 1975; Devonport and Winterbourn, 1976; Townsend and Hildrew, 1979)). Typically, the average size of ingested prey increases with size of predator, as does the variety of prey items consumed. Predaceous stoneflies tend to ingest diatoms and other non-animal items when very small. Diet changes gradually over development, first to a menu consisting primarily of chironomids, and then to a broader diet in which mayflies, simuliids and trichopterans supplement and may eventually replace midge larvae as prey (Allan, 1982a). Although some differences are reported among species and study locales, presumably reflecting differing availability of prey, any two stoneflies of about the same size, when in similar habitats, consume diets of similar species composition. Small predators tend to have less diverse diets because they do not reach sufficient size to capture those prey that are larger and more agile than midge larvae.

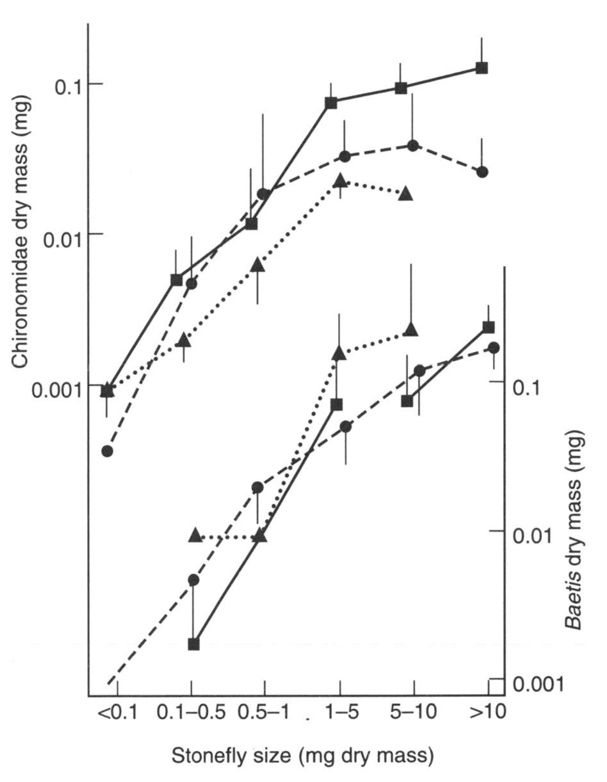

FIGURE 7.6 Average dry mass of individual prey items found in the foreguts of three species of predaceous stoneflies, as a function of size grouping of predator. Means and 95% confidence limits are shown for *Megarcys signata* (■), *Kogotus modestus* (▲) and *Hesperoperla pacifica* (●). (From Allan, 1982a.)

The average size of individual prey item eaten also increases with size of predator. By measuring head widths of ingested prey and converting those values to dry mass, Allan (1982b) showed a very similar positive relationship between prey size and predator size for several species of predaceous stoneflies and the two most common prey, *Baetis* and Chironomidae (Figure 7.6). While this result suggests a strongly size-based preference, gut analysis of field-caught animals cannot easily distinguish between two possible explanations. One is size selection on the part of the predator. Another possibility, however, is that prey and predator are both small in some seasons, larger in others, and juxtaposition of life cycles accounts for the pattern depicted in Figure 7.6, which was obtained by pooling data over an entire year.

Analysis of gut contents typically reveals a good correlation between what is eaten and what is available, as also was seen for fish. In a small and relatively species-poor stream in southern England, the caddis *Plectrocnemia conspersa* and the alderfly *Sialis fuliginosa* during summer consume prey roughly in proportion to their abundance (Hildrew and Townsend, 1976). Similarly, the rank order of prey taxa in the diet of large *Hesperoperla pacifica* is similar to the prey's rank order in the benthos (Allan and Flecker, 1988).

There is some evidence that prey availability is such a decisive factor that it may override differences between predators in foraging mode. The net-spinning *P. conspersa* and the more mobile *S. fuliginosa* exhibited considerable overlap in habitat use and diet, although the former consumed more terrestrial items, large stoneflies and small chironomids, which apparently were more easily trapped in the net of *P. conspersa* (Townsend and Hildrew, 1979). Thus, just as considerable diet partitioning can occur in closely related species (Georgian and Wallace, 1981), much overlap may occur between predators of very different foraging mode.

Finally, field studies suggest that predators influence their diet through their choice of where to hunt, referred to as patch use. Predators appear to aggregate in patches of high prey density, based on correlative field studies (Hildrew and Townsend, 1976, 1982; Townsend and Hildrew, 1978; Malmqvist and Sjöström, 1984). Individuals of *Notonecta hoffmanni* were observed to shift habitat use rapidly in the field (isolated pools of a California stream) and the laboratory in response to changes in prey availability (Sih, 1982b). The polycentropodid, *Plectrocnemia conspersa*, is more likely to place its nets where prey have recently been captured and to abandon sites after some time has elapsed since a previous capture (Hildrew and Townsend, 1980). The observed aggregation of these predators and their prey in nature apparently is due to predators staying where capture rates are high and leaving when they are not. The leech, *Glossiphonia complanata*, foraging for snails in the laboratory spent more time than expected by chance in patches containing prey compared with empty patches, although it did not discriminate between high *versus* low prey patches (Brönmark and Malmqvist, 1986). Obviously, a predator that tends to stay where capture success is high and depart when capture rate declines will result in some tendency for predators to aggregate with their prey.

In contrast, Peckarsky and Dodson (1980a) found that predaceous stoneflies were no more likely to colonize cages containing high prey densities than cages of few prey. Peckarsky (1985) argued that the absence of any aggregative behavior in these predators is explained by the ephemeral nature of prey patches, since highly mobile potential victims like *Baetis* would rapidly disperse. Area restricted search may not be effective in those circumstances where the predator's arrival prompts the prey's departure.

(b) Factors affecting predatory behavior

The number of prey consumed per predator increases with prey abundance because of increased encounter rates, and eventually reaches a maximum due to the limitation imposed by handling time on prey consumption (Figure 7.2). Comparison of functional response curves for small *versus* large IVth instar backswimmers (Figure 7.2b) demonstrates that feeding rate increased with predator size (Fox and Murdoch, 1978). In addition, prey consumption may decrease or cease prior to emergence, at molting (Malmqvist and Sjöström, 1980) and with satiation.

All indications at present are that experience and learning play little or no role in the predatory behavior of lotic invertebrates. Hunger level did however influence which prey were consumed by *Hesperoperla pacifica* offered a choice between the mayflies *Baetis bicaudatus* and *Ephemerella altana* (Molles and Pietruszka, 1983). The former is a soft-bodied, agile swimmer; the latter has a spiney and rigid exoskeleton but is slow and clumsy. Starved stoneflies ate mostly *E. altana*, while satiated stoneflies ate both prey in about equal numbers. When fresh dead prey were offered to starved predators, however, a preference for *Baetis* was evident.

The proposed explanation was that starved predators attacked both prey about equally, but with increasing satiation began to restrict their attacks only to *Baetis*. Since *E. altana* was the easier of the two species to capture, starved *H. pacifica* appeared to prefer them equally with *Baetis*.

Molles and Pietruszka's demonstration that *H. pacifica* differentiated in its attack behavior between two potential prey species, and furthermore that hunger level influenced the tendency to attack certain prey, is clear evidence of active preference on the part of a lotic invertebrate predator. Allan and Flecker (1988) also found that this stonefly rarely attacked ephemerellid mayflies. Among other prey, however, attack rates were quite similar. Laboratory preference experiments using the stonefly *Megarcys signata* extend the above results of species selection (Peckarsky and Penton, 1989). The preference rank among several mayflies, all of which have been observed in gut analysis, was *Baetis* ≫ *Cingymula* = *Epeorus* > *Ephemerella*.

Selection of prey on the basis of size also has been demonstrated in stoneflies using both *Baetis* and black fly larvae as prey (Allan, Flecker and McClintock, 1987a, b). The outcome depends on the size class of predator as well; small *H. pacifica* preferred small prey while large predators preferred larger prey. Allan, Flecker and McClintock (1987a) argued that preference by invertebrate predators generally is a bell-shaped function of prey size, but laboratory experiments typically offer predators only a subset of possible prey sizes and hence reveal only part of the preference curve. A plot of a preference index *versus* relative prey size (prey dry mass:predator dry mass) from the choice experiments of Allan, Flecker and McClintock indeed indicates a hump-shaped preference curve (Figure 7.7). However, it is unlikely that the explanation rests solely with the predator's attack preference. When the mechanistic basis for non-random prey consumption has been examined in detail (e. g. Molles and Pietruszka, 1983; Allan, Flecker and McClintock, 1987a,b; Peckarsky and Penton, 1989), it usually is best explained as passive preference based on encounter rates and capture probabilities. As with vertebrate predation, preference cannot be fully understood without careful consideration of factors influencing prey vulnerability.

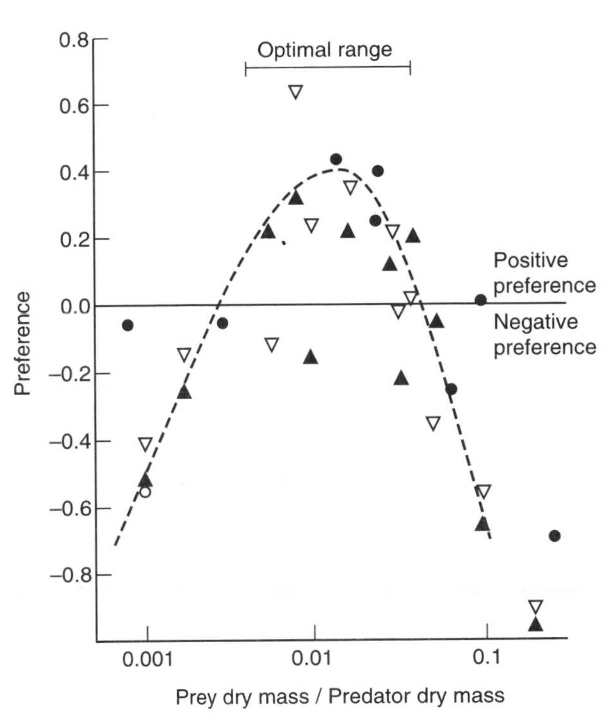

FIGURE 7.7 Prey preference as a function of relative prey size. Data are from experiments of Allan, Flecker and McClintock (1987a,b) where four size classes of prey were offered to stoneflies of a given size. ●, *H. pacifica* and *Prosimulium*; ▲, *H. pacifica* and *Baetis*; ▽, either *M. signata* or *K. modestus* and *Baetis*.

The possibility that non-random feeding results in maximizing energy consumption has received little study with invertebrate predators in running waters. Habitat shifts by *Notonecta* in response to food availability led to an

TABLE 7.1 Prey tactics that reduce the risk of predation from either vertebrate or invertebrate predators. Only those defenses thought to be relevant to stream-dwelling macroinvertebrates are included (From Allan, Flecker and McClintock, 1987b)

Risk	*Prey tactics*	
Reduce encounter rates	Spatial segregation, including use of refuges Temporal segregation (diel or seasonal) Low movement rates Reduced conspicuousness (to visual, mechanical or chemical cues)	
	Precontact	Postcontact
Reduce attack and capture rates	Detect and flee	Morphological (e.g. armor) Behavioral (aggressive retaliation, thanatosis) Chemical defenses (not documented)
	Aposematism	
	Warning/startle behaviors	

improved (but not maximal) feeding rate (Sih, 1982a). However, Allan and Flecker (1988) saw little indication that the diet of field-caught *H. pacifica* favored prey on the basis of energy reward.

7.1.3 Prey vulnerability

Characteristics of prey that act as defenses against predators can be separated into those that operate regardless of the physical proximity of predator and prey and serve to reduce the likelihood of initiation of an attack, *versus* those that function to foil attack and capture. Edmunds (1974) named these primary and secondary defenses, respectively, and several authors have reviewed their role with regard to invertebrate predators (Jeffries and Lawton, 1984; Peckarsky, 1984; Sih, 1987). Some defenses may have evolved specifically in response to predation, and others may be consequences of body plan and lifestyle. Anti-predatory behavior can be flexible or fixed, and may deter most types of predation or only certain predators. There may be substantial cost in lost opportunities in foraging, growth or reproduction, and these are discussed in a later section on indirect effects. Clearly, anti-predator adaptations are diverse and widespread.

(a) Primary defenses

The close correspondence between the relative abundance of prey items in the diet of predators and in the environment indicates that ingestion rates depend substantially on encounter rates. Thus, traits that reduce the likelihood of encounter and detection can be effective defenses. These include sensory capabilities that allow the prey to detect approaching predators, characteristics that promote spatial and temporal separation between predator and prey, and those that make the prey inconspicuous (Table 7.1).

It seems likely that any sensory modality used in the detection of prey also might be used in detection of predators. Vision is probably the most common way that fish perceive their predators, as do some invertebrates (e. g. the crayfish *Orconectes*; Stein and Magnuson, 1976). *Baetis* orients to foraging stoneflies with its caudal cerci, and apparently detects mechanical wave disturbances that signal when to flee (Peckarsky, 1987). The presence of a predaceous stonefly contained within a small cage supresses colonization of that patch of substrate (Peckarsky, 1985), so chemical cues also can be used to reduce the likelihood of an encounter.

Refuges may be absolute, rendering the prey invulnerable, but more commonly they serve to reduce the likelihood of encounter and detection. Most stream-dwelling organisms show a strong affinity for crevices and cover, and some construct burrows in softer sediments or live an entirely buried existence (e. g. *Dolania*; Tsui and Hubbard, 1979). The negative phototaxis that exists in many species of insects (Williams, 1981) serves to keep them within the substrate, reducing their chances of detection by fish, while the heterogeneity of that habitat probably also acts to reduce encounters and tactile detection by invertebrate predators. Furthermore, this affinity for cover may vary with changes in risk of predation. For example, small minnows (*Semotilus atromaculatus* and *Rhinichthys atratulus*) displayed greatest avoidance of stream channels containing predators when cover was lacking (Cerri and Fraser, 1983).

The definition of refuges may be broadened to include anything that favors spatial or temporal segregation of prey from predator (Table 7.1). The nocturnal periodicity of downstream drift is a clear example of temporal separation over the 24 h cycle, and it seems very probable that this is an evolved response to a greater risk of fish predation during the day (chapter 10). The synchronous emergence of adult insects probably 'swamps' predators with more prey than they can manage, while emergence early in the summer effectively removes the largest and therefore most vulnerable individuals from streams prior to the period when water temperatures are warmest and fish feeding rates would be expected to be highest. The aggregation of black fly larvae on stone surfaces, exposed roots and other trailing objects in very fast currents may not be an evolved response to predation yet may result nonetheless in effective spatial isolation.

Other primary defenses include any that reduce a prey organism's risk of visual, mechanical or chemical detection. Aquatic invertebrates are mostly drab and dingy in comparison to occupants of the terrestrial and marine worlds (Hutchinson, 1981; Pennak, 1985). Admittedly, avoidance of predation is only one possible reason; other explanations include the uncommonness of herbivory on higher plants from which secondary compounds may be obtained, and the fact that most invertebrates (although not fish) lack adequate visual acuity for epigamic selection to operate (Hutchinson, 1981). However, crypsis almost certainly is a common evolutionary response to predation. Some fishes are magnificently camouflaged, such as the South American leaffish *Monocirrhus* (Lowe-McConnell, 1987). In case-inhabiting caddis larvae, crypsis may be enhanced by the case itself. Otto and Svensson (1980) showed that cases constructed of leaf material were effectively camouflaged on a leaf-covered bottom, and suggested that case enlargement with stones and twigs further serves to make the caddis too large to be easily consumed by small fish. Mineral cases were especially resistant to breakage and, apparently for this reason, also were less prone to attack by trout.

Movement increases the likelihood that prey will be visually detected by fish, and *Gammarus pseudolimnaeus* reduces its movements in stream channels in response to chemical stimulation from 'essence of rainbow trout' (Williams and Moore, 1985). Movement also influences encounter probabilities. Foraging stonefly larvae encountered mobile prey at a substantially higher rate than sedentary prey (Allan, Flecker and McClintock, 1987b). Cooper, Smith and Bence (1985) provide evidence to support their prediction that evasive prey should be more vulnerable to sit-and-wait predators, while sedentary prey should be at greater risk from searching predators.

(b) Secondary defenses

Once encountered, an individual can reduce its probability of being captured in a number of ways, which can roughly be divided into precontact and postcontact defenses. Common precontact defenses include the ability to detect and flee before capture, at which *Baetis* is very accomplished (Peckarsky, 1980, 1987), and aggregation. Highly aggregated taxa such as black fly larvae may benefit as 'selfish herds' in which the risk to any one individual is reduced by its many neighbours (Hamilton, 1971), although there appears to be no evidence that any invertebrate has evolved aggregative behavior for this purpose. Schooling in fish is an effective precontact defense (Ginitz and Larkin, 1976; Morgan and Godin, 1985) that seems clearly an evolved response to fish predation. Fish also may subtly communicate their alertness, thus reducing the likelihood of an attack.

Postcontact defenses include many of the best known examples of anti-predator adaptations. Morphological defenses include spines such as occur in sticklebacks and many catfish, heavy exoskeletons with stout projections such as exhibited by ephemerellid mayflies (Peckarsky, 1987) and larger stoneflies (Otto and Sjöström, 1983), the operculum of prosobranch snails (Kelly and Cory, 1987), and so on. However, freshwater snails cannot rival their marine relatives in the extent of morphological defenses against enemies. Vermeij and Covich (1978) explain this lesser degree of evolved defenses on the basis that predation in freshwater has acted with less intensity and for less time.

Behavioral defenses include struggling to escape, such as the violent shell shaking exhibited by *Physa fontinalis* when its mantle fringe comes in contact with certain leeches and triclads (Townsend and McCarthy, 1980), feigning death (thanatosis), which occurs in stoneflies attacked by trout (Otto and Sjöström, 1983) and aggressive retaliation. Large crayfish apparently are able to defend themselves against smallmouth bass, although small individuals remain vulnerable (Stein and Magnuson, 1976).

The morphological defense of aposematism, and warning/startle behaviors, also need to be included but it is unclear whether they should be classified as pre- or postcontact. Neither is very common in the fauna of temperate running waters, nor are chemical defenses, although that may reflect inadequate knowledge.

7.2 Predator control of prey distribution and abundances

It is widely held that predation exerts an important influence over aquatic communities. This view has come about partly because predation is ubiquitous, and partly because extensive research carried out in ponds and the marine intertidal provides strong support. Allen (1951) forcefully raised the question of the apparent importance of top predators in running water by his calculation that trout in the Horokiwi stream, New Zealand, could consume up to 150 times the standing stock of invertebrates in a year. Subsequent research has somewhat modified Allen's original calculations, and we still lack sufficient precision in measurement to have real confidence in our estimates of how much food is produced *versus* that consumed in a stream. Nonetheless, 'Allen's paradox' of consumption apparently exceeding production remains unresolved, and at least should imply that lower trophic levels are strongly affected by higher trophic levels.

Predation affects the biological community directly and indirectly (Kerfoot and Sih, 1987). A direct effect is a reduction in prey abundance or biomass. Indirect effects include non-lethal but nevertheless significant alterations in the spatial and temporal patterns of prey activity and distribution, and in life histories of prey species. In addition, predation from a top predator can cascade down the food chain, for example, when predation reduces the abundance of a herbivore thereby allowing algal biomass to increase. Indirect effects can ramify further, depending on the strength of interactions among species connected by food web pathways. Strong predation thus can be an important force affecting the structure of the entire community, a topic we shall return to in chapter 11.

7.2.1 Direct effects on prey populations

(a) Comparisons and natural experiments

Naturally occurring variation in predation intensity can be examined for evidence of corresponding changes in prey populations. The simplest comparison utilizes otherwise similar stream habitats that apparently differ only in the presence or absence of a major predator. A few such 'natural experiments' (cf Diamond, 1986) have been reported. An amphipod that was common in the headwaters of a stream in the Silesian Beskydy Mountains became very scarce within a scant 100 m, coincident with a high weir and the occurrence of large numbers of brown trout (Straskraba, 1965). Waterfalls can be barriers to the upstream spread of fish, but in contrast to Straskraba's result, neither Jacobi (1979) nor Reice and Edwards (1986) found any changes in invertebrate abundances on comparison of sections above and below waterfalls in North American trout streams. Large, pool-inhabiting fishes become restricted to stream sections with sufficient water under low-flow conditions, and by chance some pools contain fish while otherwise suitable pools do not. The distribution of piscivorous bass and algivorous minnows (Power *et al.*, 1985), described in more detail below, demonstrates that under these conditions predation can have strong effects on population densities over two trophic levels.

Situations where a predator species is either present or absent may not occur routinely, but gradation in predation intensity probably is common. With sufficient data over a range of conditions, it should be feasible by use of regression statistics to ask whether predator and prey populations are inversely related. For example, samples collected from three streams of contrasting acidity (pH values of 4.8, 5.2 and 6.1) in southern England demonstrated an inverse correlation between fish and predaceous invertebrates (Hildrew *et al.*, 1984). Bowlby and Roff (1986) examined 30 stream sites in southern Ontario, Canada, to determine if the presence or absence of piscivorous fish influenced the biomass and abundance of non-piscivorous fish, predaceous invertebrates or non-predaceous invertebrates. Their intent was to contrast the 'top-down' hypothesis that predators at the top of the food chain limit lower trophic levels with other explanations, including 'bottom-up' food and habitat limitation. Sites with piscivores indeed exhibited a lower biomass of non-piscivorous fish but higher biomass and abundance of benthic invertebrates, and so the top-down hypothesis was favored.

A major limitation of natural comparisons is the obvious possibility that additional, unmeasured factors are responsible for observed differences so that correlation falsely implies causation. Usually it is difficult or impossible to resolve this satisfactorily. For instance, Bowlby and Roff also found that invertebrate abundance was positively associated with extent of riffles whereas fish biomass showed the opposite trend. Possibly this habitat relationship is the true underlying cause of the inverse relation between fish and invertebrates. Experimental manipulation of one or another population, where feasible, can be an extremely valuable alternative or supplement to natural comparisons.

(b) Experimental manipulations

The abundances of fish and invertebrate predators have been manipulated in field experiments designed to test their influence on prey populations. Zelinka (1974), Allan (1982b) and Reice (1983) each reported little or no change in benthic populations in response to artificially lowering the abundance or otherwise excluding stream fishes. Zelinka and Allan both used large sections of stream for their studies, whereas Reice used small wire baskets filled with substrate with or without tops to prevent or allow

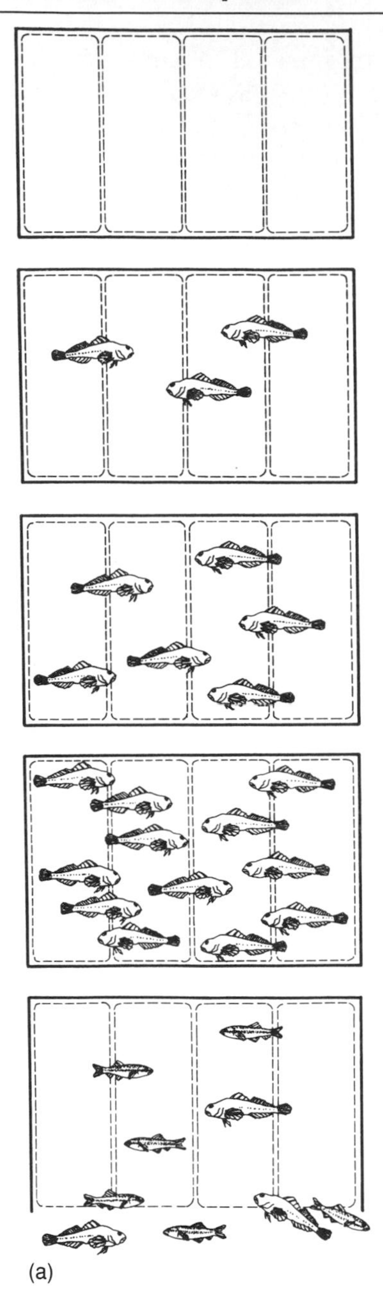

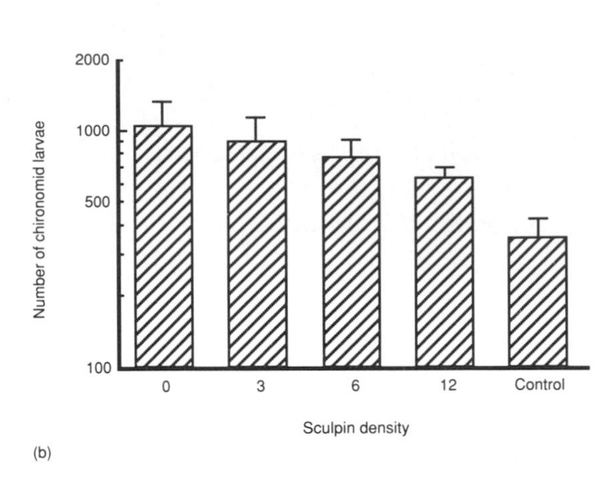

FIGURE 7.8 (a) The experimental design used by Flecker (1984) to investigate the effect of fish predation on the benthic invertebrates of a West Virginia stream. (b) Abundance of Chironomidae in each of the experimental treatments.

entry of fish. Using a field enclosure design, Flecker (1984) documented a reduction in numbers of some benthic invertebrates in response to a guild of invertivores, primarily sculpins (*Cottus*) and dace (*Rhinichthys*). Baskets of substrate were placed in enclosures containing 0, 3, 6 or 12 sculpins, while open cages were used as an additional treatment permitting free access by fish and therefore natural levels of predation. Chironomidae and the stonefly *Leuctra* showed a significant reduction in abundance with increasing intensity of fish predation (Figure 7.8), while other insect taxa were unaffected. Enclosure of small creek chub (*Semotilus atromaculatus*) within 0.5 m^2 areas of a warm-water, soft-sediment stream resulted in strong reductions in total invertebrate abundance (Gilliam, Fraser and Sabat, 1989). Oligochaetes and isopods were strongly affected while midge larvae and clams showed no response.

Peckarsky (1985) and Peckarsky and Dodson (1980b) evaluated the role of predaceous stoneflies using cages (30 × 20 × 10 cm) filled with substrate and buried just below the surface of the streambed. Prey were able to immigrate or emigrate freely but the predator was either present or excluded. Based on counts after 3–7 days of colonization, the presence of a stonefly consistently reduced prey abundance (and much of the effect can be generated by a predator restrained by a tiny cage within the cage, thus indicating that prey chemically detect the stonefly). Although the prey species list did not change, relative abundances were altered, especially in her studies of New York streams. Taxa that showed the greatest density reduction were abundant in the predator's diet, so direct predation and prey avoidance behavior probably both contributed to this result.

Other enclosure experiments (7.5 cm diameter cups) in a small stream in Alberta, Canada, showed that *Kogotus nonus* could depress the densities of *Thienemaniella* and some additional orthoclad midge larvae (Walde and Davies, 1984b). Other prey were unaffected, and so the relative abundances of the prey assemblage changed with predator density.

7.2.2 Indirect effects of predation

(a) Non-lethal effects of predation

Non-lethal effects include injury, restrictions in habitat use, foraging behavior or the timing of activity, and changes in life histories induced by the presence of a predator. The consequence presumably is a reduction in fitness due to sacrificed potential for growth and reproduction, although this can be very hard to demonstrate. In some cases these responses are flexible and manifested only in the presence of threat; in other instances the response is fixed and occurs whether or not the predator is present.

Rather little is known about the frequency of non-fatal injury. Field collections not uncommonly include larval insects where one leg is missing or much smaller than the others, indicative of injury and regeneration. Such injuries in zygopterans are believed to reflect intraspecific aggression rather than predation (Baker and Dixon, 1986), but that does not seem a likely explanation in mayflies. Nilsson (1986) reported that 13% of the nymphs of *Leptophlebia vespertina* in some unpolluted Swedish streams possessed malformed legs, and suggested that unsuccessful predation might be one cause. The survival of nymphs that had lost legs was reduced, compared with intact nymphs, but some did emerge as adults.

Various fish species, primarily in the tropics, have adopted the habit of eating particular parts of living fishes (chapter 6). Scale-eating is perhaps the most widespread, but in addition fins, gill filaments, skin, chunks of the body and even eyes may be consumed (Sazima, 1983). Here it seems that the potential effect of non-lethal injury may be substantial.

(b) Behavioral shifts in habitat use or foraging activity

The presence of enemies commonly results in restricted habitat use as potential prey shift their location on encountering a predator. Cooper (1984) found that adult water striders (Gerridae) occupied all areas of stream pools when trout were absent, but only the margins of pools containing trout. When trout were removed from some pools and transferred to others lacking trout, gerrid distribution adjusted accordingly. Harassment by trout of any gerrids venturing away from pool margins was the apparent cause. In autumn, female gerrids from trout pools weighed less than their counterparts, suggesting that lost feeding opportunity translated into reduced growth. Similarly, juvenile crayfish *Orconectes propinquus*, which are most sought after by predaceous fishes, avoid open, sandy substrates where larger crayfish are abundant (Stein and Magnuson, 1976).

The demonstration that animals sacrifice foraging opportunities to minimize predation threat raises intriguing questions about their ability to evaluate risk *versus* reward. In *Notonecta hoffmanni*, early instars are at risk from cannibalistic adults in the order (Sih, 1982b): I > II > III > IV = V = no risk. Sih used a circular aquarium to establish a central region of high reward and a peripheral region of low reward. Adult backswimmers concentrated in the central region, so early instars were presented

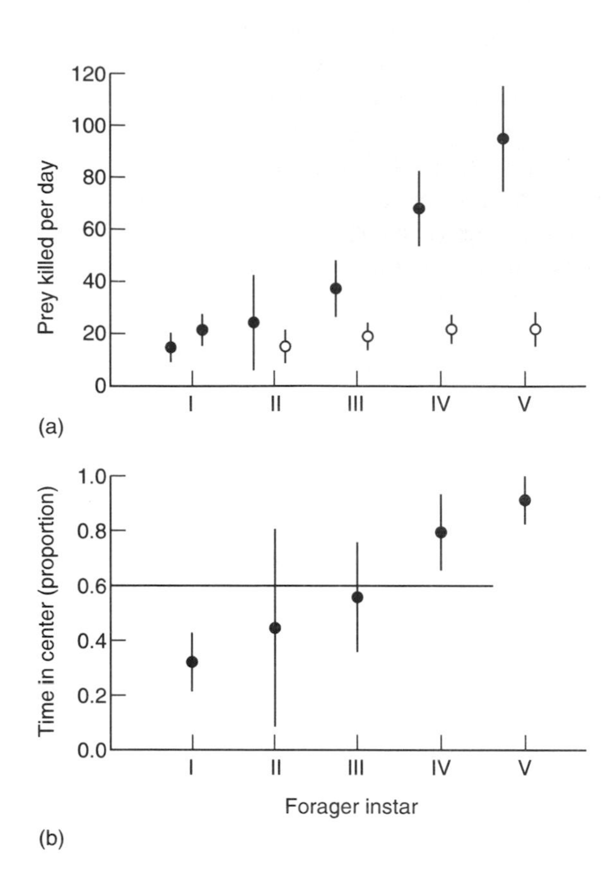

FIGURE 7.9 (a) The number of prey killed per day by *Notonecta hoffmanni* foraging in either high food–high risk (●) or low food–low risk (○) regions, as a function of instar. Values are means ± 1 standard deviation. (b) The proportion of time that a particular instar spent in the high food–high risk region when given a choice between the two regions. Values are means ± 95% C.L. (From Sih, 1982b.)

with a choice between high risk and high reward, *versus* low risk and low reward. Results demonstrate a neatly graded foraging response among instars in regard to changing risk and reward (Figure 7.9).

A similar experiment was conducted with minnows in an artificial stream where the presence or absence of predators, cover, and food were manipulated (Fraser and Cerri, 1982; Cerri and Fraser, 1983). Adult creek chub were the predators, while juvenile chub and blacknose dace *Rhinichthys atratulus* were the prey. Avoidance of risk was clearly demonstrated: prey avoided patches with predators, and were more likely to take risks during the day and when structure was present. Offered a choice among four alternatives (food *versus* no food cross-classified with risk *versus* no risk), juvenile *R. atratulus* showed no indication of balancing risk against reward. Instead they appeared to choose a patch based on food, and reduce their use of a patch proportionate to predation risk and independently of reward.

Juvenile coho salmon, *Oncorhynchus kisutch*, feed on stream drift by making short excursions from a holding position. Using house flies as prey and a model of a rainbow trout as threat, Dill and Fraser (1984) asked whether risk reduced foraging, and whether the reduction was proportionate to risk. Exposure to the model trout reduced reaction and attack distances and shortened attack time compared with young salmon foraging in the absence of threat (Figure 7.10). The investigators then varied the frequency with which they presented the model before the salmon, thereby varying the level of risk, and found that attack distance decreased proportionately (Figure 7.11). Moreover, responsiveness to the model was reduced by higher hunger levels and the presence of a competitor. This behavioral flexibility evidently allows young coho salmon to adjust their foraging in a quite complex fashion.

(c) Fixed or inflexible responses

Many adaptations to reduce risk of predation act continuously, regardless of the presence of enemies. If some cost is incurred, such as greater metabolic demands or lost opportunities, these may be viewed as indirect effects of predation. For instance, caddis cases of mineral material presumably are more energetically expensive to construct than are organic cases, but also offer greater protection (Otto and Svensson, 1980). In *Orconectes* crayfish, adult foraging is aperiodic whereas juveniles, the more vulnerable size to

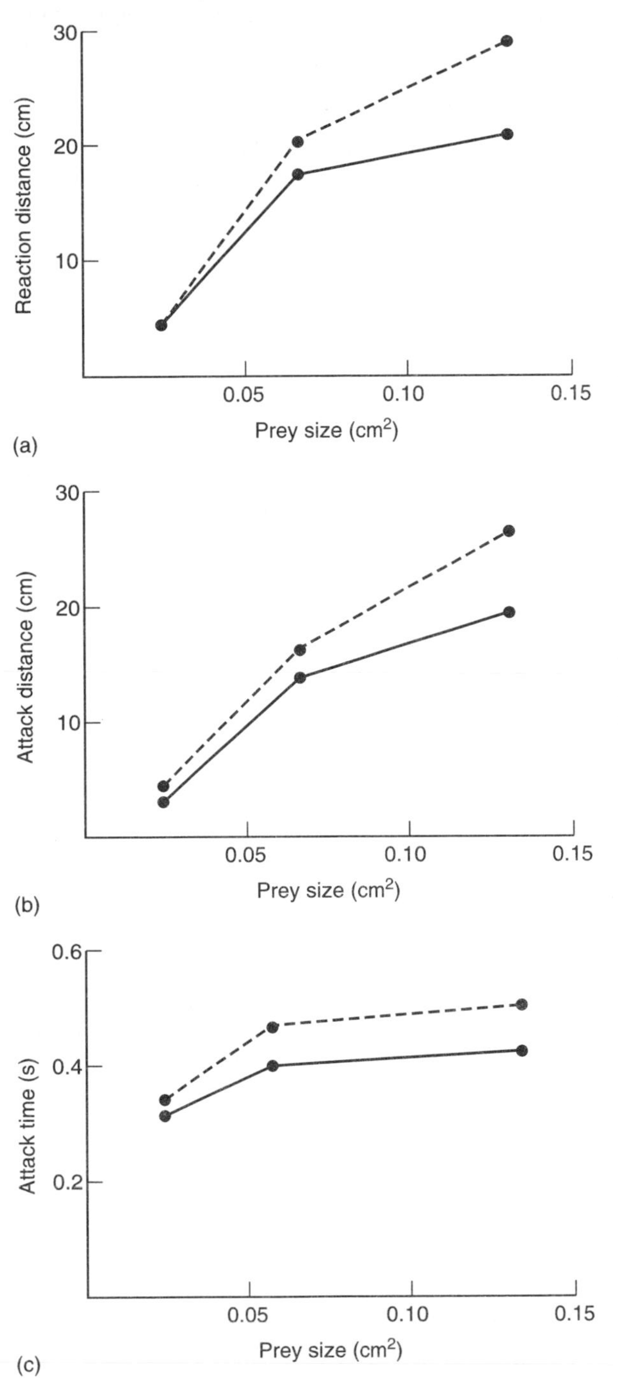

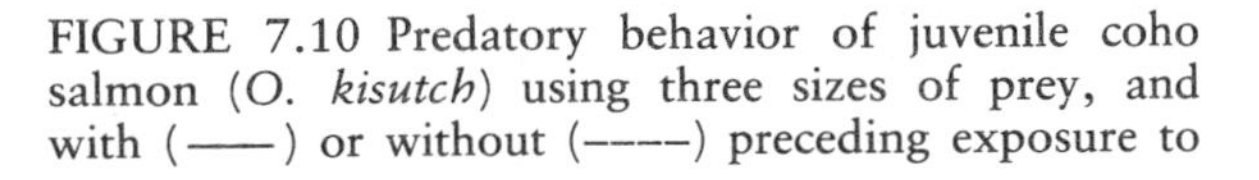
FIGURE 7.10 Predatory behavior of juvenile coho salmon (O. *kisutch*) using three sizes of prey, and with (——) or without (----) preceding exposure to a model of a predator. Means ± 1 standard error. (From Dill and Fraser, 1984.)

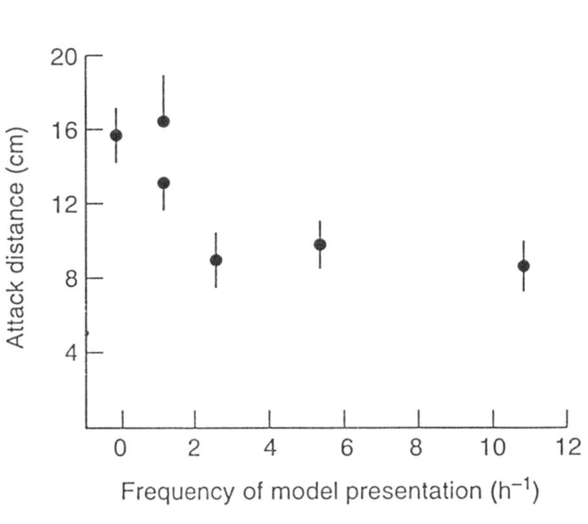

FIGURE 7.11 Mean attack distance (± 1 standard error) of coho salmon responding to medium-sized flies, as a function of frequency of presentation of a model rainbow trout. (From Dill and Fraser, 1984.)

fish, must forego daytime feeding (Stein, 1979). The nocturnal periodicity of downstream drift may have a similar basis (chapter 10). Other examples include subdued epigamic coloration and display in high risk environments (Strong, 1973; Endler, 1983), and morphological defenses such as armor and spines. All of these adaptations presumably result in some cost to the organism, but quantification of the sacrifice is extremely difficult in most instances.

(d) Effects on life histories

The presence of predators can influence the growth and fecundity of prey species by restricting feeding activities or through chemically induced life-history changes. When nymphs of the mayfly *Baetis* were reared in chambers, mayflies with predators matured at significantly smaller body sizes and produced less egg biomass compared with those reared in the absence of predators (Peckarsky *et al.*, 1993). In all likelihood, encounters with predators cause significant reduction of reproductive fitness in natural populations as well. The snail, *Physella virgata virgata*, exhibits a dramatic life-history alteration in response to a chemical released by crayfish (Crowl and Covich, 1990). Snails reared in water conditioned by crayfish grew rapidly to a size of about 10 mm before much reproduction occurred, whereas snails not exposed to the cue began reproduction at 4 mm. Evidently this is a fixed response in which the snail's rapid growth permits it to achieve a size where predation risk is lower.

(e) Cascading trophic interactions

When a predator at or near the top of the trophic web has a pronounced effect on the abundance or distribution of its prey, the potential exists for the influence of that predator to extend not just to its immediate prey, but also to ramify elsewhere through the food web. In systems where algal-grazing fish are abundant, the distribution of algae can be determined by the activities of piscivorous birds (Power, 1984a, b) or fish (Power and Matthews, 1983; Power *et al.*, 1985). *Campostoma anomalum*, the stoneroller, is an algivorous minnow occurring in the central USA. In small midwestern streams during low flow, pools separated by shallow riffles typically exhibited a striking complementarity between the stoneroller and piscivorous bass (*Micropterus*). In addition, algal abundance varied inversely with *Campostoma*. The influence of stoneroller grazing on algae was demonstrated by removing bass from a pool, dividing it longitudinally, and adding *Campostoma* to one side. Algal standing crop increased dramatically on the side lacking stonerollers (Figure 7.12). Addition of bass to *Campostoma* pools immediately resulted in these minnows restricting their grazing to shallower regions, demonstrating that habitat shifts by grazers in response to their own predators could in turn influence algal distribution.

Further studies by Power (1990, 1992) demonstrate that fish produce strong cascading effects on biota associated with boulder–bedrock substrates in pools of a northern California river during summer baseflow. By suppressing densities of damselfly nymphs and other small predators, fish released algivorous chironomids from predation (Figure 7.13). In the presence of fish, chironomid grazing reduced filamentous green algae to low, prostrate webs, whereas an upright algal turf developed when fish were excluded.

Studies of predation in streams clearly include examples where strong biological interactions are apparent and the consequences cascade through trophic webs, and others where the effects of predation are more limited. Several explanations have been offered for this diversity of results. They include adaptation to predation past, the method of study, and differences due to the specific organisms and environment.

The importance of predation in the evolutionary past and consequent adaptation to minimize risk was argued by Allan (1982b) and Thorp (1986). Unlike ponds, which harbor certain taxa found mainly in the absence of fish and rarely in their presence, fish predation may be a constant threat to the fauna of running waters. A fauna adapted always to live in the presence of predators might not be drastically reduced by their introduction, nor are there species that will increase dramatically if top predators are removed.

A number of approaches have been used to assess the importance of predation, and it is plausible that results depend partly on method. Natural comparisons have their advantages, especially in providing contrasting situations on a scale that would not be feasible or ethical to create artificially, as well as in the very fact that they are natural (Diamond, 1986). Typically, however, the contrasted areas differ in many fac-

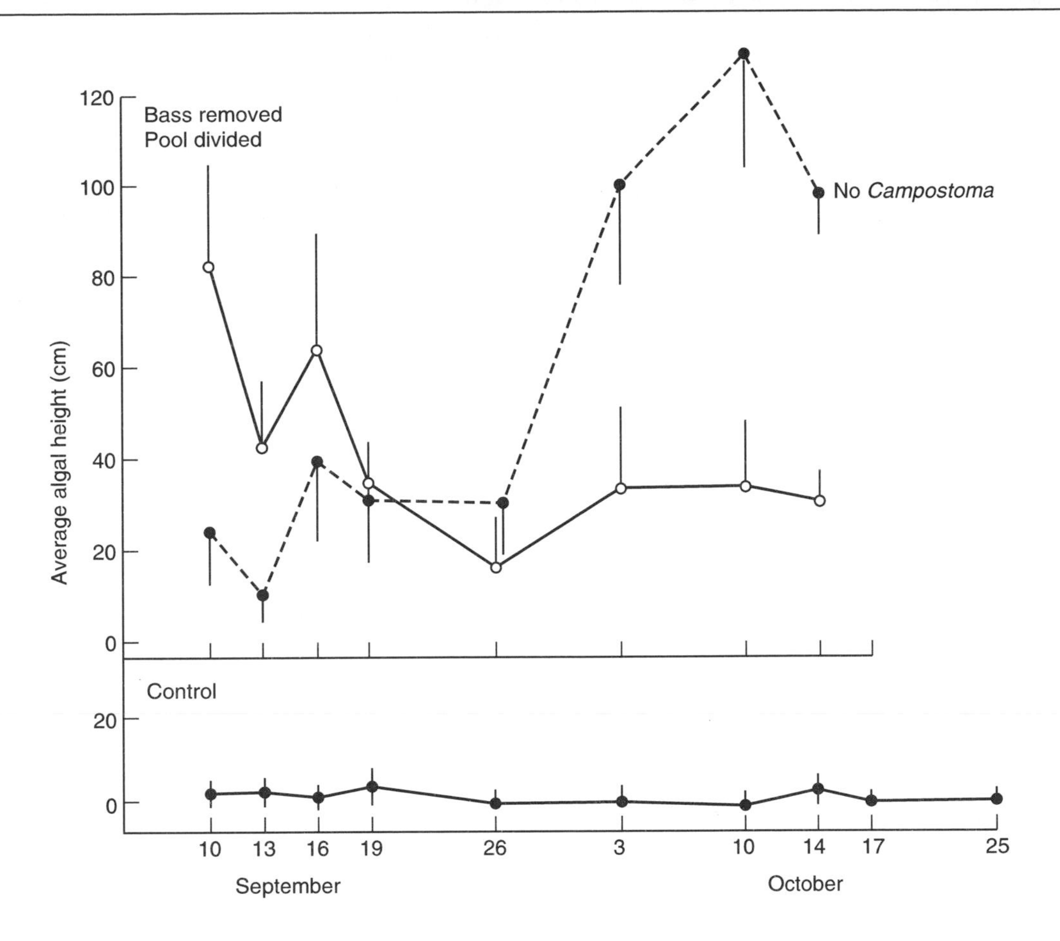

FIGURE 7.12 Algal standing crops on two sides of a pool from which bass were removed. The pool was divided in half, and *Campostoma* added to one side only. A control pool contained a school of *Campostoma*. Means ± 2 standard errors. (From Power, Matthews and Stewart, 1985.)

tors in addition to the one of interest, and so ascribing causation to a single factor is chancy. Field experiments in principle allow greater confidence in ascribing causation, but many aspects concerning how the study is designed and carried out can influence results and interpretation. Choices must be made on such controversial points as the appropriate experimental control, replication (Hurlbert, 1984), the physical size of the experimental unit, and whether the study population is added where it was absent, or removed from where it was present (Allan, 1984; Walde and Davies, 1984; Hixon, 1986). The rate at which prey are replenished within substrate patches subjected to different predation treatments is likely to be affected by mesh size of cages or physical size of the experimental unit. From their review of predation studies, Cooper, Walde and Peckarsky (1990) conclude that predator impacts are less apparent when prey replenishment rates are high. This inference likely applies to natural streams as well.

Lastly, it is entirely plausible that predators exert their greatest influence under particular circumstances that are not universal. Predation by roach and steelhead produced strong cascading effects on pool-dwelling biota of a northern California river (Power, 1990) but not

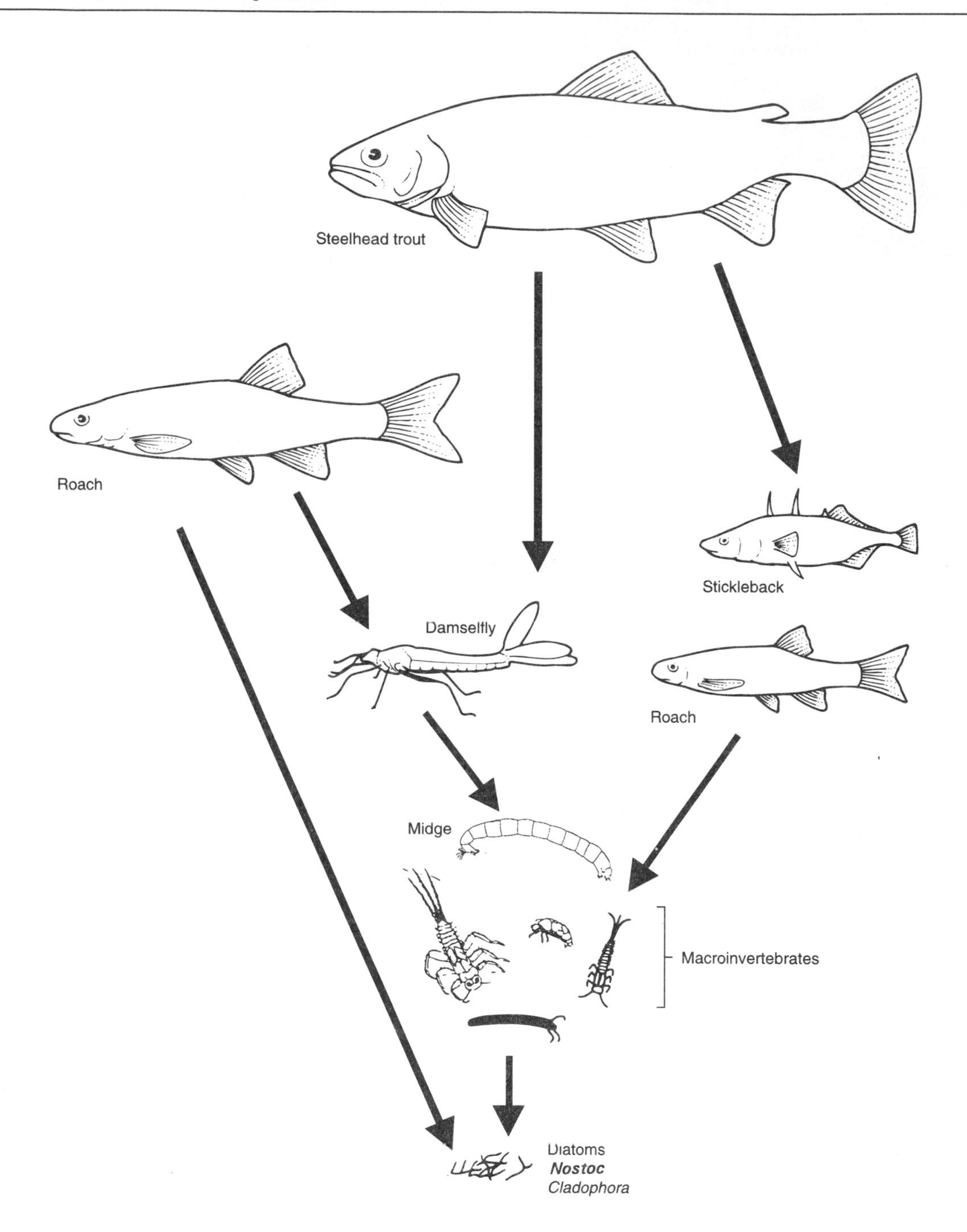

FIGURE 7.13 Diagrammatic representation of the trophic cascade described by Power (1990) on boulder–bedrock substrates in pools of a California river. Fishes include steelhead trout (*Oncorhynchus mykiss*), two size classes of the roach (*Hesperoleucas symmetricus*), and the stickleback (*Gasterosteus aculeatus*). Lestid damselflies fed on midge larvae and a number of other aquatic insects. In turn, these insects grazed a periphyton turf consisting of filamentous green algae (*Cladophora*), diatoms, and the cyanobacterium *Nostoc*.

on gravel-dwelling biota (Power, 1992), evidently due to the heterogeneity of the latter habitat. Habitat complexity and spatial refuges probably contribute greatly to the survival of prey (Macan, 1977; Thorp, 1986), and the benthos of streams surely is a heterogeneous environment. An additional characteristic of flowing water is the tremendous rate of replenishment of populations at any particular point by downstream drift. Habitat complexity and high rates of prey renewal both diminish the capacity for predators to influence their prey. At one extreme, perhaps, is a high gradient trout stream, where loose cobbles provide a complex habitat, trout feed primarily upon drift, the replenishment of populations by immigration from upstream is great, and the availability of suitable pool habitat potentially may limit numbers of trout. The fauna of pools, especially in those situations where refuges are scant and the prey conspicuous, may represent the opposite extreme where predation effects are most apparent.

7.3 Summary

Predation, used here to refer to the consumption of animal prey, is a widespread and potentially important process affecting the biota of running waters. It is the primary feeding role of many species. In addition, accidental consumption of animal prey may be relatively common even among species not normally thought of as predators.

The diet breadth of predators ultimately is determined by particular morphological and behavioral specializations, and the habitat occupied. Careful studies of feeding habits generally reveal a mix of opportunism and selectivity. However, what often is referred to as choice on the part of the predator in reality is determined by differential vulnerability among prey. Size, mobility, and day *versus* night activity are just some of the factors that govern a prey species' risk of being eaten. Predators differ widely in their feeding habits, including prey capture technique, sensory modality used, attack tactics and so on. Clearly the evolutionary pressures for predators to improve their abilities at prey capture and for prey to reduce their vulnerability to predators lock these natural enemies in a never-ending struggle.

Predation effects extend far beyond a simple accounting of numbers of prey consumed. Predation risk may force prey to restrict their foraging to certain times or places or to develop elaborate defense mechanisms, perhaps resulting in less growth or reduced fecundity. The influence of a predator on its prey might in turn benefit the trophic level below or a competitor of that prey and thus ramify through the food web. Strong cascading effects have been documented in some studies and found lacking in others, apparently because of differences in the predator–prey linkage or environmental conditions. Herbivory, while in principle very similar to predation, is sufficiently different that it deserves separate treatment, and we now turn to that topic.

Chapter eight

Herbivory

The importance of autotrophs to lotic food webs and the complexity of their interrelationships with herbivores were largely unappreciated until roughly the late 1970s. An emphasis on detrital energy pathways (chapter 5) and on abiotic control of autotrophs may have contributed to a comparative neglect of herbivory. However, with the growing realization that the extent of primary production in running waters likely has been underestimated (Minshall, 1978) there also is recognition that, even in woodland streams, herbivory can play a substantial role. Some 50 of the 75 common taxa in Linesville Creek, Pennsylvania, contained substantial amounts of diatoms and algae in their diet, and roughly one-fifth of the invertebrate biomass was believed to be supported by periphyton (Coffman, Cummins and Wuycheck, 1971). In Bear Brook, New Hampshire, where algal production was estimated to contribute less than 1% of the total energy inputs (Fisher and Likens, 1973), a moderately abundant caddis larva was found to depend heavily on consumption of periphyton for most of its spring growth (Mayer and Likens, 1987).

The field of plant–herbivore interactions has advanced rapidly in recent years. In running waters, the link between periphyton and their grazers has seen the greatest progress, primarily with invertebrates, but grazing by fishes and amphibians also has been shown to be important. Neither the river phytoplankton nor macrophytes are thought to experience strong grazing pressure very often, although this may be due to inadequate study. We will begin with a detailed consideration of the periphyton–grazer interaction, because it is best understood, and briefly consider herbivory with regard to plankton and macrophytes at the end of the chapter.

8.1 Periphyton–grazer interactions

Periphyton, comprised mainly of diatoms, green algae and cyanobacteria, are found almost everywhere in running waters (chapter 4). The extent of herbivory varies with periphyton growth form and differs among the major taxonomic groups, for reasons that we shall consider shortly, but it appears that virtually all serve as food for some grazing animals.

8.1.1 Preference and vulnerability

As was true in predator–prey relations, preference is not due solely to choices made by the consumer. A herbivore's diet is the result of characteristics of both the herbivore and the periphyton assemblage. Important traits of the former include size and motility, morphological specializations for feeding, and digestive capabilities. Differential vulnerability among periphyton species is influenced by ease of harvest, palatability and nutritional value. Reproductive traits and chemical defenses might also be important, but evidence is scant concerning these last possibilities (Gregory, 1983).

Herbivory as a feeding role was described in chapter 6, but the variation within this category deserves added emphasis. Morphological specialization of invertebrates includes the blade-like mandibles of glossosomatid caddis larvae, the rasping radula of snails, chewing mouthparts in some mayflies and brush-like structures in others, piercing mandibles in hydroptilid caddis larvae, and so on. Too strict a reliance on functional group classifications might lead to the view that herbivory is confined to scrapers. Collector–gatherers surely consume loose algae along with microbes and detritus (Lamberti and Moore, 1984), and shredders benefit from the presence of an attached flora growing on the surface of fallen leaves (Mayer and Likens, 1987). Drifting diatoms and algae also are captured by suspension feeders, especially those taxa possessing fine sieving devices (philopotomid caddisflies, some chironomids and black fly larvae), and even the relatively coarse meshes of most hydropsychids retain some diatom and algal cells. Indeed, within the North American insect fauna, grazing has been noted in at least six orders and 38 families (Merritt and Cummins, 1984). Moreover, the composition of a herbivore's diet changes with many factors, including age, season, food availability and location (Lamberti and Moore, 1984). As a consequence, generalizations concerning invertebrate feeding roles should be taken as only a rough guide to the occurrence of herbivory.

Herbivorous vertebrates also exhibit a variety of feeding specializations that influence their impact as grazers. Loricariid catfish and some amphibian tadpoles have morphologies well suited to scraping periphyton from substrates, while limiting their ability to be selective on a fine spatial scale (Hynes, 1970; Power, 1983). The guppy also scrapes attached algae from surfaces, but more delicately, using its extended lower jaw (Dussault and Kramer, 1981), and thus is capable of feeding on smaller patches. Quite different feeding mechanisms are illustrated by the blacknose dace (*Phoxinus cumberlandensis*), which consumes both single-cell and filamentous autotrophs by ingesting substantial quantities of sand (Starnes and Starnes, 1981), compared with the grass carp, which can clip grass as neatly as a pair of shears (Rottman, 1977).

Considering the greater body size of vertebrates relative to invertebrates, it is hardly surprising that herbivorous vertebrates tend to eat larger growth forms and occasionally consume vascular plants, whereas invertebrate grazers primarily eat smaller components of the periphyton. Large invertebrates such as crayfish do harvest macrophytes directly (Lodge, 1991), but most invertebrates cannot. The alternative, piercing and sucking cell fluids, is known only in hydroptilid caddis larvae, and the rarity of this adaptation among freshwater insects is a puzzle (Hutchinson, 1981). However, the scarcity of vertebrate taxa that earn a living by scraping periphyton from substrates is at least partly attributable to the seasonality of this food supply at mid to high latitudes. Algae-scraping fishes are found primarily within lineages that possess mouthparts specialized for benthic feeding, and are most common in tropical streams where periphyton biomass is likely to be substantial throughout the year.

Just as animals differ in their mode of feeding, members of the periphyton differ in a number of ways that affect overall vulnerability and their availability to particular herbivores. Benthic autotrophs vary substantially in growth form and mode of attachment as well as in overall size (Figure 4.1), and this must affect vulnerability to particular kinds of grazers. For example, field manipulations of grazer densities in a California stream established that the mayfly *Ameletus*, with collector–gatherer mouthparts, was most effective with loosely attached diatoms. In contrast, the caddis, *Neophylax*, with stout, heavily sclerotized mandibles was effective against tightly adherent diatoms (Hill and Knight, 1988).

Filamentous algae apparently are difficult for grazing insects to harvest or digest, and so they are consumed principally as new growths (Lamberti and Resh, 1983; Dudley, Cooper and Hemphill, 1986). To the snail, *Lymnaea*, however, possessing both a radula for their harvest and a gizzard for their mechanical comminution, filamentous green algae provide a very satisfactory diet (Calow, 1970).

The assimilation efficiencies of herbivore–detritivores fed different diets is a useful measure of the wide range of nutritional value of various foods. Based on a review of 45 published values of assimilation efficiency for 20 species of aquatic insects, the range of assimilation efficiencies was 70–95% on a diet of animal prey, 30–60% for a variety of algal and periphyton diets and 5–30% on a diet of detritus (Pandian and Marian, 1986). Considerable variation in assimilation efficiency can occur even for a single species feeding on periphyton. Assimilation efficiencies for the snail, *Juga silicula*, were as high as 70–80% when first added to laboratory streams, but values declined during the course of the study to as low as 40% (Lamberti *et al.*, 1989). This coincided with a shift in composition of the periphyton, from diatoms and unicellular green algae to filamentous green algae and cyanobacteria. The decline in assimilation efficiency could be the result of cell senescence and other changes in physiological condition, or caused by a decline in nutritional value due to successional changes in the periphyton assemblage.

The wide range of assimilation efficiencies observed with periphyton diets is at least partly due to their structural and biochemical characteristics. Variation in protein and lipid content and in cell wall thickness likely are responsible for differences among autotrophs in their nutritional value and palatability. Too high a ratio of carbon to nitrogen is thought to make a poor diet, indicating a high cellulose and lignin content and a low protein content. Russell-Hunter (1970) suggests that C:N ratios should be less than 17:1 for animal utilization. On this basis, members of the periphyton appear to be generally suitable (C:N is 4:1 to over 8:1), whereas aquatic vascular macrophytes appear to be nutritionally less adequate (C:N is 13:1 to over 69:1) (Gregory, 1983). Based on a correspondence across sites between periphyton characteristics and gastropod growth, McMahon (1975) concluded that low C:N ratios and high organic content are indicative of a superior diet. Variation in the nitrogen content of diets was an extremely effective predictor of assimilation efficiency for the 20 taxa of aquatic insects reviewed by Pandian and Marian (1986).

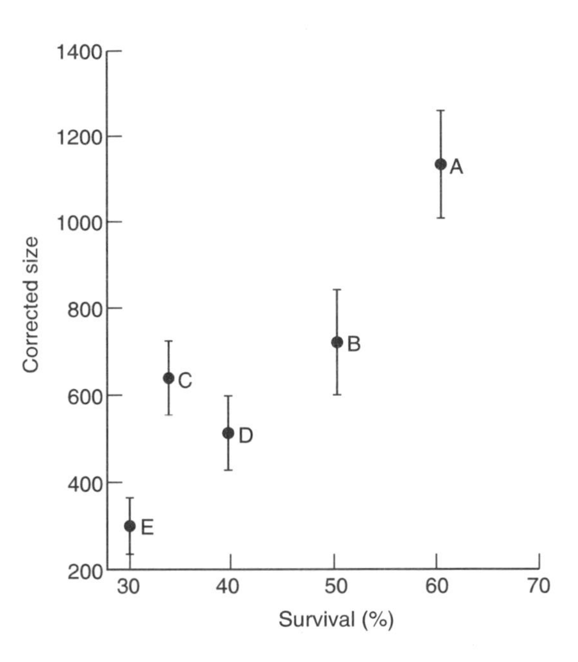

FIGURE 8.1 Growth and survival of larval black flies reared on various diets (A, diatoms; B, thin-walled Chlorophyta; C, firm-walled Chlorophyta; D, bacteria; E, unconditioned leaf litter). Final size (mean ± 95% confidence limits) of larvae of *Simulium verecundum*, *S. vittatum* and *Prosimulium mixtum* at the end of feeding trials was determined from the area of a projected image. (From Thompson, 1987.)

Even single-celled organisms vary in the degree of protection afforded by their outer walls. Larval black flies fed suspensions of algae, bacteria and detritus displayed higher growth and survival on *Chlamydomonas*, a green alga with thin cell walls, compared with several green algae with more robust cell walls (Figure 8.1).

Lipid content is another variable likely to influence the nutrition and development of herbivores. Most insects are unable to synthesize polyunsaturated fatty acids and sterols, indicating that the lipid content of their diets is important to food quality. Intense grazing by a snail and a larval caddisfly in laboratory streams altered the fatty acid composition of the periphyton, suggesting that grazing may have been responsive to this aspect of diet quality (Steinman, McIntire and Lowry, 1987b). Cargill *et al.* (1985) showed that specific fatty acids were critical dietary components to a detritivorous caddis larvae, *Clistoronia magnifica*. More detailed studies of the influence of lipid content clearly are needed.

Cyanobacteria are considered to be a poor food supply for freshwater plankton feeders (Porter, 1977). Cyanobacteria may have a high protein content, but other attributes, including a polymucosaccharide sheath rendering cell walls resistant to digestion, perhaps toxins, and a filamentous growth form all detract from their value as food. This might be expected to apply to periphytic growths of blue–green algae as well, but the evidence from lotic grazers is mixed. For example, in laboratory feeding trials the mayfly, *Tricorythodes minutus*, ate and assimilated two cyanobacteria, *Anabaena* and *Lyngbya* (McCullough, Minshall and Cushing, 1979), whereas *Asellus* and *Gammarus* would not consume *Phormidium* (Moore, 1975). Yet, as discussed in more detail shortly, orthoclad midges have suppressed blooms by *Phormidium* and *Oscillatoria* in outdoor channels (Eichenberger and Schlatter, 1978). Because studies of cyanobacteria have used primarily filamentous forms rather than colonies or single cells, consumption of the latter is little known.

Studies of utilization of cyanobacteria by fish are similarly mixed. Filamentous algae are a major component of the diet of *Ictalurus nebulosus*, the brown bullhead, which was able to assimilate the blue–green alga, *Anabaena flos-aquae*, more efficiently than the green alga, *Spirogyra* (Gunn, Qadri and Mortimer, 1977). It may be significant that *Spirogyra* has the lower protein content of these two foods.

It is difficult to specify whether the wide range in the reported nutritional quality of cyanobacteria is due to variation in plant quality or differences among herbivores in their machinery for harvest and digestion. The latter is important, however. The radula and gastric mill of some snails give them an advantage over most other invertebrates. Algivorous fish have developed three methods for breaking down plant cell walls: mechanical grinding, lysis by acids that produce a stomach pH as low as 1.4, and cellulase enzymes derived from gut microflora (Power, 1983). Digestion also is aided by a lengthy intestine, which in herbivorous and ooze-feeding fish is typically at least one to two times fish length and sometimes much longer, thus prolonging gut residence time. Clearly, more needs to be learned both about digestive machinery and plant palatability before we will achieve a full understanding of the nutritive value of various plant and periphyton diets.

Although herbivory is considered to play a role in the evolution of structural defenses in terrestrial plants, there is no evidence that the autotrophs of running waters have evolved chemical defenses against herbivores. This may reflect inadequate study, however. It is curious that *Potamophylax cingulatus*, a versatile caddis larva, was less effective in consuming various aquatic and semi-aquatic plants compared with the leaves of some terrestrial plants (Otto and Svensson, 1981). This suggests that the former may have substantial defenses, whatever their evolutionary origin.

It is also possible that some algal species or growth forms simply are more vulnerable to a range of invertebrate consumers, but are able to compensate with a high reproductive rate. Such a positive association between vulnerability to grazing and rate of reproduction, allowing plants to be scaled along a continuum from fast-edible to slow-inedible, has been demonstrated within terrestrial plant–insect interactions (Feeny, 1976). It remains to be seen whether a similar story also applies to the periphyton.

8.1.2 Foraging behavior

Because of the importance of diet to their growth and development, we might expect herbivores to exhibit foraging behaviors that aid them in selecting nutritious food and concentrate their feeding efforts in high quality patches. In one of the few studies that directly examined diet choice in a lotic herbivore, Calow and Calow (1975) found a correspondence between rank order of food preference and the nutritive value of various algae to *Ancylus fluviatilis*. Although further evidence that herbivores exhibit behavioral preference for particular food items is limited, several studies document a clear response of herbivores to the distribution and abundance of benthic autotrophs.

The periphyton is patchily distributed, from the smallest scale of the surface of an individual substrate (Munteanu and Maly, 1981), to an intermediate scale such as from stone to stone (Jones, 1974, 1978), through larger scales such as open *versus* canopied sections of streams (Power, 1983). Although some herbivores might feed essentially at random, an ability to perceive and respond to this patchiness ought to be advantageous. In fact, non-random foraging has been established in both vertebrate and invertebrate grazers of periphyton. Detailed analyses of foraging in the caddis larvae *Dicosmoecus* (Hart, 1981) and the mayfly nymph *Baetis* (Kohler, 1984) document that, when provided with periphyton patches of differing density, these insects spend much more time in more rewarding patches than would be expected under a model of random movements. When individual *Dicosmoecus* entered a region rich in periphyton, gathering movements of the forelegs and the rate of mandibular scraping both increased. In addition, overall movement rate slowed, and individuals tended to turn back on reaching a patch boundary. As a result, time spent in rich patches was two to three times what would be expected by chance alone.

The ability to perceive spatial heterogeneity in food supply and respond by simple movement rules that tend to concentrate foraging in regions of high reward is termed 'area-restricted search'. When the periphyton attached to an artificial substrate were scraped to create a checkerboard design that covered only 20% of the substrate surface, *Baetis tricaudatus* spent 67–82% of its time in food patches (Kohler, 1984). High foraging efficiency is indicated by this result, but whether efficiency is maximal is more difficult to assess. One approach involves a calculation called thoroughness, which compares the area searched to the smallest area that circumscribes the sequence of movements. Thoroughness values for *B. tricaudatus* within a patch were high (0.74 compared with a 'perfect' 1.0), exceeding by 5–10 times the values obtained both prior to locating a patch and after exiting. Moreover, search behavior on departure from a patch was influenced by patch quality. Search intensity was much greater just after departing a high quality patch, as evidenced by high thoroughness and low movement rates, compared with patches of lower quality (Figure 8.2). This had the apparently detrimental result that patches were re-visited more often than would seem optimal (Kohler, 1984). Nonetheless, aquatic insects clearly increase their foraging efficiency compared with random search, by employing simple behavioral rules that result in area-restricted search. Provided that the animals do not alter abundances within a patch too rapidly, one might expect to find aggregations of grazers within dense patches of periphyton. However, the rate at which patches are used up, and the ability of grazers to move on rapidly in search of other rich areas, are additional complications to consider.

8.1.3 Response of herbivores to variation in the distribution and abundance of periphyton

The overall abundance of herbivorous animals ought to depend upon the biomass and productivity of autotrophs, an expectation that is

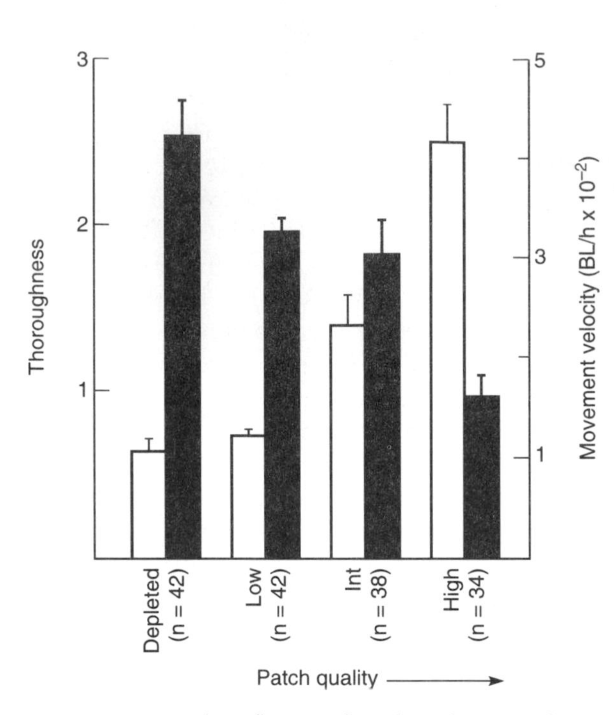

FIGURE 8.2 The influence of patch quality (periphyton cell density) on *Baetis* search behavior immediately after leaving a patch. Thoroughness (□) increases, and movement rate (■) decreases with increasing patch quality. (From Kohler, 1984.)

supported by both correlative and experimental evidence. For example, sunny, open-canopy stream sections often support greater invertebrate densities than are found in shaded stream sections (Allen, 1951; Hawkins, Murphy and Anderson, 1982; Wallace and Gurtz, 1986). The redistribution of foragers from food-poor to food-rich areas by the searching behaviors just described is one mechanism to cause such a pattern. Greater reproductive success within more productive habitats is another possible reason why areas rich in resources may also be rich in consumers.

On the other hand, if highly mobile herbivores concentrate where resources are rich, periphyton abundances will be reduced, perhaps to levels that are roughly similar among stream sections. As a consequence, a rich but crowded foraging location might offer the individual grazer no higher a rate of return than a less productive but less crowded region. A likely consequence is for the abundance and biomass of grazers to increase proportionately with autotrophic productivity, but for foraging gain per individual to be roughly constant. Power (1983) observed just this pattern in the distribution of loricariid catfish among pools in a Panamanian stream. Shaded pools were less productive and supported a lower abundance and biomass of fish compared with open pools. However, individual growth rates were similar over the resource gradient (Figure 8.3). Movements of individuals among pools in a manner similar to the finer scale foraging behaviors of *Baetis* and *Dicosmoecus* presumably result in this pattern, referred to as the ideal–free distribution.

Experimental studies show a response of animal populations to periphyton abundance, across scales ranging from individual stones, to a few square meters, to entire stream sections. Richards and Minshall (1988) conducted studies on a small scale in an alpine stream, using natural stones that were selected based on visual assessment of periphyton abundance and in some instances scraped to produce patches of various widths. Stones were replaced in the stream under glass viewing boxes, and insect presence was determined from film records. Within 1–2 days, the number of *Baetis bicaudatus* showed marked concentration in patches rich in periphyton. Moreover, baetid density among stones correlated strongly with subsequent measures of chlorophyll *a*. It seems likely that rapid recolonization via drift, and re-distribution according to the mechanisms described by Kohler (1984), explains this rapid adjustment of grazer densities to small-scale variation in periphyton abundance.

A similar response of grazers to periphyton is also apparent at larger spatial scales. Fuller, Roelofs and Fry (1986) shaded a 20 m-long riffle with black plastic, reducing chlorophyll *a* levels after four weeks to 0.1 $\mu g\,cm^{-2}$ or less, compared with values of 1–6 $\mu g\,cm^{-2}$ in unshaded

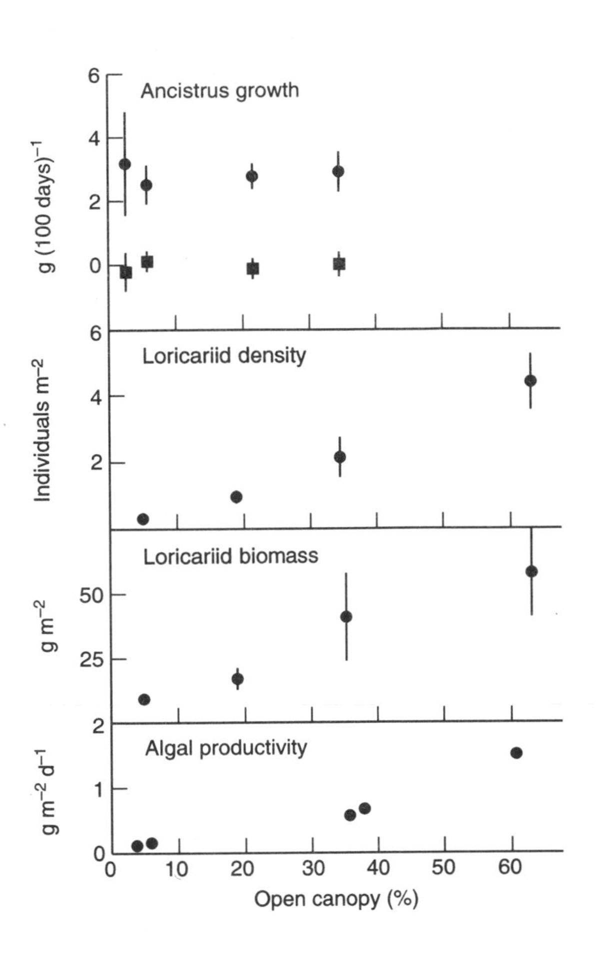

FIGURE 8.3 Evidence that the loricariid catfish, *Ancistrus*, conforms to an ideal–free distribution. Algal productivity increases in relation to openness of canopy. Density and biomass of catfish increase proportionately with algal productivity, but growth rate (●, rainy season; ■, dry season) is constant. Two standard errors are shown. (From Power, 1983).

sections. *Simulium* larvae increased slightly, and several taxa showed no apparent change, but *B. tricaudatus* was much rarer in the shaded section than in open controls. Removal of one-half of the shaded plastic resulted in chlorophyll *a* levels similar to the open stream after only 11 days, as well as much greater *Baetis* densities. Body size was smaller for those mayflies that remained in the shaded section, suggestive of reduced food availability, and also a rather surprising reluctance to depart for greener pastures. Unaffected taxa (*Simulium*, several crustaceans) presumably relied on other food sources. A similar experiment in a New Zealand stream (Towns, 1981) failed to influence any of the fauna, and was taken as evidence for the view that organic microlayers and associated microorganisms are the most important energy sources in some instances.

At an even larger spatial scale of stream sections or entire streams, grazing animals have been shown to respond to variation in resource availability (for instance, the loricariid catfish described in Figure 8.3). Wallace and Gurtz (1986) compared two small streams in the southeastern USA: one that recently had been clear-cut and a reference site. Immediately after canopy removal, *Baetis* production was roughly 18 times higher than in the reference stream, and on stable substrates the difference was even greater. Mayfly guts contained mainly diatoms and estimates of gut fullness from the clearcut stream were up to double those from the reference stream. Although cell densities were not greatly discrepant between sites, periphyton production (based on *Baetis* production and projected food consumption) was estimated to be nearly 30 times greater in the open site. Subsequent regrowth resulted in canopy closure, and after 6 years, *Baetis* was much rarer and periphyton production had dropped 10-fold. Because clear-cutting affected the entire stream, recruitment rather than redistribution is the presumed mechanism. In fact, the response of *Baetis* was much greater than occurred in other mayflies, which Wallace and Gurtz attributed to its short generation time and multivoltinism allowing a rapid increase in population size.

In sum, grazing animals respond to locations of high periphyton abundance, both by shifts in distribution and, if conditions persist for enough time, by population recruitment. The resultant increase in herbivory may eliminate any differences in periphyton standing crops, thereby obscuring any correlation between producer and consumer abundance. However, positive associations between herbivore and periphyton abundances often are observed, and because herbivores have been shown to concentrate where their food is abundant, we turn next to the issue of their effect on community and physiognomic structure of the periphyton assemblage.

8.1.4 Influence of herbivory on the periphyton

Herbivores affect the periphyton in multiple ways. As with animal predation, the influence of herbivory can be due to direct consumption or via a variety of indirect pathways, of which nutrient regeneration and physical disruption of periphyton mats are prominent possibilities. The response of autotrophs can be detected in several ways. While the standing crop or benthic biomass likely will be reduced, the photosynthetic activity per remaining cell, and turnover rate of the population, might change in the opposite direction. Effects on the composition of the periphyton assemblage also should be anticipated, but it is difficult to say precisely how assemblage structure might change. For simplicity we will consider these several responses in turn, but it is important to recognize how intertwined they really are. One also should appreciate that, inherent in this approach, is the view that the trophic level above (the herbivores) controls events in the trophic level below (the periphyton). In contrast, the preceding section considered evidence that food availability determined herbivore abundance and distribution. Whether 'bottom-up' or 'top-down' control prevails may depend on time, place and environmental circumstances, and they may not be mutually exclusive.

(a) Effects of herbivory on periphyton standing crops

An inverse relationship is expected between the abundances of benthic autotrophs and their grazers, especially if herbivory is substantial. Douglas (1958) observed this pattern for diatoms in an English stream, based on counts of *Achnanthes* cells and the caddis *Agapetus fuscipes*. One also would expect that total elimination of grazing invertebrates would allow periphyton blooms to occur. Application of pesticides has produced just this manipulation, and the response can be dramatic. When rotenone was applied to 75 m long outdoor channels to exclude orthoclad midge larvae, filamentous algae came to predominate and a shift to dominance by cyanobacteria occurred over a 2 month period (Eichenberger and Schlatter, 1978). Grazing largely prevented this occurrence in control channels, and when midge larvae were allowed to invade a third, rotenoned channel after 2 months, they caused a sharp decline in the benthic flora. Pesticide application to a small mountain stream in Japan corroborates the channel study. Most of the zoobenthos disappeared and a periphyton bloom developed with roughly an order of magnitude increase in chlorophyll *a* levels. The bloom ended coincident with recovery of the benthic fauna (Yasuno *et al.*, 1982).

Grazing studies in laboratory streams (Sumner and McIntire, 1982; Lamberti *et al.*, 1989) and in natural streams (Hart, 1981; Lamberti and Resh, 1983; McAuliffe, 1984a) have convincingly demonstrated the capacity of herbivores to reduce periphyton. Field experiments are especially compelling because they establish the importance of herbivory under fairly natural conditions. Exclusion manipulations are relatively easy with those herbivores that colonize substrates mainly by walking rather than drifting. Raising tiles on supports and constructing barriers out of vaseline and rubber bands make effective barriers to a number of grazing caddis larvae. Periphyton biomass increased some 5–20 times relative to control substrates when *Helicopsyche* was excluded (Figure 8.4; Lamberti and Resh, 1983), and cell numbers increased fivefold in a *Glossosoma* exclusion study (McAuliffe, 1984a). Because more mobile

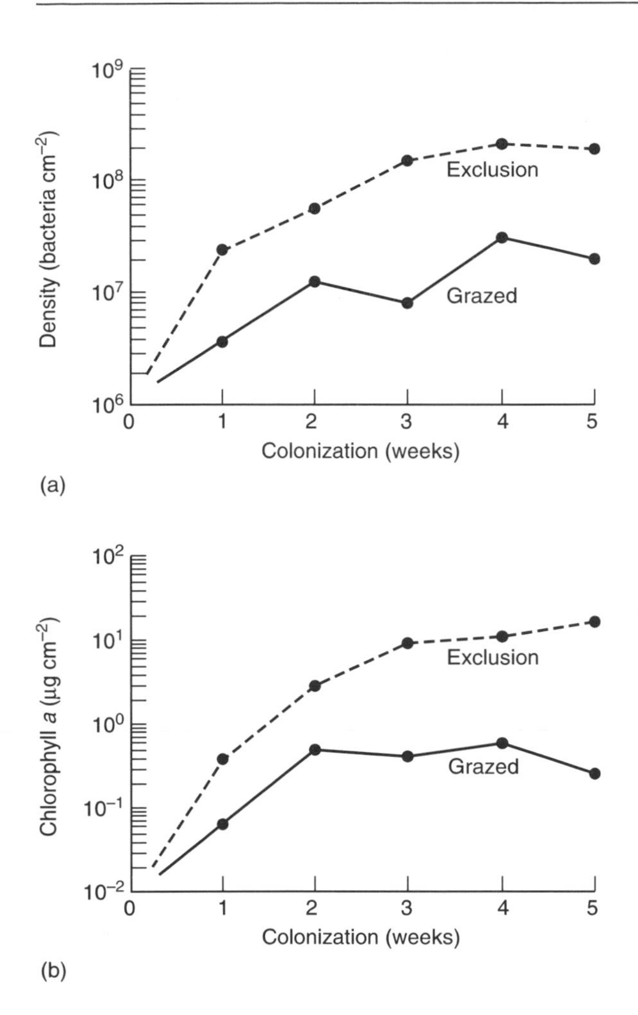

FIGURE 8.4 The influence of *Helicopsyche* grazing (——) and exclusion (----) on the time course of development of (a) periphyton populations (chlorophyll *a*) and (b) bacterial densities. (From Lamberti and Resh, 1983.)

grazers, including mayflies and midges, were unaffected by the manipulation in these two studies, the outcome suggests that these grazers were ineffectual.

The majority of studies that demonstrate a reduction in periphyton standing crop due to grazing have employed caddis larvae (above references; Hart, 1985b; Hill and Knight, 1988) and snails (Sumner and McIntire, 1982; Lamberti, Feminella and Resh, 1987; Lamberti, 1989). The implication is that effective grazers tend to be individually large, relatively slow-moving, and well equipped with scraping mandibles or a radula. Lamberti, Feminella and Resh (1987) compared the mayfly, *Centroptilum elsa*, the caddis, *Dicosmoecus gilvipes*, and the snail, *Juga silicula*, which they characterized as a browser, scraper and rasper, respectively. Laboratory streams were inoculated with algal scrapings, consumers were added at approximately natural densities 9 days later, and development of the periphyton mat was monitored for 48 days. The effect of the mayfly was slight and confined to small (less than 2 cm diameter) patches, but *Juga* had a substantial impact and *Dicosmoecus* even more so (Table 8.1). The comparatively minor effect of mayfly grazing is unsurprising, given its low biomass. In the same vein, the effect of the caddis is especially convincing, as the snail was present in greater numbers and biomass.

Although most evidence to date indicates that caddis and snails are the likely candidates to exert top-down control on temperate stream periphyton, this may reflect inadequate study. Evidence that midge larvae can suppress periphyton blooms was described above. In addition, the mayfly, *Ameletus*, stocked at realistic densities in Plexiglas chambers containing natural streambed material, caused marked reductions in periphyton standing crops, even at densities of 0.5 times ambient (Hill and Knight, 1987; Figure 8.5). The possibility also exists that entire guilds of herbivorous invertebrates have a significant effect in the aggregate, but this may be difficult to establish from studies of individual species.

Where herbivorous fish are plentiful it seems likely that they exert considerable control over benthic autotrophs. Certainly the stoneroller, *Campostoma*, in North America and loricariid catfish in tropical America can devastate algal patches. Exclusion of *Campostoma* (by the presence of a piscivorous fish) resulted in growth of filamentous algae, while introduced

TABLE 8.1 A comparison of the effects of three invertebrate herbivores on the development of periphyton in laboratory streams. Ash-free dry mass (AFDM) is a measure of periphyton biomass. (From Lamberti, Feminella and Resh, 1987)

Grazer stocking density			*Algal response*			
Species	*Numbers*	*AFDM*		*AFDM*	*Chlorophyll a*	*GPP*
Centroptilum elsa	$500\ m^{-2}$	$0.5\ g\ m^{-2}$	Slowed rate of accumulation after Day 16, eventual 20% reduction	40%	↓	Not measured
Dicosmoecus gilvipes	$200\ m^{-2}$	$2.5\ g\ m^{-2}$	Immediate reduction of algae to very low levels	>95%	↓50%	↓
Juga silicula	$350\ m^{-2}$	$4\ g\ m^{-2}$	Early effect on accumulation, eventual 50% reduction	60%	↓	25%↑

Campostoma caused rapid declines (Power and Matthews, 1983).

(b) Effects of herbivory on periphyton assemblage composition

Heavy grazing pressure has the potential not only to reduce the total biomass of periphyton, but also to alter the structural and taxonomic composition of the assemblage. One effect apparent in virtually all experimental exclusions of herbivorous invertebrates is a great increase in filamentous green algae and cyanobacteria. A reduction in large, loose, or overstory components of the periphyton assemblage and at least a relative increase in small and tightly adherent cells appears to be a general outcome of heavy grazing (Sumner and McIntire, 1982; Colletti *et al.*, 1987; Steinman *et al.*, 1987a).

Steinman *et al.* (1987a) provided a particularly detailed account of changes in the algal assemblage in response to four densities of *Juga* and *Dicosmoecus*. Ungrazed streams developed thick periphyton mats during the 32 day study. Rosettes of *Synedra* and aggregates of *Characium* became established first, then the assemblage became more heterogeneous as patches of other diatoms developed (*Scenedesmus*, *Achnanthes*, *Nitzschia*), followed by the filamentous green alga, *Stigeoclonium tenue*, and the blue–green alga, *Phormidium uncinatum*. Grazing caused severe reductions in *Characium*, whose dense aggregations apparently made it vulnerable, and in *Scenedesmus*. In contrast, *A. lanceolata* and *S. tenue* increased in representation under herbivory, the former simply because it is adnate, the latter because its prostrate basal cells are little grazed. Steinman *et al.* speculate that *S. tenue* may actually benefit from mild grazing due to removal of overstory cells, but this remains uncertain.

The snail and the caddis just described caused broadly similar changes to periphyton assemblages. However, it is equally plausible that two herbivores could have quite different effects, as Hill and Knight (1987, 1988) demonstrated in their comparison of a caddis larva (*Neophylax*) and a mayfly (*Ameletus*). Loose and adnate layers were sampled separately and *Ameletus* affected principally the former, causing declines in motile diatoms including *Surirella spiralis* and several species of *Nitzschia*. *Neophylax* affected both layers, but its major impact was due to its reduction of a particularly large, adnate diatom that comprised the bulk of total periphyton biovolume.

The most obvious mechanism by which grazers

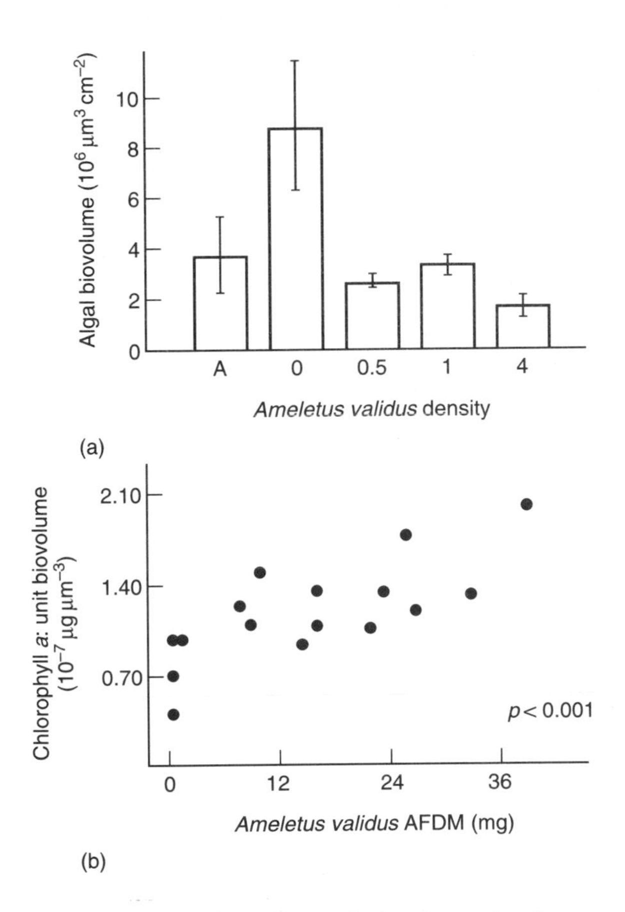

FIGURE 8.5 The effect of *Ameletus* density on periphyton standing crop and quality. (a) Periphyton abundance under various grazing conditions. A, ambient densities on streambed; 0, cages with zero density; 0.5, 1 and 4, cages with 0.5, 1 and 4 times natural densities respectively. Results were similar for chlorophyll *a* and ash-free dry mass (AFDM). Note that even low densities of grazers had a marked effect. (b) The ratio of chlorophyll *a* per unit biovolume increased significantly ($p < .001$) with *Ameletus* biomass. (After Hill and Knight, 1987.)

can alter periphyton abundance is through direct consumption. Confirming evidence is scant, but inspection of *Neophylax* gut contents revealed a sevenfold over-representation of the diatom most affected by manipulations of grazer density (Hill and Knight, 1988). Similarly, the abundance of *Nitzschia* in *Ameletus* guts was consistent with a grazing effect on this diatom, just as the under-representation of certain adnate diatoms was consistent with their relative increase under grazing. However, a number of other diatoms that decreased in response to increasing *Ameletus* grazing also appeared not to be eaten, indicating an adverse effect of the mayfly other than direct consumption. Hill and Knight (1987) found the presence of *Ameletus* to reduce silt within the loose layer, and postulated that some motile diatoms declined as a result of this physical disturbance.

There are in fact several possible mechanisms by which the activities of grazers may adversely affect some members of the periphyton, other than directly through grazing (Lamberti and Moore, 1984). The type and amount of physical disturbance and dislodgement must vary with the size, feeding and locomotory mode of the herbivore. Even non-selective removal may alter composition if members of the periphyton differ in their rates of re-colonization and growth (Sumner and McIntire, 1982). In fact, diatom colonization onto glass slides exhibits a sequence where fast colonists are gradually replaced by other species that tend to persist, evidently because of superior reproductive success once established (Oemke and Burton, 1986). Should the presence of grazers indirectly benefit autotrophs by making nutrients or light more available, members of the periphyton might differ in their rates of response. Another, perhaps unusual indirect effect is suggested by Hart's (1985b) study of *Leucotrichia*. This sessile caddis larva apparently maintains a garden of diatoms and algae close to its residence, and somehow prevents overgrowth of its foraging area by an undesirable cyanobacterial mat. Because no filaments of the blue–green alga can be found in guts, Hart suspects that *Leucotrichia* 'weeds' its foraging area to prevent overgrowth.

Studies of plankton and of the marine intertidal suggest that species diversity at the producer trophic level can be influenced by herbivory. It appears that diversity may increase or decline, depending on whether herbivore pressure falls most heavily on species of autotrophs that are competitively superior under existing environmental conditions, or the reverse (Porter, 1977; Lubchenco, 1978). Such a dual response was observed when the effects of grazing by *Juga* were examined over a range of light and nutrient treatments (Sumner and McIntire, 1982). When physical conditions favored larger, overstory taxa, grazing increased diversity and reduced dominance. When light and nutrient levels favored small, non-filamentous algae, the opposite effect occurred. Whenever grazing pressure is intense, however, the general trend appears to be towards a reduction in diversity within the periphyton assemblage (e. g. Colletti *et al.*, 1987).

Periphyton mats in laboratory channels exhibit certain distinctive trends. Mats usually develop to a much greater thickness than commonly occurs in nature, algal succession culminates in dominance by filamentous green and blue–green algae, and eventually the mat sloughs away. Usually this succession can be prevented by an abundance of grazers. Whether a similar sequence might occur routinely in streams in the absence of herbivores is uncertain; however, the aftermath of a pesticide application (Yasuno *et al.*, 1982) and the manipulative study of Lamberti and Resh (1983) certainly indicate the possibility.

In summary, evidence from laboratory and field experiments demonstrates that grazing not only can reduce total periphyton biomass, but also alter assemblage composition and structural characteristics. Both the direct effects of consumption and various indirect effects such as physical disruption and regeneration of nutrients are likely to be important. In either case, herbivore-induced changes in the amount and kinds of periphyton taxa present almost certainly will alter metabolic properties such as photosynthetic and turnover rates. Perhaps the most likely effect of herbivory is to increase production and turnover rather than decrease biomass, and we turn now to further consideration of this possibility.

(c) Effects of herbivory on primary production and plant turnover

Theoretical studies tell us that a small algal biomass with a rapid turnover rate can support a larger biomass of consumers with a much slower turnover rate (McIntire, 1973; McIntire and Colby, 1978). Estimates of grazer:algal biomass ratios of 13:1 (McIntire, 1975) and 20:1 (Gregory, 1983), and simple consideration of relative generation times, lend further support to this possibility. It is also plausible that moderate grazing enhances primary production relative to ungrazed populations due to a reduction in self-shading and nutrient depletion, thereby increasing total algal production. Thus, while an increase in herbivory is likely to cause a decline in periphyton biomass, cell-specific photosynthetic rates are likely to increase. As a consequence, areal production might be greatest under intermediate grazing pressure (Figure 8.6). This intriguing possibility remains largely untested, although some supportive evidence does exist. In laboratory stream studies, McIntire and Phinney (1965) found that highest rates of photosynthesis were attained at low biomass, which they attributed to less shading and senescence. Subsequent modeling efforts incorporated this finding, with the result that the periphyton compartment displays self-limitation. As a cautionary note, however, it might be that the effect is exaggerated in laboratory streams, where initial colonization and rapid growth, followed by senescence and sloughing, are especially pronounced.

Field manipulations of grazing intensity have confirmed that photosynthetic rates per cell or per unit of chlorophyll *a* can increase due to harvesting. Tiles grazed by *Helicopsyche* evolved five times more oxygen per unit chlorophyll *a* as compared to ungrazed tiles, and had an estimated turnover time of 8 days *versus* 16 days (Lamberti and Resh, 1983). While this result certainly indi-

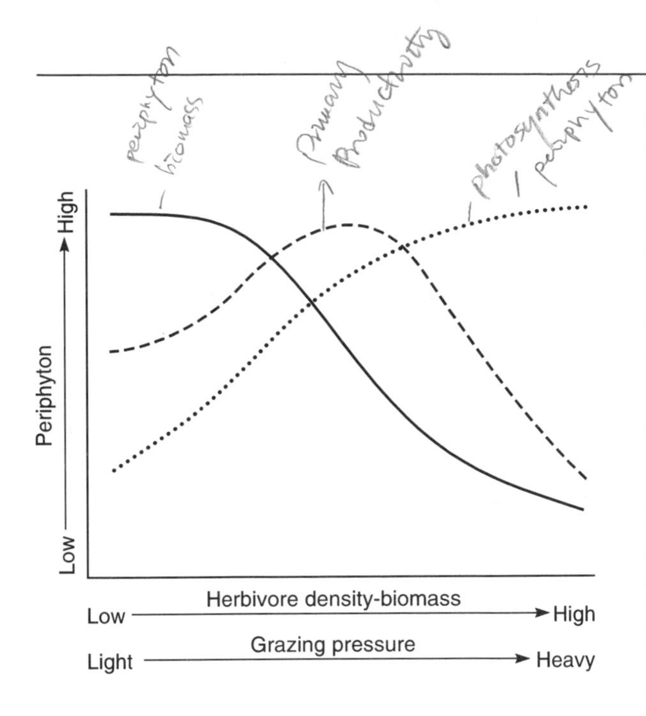

FIGURE 8.6 Theoretical expectation of the relationship between periphyton attributes and the intensity of herbivory. A decline of periphyton biomass (——) with increasing herbivory appears well supported by evidence, and a few studies also document higher photosynthetic rates per unit biomass (······) with increased grazing. The peak in gross primary production (----) at intermediate levels of herbivory is largely speculative. (Modified from Lamberti and Moore, 1984.)

cates increased periphyton productivity as a result of consumer harvesting, production per unit area remained roughly 1.75 times greater on ungrazed tiles, suggesting that increased turnover rate did not fully compensate for reduced biomass. Grazing by *Ameletus* also increased the ratio of chlorophyll *a* to cell biovolume while reducing the density of periphyton in Hill and Knight's (1987) study (Figure 8.5), although data for primary production are lacking. Grazing by *Juga* and *Dicosmoecus* each increased production per unit biomass of periphyton, but the former caused gross primary production to increase while the latter had the opposite effect (Table 8.1). This laboratory study thus supports the view that gross primary production under moderate grazing may be enhanced relative to the ungrazed condition. Possible mechanisms suggested by Lamberti *et al.* (1987) include removal of senescent cells (either by consumption or dislodgement), reduction in self-shading, and perhaps the regeneration of nutrients by grazer excretion. Clearly these are pathways deserving of further study.

8.1.5 Biotic *versus* abiotic control of periphyton populations

In the discussion of factors limiting the abundance and distribution of the periphyton (chapter 4), grazing was treated as a minor factor. Yet, herbivory has been shown to be of considerable importance in a number of instances. How can these conflicting perspectives be reconciled? One possibility is that experimental studies of grazing are few and recent. As more work is done, the importance of 'top-down' control may be strengthened further. Alternatively, it may be that the importance of herbivory depends on environmental conditions and varies considerably with time and location. Some support exists for this latter suggestion.

Certain conditions are commonly although not exclusively associated with studies where grazing effects were pronounced. A disproportionate amount of work has concerned snails and caddis larvae, either in laboratory streams during the establishment of a periphyton mat or in natural streams under low-flow, late summer conditions. Sumner and McIntire (1982) varied both abiotic conditions and the presence of grazers, and found an interaction between these factors. Unquestionably, grazing by *Ameletus* reduced periphyton populations over 3 weeks in late summer (Figure 8.5). However, winter flooding is believed to play a major role in reducing insect populations, and Hill and Knight (1987) suggest that mayfly densities may have been greater than usual during their study due to reduced flooding the previous winter. At the moment this is largely speculation, but research is needed to sort out the relative importance of environmental factors and herbivory in the dynamics of periphyton.

8.2 Herbivory on macrophytes

Herbivory has long been considered of minor significance to freshwater macrophytes (Shelford, 1918; Hutchinson, 1975; Gregory, 1983). According to Wetzel (1983), less than 10% of macrophyte production is consumed live. The tough cell walls and high lignin content that provide structural support to macrophytes are effective barriers against their ingestion and digestion. Studies of energy processing in streams have reported macrophytes to be ignored by the animal community present (Koslucher and Minshall, 1973; Fisher and Carpenter, 1976), in accord with expectations based on the low nutritive value of large aquatic plants. Thus, as discussed in chapter 5, the great majority of macrophyte production in running waters enters food chains as detritus, not as living plant tissue.

There are of course exceptions to these statements, and possibly all plants are subject to some herbivory (Gregory, 1983). Recent reviews by Newman (1991) and Lodge (1991) discuss examples of significant herbivory on living macrophytes. The majority of reports involve birds, mammals and decapods, but fish and insects also consume living macrophytes. Emergent macrophytes appear less vulnerable, presumably because of their greater investment in support tissue.

Herbivorous taxa belonging to primarily aquatic insect groups usually are ineffective grazers of higher plants (Newman, 1991). However, some invertebrates, mainly decapod crustaceans and certain insects of terrestrial origin, can reduce the biomass of submersed and floating-leaved taxa. Several studies have found crayfish to reduce the abundance of submersed macrophytes significantly, although they have no apparent effect on emergent taxa (Lodge and Lorman, 1987; Feminella and Resh, 1989). Crayfish also cause non-consumptive loss by clipping shoots of submersed macrophytes, which then float away. In northern Wisconsin lakes, *Orconectes* species altered the assemblage structure of submersed macrophytes, primarily because single-stemmed species were more vulnerable than rosulate growth forms (Lodge and Lorman, 1987).

Intriguingly, the most dramatic effects of invertebrate grazing on living aquatic macrophytes involve herbivores derived mainly from terrestrial insect lineages. These include chrysomelid and curculionid beetles, aquatic lepidopterans, and specialized dipterans (Newman, 1991). In some parts of the Ogeechee River, Georgia, leaves of the water lily, *Nuphur luteum*, are eaten by the water lily leaf beetle, *Pyrrhalta nymphacaeae* (Chrysomelidae). Wallace and O'Hop (1985) estimated that *Nuphur* leaves lasted only 17 days at a beetle-infested site, compared with over 6 weeks at another site where the beetle was absent. Some macrophytes, including the water hyacinth, *Eichhornia crassipes*, and the kariba weed, *Salvinia molesta*, can become so abundant that they present serious weed-control problems world-wide, particularly in the subtropics and tropics. A Brazilian beetle that feeds on kariba weed is one potential agent of biological control (Barrett, 1989).

Vertebrate herbivores that have significant impacts on macrophytes include waterfowl, mammals (moose, manatee) and some fishes. The grass carp, *Ctenopharyngodon idella*, is so effective a grazer that it is widely used in the control of aquatic weeds (Shireman, 1984). Although grass carp will eat many different plants, they exhibit selective feeding when offered choices among species of submersed macrophyte (Wiley, Pescitelli and Wike, 1986). However, studies of the energetics of grass carp reveal something of the limitations of a diet of living macrophyte tissue. Grass carp daily rations (in wet mass of macrophyte tissue) range from 50% (Wiley and Wike, 1986) to over 100% (Shireman and Smith, 1983) of their body mass per day, indicating that this feeding strategy is based on a high volume of material. Furthermore, the grass carp's preference among nine macrophyte species native to the midwestern USA was not correlated with measures of plant nutritive value. Instead, preferences appeared to reflect relative handing times, allowing fish to maximize 'through-put' (Wiley, Pescitelli and Wike, 1986). Grass carp also are known to have a low metabolic rate and assimilation efficiency relative to other fishes, and to require animal protein for proper growth (Wiley and Wike, 1986).

Bryophytes are consumed on occasion by a variety of invertebrates, and some, including representatives of caddis, danceflies and mites, are specialists. However, even such versatile feeders as amphipods, when offered a liverwort (*Nardia*) or nothing at all, elected to starve to death (Willoughby and Sutcliffe, 1976), and it seems unlikely that herbivory is ever a major burden of bryophytes.

The interaction between plants and herbivores may involve pathways other than direct consumption. Macrophytes influence the animal community by altering physical conditions and occupying space (Dudley, Cooper and Hemphill, 1986). The presence of large plants reduces current velocity, promotes greater settling of silt and detritus, and provides shelter for invertebrates and substrate for epiphytes (Gregg and Rose, 1982). Under conditions where plants provide virtually the only stable substrate, most of the invertebrates will be found associated with plant surfaces (e. g. Iverson *et al.*, 1985). In addition, turbidity and deposition of sediments associated with slow currents create adverse conditions for the non-burrowing bottom fauna, further favoring an epiphytic existence. In faster flowing rivers with a stable substrate, the opposite occurs, and stones usually harbor far higher numbers of invertebrates than are found on vascular plants (Gregg and Rose, 1982; Rooke, 1984).

Plant architecture apparently can favor certain groups of animals, and several studies have noted at least some differences when plants with different growth forms were compared. *Ranunculus longirostris*, with dissected leaves, trapped more fine detritus and attracted more collector–gatherers than did the broad-leaved *Potamogeton amplifolius*, which apparently was viewed simply as a flat substrate (Rooke, 1984). However, while other studies also have reported some effect of plant architecture, at present it seems that differences are not pronounced (Rooke, 1984; Gregg and Rose, 1985; Tokeshi and Pinder, 1985).

A number of studies report stimulation of macrophyte production under moderate grazing, for reasons similar to discussed under periphyton (Figure 8.6). Removal of senescent material and fertilization by fecal matter (by waterfowl especially) are likely mechanisms (Lodge, 1991). It has been suggested that some rather complex interactions take place between epiphytes, their grazers, and the macrophyte substrate. As one possibility, macrophytes might benefit epiphytes by release of DOM and nutrients, while in turn epiphytes release carbon dioxide and micro-nutrients (Wetzel, 1983). Alternatively, epiphytes can adversely affect macrophytes by competing for light and nutrients. Brönmark (1985) has shown in the laboratory that grazing snails can perceive DOM at a distance, but not the presence of epiphytic algae, and suggests that DOM release by macrophytes serves to attract grazers which in turn rid the plant of its epiphyte load.

8.3 Grazing on lotic phytoplankton

The extensive literature concerning planktonic food chains in lakes, and perhaps especially in reservoirs (Wetzel, 1983), should be a rough guide as to what to expect of river plankton. However, large rivers differ from lakes in a number of physical and chemical attributes. Rivers lack pronounced thermal stratification, contain high concentrations of inorganic particulates and nutrients, and their planktonic populations experience continual downstream displacement. Phytoplankton populations of rivers usually are lower than would be found in lakes of comparable nutrient availability (chapter 4), and zooplankton biomass also is less than would be expected based on algal abundance (Pace, Findlay and Lints, 1988). Rivers also differ from lakes in having a zooplankton where rotifers are numerically dominant and crustaceans are comparatively few (Hynes, 1970).

The scarcity of zooplankton relative to phytoplankton in large rivers likely is due to the comparatively slow population growth rates of the former. Doubling times of zooplankton may be too slow for them to 'catch up' with more rapidly growing phytoplankton populations during downstream transit. The observed scarcity of crustaceans relative to rotifers in the zooplankton of large rivers presumably has the same explanation. Rotifers reproduce at a considerably faster rate than do crustacean zooplankton (Allan, 1976), and this will result in a preponderance of rotifers wherever wash-out is high. Because population growth rates increase with temperature, warmer water and longer transit times should permit a crustacean zooplankton to develop, and this occurs in the Nile, where the temperature range is between 20 and 33°C (Dumont, 1986). As further evidence, transit times at moderate flows are two to three times longer in the White Nile compared with the Blue Nile. There is a corresponding preponderance of crustacea over Rotatoria in the White Nile, while the reverse is true of the Blue Nile (Rzóska, Brooks and Prowse, 1952).

Individuals of the crustacean zooplankton generally are much larger than rotifers, are able to consume larger algal cells, and are capable of much higher grazing rates (Wetzel, 1983). Presumably a relatively sparse river zooplankton, dominated by rotifers, therefore will exert a relatively low grazing pressure, especially with regard to larger phytoplankton cells. What is needed are comparisons of rates of grazing and rates of primary production, and unfortunately such data are scant. However, as discussed in chapter 4, there is reason to suspect that downstream export, rather than *in situ* consumption, will be the principal fate of phytoplankton production in large rivers.

8.4 Summary

A growing body of evidence indicates that herbivory is an important process in running waters. Primary production contributes significantly to consumer biomass and herbivores influence the amount, composition and productivity of the producer assemblage. The periphyton, comprised mainly of diatoms, algae and cyanobacteria, are widely distributed and often are subject to intense grazing. They are the main focus of our treatment of herbivory.

Preference and vulnerability are strongly influenced by feeding specializations of the herbivore, and structural and nutritional characteristics of the autotroph. Tightly adherent diatoms seem to be accessible primarily to snails with their specialized radula and certain caddis larvae with hardened mandibles, while the periphyton overstory is easily removed by a variety of grazers. Mobility and rapid harvest, rather than slow and thorough scraping of the substrate, might be the most effective feeding strategy for the loose overstory material. Thick cell walls and low nutritional value defend autotrophs against herbivory, but can be countered by adaptations such as the gastric mill of snails or lengthy intestine of some herbivorous fishes.

Grazers respond to local periphyton abundance by behaviors that concentrate their numbers in resource-rich areas, and eventually through an increase in abundance. Some dramatic examples are known of periphyton blooms in the absence of grazing, and of their subsequent decimation once grazers become common. A number of convincing experimental studies demonstrate that both vertebrate and invertebrate herbivory can alter the abundance and composition of periphyton assemblages, especially under low and steady streamflow conditions. While grazing may sometimes be too low to deplete producer biomass seriously, or so severe that periphyton are greatly reduced, moderate grazing might have a stimulatory effect on periphyton assemblages through reduction of self-shading, removal of senescent cells, and perhaps by more efficient nutrient regeneration. Because strong interactions between autotrophs and grazers likely develop only under favorable environmental conditions, the importance of herbivory depends on the relative strength of biotic *versus* abiotic variables at a particular place and time.

Consumption of live macrophyte tissue usually is minimal, amounting to less than 10% of annual production. However, waterfowl, some mammals and fish, and a few invertebrates can significantly reduce macrophyte abundance. Submersed and floating-leaved macrophytes appear more vulnerable to grazing than emergent taxa. Whereas aquatic insects generally are ineffective herbivores, a few insects derived from terrestrial lineages are effective biological control agents against nuisance waterweeds. Moderate grazing pressure is capable of stimulating plant growth, perhaps through fertilization with animal wastes or removal of senescent growth.

Although grazing of macrophytes occasionally is sufficient to reduce plant biomass and alter assemblage composition, macrophytes likely are more influential as structure than as a living food resource. Structural effects vary according to environmental conditions, animal functional group and plant architecture. In addition, there appear to be complex interactions involving the epiphytic algae, epiphytic grazers, and the macrophyte host, which are worthy of further investigation.

The phytoplankton of free-flowing large rivers are consumed by zooplankton and by benthic suspension feeders. Because downstream displacement usually is rapid relative to their reproductive rates, zooplankton populations of rivers tend to be sparse and dominated by rotifers rather than crustaceans. Crustacean zooplankton have longer doubling times but are capable of higher grazing rates. Provided that the residence time of a water mass is sufficient, as in long, slow-flowing rivers or impoundments, the potamoplankton can experience significant grazing pressure.

Chapter nine

Competitive interactions

The idea that competition influences the distribution and abundance of stream-dwelling organisms has been invoked routinely for more than half a century. Natural populations often are observed to differ in one or more of their ecological requirements; that is, in their niches. As competition theory came to dominate ecology after the mid 1950s, studies of ecological differences flourished, resulting in an extensive literature on resource partitioning. These examples were viewed as evidence that competition was widespread and required a minimum level of niche partitioning to permit coexistence. By the late 1970s skepticism had set in concerning the relationship between niche differences and competition past or present, causing such interpretations to fall from favor. In its place ascended a counter view that also has a long history in stream ecology, that varying environmental conditions coupled with differences in species' ecologies and life histories may play a greater role in determining species' abundances than do interactions among species. Most recently, experimental studies have convincingly established the occurrence of competition in a small number of cases. However, because such rigorous demonstrations are few, we lack any consensus concerning the frequency and strength of the occurrence of competition in flowing waters.

Many authors have offered definitions of competition and debated the validity of various types of evidence for demonstrating its existence and importance. Birch's (1957) statement suffices: "Competition occurs when a number of animals (of the same or different species) utilize common resources the supply of which is short; or if the resources are not in short supply, competition occurs when the organisms seeking that resource nevertheless harm each other in the process". Embedded within that definition are two mechanisms of competitive interaction. Exploitation competition requires the depletion of resources such that another individual is disadvantaged. Interference competition is a direct interaction usually of an aggressive nature, for instance when one individual excludes another from a preferred area.

To demonstrate the occurrence of competition is a considerable challenge. Reynoldson and Bellamy (1971) and Connell (1980) each discuss the criteria that must be met to establish a strong case. Generally the most convincing studies share two attributes: they show an adverse effect of numbers of one species upon the abundance, growth or survival of individuals of another species under reasonably natural conditions, and they provide some insight into the mechanism. Most studies fail to meet these criteria, instead simply documenting differences in resource use. While such studies must be viewed as weak evidence, for reasons discussed more fully below, they indicate the possibility of a competitive effect and provide some insight into the potential for interspecific differences in resource use to facilitate coexistence. Such studies also comprise the vast bulk of the literature on competition.

Thus we shall first consider the evidence of resource partitioning, and then the other lines of evidence that make for more convincing claims. Finally, since unrestrained competition ultimately should result in the elimination of all but the superior species, it is necessary to ask whether this occurs, and if not, why not. Physical disturbances, floods in particular, appear to be important in counteracting strong competition in a number of instances.

9.1 Distributional patterns and resource partitioning

One of the first well-documented cases of a disjunct distribution pattern involving closely related species was the triclad fauna of mountain brooks. At several locations in western Europe investigators have found an altitudinal succession, usually of three species, showing fairly clear boundaries associated with temperature and physiological performance (Beauchamp and Ullyot, 1932; Pattee, Lascombe and Delolme, 1973). Where the uppermost species was absent, presumably because it had not dispersed to that locale, the species normally confined to lower elevations was observed to extend to the headwaters. While such studies have a long history, development of ecological theory contributed greatly to their proliferation. The premise is that ecologically similar (usually, closely related) species are most likely to overlap in their critical requirements; thus, resource overlap is a measure of competition and resource partitioning is necessary for coexistence. Furthermore, niche subdivision presumably has some limit, which then must govern the number of species able to coexist in an area.

Resource overlap typically is evaluated based on similarities between individuals along three major axes: diet, habitat and time (season or time of day) when the organism is active. Enough studies have been made of resource partitioning, encompassing a variety of taxa in both aquatic and terrestrial settings, to provide grist for a general review (Schoener, 1974). Overall, habitat (including microhabitat) segregation occurred more commonly than dietary segregation, which in turn was more common than temporal segregation. Schoener also found a tendency for trophic separation to be of relatively greater importance among aquatic organisms. This last point appears to conflict with the view that many stream-dwelling animals are trophic generalists (Hynes, 1970; Cummins, 1973) and thus should not be expected to subdivide food resources. Evidence of food specialization usually comes from inspection of gut contents; thus it matters a great deal whether food items fall into distinct categories as perceived both by the animal and the researcher. Not surprisingly, food partitioning is reported more commonly from studies of grazers and predators than of detritivores. Resource partitioning between two species often involves multiple axes, and so similarity in resource use along just one axis provides an incomplete picture. Fish and invertebrates both have been studied extensively from a resource partitioning point of view, but the literature for plants and benthic algae is scant.

9.1.1 Resource partitioning within the periphyton

Few studies explicitly address competition within the periphyton. Very likely, however, this is an area where laboratory experiments have great potential. Algal dominance in artificial streams changes in response to adjustments in nutrient, light or current regime, and also undergoes succession within a particular regime. These patterns can be viewed at least as the precursors of competition studies. The tendency for filamentous green algae to dominate under high light levels is suggestive of a competitive advantage, while their scarcity under low light regimes may be due to the reduced pigment diversity of chlorophytes relative to other common stream algae (Steinman and McIntire, 1986, 1987). The longitudinal distribution of *Achnanthes deflexa* and a species of *Chlorella* in the effluent plume of a sewage outfall suggested a competitive interaction (Klotz, Cain and Trainer, 1976). Together these two taxa comprised nearly 90% of the periphyton at all sites studied; the latter was most abundant nearest the outfall, while *A. deflexa* increased with distance. When cultured alone, *A. deflexa* grew well at high concentrations of effluent, but in mixed culture *Chlorella* dominated at higher effluent concentrations and *A. deflexa* at lower concentrations, corresponding to their field distribution. Additional studies ruled out any extracellular inhibitor, leading Klotz and colleagues to conclude that the competitive superiority of *Chlorella* under high nutrient conditions was due to a combination of nutrient depletion and occupation of space.

The response of different members of the periphyton to nutrient addition also is indicative of competitive effects. In comparing the influence of nutrient availability in the water columns versus nutrient diffusing substrates, Pringle (1990) observed the diatoms *Navicula* and *Nitzschia* to dominate the upperstory and interfere with the establishment of understory taxa *Achnanthes* and *Cocconeis*. Considering the diversity of periphyton species and variety of growth forms (Figures 4.1 and 4.2), it seems likely that competitive interactions and environmental heterogeneity play a significant role within assemblages of benthic autotrophs.

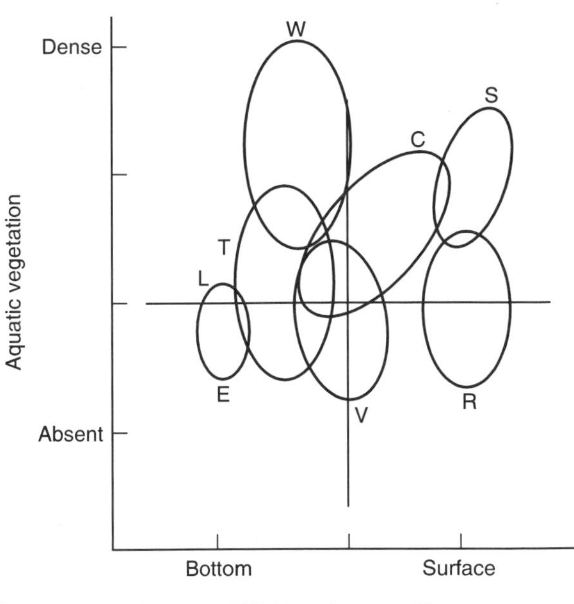

FIGURE 9.1 Ecological segregation among eight species of cyprinids in a Mississippi stream. Only *Ericymba buccata* and *Notropis longirostris* failed to separate on the axes shown, and the former was the sole nocturnal feeder in the assemblage. W, *Notropis welaka*; S, *N. signipinnis*; V, *N. venustus*; R, *N. roseipinnis*; T, *N. texanus*; C, *N. chrysocephalus*; L, *N. longirostris*; E, *Ericymba buccata*. (From Baker and Ross, 1981.)

9.1.2 Resource partitioning in fish

Many studies examine the similarity of resource use among species of fish living within a restricted area, or within one taxonomic group over a broader geographic region. In a review of some 116 such studies conducted between 1940 and 1983, primarily concerning salmonids of cool streams or small, warm-water fishes of temperate regions, Ross (1986) found that segregation along habitat and food axes was about equally frequent while temporal separation was less important. This accords well with Schoener's (1974) findings. Although resource partitioning usually was distinct, even within a similar fauna the importance of space, food and time axes varied considerably.

The neatness of ecological segregation contributes to the view that amelioration of competition must be the root cause. Many studies of darters and minnows in North America reveal a striking absence of overlap in species' distributions (Page and Schemske, 1978) or in microhabitat and feeding position at a single locale (Mendelson, 1975). Although eight species of cyprinids exhibited high spatial overlap in a

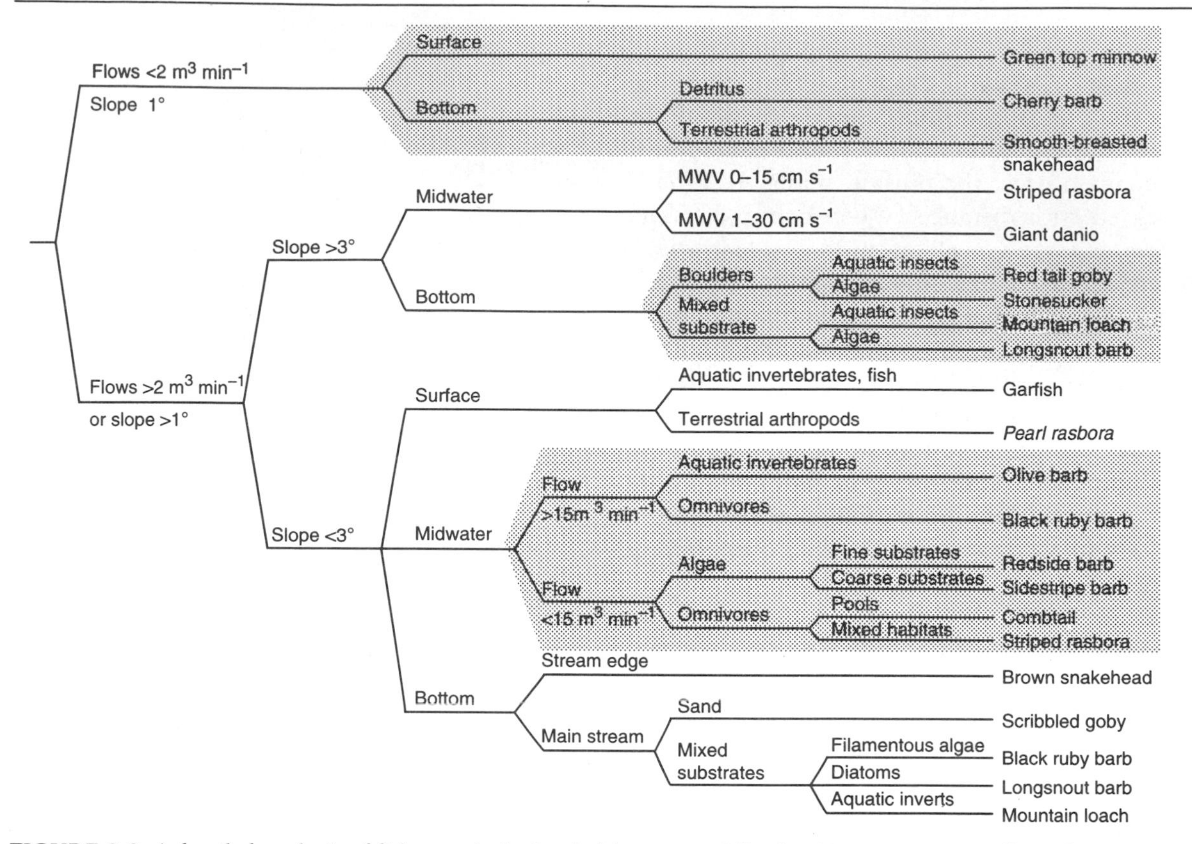

FIGURE 9.2 A detailed analysis of fish morphologies, habitat use and diet for 20 species in a small rainforest stream in Sri Lanka reveals a highly structured assemblage with low ecological overlap. MWV refers to mean water column velocity selected by fishes. (From Moyle and Senanayake, 1984.)

Mississippi stream (Baker and Ross, 1981), there was considerable microhabitat segregation with respect to vertical position in the water column and association with aquatic vegetation (Figure 9.1). Only two species failed to separate on these two axes, and one was the only nocturnal feeder in the assemblage. Moyle and Senanayake's (1984) study of an even more diverse group of fish in a small rainforest stream describes a highly structured assemblage with minimal overlap based on fish morphology, habitat use and diet (Figure 9.2).

The extent of resource sharing *versus* resource partitioning varies among studies, largely because of variation in food availability and other circumstances. Resource overlap has been reported to be high when food is scarce, because consumers have little choice as to what is available (Lowe-McConnell, 1987a). However, overlap also can be high when certain prey are very abundant, because of opportunistic feeding. Seasonal changes in diet overlap are well illustrated by Winemiller's (1989) study of nine species of piscivorous fish that were abundant in a lowland creek and marsh habitat in western Venezuela. Members of this guild exhibited substantial resource partitioning in food type, food size and habitat. Of the possible 72 species combinations among these nine piscivores, only one pair of fin-nipping piranhas exhibited substantial overlap on these three niche dimensions. For the most part diet overlap of pairs of piscivore species within their feeding guild was low (Figure 9.3). Highest overlap occurred during the wet season

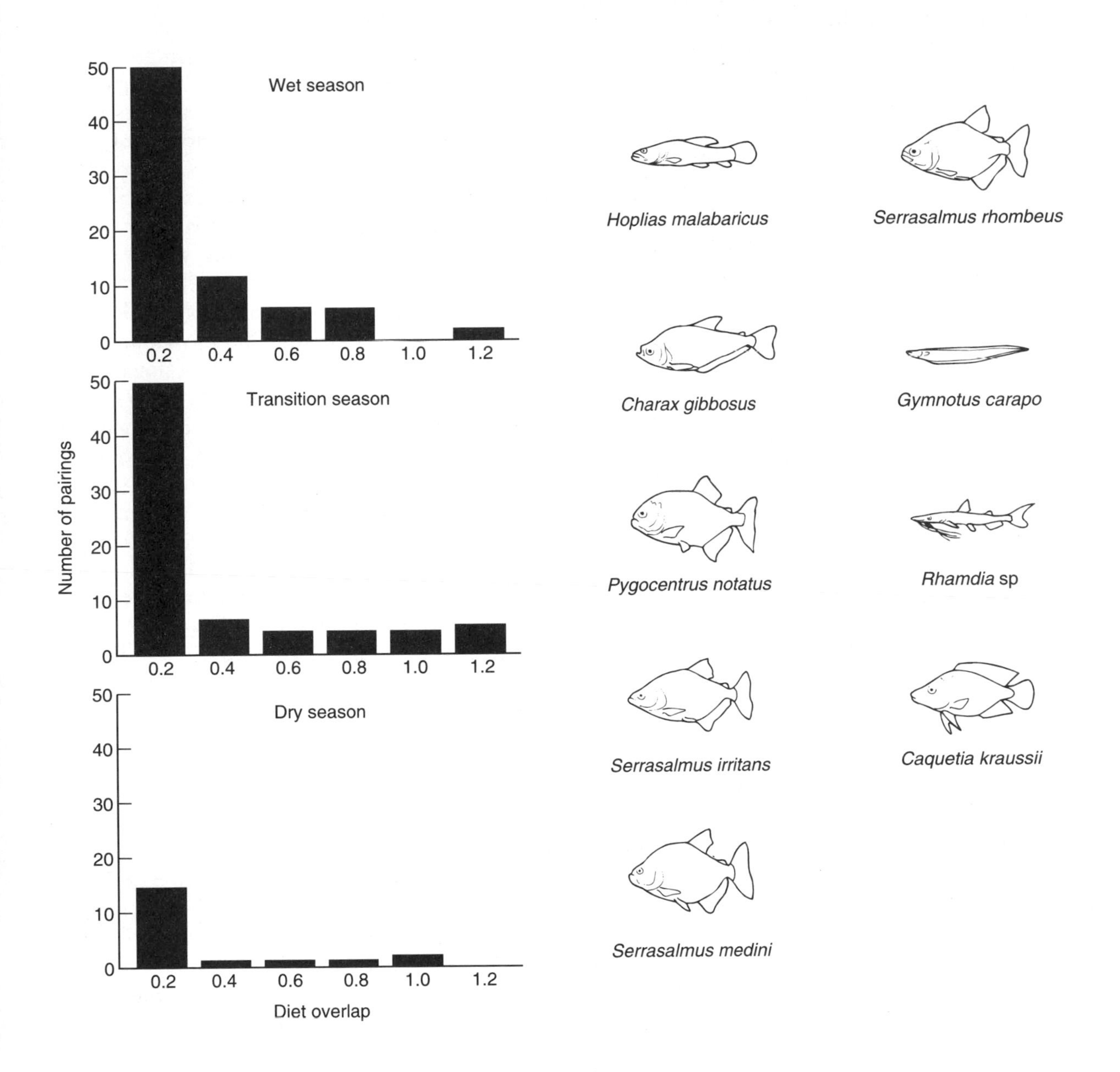

FIGURE 9.3 Frequency histograms of dietary overlap exhibited by each of nine piscivorous fish during different seasons at a lowland creek-and-marsh site in Venezuela. Wet season lasts from May to August, transition season from September to December, and dry season from January to April. Diet overlap was computed from pairwise comparisons of ingested prey after converting prey abundance to volume as an approximation of biomass. Dry season data are less extensive because not all species were present and many had empty guts. Over half of overlap estimates were less than 0.10. (From Winemiller, 1989.)

when prey were abundant, and lowest overlap occurred during the transition season when prey were least available. Thus, despite the opportunities for competition in this species-rich tropical system, food resource partitioning was widespread. Winemiller (1991) argued that the high species diversity of tropical fish assemblages is paralleled by high ecomorphological diversity (chapter 6), resulting in considerable niche partitioning and no more overlap than is common in less diverse temperate regions.

The claim that competition underlies observed partitioning of resources is most convincing when it is accompanied by evidence that one species contracts or expands along some resource axis in response to the presence or absence of a second species. For example, Baltz, Moyle and Knight (1982) found riffle sculpins to be most abundant in the upper reaches of a small California stream, while speckled dace dominated riffles downstream and the two species were inversely related over the 12.5 km section. Laboratory studies demonstrated that dace avoided areas occupied by sculpins, but also tolerated warmer temperatures, which presumably explains the downstream prevalence of dace. Although this example lacks the corroboration of a field test, such as removing sculpins from upper sections to determine if dace would invade, it certainly is stronger than field observations alone. A similar study of three species of sculpins (Finger, 1982) also demonstrated that one species occupied a different microhabitat in the field than it evidently preferred, since the presence of a congener in stream tanks caused it to shift to regions of slower current than utilized when alone. In contrast, however, two species of darters investigated for potential resource partitioning showed considerable overlap in diet and habitat use, and no shift in habitat in response to manipulations of the density of the other species (Schlosser and Toth, 1984).

(a) Range expansions

The distributional patterns of salmon and trout afford special opportunities for evaluating competition, because fishermen and managers have thrown together various combinations of ecologically similar species which, until 50 or 100 years ago, had non-overlapping ranges. So, for example, European brown trout are replacing brook trout through much of their native range in Appalachian and mid-western US streams, restricting the latter to headwater reaches (Fausch and White, 1981). In the Rocky Mountains, brook trout similarly are displacing native cutthroat trout (Vincent and Miller, 1969), and many more examples of the spread of introduced salmonids can be found (chapter 14). Moreover, interspecific aggression among salmonid species has been extensively documented (e. g. Hartman, 1965; Fausch and White, 1981; Hearn, 1987; DeWald and Wilzbach, 1992), thus lending plausibility to the view that interspecific competition is important in species replacements. In a particularly convincing study, Fausch and White (1981) measured the daytime positions of brook trout, recording 'water velocity difference' (comparing velocities at the fish's location to the highest velocity within 60 cm of the fish), depth, distance to the streambed, and amount of exposure to direct light. Brown trout, the socially dominant species in this Michigan stream, were then removed from an extensive section to determine if brook trout would exhibit ecological release. They did, shifting to resting positions that afforded more favorable water velocity characteristics and greater shade. Furthermore, this habitat shift was greatest in the larger individuals.

9.1.3 Resource partitioning in invertebrates

No general review of resource partitioning among stream-dwelling invertebrates has been written, so some discussion of individual studies is necessary. Many studies document habitat partitioning, usually at the microhabitat level, temporal separation of life cycles over seasons is frequent among the univoltine insects of temperate streams, and differences in diet are reported less often and appear principally in animals that consume easily categorized food items. An early application of the resource partitioning approach to whole assemblages of stream-dwelling invertebrates was made by Grant and Mackay (1969), whose monthly samples from five habitat categories in a small Quebec stream indicated considerable segregation among congeners within the Ephemeroptera, Plecoptera and Trichoptera. Grant and Mackay found that separation was greater over time and season than habitat, and pointed out that it resulted in temporal staggering of energy flow. In the river continuum model (chapter 12), Vannote *et al.* (1980) develop further the ecosystem consequences of this seasonal replacement of functionally similar species.

The catch-nets of caddisfly larvae were advanced by Malas and Wallace (1977) as a classic example of coexistence mediated by resource partitioning between species and between instars of the same species. *Parapsyche cardis* occupies the upper surfaces of stones at high velocities, *Diplectrona modesta* occupies under-surfaces at intermediate velocities and *Dolophilodes distinctus* occurs at the lowest velocities and also underneath surfaces. Net size decreases in the order just given, and also varies directly with instar within each species. Catch-nets of larger mesh and larger filtering area are associated with highest currents, the mean particle size found in the larvae's foreguts increases directly with mesh opening size, and so a quite convincing case can be made that food items are partitioned based on adaptation of catch-net to current regime. When a more rigorous test was made by estimating the size fraction captured by six suspension-feeding caddis larvae, as well as total availability of seston, Georgian and Wallace (1981) found no evidence that food was limiting or that resource partitioning occurred. The size fractions captured showed very high overlap, and amounted to only about 0.1% of total seston available. It now appears likely that the supply of high quality animal food may limit the production of net-spinning caddis larvae (Benke and Wallace, 1980), but competition for that resource is by no means established.

The detection of diet partitioning is easiest in animals that ingest foods that are readily categorized. Examples where gut contents reveal dietary differences include the differential utilization of diatoms by two species of grazing insects (Hill and Knight, 1988), as well as differences in food consumption by predaceous invertebrates (Townsend and Hildrew, 1979).

Microhabitat partitioning has been demonstrated in a number of studies of stream-dwelling invertebrates. Suspension-feeding caddis larvae (Malas and Wallace, 1977; Hildrew and Edington, 1979) and black fly larvae (Colbo and Porter, 1979) exhibit both macro- and microhabitat segregation with regard to current regime, position on stones, and in location downstream of lake outlets. Teague, Knight and Teague (1985) quantified the spatial distribution of five species of day-active, grazing caddis larvae with respect to water depth, velocity, rock size and roughness. Substantial separation was recorded along each axis, but especially associated with velocity and depth. Finally, Bovbjerg's (1970) study of habitat segregation in *Orconectes* crayfish is particularly instructive because it illustrates a niche expansion when a species is released from competition. Where they co-occur, *O. immunis* occupies pools and *O. viriles* is found in riffles. The former species better tolerates low oxygen concentrations and stagnant water; evidently it will enter riffles, but is displaced by the more aggressive *O. viriles*. Thus, differences in physiological tolerance and competitive ability appear to explain their habitat segregation.

Although temporal segregation apparently is less common than habitat or diet partitioning across diverse taxa (Schoener, 1974), numerous examples from temperate running waters

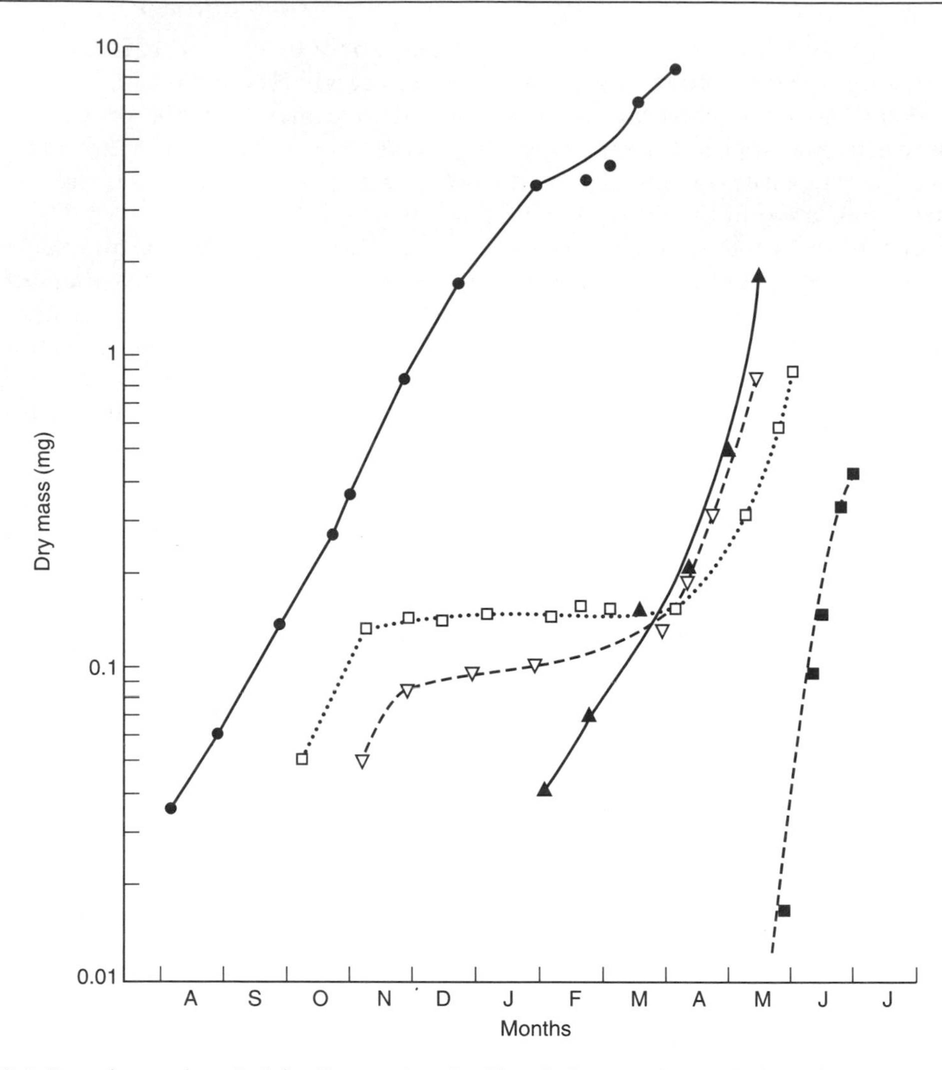

FIGURE 9.4 Larval growth period for five species of riffle-inhabiting ephemerellid mayflies in White Clay Creek, Pennsylvania. ●—● *Ephemerella subvaria*; ▲—▲ *E. dorothea*; □—□ *Seratella deficiens*; ■—■ *S. serrata*; ▽—▽ *Euryophella verisimilis*. (From Sweeney and Vannote, 1981.)

illustrate a distinct seasonal succession among closely related taxa. This has been especially well documented using *Ephemerella* mayflies (Sweeney and Vannote, 1981), stoneflies in the genus *Leuctra* and family Nemouridae (Elliott, 1987a, 1988), and a guild of scraping insects (five caddis larvae and the dipteran *Blepharicera*, Georgian and Wallace, 1983). The very neatness of the temporal separation prompted the authors of each of the above studies to suggest that amelioration of competition was the cause, albeit with some recognition that this interpretation was not fully demonstrated. For example, the periods of maximum larval growth in the five species of ephemerellids that occupy riffle habitats were sufficiently out of phase to ensure at least 10-fold size differences on any given date between individuals of the same sub-genus (Figure 9.4).

Highly synchronized, non-overlapping life

cycles among presumed competitors are not always the rule, however. Of six leptophlebiid mayflies in a New Zealand stream, only two had reasonably well-defined growth periods, and overlap of life histories was pronounced (Towns, 1983). Clearly a claim of amelioration of competition is unwarranted in this example. One way to further evaluate whether observed temporal partitioning should be attributed to competition is to ask how much overlap would be expected if species' life cycles were distributed independently of one another throughout the year, with the constraint that most growth should occur during favorable seasons. Tokeshi (1986) applied this null model approach to nine species of chironomid larvae living epiphytically on spiked water-milfoil and overlapping considerably in their diatom diet. Actual overlap of life cycles was greater than expected by chance alone, even when growth cycles within the model were constrained to the March–October time period. Since this result is the opposite of that expected in temporal partitioning, it appears that all nine species were tracking seasonal peaks in resource abundance. Competition, if it occurred, was not manifested in temporal partitioning. A null model approach could be applied *ex post facto* to other such studies, and it is possible that some will reveal greater temporal segregation than expected by chance.

This review of resource partitioning studies is relatively brief given their large number because they are considered weak evidence. However, one cannot ignore this literature because the bulk of competition studies are of this kind. On the one hand are studies that show such elegant segregation that amelioration of competition simply begs to be considered (e. g. Figures 9.1 and 9.4). On the other hand are studies that reveal indistinct segregation and overlapping ecologies (e. g. Towns, 1983; Tokeshi, 1986). Even when resource partitioning is observed, one can argue that niche segregation reflects ecological differences acquired and fixed over the species' evolutionary history (Connell, 1980). These differences cannot be interpreted as unequivocal evidence of competitive interactions in the present time; moreover the claim that past competition led to the observed differences usually is untestable. It is for these reasons that recent work has focused on more rigorous tests, typically involving experiments under fairly natural conditions, and we now consider such studies.

9.2 Experimental studies of competition

While experimental manipulations of natural populations provide especially convincing evidence of competitive interactions, laboratory studies also contribute useful information. Usually this consists of some demonstration of the mechanism by which one species affects another; for example, aggressive displacement from a location acceptable to both (Bovbjerg, 1970; Baltz, Moyle and Knight, 1982). Agonistic encounters, and thus the potential for interference competition, have often been demonstrated in laboratory studies. Salmonids are well known to behave aggressively against heterospecifics (Hartman, 1965) and maintain feeding territories against conspecifics (Dill, Ydenberg and Fraser, 1981). Invertebrates that exhibit strong agonistic behaviors include caddis (Jansson and Vuoristo, 1979; Hildrew and Townsend, 1980; Hemphill and Cooper, 1983; Hart, 1985a) and black fly larvae (Dudley, D'Antonio and Cooper, 1990) and surface-foraging hemipterans (Wilson, Leighton and Leighton, 1978; Sih, 1980). Large size almost invariably conveys an advantage, and the loser may be injured or cannibalized. In the latter instance the line blurs between competition and predation (e. g. Hemphill, 1988).

Laboratory studies of exploitative competition using invertebrates demonstrate that depression of resource levels can limit food available to others of the same species. In the laboratory, the mayfly *Baetis* departs in the drift when crowding results in depletion of algae (Bohle, 1978;

Kohler, 1985), and chironomid larvae also drift in response to increasing densities (Wiley, 1981). When *Baetis tricaudatus* and *Glossosoma nigrior* were allowed to compete for periphyton in laboratory stream channels, both strongly reduced periphyton abundances, suggesting that exploitative competition in nature was likely. *Baetis* had a greater impact on *Glossosoma* than the converse, and effects were seen mainly in terms of growth (Kohler, 1992).

Resource depletion is the presumed mechanism in exploitative competition, and so demonstration of resource decline under field conditions is evidence of the potential for competition. When Hill and Knight (1987) manipulated numbers of the mayfly *Ameletus validus* in field enclosures, algal standing crop and *A. validus*' growth both declined as its density increased (Figure 8.5). Similar results have been obtained with several caddis larvae: *Glossosoma* (McAuliffe, 1984a), *Helicopsyche* (Lamberti, Feminella and Resh, 1987), and *Neophylax* (Hill and Knight, 1988).

Hydropsychid larvae vigorously defend their nets against both conspecifics and heterospecifics. By marking third-instar larvae of *Hydropsyche siltalai* in a Swedish lake-outlet stream, Englund (1991) demonstrated that larval position within aggregations was determined by size. Smaller larvae and recent arrivals were disproportionately represented at the downstream end of an aggregation where current velocities and food availability were markedly lower. Based on other studies of hydropsychids described below, the mechanism of competition likely was aggressive interference.

An indirect demonstration of intraspecific competition in nature is evident from Otto and Svensson's (1976) experimental removal of the pupae of a caddis from the headwaters of a small stream. In the succeeding year larval densities of *Potamophylax cingulatus* initially were reduced by about half because fewer adults emerged to produce that generation. The eventual pupal population was nearly normal, however, suggesting that compensatory high survival occurred in response to reduced larval densities.

Interspecific competition among stream-dwelling invertebrates has been convincingly demonstrated primarily with grazers and taxa that occupy relatively fixed positions on stone surfaces. A study of species interactions in a Montana stream provided strong evidence that resource depression by *Glossosoma* adversely affected other species of grazers. By erecting barriers of petroleum jelly, McAuliffe (1984a, b) was able to achieve approximately a fivefold reduction in *Glossosoma* densities and a twofold increase in algal cell density. Mobile grazers such as *Baetis* were significantly more abundant in areas where *Glossosoma* was excluded. At normal densities *Glossosoma* appears able to reduce periphyton to levels where *Baetis* experiences resource limitation, and thus exploitation rather than interference is the causal mechanism. A 10-month exclusion experiment in a Michigan springbrook of very constant flow provides additional evidence of the community-wide effects of *Glossosoma*. Periphyton biomass increased substantially, as did the densities of most grazers (Kohler, 1992). Based on studies in two very different streams, several other species of grazers are strongly affected by exploitative competition from the caddis, *Glossosoma*.

Snails of the family Pleuroceridae may be competitive dominants in those headwater streams where they reach high abundances. Both *Juga silicula* in the northwestern USA and *Elimia clavaeformis* in the southeastern USA have been reported to reach densities over 500 m^{-2} and to comprise more than 90% of the invertebrate biomass (Hawkins and Furnish, 1987; Hill, 1992; Rosemond, Mulholland and Elwood, 1993). Other invertebrates in these systems potentially are strongly influenced by the presence of snails, which are effective grazers of periphyton (chapter 6). Because of their large size, individual snails also may harm other species by 'bulldozing' over substrate surfaces (Hawkins and Furnish, 1987). These authors

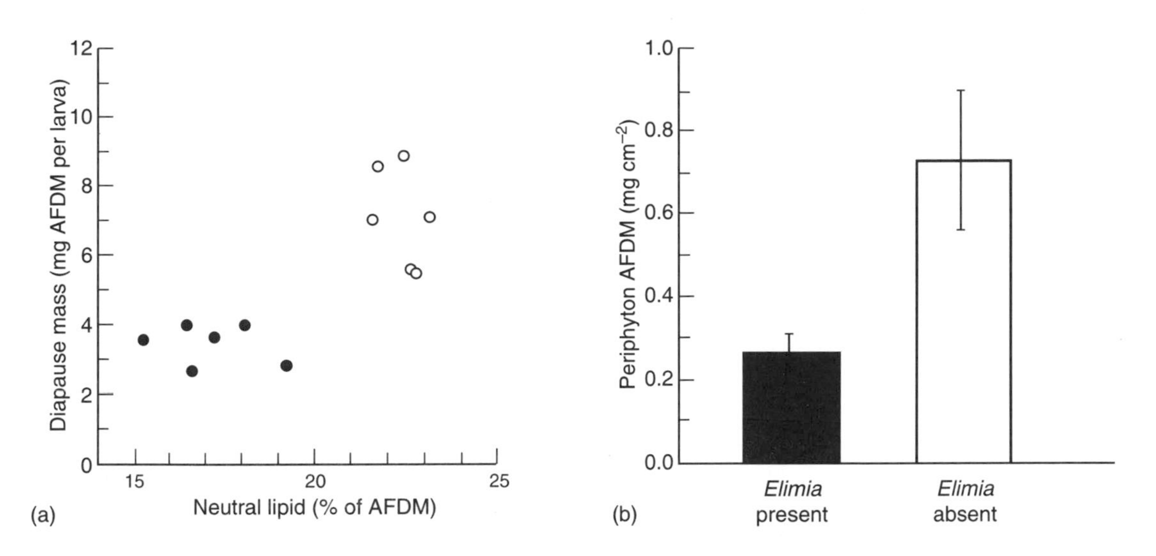

FIGURE 9.5 Comparison of six streams in the southeastern USA lacking the snail *Elimia clavaeformis* (○) and six streams where the snail was extremely abundant (●). (a) Average mass of diapausing larvae of the caddis, *Neophylax*, was higher in the absence of snails. (b) Periphyton mass also was higher in the absence of snails. (From Hill, 1992.)

showed that densities of many invertebrate taxa were inversely correlated with snail abundance, and experimental reductions in snail density resulted in higher numbers of many invertebrate taxa.

The interaction between *Elimia clavaeformis* and the caddis, *Neophylax etnieri*, in a headwater stream in Tennessee makes a strong case that the snails' influence is via exploitative competition (Hill, 1992). High dietary overlap determined from gut analysis suggested that these two grazers were competing for periphyton. Both species substantially increased their growth rates and condition (ash-free dry mass per unit wet mass) when transferred from the stream to a high quality diet in the laboratory, suggesting food limitation in nature. In a natural experiment, Hill examined periphyton abundance and *Neophylax* condition in six streams lacking *Elimia*, and in six streams where the snail was abundant. Periphyton biomass was three times greater and caddis larvae at diapause roughly twice as large as in the absence of snails (Figure 9.5). Hill could find no obvious explanation for the absence of *Elimia* from certain streams. Nonetheless, his results are consistent with a body of evidence that, when conditions permit high snail abundance, their influence on other members of the system is substantial. As Rosemond, Mulholland and Elwood (1993) document for this study system, grazing by snails also affects the species composition of the periphyton and a number of functional attributes. In those situations where they are very abundant, perhaps aided by protection from predation afforded by a hard shell, large pleurocerid snails appear to be keystone species.

Interspecific competition based upon interference mechanisms has been documented at least as often as resource-based exploitation. The hydroptilid caddis larva, *Leucotrichia*, constructs a silken case which it attaches to hard substrates, and from which larvae defend their foraging area with vigor against all intruders. In a Michigan stream, Hart (1985a) found algal standing crops to be higher inside these foraging areas than outside. Although *Baetis* ordinarily spent little time there, removal of the caddis larva led to invasion of these 'resource oases' by the mayfly, and Hart suggested its mobility might allow *Baetis* to make

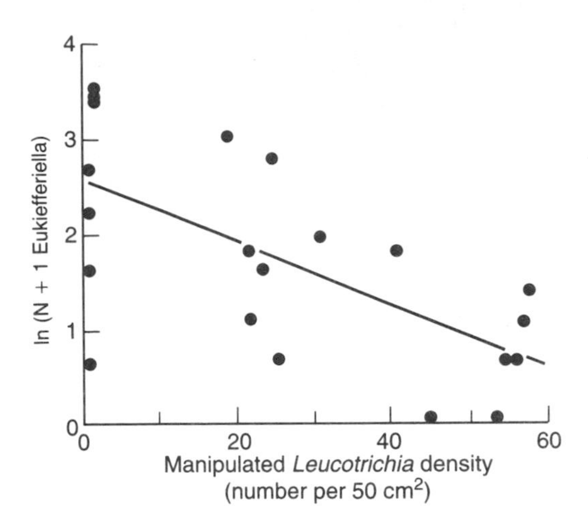

FIGURE 9.6 Numbers of larvae of the midge *Eukiefferiella* on stones from which nearly all, roughly half, or none of the *Leucotrichia* had been removed 15 days previously. (From McAuliffe, 1984b.)

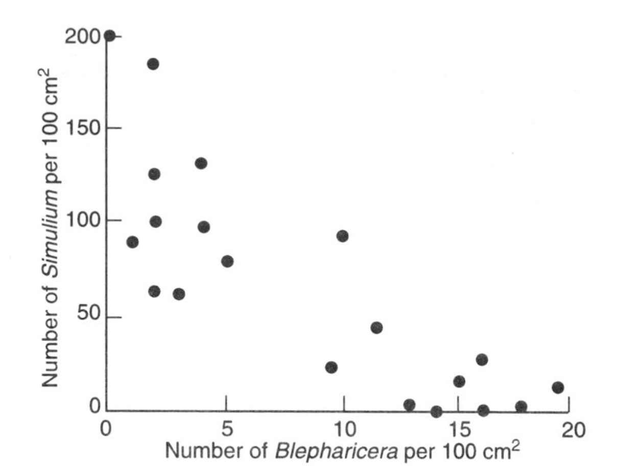

FIGURE 9.7 Relationship between densities of *Blepharicera* and *Simulium* in a California stream. (From Dudley, D'Antonio and Cooper, 1990.)

brief grazing forays into *Leucotrichia*'s feeding territories on a regular basis.

Another study involving *Leuchotrichia* revealed far-reaching effects stemming from this tiny but nonetheless doughty creature. McAuliffe (1983, 1984b) manipulated *Leucotrichia* density by removing all, half, or none of the cases from a number of stones, which then were returned to the stream for subsequent colonization. There were significant inverse effects on sessile and slowly moving taxa (Figure 9.6) including a moth larva (*Parargyractis*) and a midge (*Eukiefferiella*), but apparently more mobile species were affected as well. Additional details add to the complexity of the interaction. Although *Leucotrichia* can exclude *Parargyractis*, the former will not colonize regions covered by the silk of the latter, resulting in what McAuliffe refers to as a 'founder-conferred dominance'. Additionally, cases left by *Leucotrichia* from previous years remain on stones and facilitate the subsequent establishment of populations. This, coupled with the tendency for spates to flip small stones but not larger ones, results in a positive correlation between *Leucotrichia* density and stone size. As a consequence, not only does this particular caddis larva have a strong effect on the overall community, but its effect is greatest on larger stones and presumably in regions of less severe flooding.

Interference competition is well documented in space-limited taxa. Larvae of a netwinged midge (*Blepharicera*) and black flies (Simuliidae) compete for space on stone surfaces in swift-flowing small streams, even though the former feed on attached periphyton and the latter are primarily suspension feeders. Dudley, D'Antonio and Cooper (1990) observed a strong inverse relationship between blephariccrid and simuliid densities (Figure 9.7) in a small stream in California. Behavioral observations of larval interactions made within the stream revealed that simuliids 'nipped' at *Blepharicera* within reach, disrupting their feeding. *Blepharicera* spent significantly less time feeding in the presence of simuliids, compared with when the investigators removed all simuliids within a 5 cm radius using a fine probe. Feeding trials on natural substrates and in flowthrough chambers on the streambank showed that blephariccerids were significantly smaller when reared in the presence of black fly larvae.

Larval hydropsychids compete for superior positions to attach their silken capture nets to substrate surfaces. Both simuliids (Hemphill and Cooper, 1983) and smaller conspecifics (Englund, 1991) are displaced by a combination of mechanisms. The interactions between *Hydropsyche* and *Simulium* included predation, interference, behavioral avoidance and a change in feeding efficiency due to altered flow patterns (Hemphill, 1988).

Experimental studies of competition in streams remain too few in number to generalize with confidence. However, both intraspecific and interspecific competition have been convincingly demonstrated in a number of settings. These examples include competition for both food and space, and document the action of both exploitative and interference mechanisms. It is tempting to suggest that interference competition is most apparent in space-limited and relatively sedentary taxa, and exploitative competition in more mobile taxa, but further studies are needed to establish this. Competition usually is asymmetric, meaning that a competitively dominant species influences other taxa much more than the reverse.

The immediate effect of competition among animals commonly is a reduction in feeding opportunity. An individual may spend less time feeding because of harassment by neighbors (*Blepharicera*; Dudley, D'Antonio and Cooper, 1990), exclusion from preferred locations (*Simulium* and small *Hydropsyche*; Hemphill and Cooper, 1983; Englund, 1991), or simply experience lower resource levels (*Neophylax*; Hill, 1992). Diminished growth and reproduction are important means by which competitive interactions are expressed (Figure 9.5), and this can be hard to detect. In addition, a number of studies indicate that competition occurs episodically, depending on fluctuations in environmental conditions and in the timespan during which two species share a limiting resource. For all of the above reasons, it is likely that the full extent of competition in stream communities is yet to be discovered.

9.2.1 Mitigating influence of environmental variation

Competition cannot play a decisive role in environments so variable that populations rarely reach abundances where resources are substantially depleted (Connell, 1975). The extreme of this position holds that the particular species favored by the environmental conditions of the moment floods the region with its offspring, and at least for a time holds sway. In such a lottery system (Sale, 1977) all species are competitively equivalent, and dominance shifts according to changing environmental conditions and the lifespan of component species. While the relevance of this explanation to stream communities is uncertain, harsh and variable environments unquestionably can minimize the role of competition or limit its importance to periodic episodes. For example, the fish assemblage of a southern Great Plains river was characterized by transitory species associations and minimal ecological segregation, apparently because of great seasonal variability of the locale (Matthews and Hill, 1980). The seasonal replacement of *Simulium* by hydropsychid larvae in a California foothills stream is due partly to competition, and partly to an interaction between the life histories of the two species and year-to-year variation in environmental conditions (Hemphill and Cooper, 1983). Severe winter flooding reduced the abundance of aggressively superior hydropsychids and permitted simuliids to remain abundant for a longer time. Similarly, the overlap between simuliids and blephariccerids in the same system was at most one month in dry years but more than 4 months in wet years (Dudley, D'Antonio and Cooper, 1990). Such an interplay between biotic interactions and physical disturbance likely is common, and is discussed more fully in chapter 11. Even in the most convincing studies of competitive effects (McAuliffe, 1983, 1984a b; Hill and Knight, 1987, 1988; Kohler, 1992) the strength of competitive interactions likely will vary seasonally and spatially.

Species-specific differences in timing of recruitment of stream fishes can combine with variability in environmental disturbance to produce patterns that resemble competitive interactions. In Sagehen Creek, California, rainbow and brook trout varied inversely over a 10 year period in parallel with flood events (Seegrist and Gard, 1972). Severe winter floods destroyed the eggs of brook trout, which spawn in the fall, whereas spring floods similarly ravaged eggs of spring-spawning rainbow trout. In each instance the unaffected species increased in abundance, and effects persisted for several years. The resulting oscillations gave the appearance of shifting competitive dominance, but in fact were caused by life history differences and the timing of floods.

Differences in the timing of recruitment among warm-water fishes also provide ample opportunity for flooding to influence local relative abundances. In a small stream in Kentucky, the order of appearance of juveniles was approximately: cottids, percids, catastomids, cyprinids, centrarchids, ictalurids. The length of recruitment period for individual species varied from as little as three weeks to as long as 16 weeks (Floyd, Hoyt and Timbrook, 1984). Clearly, such variation in timing of recruitment is widespread, with the consequence that variation in timing and duration of disruption in the physical habitat can be of great importance.

Long-term environmental change brought about by human activity also can produce patterns with the appearance of competition. Over 15 years, brown trout have largely replaced brook trout in a Minnesota stream (Waters, 1983), a common observation usually attributed to competition and/or predation. In Waters' study, however, a local increase in rainfall in conjunction with housing construction led to greater flooding, erosion and siltation, resulting in habitat changes particularly detrimental to brook trout.

The above examples establish the range of the interplay between competitive interactions and environmental change. Extreme variability might override competition completely, while moderate or regular environmental change may constrain discernible interactions to particular seasons or locations. This is part of the general view, discussed more fully in chapter 11, that the strength of biological forces varies inversely with environmental harshness.

9.3 Summary

Competitive interactions among stream-dwelling species have been assessed through studies of resourcc partitioning and by experimental approaches in laboratory and nature. The many examples of resource partitioning in invertebrates and fishes demonstrate that actual overlap in resource use between potential competitors typically is ameliorated by non-overlapping spatial distributions or by subdivision of resources. Resource partitioning can occur along any of three major axes: food consumed, microhabitat use and time of activity. Not infrequently, however, careful studies of resource use among presumed competitors fail to detect any indication of resource partitioning. In addition, it is likely that some of the subtle differences in niche dimensions seen in closely related species were acquired over their evolutionary history, and have little to do with present-day interactions. For these reasons it has been argued that resource partitioning alone is not convincing evidence of a competitive interaction. Laboratory studies that demonstrate aggressive displacement of one species by another, and restriction in habitat use in the presence of the presumed competitor, strengthen the case for competition.

The introduction of an exotic species often results in displacement of native species. This is widely seen among trout, because of the numerous introductions that have occurred and the strong aggressive interactions that are observed in experimental studies.

The mechanism of competition can be exploitation of a common resource required by members of both species, such as food or space, or interference with the ability of another to make use of that resource, usually by direct aggression. Both mechanisms of competition have been demonstrated for stream-dwelling invertebrates, whereas interference competition seems most common among fishes. Resource depletion has been convincingly demonstrated with a number of invertebrate grazers, whose influence clearly limits food availability to both conspecifics and heterospecifics. This results in reduced growth, a smaller final body size, and as a consequence fewer offspring for the next generation. The larvae of some caddis, simuliids, and other space-limited taxa aggressively defend against neighbours, causing their departure or reducing their foraging opportunities. In some instances, the difference in body size is sufficient that the superior competitor becomes a predator. Competition appears to be asymmetric in most instances now reported, and perhaps further study will establish that one species commonly is the competitive dominant and the other has little effect.

Although the existence of competition is widely suggested by resource partitioning studies and convincingly demonstrated in a number of instances, no consensus has emerged on the frequency and importance of competition in running waters. This may be partly because competition can be difficult to demonstrate, and partly because it may operate episodically. Typically the loser in competition departs for other habitats, or suffers diminished growth and reproduction over an extended period, and this can be hard to detect. We do not expect competition to play a decisive role in environments so variable that populations do not become sufficiently abundant to deplete resources, and streams are notoriously variable environments. Some of the most convincing examples of competition under natural conditions involve species that co-occur for only a few months of the year, or that were studied under low-flow or otherwise favorable environmental conditions. The expression of competition, like that of other biological interactions, is strongly influenced by variation in the favorability of abiotic conditions.

Chapter ten

Drift

Stream-dwelling organisms often are transported downstream in the water column in substantial numbers. Because they have limited swimming ability and apparently are being swept downstream, the phenomenon has been called 'drift'. While most descriptions of drift focus on invertebrates, downstream transport also has been reported for taxa of the periphyton and some larval fishes and amphibians. Brown and Armstrong (1985) found that 18 of some 60 fish species in an Arkansas river exhibited larval drift. Diatoms also occur in the drift (Müller-Haeckel, 1971), providing nourishment to suspension-feeding invertebrates and resulting in rapid colonization of new habitats and artificial substrates. In this chapter however, and in most of the literature, the principal focus is on insects and crustaceans.

Drift became the subject of serious inquiry after Müller's (1954b) studies in a Swedish river led him to propose that compensatory upstream flight by ovipositing adults (the 'colonization cycle') was necessary to maintain populations. Earlier workers had noted and quantified the drift of aquatic invertebrates (Borgh, 1927; Needham, 1928; Ide, 1942; Dendy 1944; Berner, 1951), and Mottram (1932) proposed the basic idea of the colonization cycle in an obscure publication that never attracted attention. The finding by Moon (1940) that trays of substrate set on the bottom of a lake were colonized more rapidly during the night than during the day presaged the discovery by lotic ecologists that drift activity exhibits a nocturnal rhythm. However, study intensified after Müller (1954) provided a conceptual framework within which to investigate drift. Subsequent discovery of drift's nocturnal periodicity (Tanaka, 1960; Waters, 1962a; Müller, 1963) further stimulated research.

10.1 Composition and periodicity

10.1.1 Composition of the drift

As Waters (1972) points out, there is no drift fauna *per se*. Drifting invertebrates are derived from the benthos and spend very little time in the water column. It seems that any member of the benthic fauna may at some time be captured in the drift, but some taxa are particularly common. Among the insects, many of the Ephemeroptera, some Diptera (especially the Simuliidae) and some Plecoptera and Trichoptera are common components of the drift, in roughly that order. Among the crustaceans, amphipods and isopods also have been reported in the drift. However, it would be difficult to claim that any species is totally absent.

Some investigators have reported a preponderance of young animals in the drift. This appears to be a frequent result in studies of trichopterans, and may suggest that one role of drift is dispersal (Waters, 1972). However, more investigations have reported drift to be common in later rather than earlier life history stages. Waters (1972) suggests several reasons why this might be so. Absolute growth is greatest in later instars, so energy demands may lead to greater intraspecific competition, foraging activity and drift. Final instar individuals may disperse prior to emergence or pupation, in preparation for reproductive activity. Waters also raised the

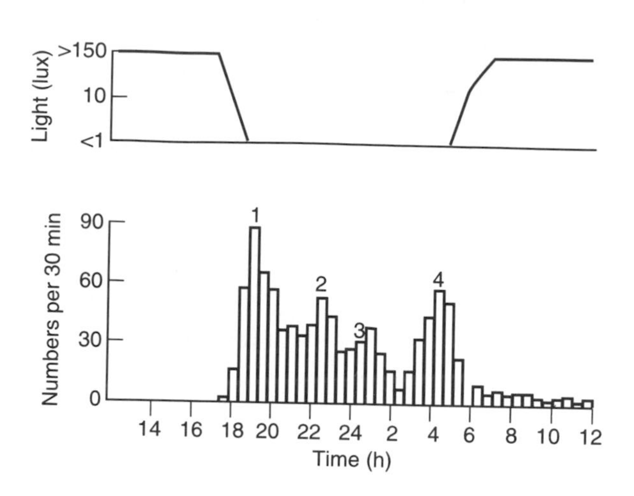

FIGURE 10.1 Diel variation in drift catches of *Baetis rhodani* at 30 min intervals over a 24 h period. Four apparent peaks are indicated (1–4). (From Elliott, 1969.)

possibility that larger organisms may protrude farther into the current, increasing the risk of dislodgement. Lastly, choice of mesh size influences the size classes sampled, and many studies probably neglect the smallest stages for this reason. Regardless of which, if any, of these explanations is correct, it is worth noting that drift activity varies over the life cycle.

10.1.2 Diel periodicity

The number of drifting animals usually is greater by night than by day, often as much as an order of magnitude greater, an observation that led Waters (1965) to suggest the following classification scheme. Behavioral drift exhibits a consistent pattern, usually a pronounced nocturnal increase. Constant drift refers to the continuous background of low numbers, most easily detected during the day. Catastrophic drift refers to the effects of floods or other major, adverse events such as drought, high temperatures and anchor ice. Catastrophic drift also can provide a useful index of human disturbance (e. g. Hall *et al.*, 1980).

Although most taxa show a nocturnal peak in numbers, the Chironomidae usually are reported to be aperiodic, while water mites and some trichopterans are day-active. Nocturnal drift usually reaches a major peak shortly after dusk, declines through the night, and may rise to a smaller peak just before dawn (the 'bigeminus' pattern; Müller, 1965). Less frequently, the predawn peak is the larger ('alternans' pattern).

Elliott (1969) sampled diel periodicity at very close intervals in an effort to elucidate the timing and number of peaks. His observations for *Baetis rhodani* and some other taxa were taken over 5 days in March and again in July, at 30 minute intervals (Figure 10.1). Drift was greatest at night, but the increase at dusk was not always as sudden as supposed from less frequent sampling. Except for *Simulium* species, which showed the alternans pattern, drift followed the more typical bigeminus pattern. Elliott demonstrated that the number of observed peaks depended on the number of sampling periods, and suggested that there were usually four, although he allowed that more frequent sampling might reveal more peaks.

Light level clearly serves as the time signal (Zeitgeber) for behavioral drift. Müller (1965, 1966) manipulated the light–dark (LD) cycle in the field by means of lamps and opaque plastic. In the former study, two species of *Baetis* tracked LD cycles as extreme as 23:1, which concentrated all drift into 1 h (Figure 10.2). Continuous light resulted in virtually no drift and no rhythm, whereas natural drift patterns persisted for 8 days in artificial darkness. In Müller's (1966) study, *Gammarus pulex* responded similarly, cueing to such unusual light regimes as LD 2:22 and 20:20. In constant darkness, drift exhibited one further cycle and then became arrhythmic. Holt and Waters (1967) obtained similar results with *B. vagans* and *G. pseudolimnaeus*; like Müller, they concluded that the exogenous signal of light overrode any endogenous rhythm, if one existed. Indeed, the constant light of polar summers extinguishes drift rhythm entirely

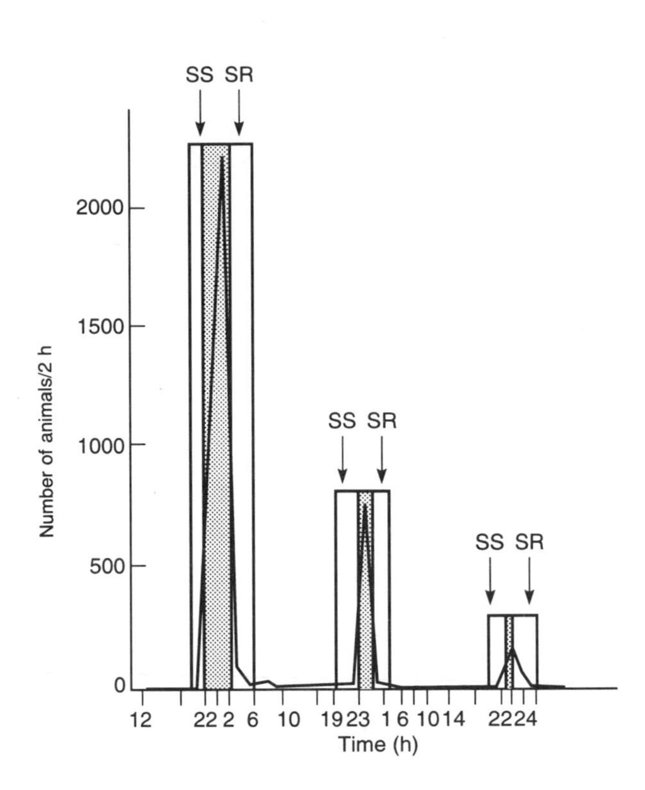

FIGURE 10.2 Drift activity of *Baetis* nymphs in artificially shortened nights. SS and SR denote natural sunset and sunrise, but artificial lights reduced the period of darkness to 4, 2 and 1 h, respectively (shaded regions). (From Müller, 1965.)

(Müller, 1973), and moonlight also may suppress drift (Anderson, 1966).

There has been considerable effort to determine whether a light threshold exists, and whether behavior has an endogenous component. Correlations between the onset of drift and darkness provide a number of estimates of the light level below which drift commences: 0.1–1 lux (Tanaka, 1960), around 1 lux (Holt and Waters, 1967), and 1.58 lux (the average of many correlations by Chaston, 1969, 1972). Müller (1965) found that a light level of 5 lux maintained throughout the night totally suppressed drift; levels of 1 and 2 lux did not prevent a nocturnal peak but reduced it 5–10-fold. While these results appear consistent, they are contradicted by the finding that bright moonlight (about 0.2 lux) may (Anderson, 1966; Bishop and Hynes, 1969a) or may not (Chaston, 1972) suppress drift. Additionally, Elliott (1969) concluded from field observations that drift did not increase suddenly at dusk, but instead increased gradually as light fell from 60 to less than 1 lux (see Figure 10.1). Holt and Waters (1967) make the sensible point that animals on the stream bottom, in varying amounts of shadow, probably detect different light levels and so may not respond in perfect synchrony. Thus field studies do not provide an unambiguous threshold, but do agree that light levels above approximately 1 lux reduce drift substantially.

Laboratory studies indicate that variation in wavelength has little effect (Bishop, 1969; Chaston, 1969), and that an even lower threshold (approximately less than 0.01 lux) may be detected under controlled conditions. Variable light penetration due to depth and turbidity, and shadows due to irregularities of the substrate, help to account for higher apparent thresholds in natural studies. However, some aspects of this exogenous signal appear to be insufficiently studied. Possibly the rate of change of light intensity is important, as has been demonstrated for light-dependent movements in vertically migrating zooplankton (Ringleberg, 1964), and apparently no one has determined whether physiological state, such as hunger level, influences the light threshold.

The possibility of an endogenous rhythm has been indicated by several workers, most convincingly by Elliott (1968, 1970a). In the former study, individuals took positions on the tops or bottoms of stones according to a negative phototaxis, and so presence on the tops of stones was exogenously controlled in five species of mayflies. Movement of the nymphs appeared to be an endogenous activity, however, which continued on the original cycle in complete darkness for as long as one week. In the latter study, four species of caddis larvae were solely under exogenous control, three being night-active and one being day-active. Activity shifted developmentally in another species of caddis larva, *Potamophylax luctuosus* (Lehmann, 1972). Early instars showed no avoidance of light or preference for nocturnal *versus* diurnal activity, later instars were negatively phototactic but lacked an endogenous rhythm, and final instar larvae were strictly nocturnal and maintained an endogenous rhythm in total darkness. Field studies employing continuous darkness have reported retention of drift periodicity for one day in *G. pulex* (Müller, 1966) and at least one week in *Baetis* (Müller, 1965).

The drift of larval vertebrates (Brown and Armstrong, 1985; Bruce, 1985) also exhibits nocturnal periodicity, but drift of diatoms usually is diurnal. In diatoms, daytime drift often coincides with photosynthetic activity and cell division (Müller-Haeckel, 1971), and these are potential mechanisms inducing detachment of cells. However, an endogenous drift rhythm, which persisted for 3 days in constant darkness, has been demonstrated in *Synedra ulna* (Müller-Haeckel, 1971).

10.2 Drift and downstream displacement

Drift samples are obtained by suspending a net in the water column over some time period. Measurements of total discharge and flow through the net allow one to calculate the two principal metrics, drift density and drift rate (see Elliott, 1970b; Allan and Russek, 1985, for details). Drift density is expressed as numbers per volume of water filtered, usually standardized to a volume of 100 m^3. Drift rate is the number passing a point during a time period, usually the number for the entire stream per 24 h. Drift densities typically are between 100 and 1000 per 100 m^3 (Bournard and Thibault, 1973; Armitage, 1977). Drift rates can be very high, in the range of 10^4–10^5 per 24 h for small trout streams, and may be higher yet in large rivers. Berner (1951) calculated that in 24 h, some 64 million organisms weighing about 200 kg passed beneath the Boonsville bridge on the Lower Missouri.

The magnitude of downstream drift has perplexed researchers since it first was quantified, because of the imagined depletion of headwaters caused by such enormous export of organisms. Most research on drift has in some way been motivated by the apparent dilemma caused by such massive, one-way migration. Efforts have been made to compare the drift past a point to benthic densities, and to determine how far individuals drift, in order to gain some appreciation of the scope of the problem.

10.2.1 Downstream movements

The abundance of organisms in the drift has been expressed as the percentage of the bottom fauna that is in the water column above a unit area of bottom at an instant in time. Elliott (1967), Ulfstrand (1968) and Bishop and Hynes (1969a) report this percentage to be a very low value, usually less than 0.01%, and rarely greater than 0.5%. As Waters (1972) points out, however, the cumulative effect nonetheless is great, with 24 h drift over a unit area reaching 10 or even 100 times benthic density. Waters re-examined Ulfstrand's data to conclude that daily drift at a sample station in a Lapland river was equivalent to the entire estimated population upstream for 30–400 m, depending on the species. Drift rates of Ephemeroptera in a Rocky Mountain stream were found to be very high, in the vicinity of 2–8 × 10^5 per 24 h, in a small trout stream some 4–6 m wide and 30 cm deep (Allan, 1987). Instantaneous ratios of drift to benthos were in the range 0.2–0.5%, yet the cumulative drift over 24 h was estimated to be equivalent to the entire benthic population for 40–150 m upstream.

It is not immediately apparent how to reconcile the results of these two calculations. The instantaneous method assumes that immigration equals emigration per unit area of substrate, which seems reasonable and has been verified by Waters (1962b). In fact, calculations incorporating estimates of transport distances, per cent entering drift, and number of trips per night (e. g. Elliott, 1971b) make it probable that only a tiny fraction of the benthic population is suspended in the overlying water column at any instant in time. Nevertheless, the cumulative calculation, particularly when extended over weeks or months, appears to re-assert the basic conundrum of what prevents a virtual denuding of upper reaches. In fact, depletion of headwaters is almost never observed (for an exception, see Madsen and Butz, 1976).

Estimates of drift distance establish that animals traverse a substantial length of stream. Waters (1965) provided the first, albeit crude measure of this by blocking drift across an entire stream, and then sampling downstream at various distances to determine at what point drift abundances returned to normal. After 38 m, over two riffles and a pool, drift was still less than 100% of the expected value. Subsequent work by McLay (1970), Elliott, (1971b) and Larkin and McKone (1985) establishes the adequacy of a simple physical model which assumes that animals return to the substrate according to an exponential decay function:

$$N_x = N_o e^{-Rx} \qquad (10.1)$$

N_o is the number of animals entering the drift, N_x is the number still in the water column x meters downstream from the point of entry, and R is a constant that reflects the rate of return to the substrate. The mean distance traveled by a drifting organism ($\bar{X}$) is estimated by:

$$\bar{X} = 1/R \qquad (10.2)$$

Using this approach one can estimate the mean distance journeyed by the furthest traveling X_p% (e. g. 10% or 50%) of the population (Figure 10.3).

How far animals travel varies with the species, water velocity, and perhaps also with turbulence, although that has yet to be investigated. Elliott developed statistical relationships for particular taxa of the form:

$$R = aU^{-b} \qquad (10.3)$$

$$\bar{X} = cU^{d} \qquad (10.4)$$

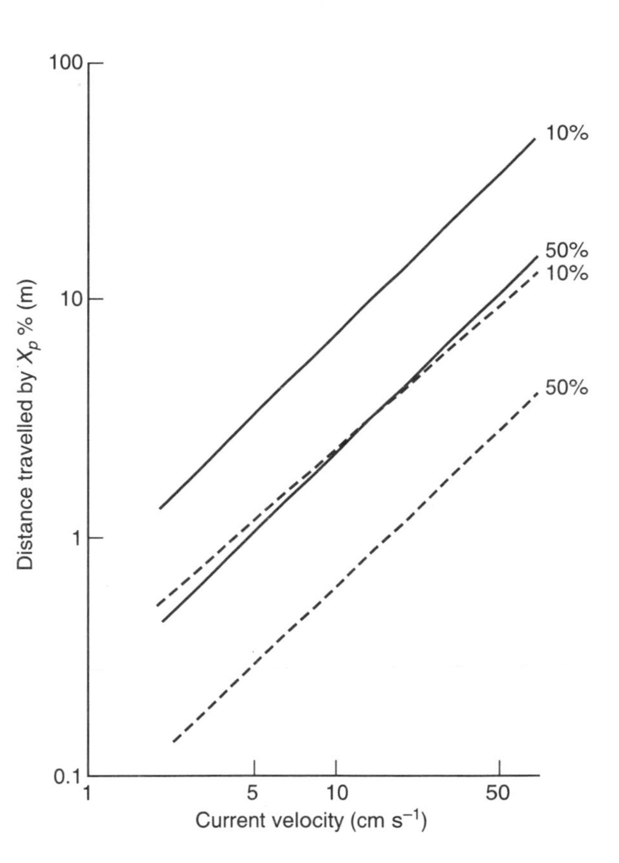

FIGURE 10.3 The distance traveled by drifting invertebrates as a function of water velocity, from Elliott's (1971b) equation 12 and Table 6. The solid lines depict taxa that regained the bottom no more rapidly alive than dead at all current speeds, or performed slightly better than this at slow currents only (a flatworm, chironomid and simuliid larvae, riffle beetles, several stoneflies and caddis larvae, and a heptageniid mayfly). The dotted lines depict *Baetis rhodani* and *Gammarus pulex*, which were the most adept at regaining the bottom. *Hydropsyche* spp. and *Ecdyonurus venosus* were intermediate. Note that $(100 - X_p)$% of individuals have settled out of the drift in the specified distance.

where U is water velocity and a, b, c and d are fitted parameters.

For those taxa with no special ability to regain the bottom (drift distances are similar for live and dead specimens), mean drift distances are of the order of 10–20 m at moderate (30–60 cm s^{-1}) current speeds. Comparable figures for the agile *Baetis* are much less, 3–6 m on average. While multiple trips clearly could result in animals traveling enormous distances, Larkin and McKone's results indicate that individuals spend much more time on the stream bottom than in the water column, even while in the mode of moving downstream. In one instance, after disturbing the substrate within a very uniform spawning channel, individuals in the drift were expected to reach a net some 5 m downstream in 11 s if they traveled the entire distance in a single trip at the current speed of 47 cm s^{-1}. In fact, in the first 7.5 min after animals were dislodged, the catch at that net was about ten times natural, in the next 7.5 min it was roughly five times natural, and thereafter it was less than natural. Clearly, animal drift velocity measured over this distance was much less than current velocity, suggesting that animals drifted a short distance, regained the bottom, re-entered the drift once again, and so on. Nonetheless, total distances traveled per animal per hour or night still are a mystery, and one cannot deny the possibility that downstream displacement may be very large. Larvae of *Pycnopsyche guttifer* that were marked with acrylic paint or fish tags and released in pools later were recaptured some 400–700 m downstream on some occasions, and a few traveled 1.5 km in 20–25 days (Neves, 1979). Other species of *Pycnopsyche* have shown substantial downstream displacement due to catastrophic drift during spring thaws (Mackay and Kalff, 1973).

Since a regular alternation of riffles and pools typifies many streams (Leopold, Wolman and Miller, 1964) and pools serve as depositional regions, one might imagine that the settling out of drift in pools serves as a major brake on downstream transport. Few have studied this and there is little to support such a contention. Dendy (1944) showed that stream drift entering a lake could result in animals distributed up to 5–10 m beyond the stream mouth, and rarely some distance further, so animals do not appear to immediately settle out of the water column. Bailey (1966) claimed that pools trapped drift, but his data were limited. Waters (1962b) made a careful study of drift over a sequence of two riffles and two pools. Drift out of the pools was about 20% less than drift off the riffles, suggesting some trapping of drift in pools, but also that the majority pass through. In fact, since the species studied (*B. vagans*) is a riffle-dwelling species, these results may be explained by suggesting that animals exited the drift equally in both habitats, but entry to the drift was reduced in the pools. Thus, this study and others by Elliott (1967) and Campbell (1985) give little support to the proposition that pools trap drifting individuals.

10.2.2 Upstream movements

As mentioned, depletion of upstream populations, which might be expected to result from drift throughout a season, has rarely been demonstrated. Either the amount of downstream displacement has been exaggerated and upstream movements compensate, or those transported downstream represent a surplus. Both aquatic and aerial stages have been shown to move upstream, and the initial establishment of populations surely requires such dispersal. Upstream movements have been shown to contribute substantially to the colonization of temporary streams by invertebrates (Williams, 1977; Gray and Fisher, 1981). No such active upstream migration is possible for diatoms and other members of the periphyton, but passive transport by other organisms on their body surface or internally (Solon and Stewart, 1972) clearly has the same effect. Introduced species of amphipods have demonstrated impressive ability to disperse once major barriers have been surmounted. *Crangonyx pseudogracilis* apparently was transported from North America to England, where it spread fairly rapidly by means of canals (Hynes, 1955). Several gammaridean amphipods of the Ponto-Caspian region have spread northwest into Germany, also via canals (Jazdzewski, 1980). Several instances of upstream migrations have been reported, some of them rather spectacular. Minckley (1964) observed a mass upstream migration of *Gammarus bousfieldi*, ascending riffles by using turbulence and eddies, and proceeding mainly along the banks. Neave (1930) reported a substantial upstream migration by the mayfly, *Leptophlebia cupida*, which covered 1.6 km at approximately 200 m day^{-1}.

Positive rheotaxis will lend an upstream bias to movements within the substrate, and this behavior is found in most major invertebrate groups. Possible functional interpretations of pronounced upstream movement include searching for unexploited resources, and for suitable emergence, pupation, and mating sites, avoidance of unfavorable abiotic conditions, and dispersal *per se* (Söderström, 1987).

Several investigators have attempted to quantify upstream movements using directional traps. Buried nets (Bishop and Hynes, 1969b; Elliott, 1971a; Bird and Hynes, 1981), traps open at one end (Waters, 1965; Williams and Hynes, 1976) and inclined troughs (Hultin, 1968; Otto, 1971) have all documented upstream movements by amphipods and a variety of insect taxa. Laboratory studies also are supportive (e. g. Hughes, 1970; Mackay, 1977). In an especially thorough study, Elliott (1971a) found that during most of the year, upstream movements were disproportionately concentrated among small rather than large individuals, in traps containing small rather than large stones, and near banks. However, large *Baetis rhodani* in April and June tended to move up in midstream among large stones. While movements by day were substantial, both *Baetis* and *Gammarus* showed distinct nocturnal peaks in movement activity. Unlike the findings of Hultin, Svensson and Ulfstrand (1969) and Otto (1971), movements were not confined to pre-emergent nymphs and mature larvae. Elliott also found no indication that mate searching played a role in upstream movements, unlike Lehmann's (1967) conclusions for gammarids.

On the critical question of compensating for downstream displacement, upstream movements appear inadequate. During winter, Elliott (1971a) estimated upstream movements to be about 30% of drift. During the high drift periods of spring and summer, the estimate fell to 7–10%. Similar calculations by other investigators are in the same range (1.6–14.9%, Bishop and Hynes, 1969b; 2.1–15.2%, Bird and Hynes, 1981). These results suggest that upstream movement by aquatic stages is inadequate to counter-balance the downstream displacement of drift, although it probably ensures that populations persist in upstream reaches. Measuring upstream movement rates is more difficult than sampling drift, however, and it is possible that trap estimates are too low. Animals might exit these traps as readily as they entered, and traps typically have been emptied rather infrequently (every 24 h, as compared with 2 h for drift nets). For those taxa that rarely drift, relatively little upstream movement may be required to offset downstream transport. Laboratory and field investigations of the stonefly, *Diura bicaudata*, indicated that drift was minimal and upstream movements were sufficient to maintain the population (Schwarz, 1970). *Isoperla goertzi*, in contrast, exhibited more drift and less upstream movement by nymphs, and Schwarz concluded that upstream flight by adults was necessary to counteract a steady loss during nymphal life.

When Müller (1954) originally proposed the colonization cycle, only a few anecdotes supported the claim that adult insects tended to fly in an upstream direction. Subsequent work now establishes that aerial stages of many aquatic insects exhibit flight directed upstream.

Extensive sampling using bidirectional Malaise traps on the River Ammarån in northern Sweden documented upstream migration in several groups, including the Ephemeroptera, Plecoptera, Simuliidae and Trichoptera (Roos, 1957). Mayflies typically experience two molts on emerging from the water, first from nymph to a non-reproductive aerial form, the sub-imago, and with a subsequent molt enter the imago or reproductive stage. Roos observed upstream flight to be most pronounced in female imagos of *Baetis*, and also in females carrying mature eggs in the trichopterans *Cheumatopsyche lepida* and *Rhyacophila nubila*. In contrast, subimagos of mayflies and female caddis whose eggs were not mature showed no flight direction, or flew downstream. Roos suggested that the tendency to fly upstream was confined to ovipositing females, and pointed out that trap collections must make this distinction in order to collect meaningful data.

Perhaps the strongest case against a colonization cycle was made by Elliott's (1967) study of a small English stream. He found no directional upstream flight in Ephemeroptera or Plecoptera; Trichoptera moved down when the wind was strongly downstream, and upstream when the wind was weak. Since he was unable to observe any population depletion in the upper reaches, while adults in fact emerged in upstream regions, Elliott concluded that a colonization cycle was neither necessary nor present. Elliott opined that wind direction was likely to be the prime determinant of flight direction in weak fliers like the Ephemeroptera, and perhaps a stimulus to upstream flight whenever topographic conditions caused wind to blow up-valley in the evening.

Subsequent studies now confirm that upstream flight by adults occurs in a variety of aquatic insects (Madsen, Bengston and Butz, 1973; Svensson, 1974; Müller, 1982 and references therein). In the Ängerån estuary in north-central Sweden, a number of mayflies and stoneflies drift into brackish water, where they survive but their eggs do not. An upstream migration clearly is necessary, and a directional Malaise trap at the river mouth revealed that 70–98% of captured insects were moving in an upstream direction (Lingdell and Müller, 1979; Müller and Mendl, 1980). In another study (Müller, 1982), directional traps on two rapids in the River Dalälven showed predominantly upstream movements (Figure 10.4). Unfortunately, data were not analyzed by sex or whether females were gravid, nor were wind direction and strength noted.

Svensson (1974) compared the magnitude of upstream migration *versus* downstream drift for *Potamophylax cingulatus* (Figure 10.5), and concluded that the former was much smaller than the latter. However, when one considers the number of eggs carried by each female, and the likelihood that a greater fraction of larvae will survive where densities are low, it seems plausible that a relatively small amount of upstream movement by adults can offset quantitatively greater downstream drift. Otto and Svensson (1976) conducted an imaginative test of compensatory survival in the offspring of upstream migrants of adult *P. cingulatus*. Larvae pupate on the underside of large stones, and the investigators were able to remove by hand nearly every pupa from the upper reaches of a small Swedish stream. Compared with control sections and pre-removal conditions, larval populations in the following year were reduced by more than half. From knowledge of survivorship of *P. cingulatus*, Otto and Svensson concluded that upstream migration provided 29–46% of first instar larvae of the following year's population. Furthermore, high survival, attributed to reduced intraspecific competition, resulted in a pupal population fully 80% of that expected in the absence of the depletion experiment. Clearly, this population can endure substantial losses to drift without very marked depletion in the subsequent year, and adult females that fly upstream may obtain exceptional reproductive benefits.

While no study provides a truly satisfactory

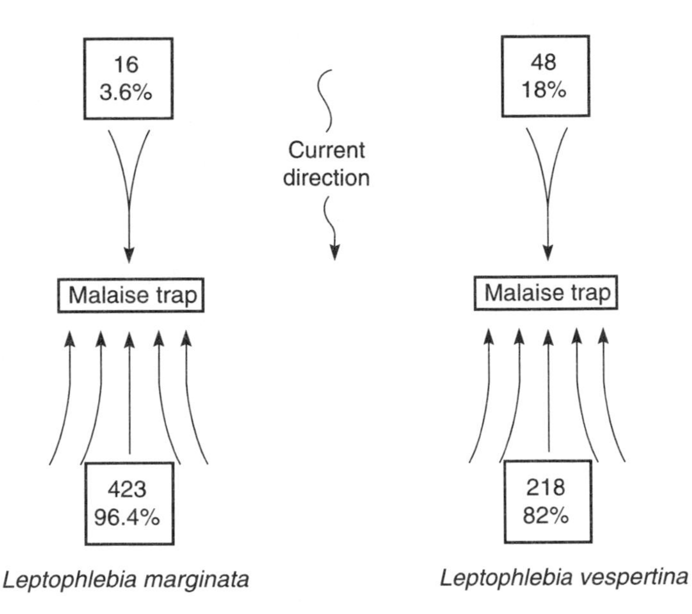

FIGURE 10.4 Flight directionality of two mayflies in the Gysinge rapids of the River Dalälven, showing predominantly upstream flight. (From Müller, 1982.)

resolution, evidence and logic suggest the following interpretation. Since depletion of upstream populations rarely occurs, it appears to be unnecessary, from a population or ecosystem perspective, to postulate massive upstream movements. Nor does the admittedly provisional evidence support the claim that upstream movement by either aquatic or aerial stages is sufficient to compensate for the apparently great downstream displacement engendered by drift. Thus, from the population perspective, it seems reasonable to presume that drift really is a net loss of individuals, but not to the extent that populations are unable to maintain themselves. From an individual perspective, however, there clearly can be benefits to upstream movements, perhaps most dramatically demonstrated by the examples of the Ängerån estuary, and of *P. cingulatus*. Individuals that tend to move upstream will reduce their risk of eventually being flushed from the system, and also may attain greater reproductive fitness.

10.3 Functional basis of drift

Throughout the study of drift there has been considerable interest in questions of ultimate causation: why drift exhibits a strong, nocturnal periodicity, why organisms enter the drift and whether entry is purposeful, and what role drift plays in the biological functioning of streams. Early studies focused on crowding and accidental dislodgement as primary causes of drift entry. Müller's (1954) explanation places a strong emphasis on life cycle events, including downstream dispersal by early instars and compensatory upstream flight by adults. Waters (1961, 1965) suggested that drift represented excess production, removing surplus animals from the benthos and making them available to higher trophic levels, especially drift-feeding fishes. A third perspective is to consider the costs and benefits to an individual organism that drifts compared with another that does not. Granted, individuals might not always be able to influence whether they drift. However, agile swimmers

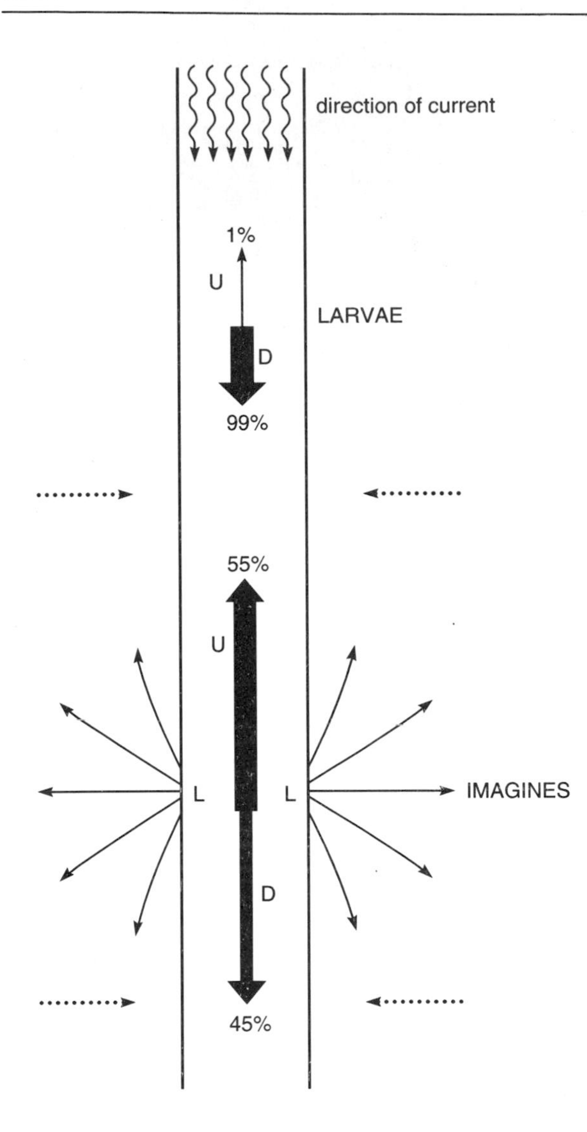

FIGURE 10.5 Movements of the trichopteran, *Potamophylax cingulatus*, in a south Swedish stream. U, upstream-moving larvae and upstream-flying adults; D, downstream drifting larvae and downstream-flying adults. Some lateral dispersal (L) of adults also took place, and lateral immigration of adults may have occurred (broken arrows). (From Svensson, 1974.)

such as *Gammarus* and *Baetis* appear to have considerable control over their movements, and the widespread nocturnal periodicity of drift has long been viewed as evidence of some sort of behavioral control. Recent studies that examine drift from the perspective of natural selection operating on individual organisms provide evidence that, at least in some taxa, drift may be best interpreted in a behavioral context based upon foraging opportunities and predator avoidance (Allan, 1978; Kohler, 1985; Flecker, 1992a).

10.3.1 The cause of nocturnal periodicity

While a great deal of effort has gone into quantifying behavioral drift and examining the influence of light level and circadian rhythms on its periodicity, less effort has been directed toward asking the obvious question, why is drift nocturnal? Perhaps this is because the answer also seems obvious, namely, avoidance of predators that rely on vision for prey capture. In addition, the risk of predation as an explanation for nocturnal drift is difficult to test directly. Simply placing the organisms in a predator-free environment may not elicit an immediate shift in activity, because nocturnal drifting almost certainly is encoded genetically. Indeed, numerous laboratory studies of drift show that nocturnal periodicity is retained in a predator-free environment (e. g. Elliott, 1968; Bohle, 1978).

One approach to testing the predation hypothesis is based on two properties of trout predation that may be general to many species of drift-feeding fish. Brook trout feeding on *Baetis* were found to feed preferentially on larger prey, and to exhibit the greatest size-selectivity when feeding during the day (Allan, 1978). As a consequence one would expect the risk of drifting during the day to be greater for large invertebrates than for small ones. Using the ratio of nocturnal:diurnal drift density for various size classes of *Baetis*, Allan found that nocturnal drift was most pronounced in the larger size classes. Thus, relative vulnerability to fish predation appears to be a useful predictor of the strength of nocturnal drift behavior. Skinner (1985) obtained similar results for *Baetis* in an Idaho stream, as did Andersson *et al.* (1986) and Newman and

Waters (1984) in studies of *Gammarus* drift.

This size-dependent version of the risk of predation hypothesis has some generality for explaining reported exceptions to nocturnal drift periodicity. Water mites tend to drift by day (Elliott and Minshall, 1968), are small, and appear to suffer little trout predation (Allan, 1981). The Chironomidae usually are reported as aperiodic; they tend to be small and often are under-represented in fish diets. Cowell and Carew (1976) found that the midge *Polypedilum haterale* changed from aperiodic to nocturnal drift when final instars predominated. Anderson (1966) demonstrated that moonlight suppressed drift in an Oregon stream by reducing the drift activity of the larger individuals to approximately daytime levels.

A recent, detailed study of drift by members of the meiofauna complicates this explanation, however. Palmer (1992) documented strong nocturnal drift periodicity in copepods, oligochaetes, rotifers and early instar chironomids, all of which are well below the size where risk of predation is expected to constrain drift activity to nocturnal periods. Moreover, these taxa dwell mainly within the sediments, which makes active drift entry much more plausible than passive entrainment.

Strong evidence that predation risk determines nocturnal periodicity in macroinvertebrates is provided by comparisons between locations differing in their fish populations. Drift-feeding fish do not occur at high elevations in the Andes but are abundant in foothill streams, a pattern that Flecker (1992a) exploited to show that fish predation governs diel periodicity. Drift was aperiodic or diurnal in streams lacking drift-feeding fish, a finding also reported by Turcotte and Harper (1982). Along a gradient of increasing predation intensity, the night:day drift ratio for a number of mayfly taxa increased dramatically, with differences sometimes more than 100-fold (Figure 10.6). To address the concern that fish feeding was the cause of low daytime catches in drift nets, Flecker constructed 10 m long by 1 m wide fish exclusion cages in the channel of a river containing abundant drift-feeding fish. Nocturnal periodicity was unaffected by the absence of fish, establishing that direct consumption does not explain low daytime drift catches. Lastly, in some Andean streams trout from local hatcheries had established resident populations. In these streams *Baetis* drift was nocturnal, suggesting a rapid evolutionary response to novel predation risk. Clearly, predation by drift-feeding fishes has considerable support as the explanation for the diel timing of drift. The reasons why individuals enter the drift are, however, still debated.

10.3.2 Why organisms enter the drift

On 4 May 1952, Müller had 150 m of a streambed cleared by caterpillar tractors and new gravel laid down. Eleven days later, he estimated the total benthic density in that section at 4 158 000 organisms. His calculation of total drift rate over 11 days was of the same order of magnitude, demonstrating the enormous colonization potential of drift and forcefully raising the question of why it occurs. Accidental dislodgement is of course one answer, and both the colonization cycle and excess production explanations also invoke crowding and resource depletion. From an individual perspective, drift may be an efficient means of reaching new foraging locales, constrained to hours of darkness by predation risk.

The colonization cycle emphasizes the downstream displacement of a cohort of individuals over their life cycle, and the compensatory role that upstream flight plays in maintaining the population. Thus it requires evidence that numbers in upstream reaches decline over time, and that adults make directional, upstream flights. The former expectation has little support, and while upstream flight unquestionably has been demonstrated (Figure 10.4), some careful studies have found little supporting evidence (e. g. Elliott, 1967). Critics of the colonization cycle

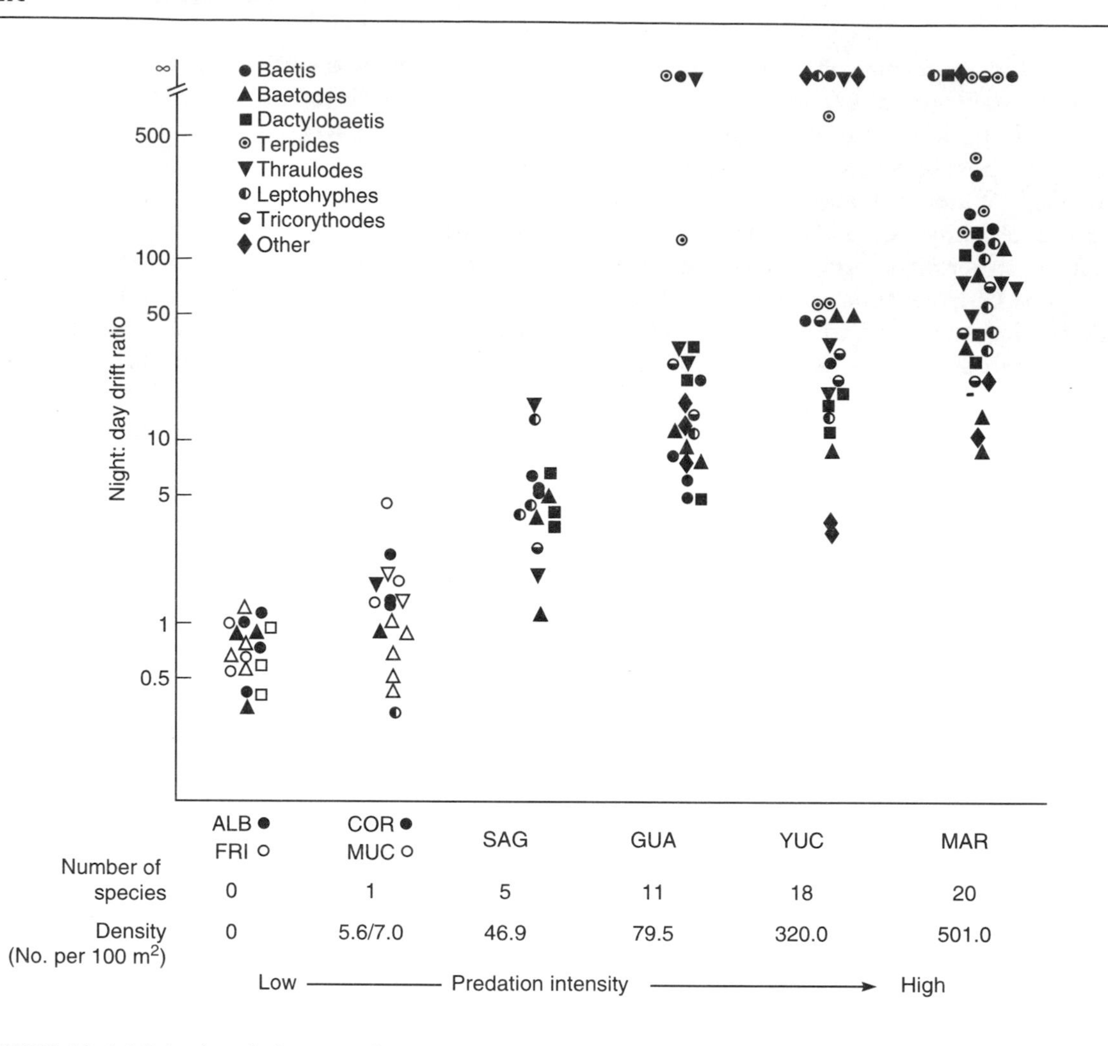

FIGURE 10.6 Night:day drift ratio of mayfly drift densities from a series of streams in the Venezuelan Andes representing a gradient from low to high predation. Note that drift is greater by day in high elevation streams lacking drift-feeding fish (Rio Albarregas [ALB] and Quebrada La Fria [FRI]) compared with nearby streams containing introduced trout (Qda. Coromoto [COR] and Qda. Mucunutan [MUC]). Other rivers are Rio Saguas [SAG], Rio Guache [GUA], Rio La Yuca [YUC] and Rio Las Marias [MAR]. (From Flecker, 1992a.)

also point to gammarid amphipods, which lack an aerial stage yet may drift in abundance.

The production–compensation model (Waters, 1961) differs from Müller's explanation in that downstream drift is presumed to represent production in excess of carrying capacity. Under this model, upstream flight is no longer required and entering the drift is disadvantageous. Drift is expected to be low when populations are below their carrying capacity, and to rise sharply as population numbers approach or exceed the food resources needed for their support. One widely cited piece of supporting evidence concerns the observed recovery of benthos and drift in a Maine stream after aerial spraying of an insecticide to control spruce budworm ravaged aquatic invertebrates as well (Dimond, 1967). As benthic populations recovered, drift initially was very low, then increased curvilinearly to previous levels (Figure 10.7), suggesting a

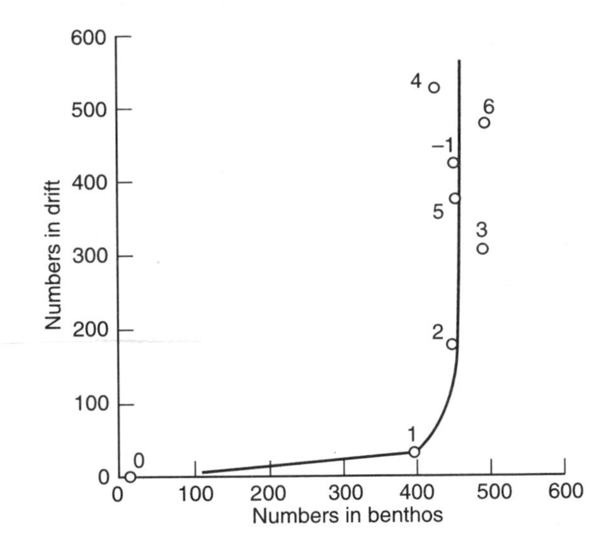

FIGURE 10.7 Relationship between total number of insects in drift samples and numbers in the benthos in streams affected by DDT-spraying in year 0. After 3–4 years, these densities were similar to the pre-treatment year (indicated by −1). (From Dimond, 1967.)

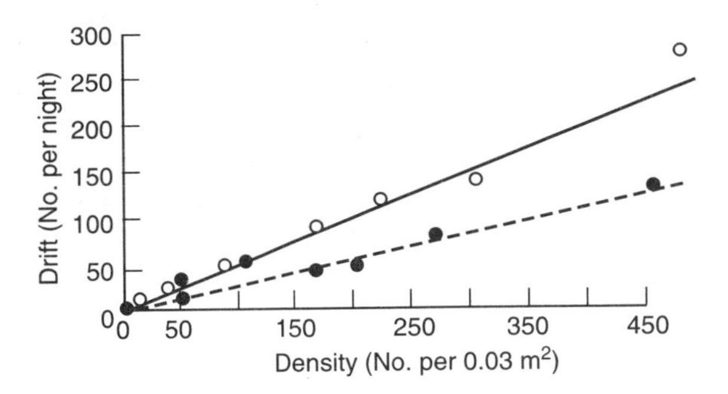

FIGURE 10.8 Relationship between benthic density and numbers in night drift for the mayfly *Ephemerella needhami* in artificial stream channels containing stones with low (○, ——) *versus* high (●, ----) periphyton densities. (From Hildebrand, 1974.)

density-dependent relationship. However, the shape of the curve is highly dependent on only a few data points.

Several studies have examined the correlation between drift and benthic densities to determine if the relationship is curvilinear, an expectation under the production–compensation model. Measuring drift out of stream channels containing differing insect densities, Hildebrand (1974) found drift to be simply a constant proportion of benthic density (Figure 10.8), a finding that contradicts Waters' model. Waters and Hokenstrom (1980) compared measurements of drift, standing crop and annual production in a long-term study of a 100 m section of a Minnesota stream over 5 years. Apparently due to siltation, all variables declined over the study, and production and drift were correlated. However, if drift were a constant percentage of production and played no compensatory role (as in Figure 10.8), a correlation still would be expected, so this type of study is inconclusive unless it is detailed enough to detect deviations from linearity.

While the explanations of drift offered by Müller and Waters contain some elements of an individual perspective (i. e. one that evaluates the benefits and costs to the drifting individual), they primarily reflect a population and ecosystem focus. Elliott (1967, 1968) was amongst the first to focus on a mechanistic interpretation of individual behavior. Field observations and laboratory experiments suggested that individuals were under stones and comparatively inactive during the day, and on stone-tops and actively moving by night. He concluded that an accidental explanation of drift was most plausible, in which dislodgement resulting from nocturnal movements explained drift periodicity, and competition for food and space might influence the likelihood of dislodgement.

A central issue in the study of drift is whether drift entry is an active process in which the drifting animal in some sense purposefully enters the water column, or whether entering the drift always is an accidental and passive phenomenon. Active drift entry implies some benefit to drifting. At first glance, passive drift implies at best no benefits to the individual and is likely to be detrimental, resulting in lost foraging opportunity or exposure to predators. Unfortunately the active *versus* passive explanations are not easily separated. For example, movements by individuals from underneath to the tops of stones and over stone surfaces are likely to be active behaviors associated with foraging, yet result in dislodgement purely as an accidental consequence of position and activity. On the other hand, entering the drift could be a purposeful movement to desert unprofitable foraging locations and seek patches that are richer in food resources. It is confusing to refer to the former mechanism of entering the drift as passive and the latter as active, because both are due to foraging activity. By examining the behaviors of individual animals, including day *versus* night position and activities and the influence of environmental variables such as light level and hunger, a number of investigators have attempted to sort out these conflicting interpretations of why benthic invertebrates enter the drift.

Accidental drift entry is consistent with the expectation that animals spend the daylight hours in refuges (usually under stones), perhaps feeding but probably inactive. Since their food resources including diatoms and organic layers are found mainly on the upper surfaces of stones, animals must move onto the substrate surface at night to feed and move about in search of profitable feeding patches. Several lines of evidence support this view.

Laboratory experiments document a negative phototaxis in a variety of aquatic insects. Hughes (1966) found that *Baetis harrisoni* and *Tricorythodes discolor* both preferred shade to bright light, although *Baetis* tolerated higher light levels before responding. Hughes described the behavior as a kinesis where animals simply kept moving while in high light and ceased movements in dim light. The phototactic behavior of five mayflies (Elliott, 1968) and four caddisflies (Elliott, 1970a) in a laboratory stream was negative in all but the caddis, *Anabolia nervosa*, as exhibited by the animals' choice of the underside rather than the tops of stones.

An endogenous activity rhythm, corresponding to the hours of darkness, was demonstrated in five species of mayflies studied by Elliott (1968). All exhibited an endogenous increase in activity during the night. Furthermore, in constant darkness the rhythm persisted whereas positioning behavior became unrelated to the 24 h cycle. However, of four species of caddis larvae studied, none showed an endogenous rhythm (Elliott, 1970a). It appeared that only exogenous control accounted for the nocturnal activity of three species and the diurnal activity of the fourth.

A greater number of individuals usually can be seen on the tops of stones at night compared with during the day. Elliott (1968) counted the numbers of *Baetis rhodani* and *Ephemerella ignita* at night, using a red flashlight, and found that the numbers in view showed the same pattern of nocturnal increase as did drift, although drift was more variable through the night than numbers were on the substrate surface. Campbell (1980) collected the large, carnivorous siphlonurid, *Mirawara purpurea*, in greater numbers in the upper layer of the benthos by night than by day, and surface abundance also increased on an overcast afternoon.

Observations of gut contents suggest that more feeding activity occurs during the night than during the day (Chapman and Demory, 1963). *Paraleptophlebia* appeared to feed on periphyton by night and detritus by day, implying a corresponding shift in microhabitat use from the tops to the undersides of stones. Meier and Bartholomae (1980) measured gut fullness using an ocular micrometer and obtained the lowest values just before dark and the greatest rate of filling just after dark. Ploskey and Brown (1980) compared the dry mass of *Baetis* guts from the drift before and after dark, and from the benthos. Three hours after sunset, guts were roughly 50% heavier in drifting nymphs compared with just before sunset.

Increasing drift rates, but more-or-less constant drift densities with increasing discharge, have been interpreted to imply that the probability of dislodgement is a function of current. Elliott (1967) demonstrated this for a Dartmoor stream and Allan (1987) observed a similar pattern in a Rocky Mountain stream. However, drift distance increases with increasing current velocity (equation 10.4), and this could result in a positive relationship between discharge and catch without any increase in drift entry.

While the above evidence appears persuasive, and so drift entry almost certainly is determined in part by diel changes in foraging movements and the resultant accidental dislodgement, the possibility that individuals actively enter the drift is not without support. By elevating trays of substrate above the stream bottom, Townsend and Hildrew (1976) determined which taxa most readily reached new habitats via the drift. Because current velocities in this small English stream were low, generally less than 15 cm s^{-1}, passive dislodgement did not seem likely to be important. Drift was substantial, however, and different taxa varied greatly in their capture rates in suspended *versus* benthic substrates, an index of drifting *versus* crawling. Townsend and Hildrew argued that the greatest tendency to drift was exhibited by those taxa whose food was patchily distributed (several predators and a detritivore that fed in leaf packs), whereas another species that fed on more uniformly distributed surface bacteria drifted little.

Laboratory studies have reported reduced drift and walking movements when food is plentiful compared with when food is depleted. Presented with an experimental range of algae on stone surfaces, *Baetis rhodani* exhibited no nocturnal periodicity in feeding or in occupation of the tops of tiles, although drift was nocturnal and inversely related to food abundance (Bohle, 1978). The implication was that animals drifted, primarily by night, to exit the habitat after food was depleted but not as a consequence of feeding activity. Kohler's (1985) study of *B. tricaudatus* extends these findings. Nymphs fed *ad libitum* were more active and more common on stone-tops by night than by day. Starved nymphs showed little day–night variation in activity or position. Their behavior during the day was similar to that of fed nymphs at night, leading Kohler to predict that if drift entry was passive, diurnal drift of starved nymphs should approximate nocturnal drift of fed nymphs. In fact, virtually all drift was nocturnal and did not correspond to diel patterns in positioning or activity. High food levels reduced drift, and only in the absence of food was day drift detectably greater than zero. Kohler concluded that the relationship between food availability and food demands determined purposeful drift entry, subject to constraints mainly due to predation risk.

Studies of benthic distributions have on occasion found evidence that invertebrates can be present and even common on stone surfaces during the day. Collecting from the tops and bottoms of colonized bricks and natural stones, Kovalak (1976, 1978) reported no convincing evidence of a shift to upper stone surfaces at night, and remarked on the unexpected occurrence of substantial numbers on stone-tops by day. In a similar study with tiles colonized for 4 weeks, Kohler (1983) also found no significant differences, although there was a tendency for the proportions on top to be higher just after dark. Kohler also observed differences in behavior among species: in general, *Baetis* was more common on tops of stones, whereas other mayflies (*Paraleptophlebia*, ephemerellids and heptageniids) were more common on the bottom of stones. Graesser and Lake (1984) studied epibenthic activity in an Australian stream by scrubbing the tops of 10 stones at 4 h intervals. Peak densities were on stone-tops by day, while peak densities in drift occurred at night. This negative correlation occurred in all 33 taxa studied, significantly in 16 of the 33. Graesser and Lake suggested that lower numbers on stone-tops by night reflected the presence of so many

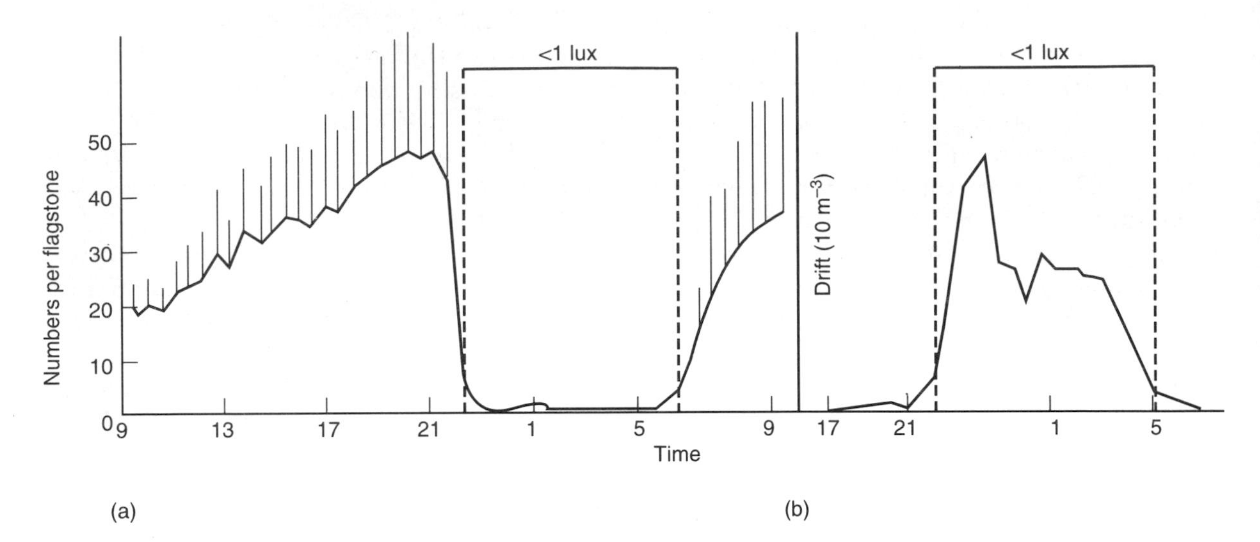

FIGURE 10.9 Numbers of *Baetis* (almost entirely *B. buceratus*) (a) on the surface of flagstones over a 24 h period (mean ± 1 standard deviation), observed with a light-amplification device and (b) in drift catches over the same time period. Darkness is indicated as light less than 1 lux. (From Statzner and Mogel, 1985.)

animals in the drift. However, based on reports that the amount in the drift at any one time usually is much less than 1%, it seems more plausible to interpret these results as counter-evidence to accidental drift entry.

Direct behavioral observations of insect activity on substrate surfaces in the field have been made using red filters on flashlights, light amplification devices, and infrared-sensitive cameras. Results generally confirm a greater presence and activity of aquatic insects on the substrate surface at night (Allan, Flecker and Kohler, 1990), but greater abundance by day also has been reported. Using a light amplification technique, Statzner and Mogel (1985) observed low densities of *Baetis buceratus* on substrate surfaces at night and high numbers during the day, while drift typically was nocturnal (Figure 10.9). Using red lights and visual observations, Allan, Flecker and McClintock, (1986) found several species of mayflies to be primarily diurnal in abundance and active on stone-tops, while drift as usual was nocturnal. While it was not possible to see feeding movements of the mouthparts of *Baetis*, Allan and colleagues observed steady sweeps of the labial palps in the heptageniid *Cinygmula*, which clearly engaged in at least some feeding by day. Furthermore, whereas *Baetis* was relatively difficult to disturb, both *Cinygmula* and *Rhithrogena* came into view on rock surfaces only after the observer remained still for at least 5 min, and would immediately disappear under the stone at the smallest disturbance. Possibly the common observation of stream biologists that aquatic insects are under stones during the day reflects this behavior, and not their usual position.

Lastly, active drift behavior may be an escape response to predators or aggressive conspecifics. Hildrew and Townsend (1980) showed that aggressive encounters between caddis larvae for net sites resulted in escape by drifting in the loser. Walton, Reice and Andrews (1977) inferred that intraspecific competition for space caused drift in a perlid stonefly, although they did not observe interactions. Walton (1980) found that a predaceous stonefly affected the drift of one prey species (*Taeniopteryx*) but not another (*Stenacron*), Peckarsky (1980) found that some mayflies drifted on encountering predaceous stoneflies, and Corkum and Clifford (1980) reported higher drift of *Baetis* in the presence, compared with the absence, of a stonefly predator. At this time it is uncertain whether the necessary interactions (predation and interference competition) occur with sufficient frequency and a nocturnal periodicity in order to account for the magnitude of behavioral drift. However, they clearly provide another mechanism for purposeful drift entry.

10.4 Summary

Downstream transport within the water column has been reported in many stream-dwelling organisms, including diatoms, larval vertebrates and many different invertebrates. Some taxa routinely drift in great abundance, including the Ephemeroptera, some Diptera, Plecoptera and Trichoptera. The diel timing or chronobiology of drift is its best understood aspect. Nocturnal periodicity is widespread and appears to be satisfactorily explained by the risk of predation hypothesis, which also accounts for those taxa that are aperiodic or day active. The proximal cue unquestionably is daylight. Exogenous control of drift activity appears to be a sufficient explanation, although an endogenous rhythm has been demonstrated in some instances. Estimates of light thresholds still are not very precise, and perhaps this value depends on environmental or physiological variables yet to be explored.

The colonization cycle proposed by Müller, (1954) interprets drift in a life cycle context, in which larval populations move downstream, perhaps in response to crowding, and upstream flight by adults completes the process. Supporting evidence exists for the colonization cycle, especially that egg-bearing females often fly upstream to oviposit. Because of compensatory high survivorship in headwater reaches, a relatively small number of ovipositing females may suffice to re-populate depleted stream sections. However, downstream displacement of the population has rarely been documented and upstream flight does not seem to be universal. Thus, while the life cycle and dispersal aspects of Müller's explanation clearly have merit, this is only a partial explanation.

In the production–compensation model proposed by Waters (1961), drift represents production in excess of the carrying capacity of the substrate. Thus these individuals can be lost without any need for compensatory upstream flight, serving instead to support the growth of higher trophic levels. Although unquestionably, drifting individuals are food to a variety of consumers, especially those that feed from the water column, this explanation is undermined by experimental and correlative evidence that drift appears density independent.

A more mechanistic and functional explanation of drift is achieved by asking what benefits and costs accrue to an individual depending on whether or not it drifts, and by appropriate experiments and direct observations of individual behavior. Foraging movements strongly influence drift entry both directly and indirectly. Numerous studies report aquatic insect larvae to avoid the substrate surface during daylight hours, and to move to stone-tops at night to feed on periphyton and organic matter. The diel periodicity in drift can be explained from this foraging rhythm and the greater likelihood of dislodgement while feeding and traversing the substrate surface. Predation risk, greatest during the day and for larger individuals, constrains foraging activity to hours of darkness.

In addition, it seems likely that some aquatic invertebrates purposefully enter the drift as part of their foraging behavior. This is supported by evidence of drift under flows considered too low to promote accidental dislodgement, and by evidence from laboratory studies that drift increases as food resources are depleted.

Chapter eleven

Lotic communities

The forces that shape community structure are those that determine which and how many species occur together, which species are common and which are rare, and the interactions among them. Thus the topic of community structure involves a synthesis of all the environmental factors and ecological interactions influencing an assemblage of co-occurring species.

The idea that communities exhibit structure requires that assemblages be more than haphazard collections of those species able to disperse to and survive in an area. It leads us to expect that the same species, in roughly the same abundances, will be found in the same locale as long as environmental conditions do not change greatly, and that similar communities will occur elsewhere if environmental circumstances are comparable. Biotic interactions, including, but not restricted to, food web linkages, are the important processes that govern the assembly and maintenance of communities. Environmental variation has a counteracting tendency, and, if sufficently extreme, is likely to prevent biotic interactions from acting with the strength and regularity to allow any consistent community pattern to emerge. However, a moderate level of environmental variability may offset the tendency of a few superior species to win out, and thus enhance diversity. The interaction between biotic and abiotic factors is complex, and as a consequence there are a number of theoretical frameworks that attempt to explain what regulates the composition of communities.

Studies of community structure and biotic interactions often assume that observed pattern is the deterministic outcome of local interactions, without regard for such larger scale processes as dispersal, speciation and unique historical circumstances (Ricklefs, 1987). However, the species pool of a region is important to any consideration of local community structure because it determines which species are available to colonize any given locale, and thus influences the make-up of the community. In addition, consideration of regional diversity reminds one that the long-term persistence of any individual species usually does not depend on its survival in any one local community. Separate populations of a species may exhibit different trends in different locales, with the consequence that dispersal permits a long-term regularity on a larger scale that is not apparent by detailed investigation on a finer scale. Such a perspective lessens the need for equilibrium-enhancing interactions: regional processes of immigration and emigration may contribute some of the buffering against extinction that otherwise must be attributed to biotic factors.

11.1 Local and regional diversity

11.1.1 Species richness at the local scale

A complete survey of all the species within a single community is an extremely challenging task. Such studies are unusual partly because the

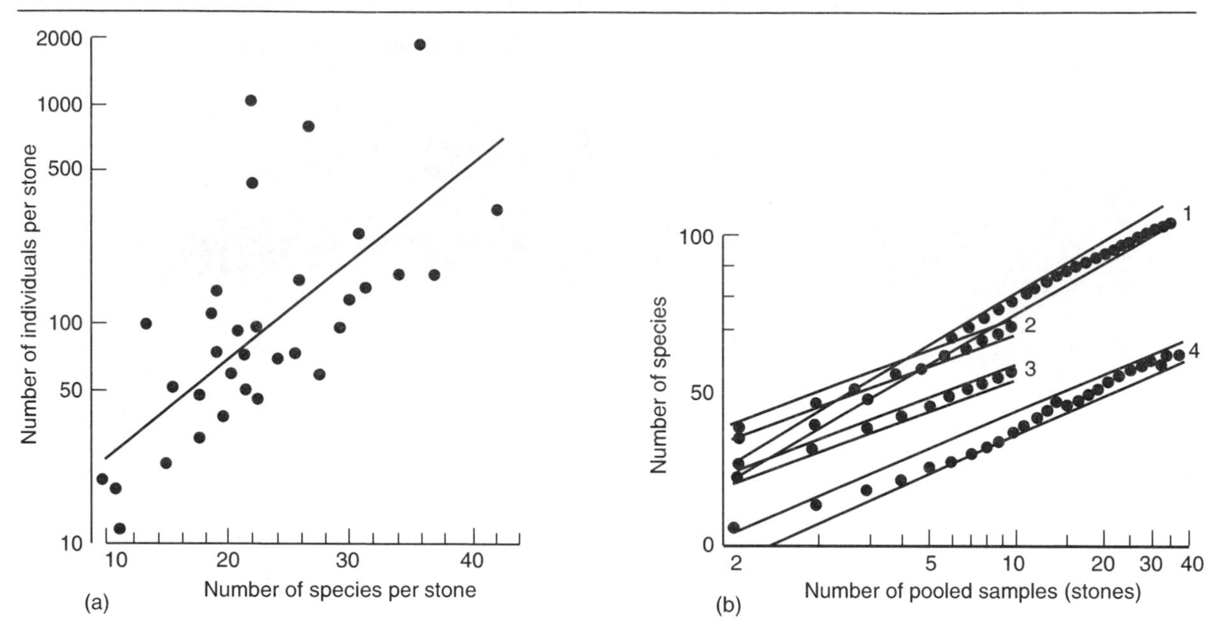

FIGURE 11.1 The number of species collected increases with the size of the sample, illustrated by Kuusela's (1979) study of the fauna on individual stones in a large Finnish river. (a) The number of species found on a single stone increases with the number of individuals ($r = 0.71$). (b) The cumulative number of species collected increases with the logarithm of cumulative number of stones sampled; 95% confidence limits as illustrated. 1, River Lestijoki, Finland (Kuusela, 1979); 2, Vaal River, South Africa (Chutter and Noble, 1966); 3, Lytle Creek, Utah (Gaufin, Harris and Walter, 1956); 4, Rio Java, Costa Rica (Stout and Vandermeer, 1975).

taxonomic knowledge of many groups is inadequate, and partly because the exhaustive compilation of a species list is rarely a priority. It is more usual to find either a detailed study of a single taxon, or an ecological investigation where the focus is on the more common species, while difficult to identify taxa are lumped, often at the family level. Nonetheless, it is apparent that species richness is affected by a number of variables. The estimated number of species always depends on the size of the sample collected. More species are found in large rivers than small streams, apparently because the spatial area and habitat diversity are greater in larger systems. Historical differences between regions in rates of speciation and rates of extinction will influence the size of the species pool, and hence have an impact on local diversity.

The importance of sample size is nicely illustrated by a study of macroinvertebrates collected from individual stones in a large rapids (20–40 m wide, approximately 1 km in length) of the River Lestijoki, Finland (Kuusela, 1979). The number of individuals per stone was positively correlated with the number of species per stone (Figure 11.1a). In addition, the cumulative number of species increased with the logarithm of the cumulative number of stones sampled; that is, at a decelerating rate. The latter relationship has been reported from streams of widely different regions (Figure 11.1b), and clearly illustrates the dependence of local species richness on sampling effort.

The area of environment from which the sample derives also has a profound effect on the number of species obtained, as has been demonstrated most clearly using collections from such discrete units as islands (MacArthur and Wilson, 1967) and lakes (Barbour and Brown, 1974). The relationship typically is

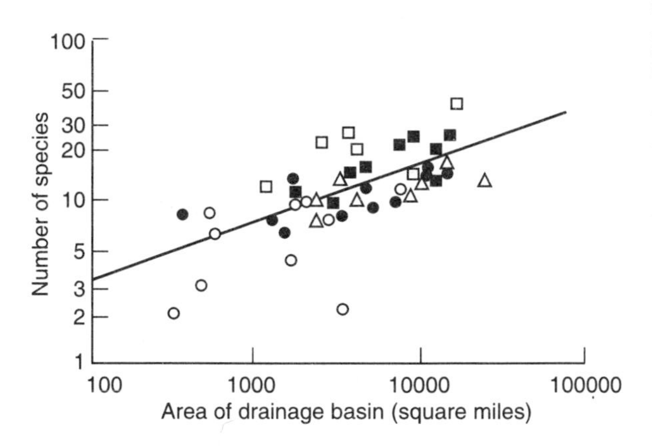

FIGURE 11.2 Species–area relationship for freshwater mussels (Unionidae) from 49 rivers draining the North American Atlantic coast between southern Canada and the eastern Gulf of Mexico. ●, Northern Atlantic slope rivers; △, middle Atlantic slope rivers; ■, southern Atlantic slope rivers; ○, peninsular Florida rivers; □, eastern Gulf slope rivers. (From Sepkoski and Rex, 1974.)

log-linear, according to the equation:

$$S = cA^z \tag{11.1}$$

where S is species richness, A is area of habitat, and c and z are parameters determined from the data. The slope parameter z quantifies the rate of increase in number of species with area studied, and frequently falls between 0.2 and 0.4. Several studies document that equation 11.1 applies to rivers, although estimates of z vary widely. Freshwater mussels fall well within the expected range, with $z = 0.32$ (Figure 11.2).

The main cause underlying thc spccics–area relationship probably is that larger rivers include a greater habitat diversity due to their greater habitat area. In addition, a larger habitat volume will contain more individuals, and based on Figure 11.1, this in itself is likely to result in the inclusion of more species. Finally, many species that occur in larger rivers appear unable to live in small streams. This is indicated by studies of species distributions, which typically show more addition than replacement as one proceeds downstream along a river's length (Illies and Botosaneanu, 1963; Horwitz, 1978), and by a positive correlation between fish size and river size (Hynes, 1970).

Another very general finding from species collections is that a few species are common and most are quite rare. In the first of an extensive series of studies on productivity at Schlitz, Germany, some 52 000 insect specimens were trapped while emerging from a stream flowing underneath an 11 m^2 greenhouse (Illies, 1971). A total of 148 species was collected, but the 15 most abundant species contributed 80% of the total number of individuals (Figure 11.3).

Why, in general, a few species should be very common and many species quite rare is subject to unresolved debate (May, 1984). However, it has the consequence that collection of few samples will include most of the common species, while further sampling effort will continue to produce additional species almost (but not quite) indefinitely. This underlies the relationship between sample size and local species richness (Figure 11.1), which in turn influences the amount of sampling effort necessary to characterize a system.

Species diversity sometimes refers just to the number of species, also known as species richness, but it also is used more generally to describe other attributes, including the relative abundance of species in the collection. One commonly employed statistic derived from information theory is:

$$H' = -\sum_{i=1}^{S} (p_i \log p_i) \tag{11.2}$$

where S is the number of species in the collection, and p_1 is the relative (proportional) abundance of the ith species (Pielou, 1975). High values occur when species richness is high and most species are equally abundant in the collection. The latter is called 'evenness' because an assemblage will appear most diverse when its

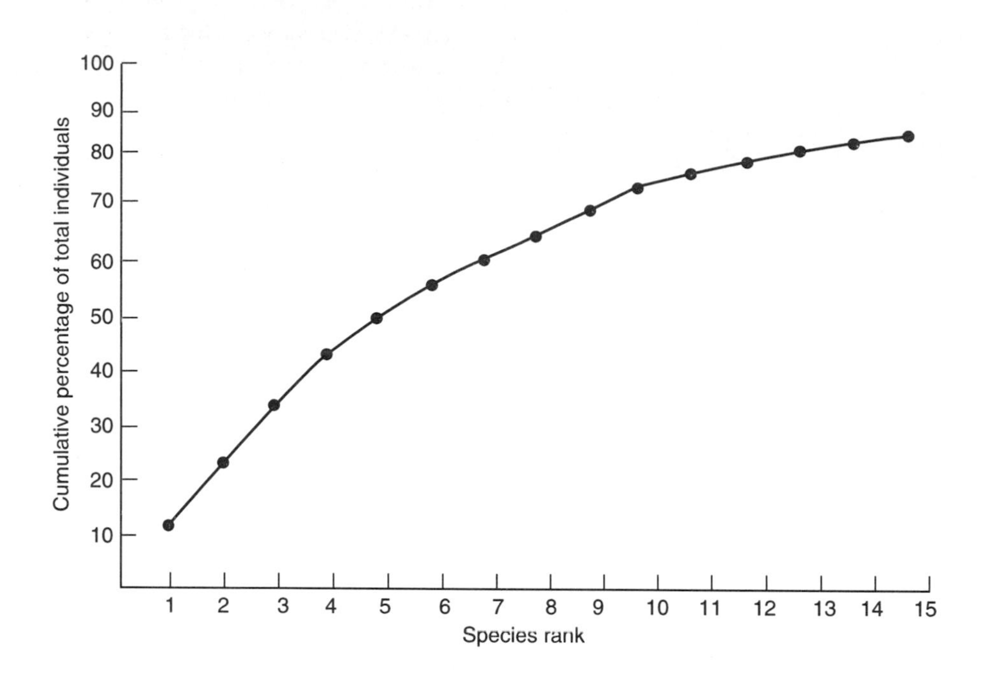

FIGURE 11.3 The cumulative percentage of total individuals collected that are contributed by the 15 most common species. (From emergence collections of Illies, 1971.)

individuals are evenly distributed among all species.

Diversity statistics have received much use in environmental assessment (Wilhm and Dorris, 1968), because it is often found that in polluted waters one or a few species are very strongly dominant, and others scarce or absent. Unfortunately this approach has several drawbacks, including the dependence of species richness on sample size and, at least in unpolluted systems, the relatively invariant pattern of relative abundances. Probably it is more useful to provide separate estimates of species richness and evenness. For example, Hawkins, Murphy and Anderson (1982) compared dominance–diversity curves for the invertebrate taxa of canopied *versus* open (due to clear-cutting) sections of some Oregon streams. Greater dominance and less evenness characterized the open site, as expected in disturbed habitats.

11.1.2 Local *versus* regional diversity

The number of species that occur within a region clearly sets an upper limit on the number of species that comprise any particular biological community, whereas local environmental conditions largely determine what subset of the species pool is represented. Often it seems that the species richness of a local community is half or less of the species pool. In a survey of the invertebrate taxa from 34 stream sites that varied in pH, the number of taxa decreased in response to increasing acidity (Hildrew *et al.*, 1984). In fact, the total species pool decreased as streams became more acid, and the average number of species at a site was a steady 41–48% of the available species pool (Figure 2.8b).

Because ichthyologists have been collecting and describing the fishes of North America and Europe for several hundred years, the fish species and their distribution are well known (Hocutt and Wiley, 1986). North America with its greater size and physiognomic diversity contains about three times as many fish species as Europe, over 700 *versus* 250 (Briggs, 1986). Regions within North America differ greatly, however. All of Canada and Alaska contain some 180 species of fish, considerably fewer than the rich Mississippi basin where most of the major adaptive radiations in North America have occurred. The Tennessee and Cumberland Rivers drainage realm alone includes some 250 species of fish (Starnes and Etnier, 1986). Species richness declines from east to west across the USA (Fausch, Karr and Yant, 1984), due in part to major differences in extinction rates during the Pleistocene. As a consequence, the western USA contains only about a quarter as many fish species as the east.

Not surprisingly, numbers of fish species at a site also differ between regions. In the species-rich midwest, as many as 50–100 species may occur at a particular location, although reports in the range of 20–50 species are more common (Horwitz, 1978). In some studies that will be examined in more detail later in this chapter, at any one time investigators found 23 fish species in a single riffle of an Indiana stream (Grossman, Moyle and Whitaker, 1982), 47 species from several sites in an Ozark stream (Ross, Matthews and Echelle, 1985), and only seven species in a small California stream (Moyle and Vondracek, 1985).

The number of invertebrate species at a site, and trends across regions, are reasonably well known for some thoroughly studied groups, but not for the entire assemblage. Because of the challenge of collecting and identifying all of the invertebrates, particularly dipterans and many members of the meiofauna, most ecological studies work with a restricted subset of the invertebrate fauna, identifying some taxa only to family, order or even phylum. In a few instances, however, investigators have attempted to catalog all invertebrates at a site. Sixmile creek, Pennsylvania, has in excess of 300 identified taxa of aquatic insects (E.C. Masteller, personal communication), and over 550 insect species have been catalogued from Upper Three Runs Creek, South Carolina (J.C. Morse, personal communication). An intensive effort of almost two decades in a small German stream, the Breitenbach, provides an unusually complete list (Table 11.1). It includes 1044 invertebrate species, a number that is likely to increase with further studies, but probably not greatly. It is difficult to say what fraction derives from habitats other than the stream, such as a small impoundment and other standing-water habitats, but probably less than one-third (P. Zwick, personal communication). This compilation indicates that the greatest invertebrate diversity is located in a few groups, including several minute, interstitial phyla (Nematoda, Rotatoria, Annelida and Turbellaria) and the highly diverse Diptera, especially the midge family Chironomidae. Despite the scarcity of complete inventories of invertebrates, there can be little doubt that, at least in temperate streams, the number of invertebrate taxa exceeds the number of fish species by an order of magnitude.

Studies of the number of diatom species that colonize glass slides suspended in the current also suggest that local species richness is proportional to regional species richness (Patrick, 1975). In two species-rich streams of the eastern USA, between 79 and 129 diatom species colonized glass slides; in two species-poor streams on Dominica, West Indies, the range was 46–61 species. Comparisons of species-rich and species-poor streams in the USA gave a similar result: 160 species were collected on slides in a stream where the species pool was about 250; fewer than 30 were found in a stream with about 100 total species.

Patrick's studies also demonstrate that invasion rate affects local species richness. Fewer

TABLE 11.1 Provisional faunal list of the Breitenbach, Schlitz, Germany (courtesy of Dr Peter Zwick)

Group	Species	Insecta	Species	Diptera	Species	Chironomidae	Species
		Odonata	1				
Insecta	642	Ephemeroptera	16				
		Plecoptera	18				
Turbellaria	50	Megaloptera	2	Tipulidae	30		
		Planipennia	2	Limoniidae	86		
Gastrotricha	6	Coleoptera	70	Ptychopteridae	2		
		Trichoptera	57	Psychodidae	35	Tanypodinae	15
Nematomorpha	1	Diptera	476	Chironomidae	152	Diamesinae	8
				Dixidae	4	Orthocladiinae	88
Nematoda	125			Culicidae	2	Chironominae	41
				Tahumaleidae	3		
Rotatoria	106			Ceratopogonidae	61		
				Simuliidae	10		
Mollusca	12			Rhagionidae	2		
				Empididae	21		
Annelida	56			Dolichopodidae	50		
				Tabanidae	8		
Crustacea	24			Stratiomyidae	3		
				Ephydridae	2		
Hydracarina	22			Syrphidae	5		
Total	1044 species						

species of diatoms were found on glass slides suspended in a flow rate of $1.5\,l\,h^{-1}$ compared with a flow rate of $650\,l\,h^{-1}$. The underlying biology responsible for this result merits further investigation, but certainly must involve a high colonization rate counteracting the presumably marginal suitability of existing conditions for some species.

Local conditions surely influence the subsample of regional diversity that occurs at a site. In general, we anticipate a greater number of co-occurring species whenever the heterogeneity in environmental conditions also is large. This is widely known as Theinemann's (1954) first law, although Macan (1974) points out antecedents. Supporting evidence comes from studies that demonstrate a positive correlation between measures of habitat diversity and numbers of species at a site (e.g. Gorman and Karr, 1978), and the habitat-specificity of particular organisms. Mackay's (1969) separation of her invertebrate collections into five distinct habitat types clearly demonstrated that total diversity within a small stream section was enhanced because different habitats possessed different assemblages (Table 3.4). Moreover, species richness was lower in the more physically uniform habitats (sand *versus* gravel, for instance).

Environmental variability plays a double-sided role, however. Conditions that are extremely variable, or otherwise deviate greatly from the norm, generally support very few species (Thienemann's 'second law'). This can be seen in the decline in species richness with increasing acidity (Figure 2.8), and from the many studies where extreme conditions, whether natural or due to human influence, support a diminished biota.

11.1.3 Evolutionary history of a region

Geographically distant regions typically have their own distinctive flora and fauna, attesting to chance differences in the establishment and diversification of particular taxa, and the local interplay of environmental and biological forces that direct evolutionary change. Thus, even if environmental conditions appear very similar, as they do in mountain streams almost everywhere, communities that are assembled from distinctly different species pools are unlikely to be identical.

Comparative studies of fish taxocenes clearly show that history and biogeography can influence the taxonomic and ecological diversity of a region. The Nida River in south-central Poland, and the Grand River in Ontario, Canada, are two river systems that exhibit similar gradients from the headwaters downstream and occupy similar climates. Thus they might be expected to support a similar number of species, filling roughly similar ecological roles. There were in fact many similarities (Figure 11.4), due largely to the abundance of cyprinids in both, and comparable species in the Esocidae, Cottidae and Gasterosteidae (Mahon, 1984). There also were prominent differences, including the diversity in North America of Centrarchidae and Ictaluridae, and radiation within the genera *Notropis* and *Etheostoma*. The Grand River contained more species overall, especially in its smaller streams. In addition, more species in the Grand River were primarily stream dwellers, whereas the Nida possessed a greater proportion of large species that were only occasional stream dwellers.

Explanations for such differences always are speculative. Mahon (1984) favors the possibility that the success of the lentic specialists (Centrarchidae and possibly Ictaluridae) in North America closed out the migratory option typical of the larger cyprinids of Europe, and favored species that formed resident populations in small streams. Possibly this accounts for the greater North American diversity within smaller species. A central point from this comparison is that while species richness may increase downstream in a like fashion in quite different stream systems, dissimilarities may be as prominent as similarities in terms of faunal composition and various ecological attributes of the taxocene.

Of all the geographic patterns that have been described for species richness, a general trend towards greater numbers of species in the tropics is perhaps the most predictable. It unquestionably is true for the fishes of running waters (Figure 11.5). The slope of the species-area relationship is 0.15 in northward-flowing Russian rivers, and 0.24, 0.43 and 0.55 for European, African and South American rivers, respectively (Welcomme, 1979), indicating a trend toward a more rapid increase in species richness with increasing river size at low latitudes. Despite the inadequate state of taxonomic knowledge, in excess of 3000 species of freshwater fish are estimated to occur in tropical South America, primarily in riverine habitats (Moyle and Cech, 1982; Lowe-McConnell, 1987). This greatly exceeds the roughly 700 species found in the lakes and rivers of temperate North America.

It therefore is surprising that no one has yet established convincingly that a similar latitudinal trend exists in numbers of invertebrate species. Patrick (1964) suggested that, except for the fish, the biota of streams might be an exception to the general trend toward more species in the tropics. Because tropical studies are few and taxonomic knowledge is limited, the absence of a latitudinal trend in invertebrate diversity is far from established. Nonetheless, temperate biologists who have sampled tropical streams have been struck by the appearance of equivalent or reduced diversity in the latter, at least at the within-habitat scale (Patrick, 1964; Illies, 1969). Only Stout and Vandermeer (1975) have reported a greater species richness for tropical invertebrates, and their conclusion rests on the extrapolation of the species richness *versus* sample size curve (Figure 11.1b). Their 30 stone samples contained similar or fewer taxa than typically would be found in a temperate stream (although it is possible that they under-estimated the true number of species due to incomplete taxonomic knowledge). Because the tropical samples differed greatly from stone to stone, species richness increased rapidly with increasing sample size, and by

FIGURE 11.4 Some of the fish species occupying small drainage basins (300 km^2 or less) in (a) the Grand River system, southern Ontario; and (b) the Nida River system, south-central Poland. (From Mahon, 1984.)

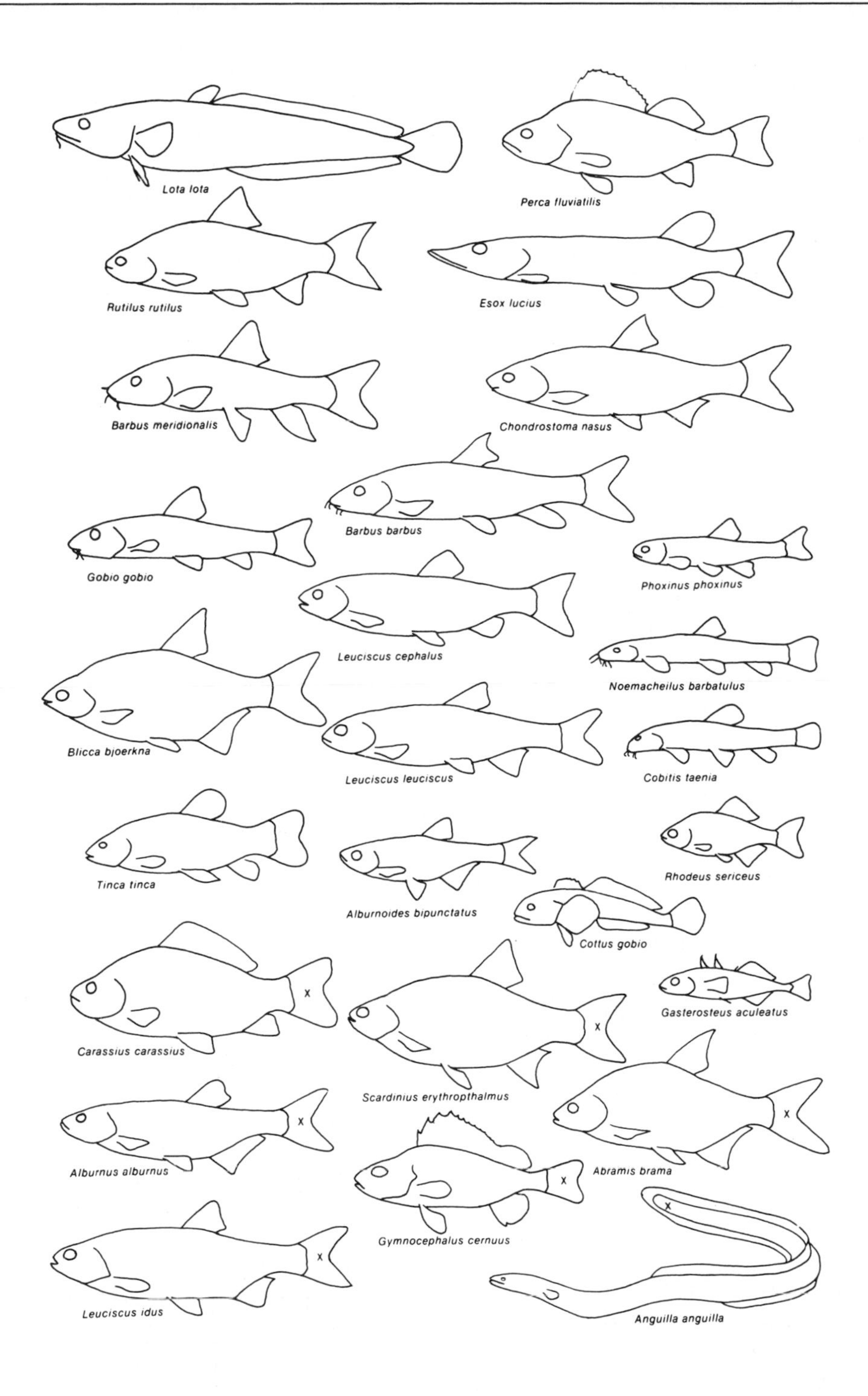

FIGURE 11.4 (*cont.*)

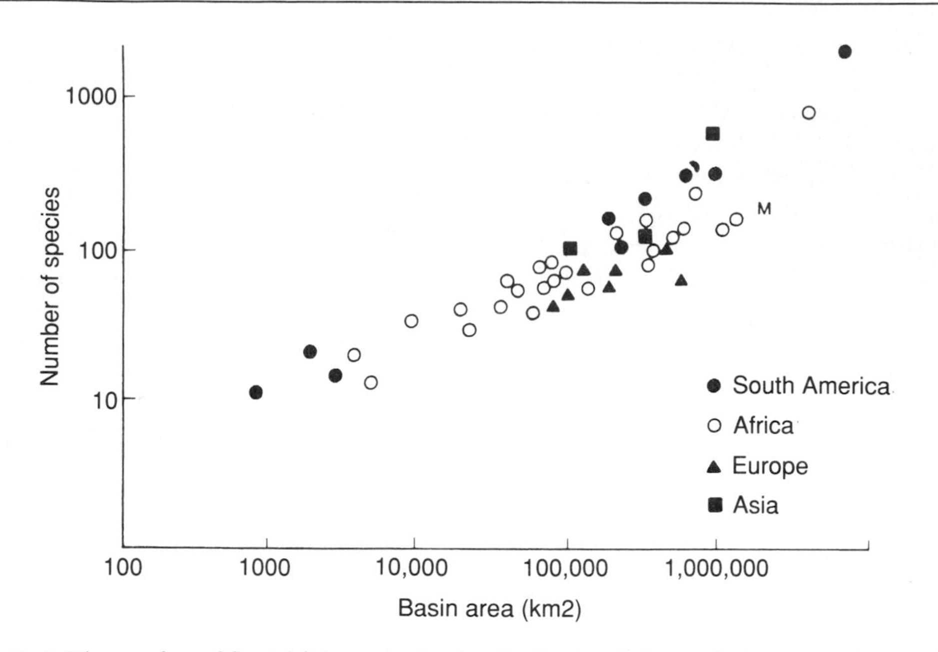

FIGURE 11.5 The number of fluvial fish species in river basins in relation to drainage area (a correlate of size and habitat complexity of a river system). The slope of the species–area relationship is 0.15 in northward-flowing Russian rivers, and 0.24, 0.43 and 0.55 for European, African and South American rivers, respectively. ●, South America; ○, Africa; ■, Asia. (Data from Welcomme, 1979; redrawn by Smith, Stearley and Badgley, 1988.)

extrapolation resulted in a theoretically more diverse assemblage. However, tropical sites are more diverse in Stout and Vandermeer's study only in their projected number of species, and not in their actual collections.

In summary, there are numerous factors that contribute to some areas being relatively rich in species while other areas are less so. A first level of explanation must take into account regional diversity, which is influenced by historical and biogeographical events; intensity of the sampling effort at both within and between-habitat levels; and the variability of the physical environment and the habitat. There is also good reason to believe that interactions between species, which in turn probably vary in their intensity depending upon environmental conditions, play a major role in determining local species richness. This provides the link between the topics of species diversity and community structure, and we turn now to the latter.

11.2 Community structure

Whether communities constitute a tightly interwoven network of strongly interacting species or are merely loose assemblages that persist in an area because the individual species are adapted to the particular environment, is a debate of long standing (Strong *et al.*, 1984; Diamond and Case, 1986). The former idea is associated with the view that biotic forces act to maintain communities at or near equilibrium. The opposing view is that variable and unpredictable abiotic forces are pre-eminent, and the species that are found together are simply those that happen to be favored by the environmental conditions of the moment.

How the interaction between abiotic and biotic forces influences ecological communities has been formulated into a number of similar but subtly different frameworks. One viewpoint holds that local environments vary from harsh to benign (Peckarsky, 1983), with a corresponding shift in the relative importance of abiotic and biotic forces. Although the implication of a continuous scale obviously permits one to be somewhere in the middle, this framework has often resulted in vigorous debates between advocates of one or the other extreme.

A second perspective emphasizes the importance of biological interactions, and especially competition, in structuring communities. It is based on the expectation that in very constant environments, strong biological forces permit only a few biologically superior species to maintain populations. Some moderate level of physical disturbance prevents dominance by a small number of the most specialized and effective competitors and allows other species, including those that are rapid colonizers but easily displaced, to coexist (Connell, 1978; Yodzis, 1986).

A third framework, referred to as patch dynamics, shares some aspects of the preceding models but places more emphasis on the dispersal ability of organisms and a shifting mosaic of environmental conditions. It is based on the ever-changing environmental circumstances within a stream section and the considerable ability to colonize and rapidly reproduce that characterizes most of the flora and fauna of running waters (Patrick, 1975; Townsend, 1989). Under the patch dynamics model, the fluctuating nature of running water environments permits more species to co-occur than would be true if conditions exhibited greater constancy. It also confers some regularity to pattern because environmental circumstances are predictable in the aggregate even though they are unpredictable for any given place and time.

The emerging view is that no single model adequately describes all communities (Townsend, 1989). Some are more strongly influenced by biotic processes, others by abiotic processes, and the applicability of each of these theoretical frameworks is situation-specific. We will briefly review evidence for the separate importance of biotic and abiotic processes within lotic communities, and then turn to an evaluation of the several conceptual models that describe the interaction between them.

11.2.1 Frequency and strength of biotic interactions

It is the central premise of an equilibrium view that observed pattern is the deterministic outcome of local processes, and in particular is the result of interactions among species. This requires that competition, predation and herbivory act with sufficient force and frequency to maintain regularity in the composition and relative abundances of species within some reasonably defined area. To the extent that biotic interactions are found to be strong and communities to persist with little change, the equilibrium view is upheld.

While evidence of resource and habitat partitioning among ecologically similar species of running waters is widespread (chapter 9), competitive effects have been documented convincingly in a relatively small number of studies. Competition appears to occur episodically: during certain seasons, when food is in short supply, or when favorable conditions permit populations to reach high densities. Existing evidence does not permit us to specify how frequently competition is expressed within the communities of running waters, and so we should be cautious about theories that require competition to be widespread and strong.

Linkages between trophic levels, including predation and herbivory, unquestionably are ubiquitous for these are the connections that comprise food webs. Predators (chapter 7) and grazers (chapter 8) have often been shown to reduce the numbers and alter the relative abundances of the assemblage of species on which they feed. A variety of indirect effects has also been implicated, including prey shifts in habitat use in order to minimize predation risk, and stimulation of algal growth due to the removal of senescent cells. Moreover, consumers have been shown to affect other species that are one or more links removed from the immediate prey in the food chain. These trophic cascades occur because the reduced abundance or feeding activity of the immediate prey in turn reduces predation pressure for those organisms upon which it feeds.

Strong predation effects appear to be associated with certain conditions. High rates of immigration and emigration are likely to minimize local control, and predation effects have been least apparent in fast-flowing streams where drift and dispersal are substantial (Cooper, Walde and Peckarsky, 1991). Vulnerability of prey to predators appears to be a second factor governing the force of the interaction. Wherever the prey are conspicuous or lack a refuge, predation is likely to be most effective, as Power (1992) demonstrated in contrasting the impact of predaceous fish in riffles *versus* boulder-pools of the same river.

Studies of herbivory likewise establish that strong biotic interactions can have far-reaching effects. Periphyton taxa that are especially vulnerable, meaning those that are large and accessible (Figure 4.1), species of grazers that are particularly effective because of their mouthparts or feeding rates, and benign environmental conditions such as low flows during late summer are all associated with enhanced 'top-down' control at the autotroph–herbivore interface. Not surprisingly, these conditions are similar to those that favored strong predation effects as well.

11.2.2 Quantification of environmental variability

Rivers and streams appear to be physically harsh environments. Organisms frequently are at risk of being swept away in the current. Floods scour the substrate and re-shape the channel. Droughts cause temperature and oxygen stress, and perhaps a total dewatering. Thus it often is suggested that abiotic factors might have an especially great influence in running waters compared with other environments, or in particular lotic settings relative to others. Testing such assertions is difficult, however, for at least two reasons. First, we need some way to quantify the relative harshness of different environments. Second, it can be argued that because the biota generally are adapted to the conditions they normally experience, small deviations might be devastating to species adapted to very predictable conditions whereas large fluctuations might be the norm for species adapted to variable conditions.

Although many environmental variables fluctuate widely, much attention has focused on discharge regime. Variation in flow is relatively easy to quantify and appeals to our common sense as an important abiotic force. Discharge volume at a site can vary widely on time scales of days, years and decades; and among sites because of differences in drainage basin area, climate, vegetation and physical characteristics such as soils and topography (chapter 1). Some of this variation is reflected in the hydrographic pattern typical of a climatic or geographic region. In the Rocky Mountain region of the western USA, for example, most precipitation falls as snow, and runoff occurs with highly predictable timing during spring melt. In contrast, the Pacific Northwest receives much of its precipitation as winter rain, resulting in an irregular recurrence of hydrograph peaks throughout the winter months. And of course some regions receive precipitation inputs fairly evenly through the year, whereas erratic storms typify other areas (Figure 11.6).

Poff and Ward (1989) attempted to characterize these differences in flow regime using hydrologic data from 78 stream gages across the lower USA. Using a number of measures expected to characterize flood frequency, seasonal flood predictability, flow intermittency and overall flow variability, they identified nine categories of flow regime (Figure 11.7). Although these nine flow regimes are not linearly ranked in terms of harshness, one can at least compare regions of differing streamflow vari-

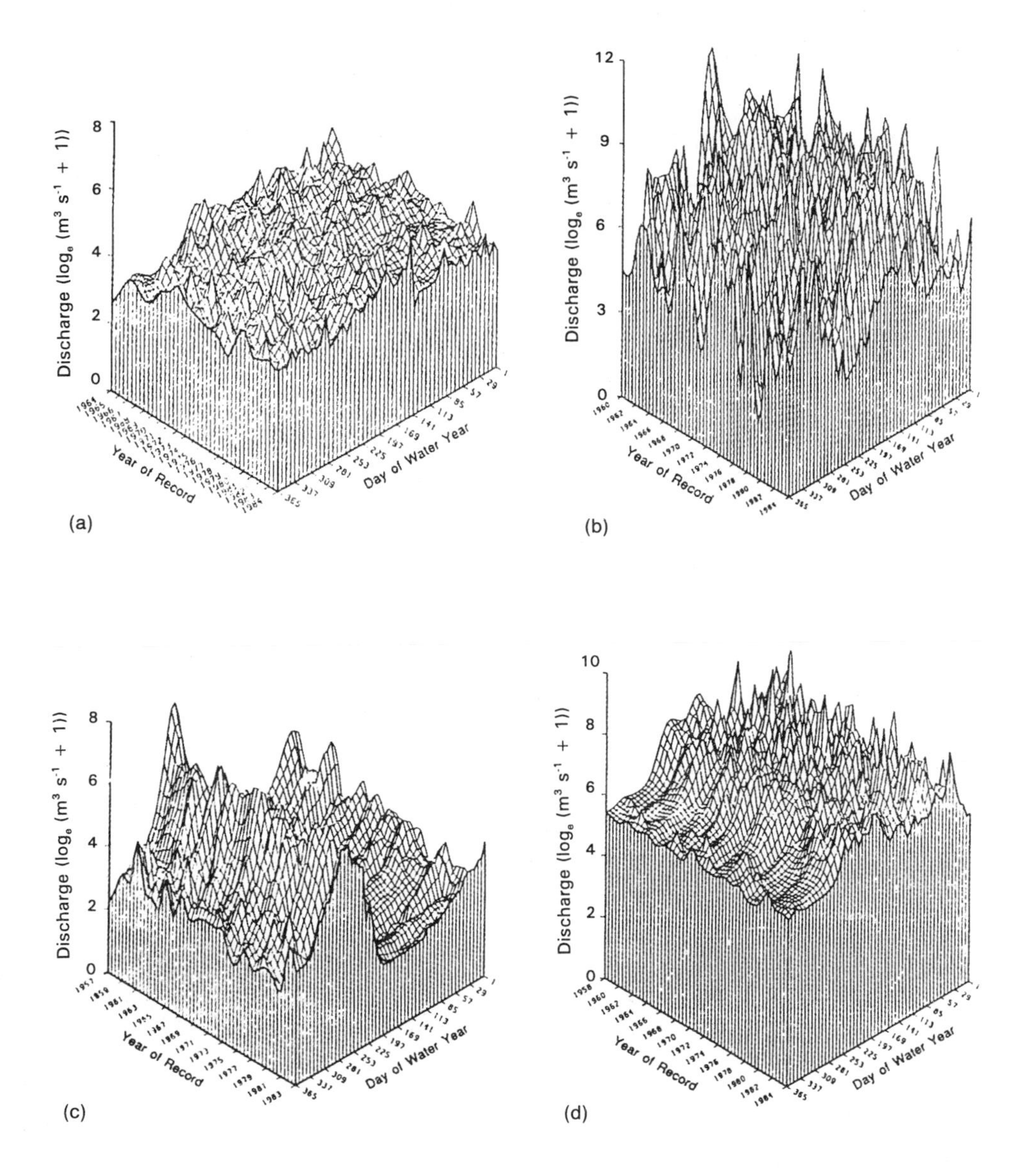

FIGURE 11.6 Patterns in streamflow variability based on long-term, daily mean discharge records for four different stream types. Water year is 1 October–30 September. Three-dimensional graphs portray within-year and between-year variation. (a) 'Mesic groundwater', Augusta Creek, Michigan; (b) perennially flashy Satilla River, Georgia; (c) 'snowmelt', Colorado River, Colorado; (d) 'winter rain', South Fork of the McKenzie, Oregon. These represent four of the nine flow categories of Poff and Ward (1989).

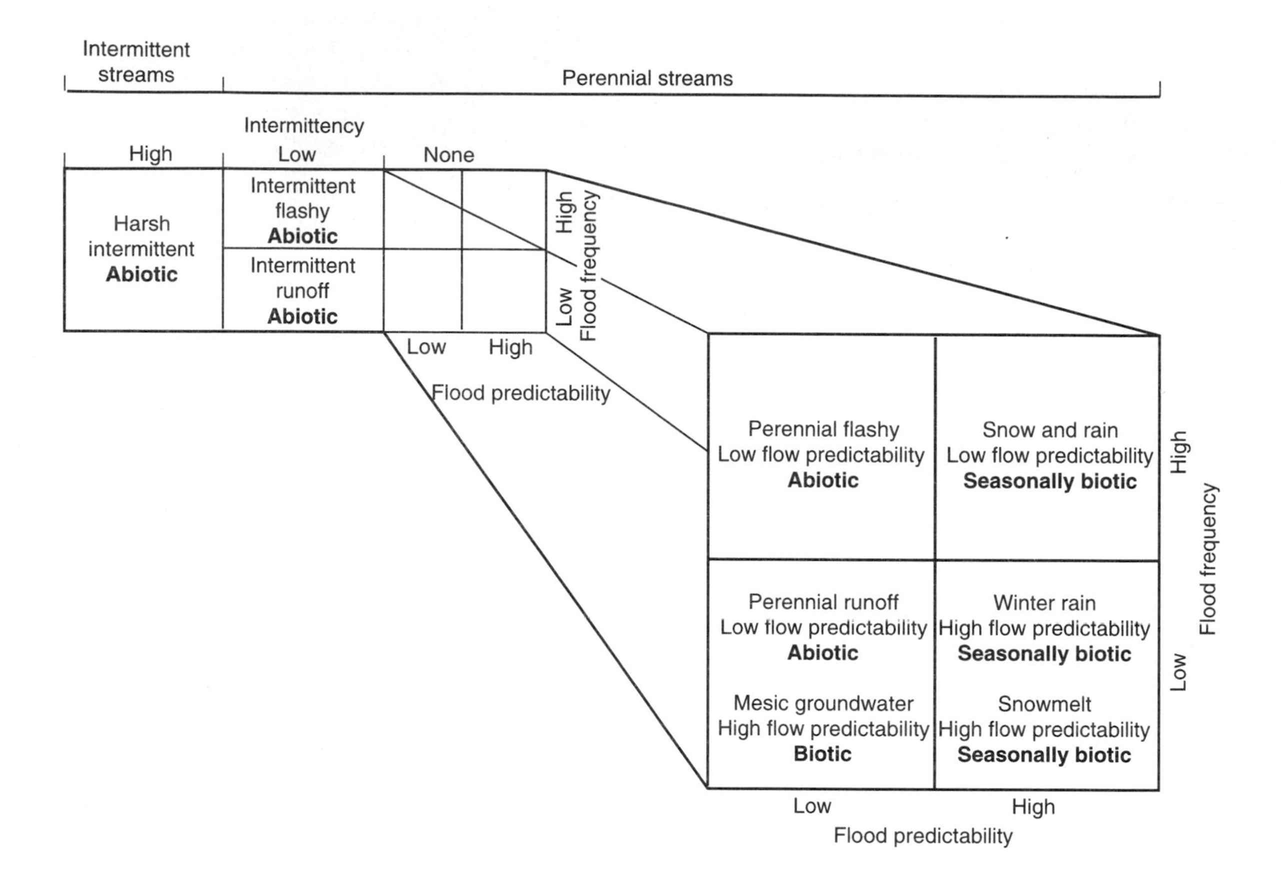

FIGURE 11.7 Conceptual model of nine stream types based on several temporal measures of discharge regime. The degree of intermittency is the first classification variable. For streams of low intermittency and for perennial streams, flood frequency provides additional separation. For perennial streams, flood predictability also must be considered. Names ('winter rain') are indicative of environmental conditions resulting in hydrographs of a particular class. Additional descriptions (abiotic, seasonally biotic) are Poff and Ward's (1989) predictions of the relative contributions to community structure of abiotic and biotic processes for each stream type.

ability. When some 43 stream locations in Minnesota and Wisconsin were divided into those with more or less variable flow regime, some expected differences in their respective fish assemblages were confirmed (Poff and Allan, 1995). Fishes from the more variable sites contained a higher proportion of species that were trophic generalists and tolerant of silt and disturbed environments.

Seasonal variation in flow might be expected to cause a regular alternation in the importance of abiotic and biotic factors. This is clearly seen in Flecker and Feifarek's (1994) study of aquatic insects of rivers at the foot of the Venezuelan Andes, which experience frequent flash floods during the rainy season and droughts during the dry season. Total abundances of benthic insects exhibited a strong negative relation with average monthly rainfall, used as a surrogate of flow because no stream gages were

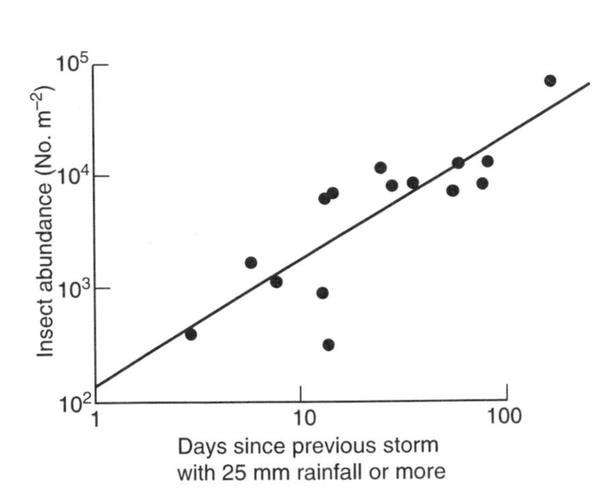

FIGURE 11.8 The numbers of aquatic insects in a river of the Andean foothills subject to a pronounced dry season and frequent floods during the rainy season. Time elapsed since the most recent rainfall event over 25 mm is used in lieu of stormflow data ($r = 0.82$). (From Flecker and Feifarek, 1994.)

available. Presumably due to colonization and recruitment, numbers recovered during flood-free periods, resulting in a strong positive relationship between insect abundances and time elapsed since the last storm (Figure 11.8).

11.2.3 Interplay of biotic and abiotic factors

The preceding material provides ample evidence that biotic and abiotic factors each vary in magnitude among different taxonomic groups and across space and time. The following examples illustrate how an interplay between physical and biological forces can facilitate species coexistence when only one species might persist in the absence of this interaction.

Throughout most of Arizona, the introduced mosquitofish, *Gambusia affinis*, has replaced a native poeciliid, the Sonoran topminnow, *Poeciliopsis occidentalis*, largely through predation on juveniles (Meffe, 1984). In mountainous regions subject to extreme flash floods, however, long-term coexistence of the two species results from the native fish's superior ability to avoid downstream displacement. Hydropsychids and simuliids exhibit a similar interaction seasonally in coastal Californian streams (Hemphill and Cooper, 1983). On hard substrates in fast-flowing sections, black fly larvae are more abundant in spring and early summer, while caddis larvae predominate thereafter. Winters of high discharge lead to greater numbers of simuliids, and winters of low flow lead to higher densities of hydropsychids. By scrubbing substrate surfaces with a brush, Hemphill and Cooper showed that disturbance benefitted simuliids because they were the more rapid colonizers, whereas caddis larvae were superior at monopolizing space on rock surfaces. As time passes since the last disturbance, hydropsychid larvae gradually replace simuliids due to their aggressive defense of net sites.

It is apparent that biotic and abiotic factors are difficult to discuss separately, because often the best evidence for each arises from their interaction. It is also clear that the interplay between biotic and abiotic factors takes many forms and has been formulated into a number of models of community structure, with both overlapping and some distinctive elements. We shall now more closely examine three conceptual frameworks that have received much discussion by lotic ecologists.

The harsh–benign framework is based on the expectation that biotic processes are not able to convey regularity and structure to communities when environmental conditions fluctuate widely and unpredictably. This is well supported by studies such as those of Hemphill and Cooper (1983) and Meffe (1984) just described. Indeed, most of the convincing examples of strong biological interactions described in chapters 7–9 were obtained under relatively benign conditions, such as low and steady summer streamflows. Biological interactions often are most pronounced when populations are most numerous, and as Figure 11.8 illustrates, frequent flood episodes result in sparse populations.

Regularity in assemblage structure appears to be inversely correlated with degree of environmental variability, as expected under the harsh–benign hypothesis. Fish surveys made 9 years apart allowed Ross, Matthews and Echelle (1985) to contrast two streams, one of which was considered physically more rigorous based on seasonal and year-to-year variation in flow regime, maximum summer temperatures and frequency of dewatering. Species persistence was high and similar in both, but the difference between the collections 9 years apart was greater at the more variable site. A comparable invertebrate study from surveys made 8 years apart in a number of small streams in southern England revealed substantial change in both species composition and abundance ranks (Townsend *et al.*, 1987). Highest persistence in assemblage composition occurred at headwater sites characterized by low discharge and cool temperature regimes, and a low but unchanging pH.

Providing an unambiguous measure of environmental harshness remains the greatest difficulty in evaluating the harsh–benign framework. It describes a real polarity between the strength of abiotic and biotic processes, and can be applied usefully to comparisons between seasons or locales that obviously differ in the extent of environmental variability. However, until better and more objective means are developed to characterize the strength of environmental variation, it will be difficult to extend this framework beyond individual comparisons.

Competition-based models of community structure occur in a number of guises (Yodzis, 1986; Townsend, 1989). In niche-controlled communities, coexistence is attributed to species differences along resource or habitat axes, and abiotic factors are of little import. Once environmental variability is allowed to interact with competition, several possibilities arise. If competitive ability is hierarchical and inversely correlated with colonizing ability, one expects a predictable succession in which good colonizers are gradually replaced by good competitors. Interestingly, this model predicts greatest diversity under conditions of intermediate disturbance, because good competitors and good colonizers both would be represented (Connell, 1978). A more stochastic version results if all species have roughly similar competitive and colonizing abilities. When all species are equally good colonizers and competitors, chance establishment determines assemblage composition and competition maintains it until the next disturbance. This was referred to as lottery competition in chapter 9.

It is likely that niche-controlled communities are rare in running waters, because environmental variability is widespread and existing evidence does not indicate that competition is routinely a strong force. Some evidence has been marshalled for the interaction between competiton and disturbance, however. The *Simulium–Hydropsyche* and *Gambusia–Poeciliopsis* studies are indicative of coexistence mediated by an inverse correlation between competitive superiority and ability to colonize after floods or withstand floods. A competitive hierarchy appears to explain the interactions among a shelter-building caddis larva, *Leucotrichia*, a moth, *Parargyractis*, and a midge, *Eukiefferiella*, on stone surfaces (McAuliffe, 1984b, chapter 9). Strong evidence of a successional sequence, in which initial colonists are replaced by (apparently) superior competitors, is provided by the study of Fisher *et al.* (1982) of the recovery sequence in an Arizona stream following a flash flood. Diatoms colonized and grew quickly, but were soon overtaken by slower growing but larger filamentous algae, which presumably reflects their competitive dominance.

The remaining theoretical framework is based on the patchy distribution of habitats in space and time and the high dispersal ability that characterizes much of the biota of running waters. In the patch dynamics model, disturbance frequently removes organisms and opens up space which can then be colonized by individuals of the same or different species (Townsend, 1989). The patch dynamics model emphasizes differences in life cycles and colonization ability of individual taxa, and the ever-shifting spatial and temporal mosaic of stream habitats. The assemblage exhibits regularity in its species composition because the overall mosaic of environmental conditions is predictable even if conditions within a patch are not.

It can be seen that the patch dynamic framework is quite general, encompassing aspects of models discussed previously. If early arrivals are replaced by biologically superior species, the model has much in common with those based on competition. Should environmental fluctuations interrupt the outcome of biotic interactions with sufficient frequency, communities will exist in a permanent state of non-equilibrium. Biological attributes of the species help to determine assemblage structure, but species differences in colonizing ability and other life cycle attributes are included along with direct species interactions. Although a species may disappear from a patch or even a larger stream segment, the existence of a diverse array of environmental conditions throughout the larger region makes it probable that the species persists elsewhere and will soon reappear.

Much of the evidence for the patch dynamics model is circumstantial, but compelling nonetheless. Patrick (1975) argues that, compared with their terrestrial and marine counterparts, many stream-dwelling organisms across all trophic levels possess fast life cycles, are fecund and disperse widely. These are characteristics we would expect of taxa living in an ever-changing environment in which frequent re-establishment of populations is a necessity. In addition, species unquestionably differ in their rates of dispersal, colonization and population increase, and so some taxa are more vagile than others. At the extreme, 'tramp' species may persist solely because their superior colonizing ability enables them to capitalize on habitat made available by episodic events that, while individually erratic, are in the aggregate wholly predictable.

Studies of two outdoor experimental streams in southern England lend anecdotal support to the view that some species persist because they are adept at colonizing ephemeral environments. When the channels were flooded they provided new, uncolonized habitat, and accumulated some 33–35 taxa in roughly one month (Ladle *et al.*, 1985; Pinder, 1985). On day 16, the chironomid, *Orthocladius calvus*, completely dominated both channels; by day 37 it had virtually disappeared. Remarkably, this was an undescribed species, in a region where the Chironomidae are relatively well known.

The chironomid, *O. calvus*, illustrates an extreme example of a fugitive species requiring the frequent provision of new habitat for its persistence. Fugitive and resident species are not distinct categories, however, but a continuum. Because of this, so long as environmental variability falls within the adaptive capabilities of the species present, a changing environment may often act to enhance diversity. For example, the dominant fish species in some desert streams changes between wet and dry years, depending on species' abilities to withstand drought and recolonize following population wash-out (Deacon and Minkley, 1974). One would predict that a climate that was permanently wetter or drier would lose species as a result.

Under the patch dynamics model, the longevity of constituent species contributes to the stability of assemblages in individual habitats. Most stream-dwelling organisms are short-lived, however, with lifespans usually of days to months, and only a few groups such as fish and mollusks live a few years or longer (Patrick, 1975). Two examples illustrate how longevity influences to the stability of local populations. Although the caddisfly, *Leuchotrichia*, has a 1 year life cycle, it is able to occupy cases left from the previous year, and the relative permanence of these cases contributes substantially to *Leucotrichia's* monopolization of space (McAuliffe, 1984b). In trout, a lifespan of several years and repeated reproduction likewise contributes to population stability. Elliott (1987b) compared two populations of brown trout in the English Lake District: at one site flooding was a significant cause of population loss, whereas the other population occupied a more stable environment and appeared to be regulated by its own density. Nonetheless, each population was relatively invariant, apparently because an adult lifespan of several years allowed repeated reproduction, thus buffering against failure of any year class.

A study of the benthic invertebrates of an Andean foothills stream demonstrates the utility of a patch dynamics framework to understanding community structure (Flecker, 1992b). A high rate of deposition of sediments on the stream bottom resulted in a thick layer of sediments and ooze on stone surfaces, and at approximately 10% organic content, this layer supported a diverse group of detritivores. The feeding activities of algal-feeding catfish and of an abundant detritivorous fish, *Prochilodus*, so dramatically reduced sediment accumulations that their feeding scars produced a visible mosaic of clean and silty stones on the substrate. Thickness of sediments and amount of algae were inversely correlated on the two types of patches, and each supported a particular array of aquatic invertebrates. Experimental manipulations of access to the substrate by fish established that the invertebrate community was controlled by the mixture of patch types which in turn was maintained by the interaction between silt deposition and its removal by fish (Figure 11.9).

The three principal models discussed in this chapter are overlapping, and it is likely that none is correct to the exclusion of the others. In some situations extreme environmental conditions apparently do prevent biological interactions from exerting much influence over community structure. Especially at a local scale, chance and a fluctuating environment are likely to be of considerable importance. Under relatively constant environmental conditions, however, strong biological interactions may convey considerable regularity to the composition and relative abundance of species in an assemblage. In some circumstances, competitive ability may be inversely correlated with colonizing ability, resulting in predictable alternations in species dominance depending on environmental conditions. However, the patch dynamics view, in which environmental unpredictability interacts with species-specific differences in life history characteristics and dispersal ability, may frequently play the greatest role in governing the make-up of stream communities.

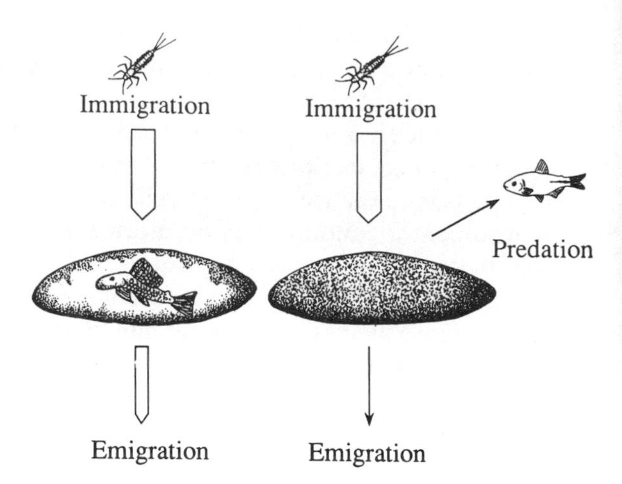

FIGURE 11.9 Patch dynamics model of the interaction between arrival and departure of aquatic insects from habitat patches, and the efficacy of consumers in influencing benthic insect densities. Predation is overshadowed by high replacement rates, whereas grazing catfish and the detritivore, *Prochilodes*, reduce habitat quality causing high emigration rates. (From Flecker, 1992b.)

11.3 Summary

To say that communities have structure implies that the species composition of a given locale is relatively persistent, and also that similar environments in different regions contain similar assemblages of species. It also should be possible to demonstrate that observed pattern is strongly influenced by the biological attributes of species and by biotic interactions among species. An alternative view holds that local assemblages simply contain the subset of species that is favored by the environmental conditions of the moment, which change continually and with little predictability. Between these two extremes fall a number of conceptual frameworks to explain the makeup of communities. Some interplay between biotic and abiotic forces is central to these models. In addition, it is useful to recognize that each local community is strongly affected by interchange with surrounding communities that comprise the species pool of the larger region.

The number of species observed in a local community increases with the size of the species pool, the size and heterogeneity of the environmental unit under study, and the effort devoted to sampling. The regional species pool is a result of historical influences on the rates of species formation and extinction. The relationship between local and regional diversity, together with the interplay between abiotic and biotic forces at the local scale, are important to our understanding of assemblage structure.

As a whole, running water environments appear variable and often harsh, most evidently in their flow regimes. This has led some to argue that abiotic forces frequently are pre-eminent in lotic communities. In fact, evidence exists that running water environments can be highly variable in space and time, although it is difficult to quantify this variation unambiguously. There is also ample evidence that species differ in their biological attributes including resource needs, life cycles and mobility, and in their strength of interactions with one another.

The effects of abiotic and biotic factors have been combined into a number of similar but subtly different models. One view argues that environmental variation constitutes a continuum from harsh to benign. Abiotic forces are expected to be pre-eminent under harsh conditions, which may be common in running waters. A second view emphasizes the importance of biotic interactions and suggests that an intermediate level of environmental disturbance should promote both persistence and diversity because it prevents biologically superior species from excluding inferior forms. The patch dynamics framework emphasizes the rapid colonizing ability characteristic of many stream-dwelling organisms, along with the ever-shifting mosaic of environmental conditions. In this view it is the predictability of environmental conditions in the aggregate, rather than at any one place and time, that ensures some regularity and persistence in communities.

Chapter twelve

Organic matter in lotic ecosystems

The study of ecosystems includes non-living matter as well as the biota. The flux of energy, expressed as organic matter or carbon, and of materials, referred to as elements, minerals or nutrients, are principal areas of ecosystem-level investigations. Both organic matter and elements are present in organisms and in various non-living organic forms. In addition, the elements necessary for life cycle back and forth between organic and inorganic states. Whether the quantity being measured is organic carbon, nitrogen, or some other material, it exists in various states (stores) and moves among these states by various processes (transfers) that may be broadly classified as biological or physical–chemical. Ecosystem studies attempt to understand the processes that govern the movement and transformation of energy and materials from one state to another. Valuable insights into ecosystem function are gained by study of either the stores or the transfers, but both are necessary to a full understanding.

Energy pathways through consumers and the flux of organic matter through ecosystems are related topics that have been the focus of much ecosystem research. As can be seen from a simplified scheme of organic matter flux through a woodland stream (Figure 12.1), such systems are very complex. Although additional detail and complexity could be added to any part of this diagram, this simplification is essential in identifying the main topics of interest.

Organic matter enters some defined area of an ecosystem from multiple sources, or inputs. Once within those boundaries it meets with various fates. Organic matter is transformed from one state to another, respired by the consumer community, or exported downstream. Organic matter dynamics usually are strongly pulsed according to the growing season, hydrologic events, or a combination of the two. Some processes operate on a time scale of minutes to hours, while others occur on a scale from weeks to the seasons, and some are longer yet. A substantial amount of particulate organic matter resides on or within the sediments for tens to hundreds of years, entering the pool of mobile organic matter only during episodes of rare occurrence such as major storms. Because streams are especially open ecosystems, meaning that fluxes across boundaries are large, export to downstream is invariably an important fate of organic matter inputs. The amount of material present varies among stores, and the rate at which any particular material is transformed differs for different processes. Thus, comparison of these quantities and rates within the study system lends insight into their relative importance.

A useful framework for comparisons across

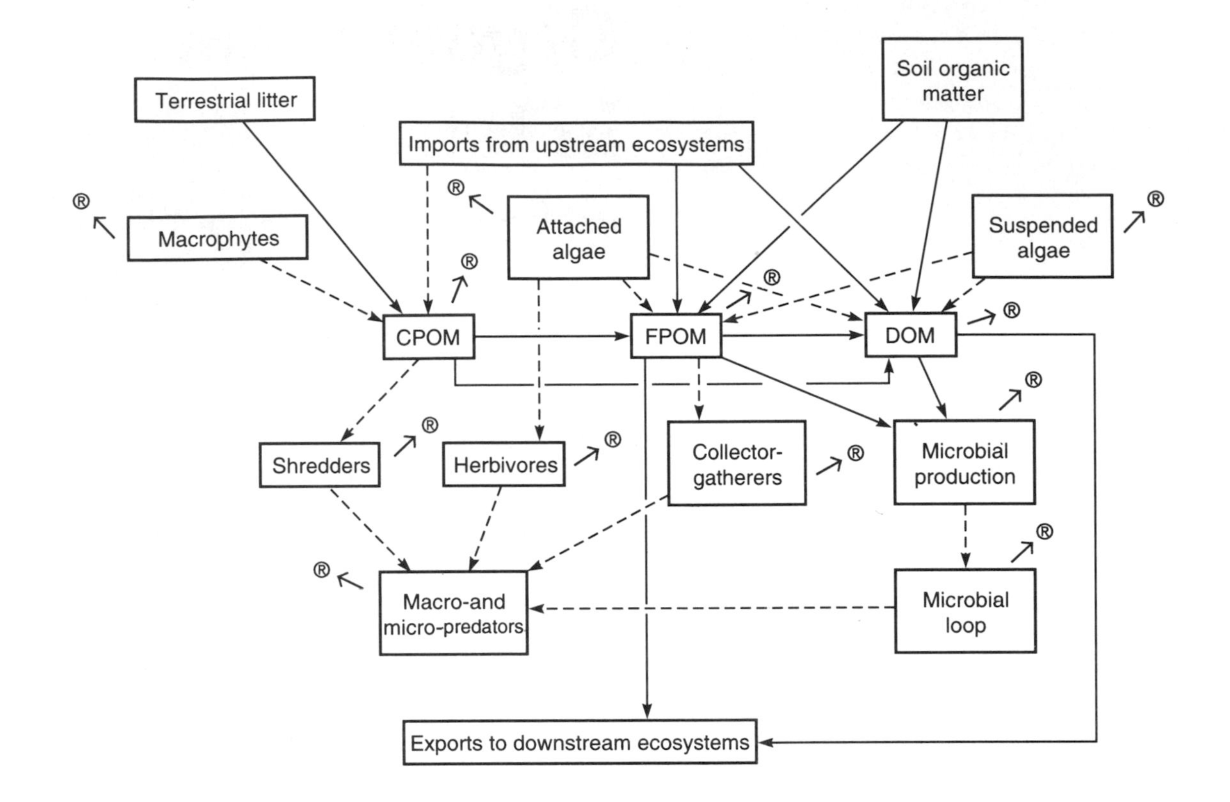

FIGURE 12.1 Simplified model of principal carbon fluxes in a stream ecosystem. Solid lines indicate dominant pathways of transport or metabolism of organic matter in a woodland stream. ® denotes mineralization of organic carbon to carbon dioxide by respiration. See Figure 12.7 for a depiction of how energy inputs change with increasing river size. Note that storage is omitted. CPOM, coarse particulate organic matter; FPOM, fine particulate organic matter; DOM, dissolved organic matter. (Modified from Wetzel, 1983.)

lotic ecosystems also is apparent in this model. The importance of various inputs ought to vary among streams and rivers because of differences in size, in the vegetation of the surrounding landscape, and in the composition of the biological community. Running water ecosystems traditionally have been thought to derive most of their energy from dead organic matter produced outside the stream channel rather than from instream primary production. Quantification of the terms in Figure 12.1 allows this question to be addressed. The fate of organic matter inputs also is likely to vary among ecosystems. In particular, the proportion of inputs that is respired by the biological community compared with that lost to downstream ecosystems, provides a comparative measure of ecosystem efficiency. Out of such comparisons among ecosystems may emerge generalizations concerning ecosystem function.

We will begin by examining the principal inputs, stores and outputs of organic matter for a section of stream. This draws on previous chapters concerning primary production (chapter 4), the inputs and transformations of DOM and POM (chapter 5), and energy flow through the food web (chapter 6), and integrates this information into an ecosystem framework. Following this description of the dynamics of organic matter, we will turn to the important instream processes that influence the transformation of organic matter from one state to another, and thus collectively influence ecosystem efficiency. Having considered both pattern and process, we will take a comparative view of ecosystems, first by comparing the magnitude of inputs and outputs calculated for different stream ecosystems, and second by examining how inputs and dynamics vary in predictable fashion with increasing river size and with characteristics of the surrounding watershed.

12.1 Dynamics of dissolved and particulate organic matter

The total pool of organic matter in a section of river is the sum of inputs resulting from instream primary production and diverse inputs of non-living organic matter. Some organic matter is autochthonous in origin, meaning that its production took place within the stream. Primary production by periphyton, macrophytes and phytoplankton constitutes important autochthonous sources. Primary production also contributes to DOM pools by extracellular release and enters both DOM and POM pools after sloughing and die-back. Microbial production based on use of instream DOM and POM can be a significant component of autochthonous production. Allochthonous material, which includes all of the organic matter that a stream receives from production that occurred outside the stream channel, often constitutes an even larger fraction of a stream's total inputs of organic matter. Some allochthonous inputs of DOM and POM have a clearly identifiable organic matter source such as plant litter of terrestrial origin, fine particles produced by the breakdown of CPOM and dissolved organics released by leaching of freshly wetted leaves. All of the above inputs are pulsed by the growing season. In addition, a considerable amount of FPOM and DOM reaches streams from surrounding terrestrial landscapes after substantial processing in the soil, and so is highly refractory (chapter 5). These terrestrial sources of FPOM and DOM likely play a minor role in the instream dynamics of organic matter, changing little in composition between entering the stream channel and reaching the sea.

12.1.1 Concentrations of organic carbon in running waters

The amount of DOM and POM in river water varies on daily, seasonal and yearly time scales; and spatially, in accord with local geology, vegetation and rainfall. From a study of 45 streams and small to medium rivers in the USA, mean annual DOC concentrations were found to be 0.7–28 $mg\,l^{-1}$. However, values of 1–4 $mg\,l^{-1}$ were common and maximum values rarely exceeded 10 $mg\,l^{-1}$. Higher values often were associated with larger streams and human disturbance. DOC concentrations tend to be high where production is high and decomposition is slow. In waters draining wetlands, high concentrations of organic acids reduce pH to 3–6, preventing rapid bacterial decay and resulting in very high DOC levels (Thurman, 1985). Table 12.1 gives an approximate guide to typical natural concentrations. (From the standpoint of the flux of energy and carbon in ecosystems, organic matter and organic carbon are used interchangeably because organic carbon compounds are the energy source within organic matter. POM (also called seston) is usually determined as mass and converted to particulate organic carbon (POC) by assuming it is 45–50% organic carbon by mass. DOC is usually determined by chemical oxidation or combustion, and converted to DOM by the reverse calculation.)

TABLE 12.1 Typical concentrations of dissolved organic carbon (DOC) in running waters and groundwater (from Thurman, 1985). Local circumstances may result in higher values than those given here

	DOC (mg l^{-1})	
	Range	*Average*
Groundwater	0.2–2	0.7
Pristine streams	1–4	2
Major rivers	2–10	5
Swamps, bogs and blackwater streams	10–30	–

The ratios of dissolved to particulate forms of organic carbon vary greatly. On a world-wide basis, DOC exceeds POC by approximately 2:1, but this depends on the type of stream and discharge regime. Based on annual means, reported DOC:POC ratios for North American streams range from 0.09:1 to 70:1 (Moeller *et al.*, 1979). Mean annual POC concentrations frequently are less than 5 mg l^{-1} (Webster, Benfield and Cairns, 1979). Stream size and watershed characteristics account for some of the variation. In undisturbed forested watersheds, mean annual POC concentrations are often less than 1 mg l^{-1} (Fisher and Likens, 1973; Naiman and Sedell, 1979). Higher POC concentrations are found in low-gradient streams flowing through agricultural or multiple-use watersheds (Malmqvist, Nilsson and Svensson, 1978; Sedell *et al.*, 1978) and especially in larger lowland rivers (Thames (Berrie, 1972); South Platte, Colorado (Ward, 1974)). In larger rivers, POC and DOC concentrations are similar, and at high discharge POC can exceed DOC (Thurman, 1985).

12.1.2 Factors influencing DOC concentrations

DOC concentrations vary seasonally due to changes in inputs between the growing and dormant seasons, and because of flushing associated with high discharge events. The strong influence of seasonality in plant production on DOC concentrations in streams is manifested in many ways. In a Pennsylvania stream flowing through woodland and meadows, extracellular release of DOC during periphyton blooms elevated stream DOC by nearly 40% on a daily basis, but only in springtime because canopy development reduced instream primary production during summer (Kaplan and Bott, 1989). Leaf fall is an important organic carbon source in woodland streams, and in Roaring Brook, Massachusetts, 42% of stream inputs of DOC during autumn were attributable to leaf leachate (McDowell and Fisher, 1976). Meyer and Tate (1983) found DOC inputs to be less in a recently clear-cut stream compared with a forested stream, apparently because successional recovery of the surrounding forest reduced DOC transport from soil water into the stream channel (Tate and Meyer, 1983). The activity of leaf-shredding invertebrates can increase the release of DOC (Meyer and O'Hop, 1983), particularly during periods of low flow (Webster, 1983).

Storms commonly elevate DOC concentrations in streamwater (Figure 12.2). The highest concentrations are found on either the ascending or the descending limb of the hydrograph (McDowell and Fisher, 1976; Meyer and Tate, 1983), depending on the pathway by which DOC enters the stream. Sub-surface water carrying DOC flushed from storage in the terrestrial system enters the stream fairly slowly. On the other hand, throughfall, leaching of POC from streambanks and side channels inundated during rising water levels, and flushing of interstitial DOC from within the streambed can occur quickly enough to be reflected primarily during the initial rise of discharge. Streamside vegetation (and its modification by human activity) strongly influences how stormwater reaches stream channels, and so is an additional variable to consider.

DOC concentrations in streams and rivers are often highest during seasons of high discharge, presumably due to flushing of interstitial and

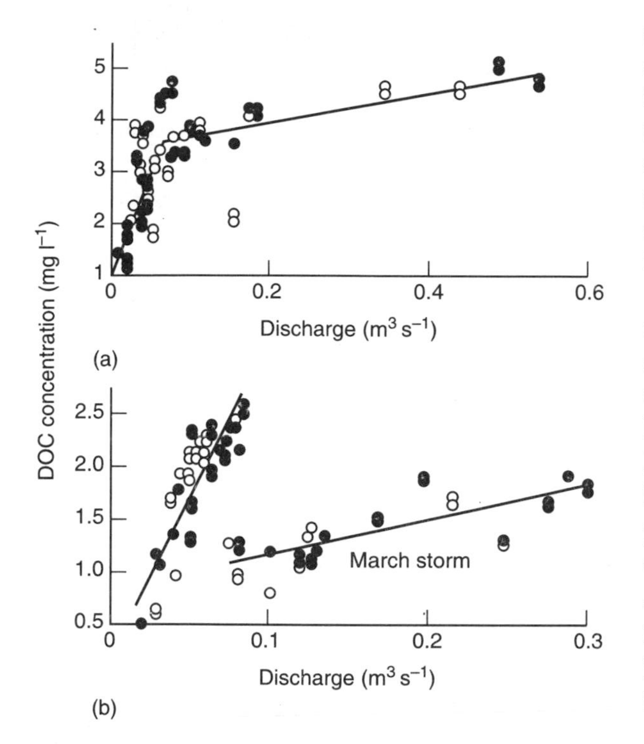

FIGURE 12.2 DOC concentration as a function of stream discharge in a forested headwater stream in the southern Appalachians. Samples were taken on both the rising (●) and falling (○) limbs of the hydrograph. Note higher values in the growing season (a) compared with the dormant season (b). The March storm followed a long wet period which presumably depleted particle supply. (From Meyer and Tate, 1983.)

soil water DOC into the channel. DOC concentrations were greatest during the spring in the Salmon River, Idaho, and exhibited a winter–spring maximum in the Kalamazoo River, Michigan, and McKenzie River, Oregon (Moeller *et al.*, 1979). In contrast, however, DOC concentrations were relatively low and constant during the spring freshet in boreal forest streams in Quebec, and higher during the productive summer and autumn (Naiman, 1982). Exceptions are also found in small streams where annual discharge variation is not great and growing season inputs coincide with low flow. In White Clay Creek, Pennsylvania, fall maxima in DOC concentrations were attributed to litter inputs and leaf leachate, and a second maximum in late winter–early spring to inputs during the thaw (Larson, 1978). In small upland streams of the southeastern USA, DOC concentrations were greatest during the growing season, whereas 65% of the discharge occurred during the dormant season (Meyer and Tate, 1983).

On a yet longer time scale, DOC concentrations can be influenced by the magnitude of discharge events during the previous decade. Comparing four watersheds of varying maturity located in the Coweeta Hydrologic Laboratory, North Carolina, Tate and Meyer (1983) found little difference among watersheds for any given year, but roughly a twofold difference in DOC export between two water-years a decade apart. A 100 yr flood occurred in the decade that preceded the lower measurements, and this presumably flushed most of the accumulated sub-surface DOC from each watershed.

12.1.3 Factors influencing POC concentrations

Like DOC, the amount of POC in suspension fluctuates in space and time. Concentrations are controlled by changes in particle availability, which is influenced primarily by seasonal changes in biological processes; and by discharge, which varies both seasonally and unpredictably. In comparison to DOC, accumulation and storage of POC in the streambed can be of considerable importance. As a consequence, POC varies with discharge to a much greater degree than does DOC, and concentrations are strongly influenced by time elapsed since the last scouring of the stream bottom.

The bulk of POC consists of very small particles, generally 100 μm or less. Median particle sizes are reported to range from 5 μm to perhaps 65 μm. Large particles in transport are usually less than 10% of fines (Gurtz, Webster and Wallace, 1980; Wallace, Webster and Cuffney, 1982), especially in larger rivers (Cudney and Wallace, 1980). Strongly pulsed leaf inputs can

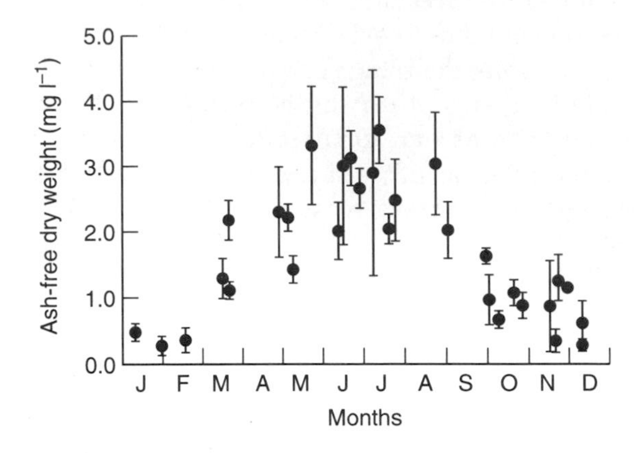

FIGURE 12.3 Seasonal variation in organic seston concentrations in a forested headwater stream in the southern Appalachians. Error bars show 15% confidence intervals. (From Webster and Golladay, 1984.)

reverse this pattern temporarily, of course. During autumn in a small New England stream, McDowell and Fisher (1976) found CPOM to comprise 60% of total POM in suspension.

Seasonal variation in POC concentrations is common, but the pattern varies among streams. In forested headwater streams of the southern Appalachians, non-storm organic seston concentrations are highest during the summer when flows are low (Figure 12.3). Lower concentrations during winter may be due to dilution, less particle generation due to biological activity, or be a consequence of wash-out during earlier fall storms. Dilution is not the sole explanation for lower winter concentrations in southern Appalachian streams, however, because transport (discharge × concentration) is highest in spring and summer (Webster and Golladay, 1984). In Bear Brook, New Hampshire, POC concentrations are highest in summer and transport is highest in winter (Fisher and Likens, 1973), and so dilution suffices to explain low winter concentrations.

Much of the POC in transport during non-storm periods results from biological activity. Thus, high values during the growing season can be attributed to high biological processing of inputs available in spring, and continuing biological activity during warm periods (Webster, 1983). However, as was discussed in chapter 5, capture of FPOM by suspension-feeding consumers typically is an insignificant factor in its dynamics.

Hydrologic events play a major role in controlling concentrations of suspended POC. Organic seston concentrations can vary by two to three orders of magnitude in response to changes in discharge (Thurman, 1985; Kazmierczak, Webster and Benfield, 1987), a much stronger response than is seen for DOC concentrations. This is true for individual storms, seasonal periods of high discharge and major flood events. Because scouring associated with high discharge depletes the available particle supply, the amount of time elapsed since the last discharge peak influences POC transport in response to a subsequent storm. Thus, the low frequency of storms during the growing season in some regions is an additional explanation for seasonally high POC concentrations.

Storms in small streams cause discharge and POC concentrations to rise and fall dramatically over a few days or even a few hours. Almost invariably, FPOC concentrations are highest on the rising limb of the hydrograph (Figure 12.4) and decline at peak flows (Bilby and Likens, 1979; Golladay, Webster and Benfield, 1987). This indicates that the major pool of POC lies in areas already wet or adjacent to the stream's wetted perimeter, where FPOM accumulated during low flows. Rising water levels and current velocities cause fine particles to be dislodged, side channels wetted, and pools and backwaters to be subjected to increased flow. Inputs of POC from outside of the stream also are likely to be greatest during the rising limb of the hydrograph, due to the erosive effects of rainfall on soil and streambank litter. During the descending limb water enters the stream principally by sub-surface flow (chapter 1), and carries little or no POC.

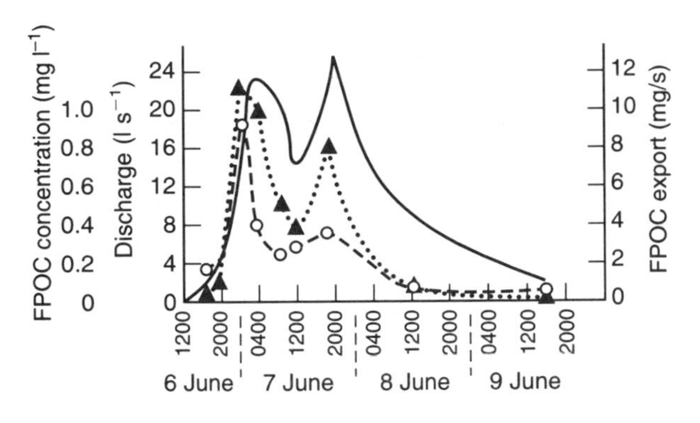

FIGURE 12.4 Changes in discharge, FPOC concentrations, and FPOC transport during a summer storm in a small forested watershed in New Hampshire. Note that FPOC concentrations peak on the rising limb of the hydrograph, indicating rapid entrainment of small particulates. A second hydrograph peak resulted in a much smaller FPOC concentration peak, evidence that wash-out rapidly depletes the available FPOC supply. —— indicates discharge; ○—○ denotes FPOC concentration and ▲---▲ denotes FPOC export. (From Bilby and Likens, 1979.)

Just as depletion of benthic POM contributes to concentration differences between rising and falling limbs of the hydrograph, time elapsed since the last storm event and seasonal differences in particle generation interact with discharge to determine seston peaks during storms. In a forested headwater stream studied by Wallace, Ross and Meyer (1982), seston concentration increased greatly with rising discharge during the first autumn storm, apparently due to the availability of abundant FPOM generated by autumn leaf fall, coupled with a lengthy prior period of low flow. A winter storm resulted in a smaller rise in seston concentrations, which Wallace and colleagues attributed to depletion of benthic FPOM by wash-out during prior fall storms.

Although discharge and organic seston concentrations correlate well for a particular stream over the range of conditions encompassing non-storm and storm periods (Figure 12.3), correlations between discharge and POM across streams during non-storm periods are weak or non-existent (Sedell *et al.*, 1978; Naiman and Sedell, 1979; Naiman, 1982). This is at first surprising, because physical models of transport of the inorganic load indicate that differences among rivers can be related to stream power, a measure of the river's transport capability that is determined primarily by discharge (Bagnold, 1966). However, transport models assume unlimited availability of transportable particles (Webster *et al.*, 1987). Their inability to predict organic seston transport during stable discharge indicates that mechanisms other than physical entrainment are limiting the amount of organic seston in transport, underscoring the importance of particle availability and depletion.

The above discussion focuses mainly on streams and smaller rivers. What of POC in larger rivers? The influence of individual storms on discharge is reduced with increasing river size. Consequently it is the wet season rather than individual storms that determines variation in particle concentration. Seston originates either from upstream sources of FPOM, by suspension of benthic deposits of particulates, or by lateral expansion as the rising river inundates side channels and, if present, a floodplain. Consequently, discharge events rather than in-stream biological processes govern seasonal changes in concentration. As is found in smaller systems, POC concentrations in large rivers are highest on the rising limb of the hydrograph, and decline thereafter (Figure 12.5).

12.2 Fate of dissolved and particulate organic carbon

Organic matter that enters the channels of streams and rivers experiences one of three fates. It is stored for some time period on streambanks and by burial within the channel, exported

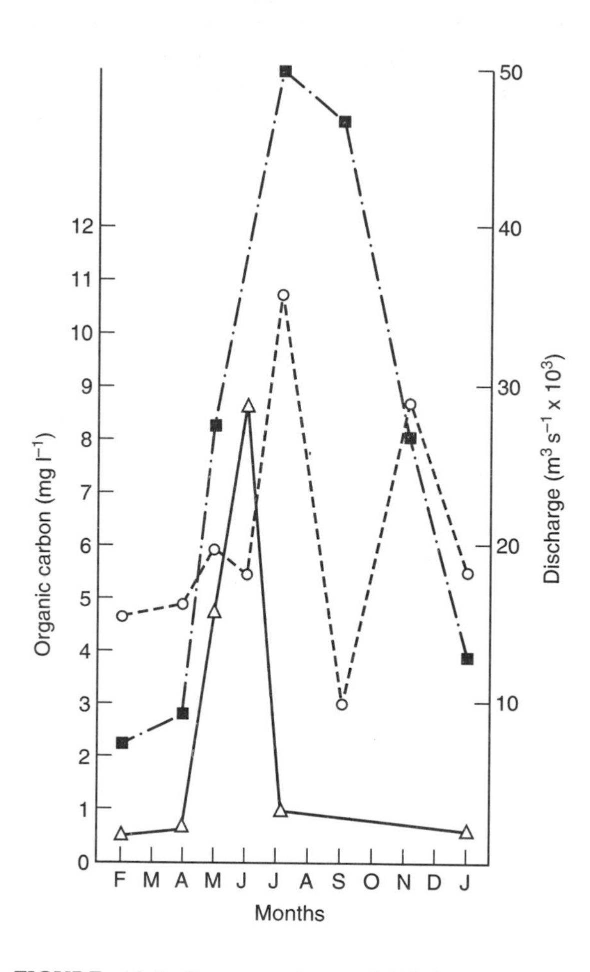

FIGURE 12.5 Concentrations of POC (△—△) and DOC (○----○) over the annual cycle in the Orinoco River, Venezuela. Note that the peak in POC concentrations is prior to peak discharge (■-•-■) and that POC:DOC ratios vary over the year. (From Thurman, 1985.)

to downstream ecosystems, or respired as carbon dioxide due to biological processes within the stream segment. Export is the fate of a great deal of organic matter. Present estimates are that 25% of carbon entering the world's rivers is processed within the system, 25% is stored as sediment POC, and 50% is transported to the oceans (Mulholland and Watts, 1982; Meybeck, 1982; Thurman, 1985). This export consists of roughly equal quantities of POC and DOC.

The material that reaches the seas appears to be highly refractory. DOC concentrations are substantially higher in river than in sea water, and decrease linearly on reaching an estuary, indicating that dilution rather than biological processing causes the decline (Thurman, 1985). Although some precipitation and microbial decay of DOC inputs take place in estuaries, as much as 65% is highly refractory and destined for ocean sediments (Ittekkot, 1988). Based on existing knowledge of DOC composition, and the relative constancy of DOC and POC in any particular river along most of its length, this material likely is refractory when it first enters the stream channel, or whatever fraction is labile is very rapidly utilized. For example, both POC and DOC were constant in composition over great distances along the Amazon mainstem (Ertel *et al.*, 1986; Hedges *et al.*, 1986). This supports the view that export is the dominant fate of DOC and POC in running waters. On the other hand, even medium and high molecular weight DOC supported some bacterial growth in the Ogeechee River (Meyer, Edwards and Risley, 1987); thus even organic carbon in transit is subject to some biological processing.

Clearly, if the dynamics of organic matter in running waters is to have relevance beyond the role of rivers as conduits of carbon from the land to the sea, there must be opportunities for transformations to take place within the river channel. This depends partly on the abiotic and biotic processes that affect POC and DOC, and partly on mechanisms that retard downstream transport and thus enhance the likelihood that instream processes will come into play.

12.2.1 Export of DOC and POC

The amount of total organic carbon (TOC) exported by a river system equals discharge multiplied by water column concentrations of

DOC and POC over the annual cycle. Because concentrations and discharge often are correlated, seasonal patterns of TOC export are strongly pulsed. Fluvial export of TOC is lowest in regions of low productivity such as deserts and tundra, and greatest in regions of high productivity such as the wet tropics (Thurman, 1985).

DOC and POC differ significantly in how they are transported. DOC moves continuously downstream in pace with water molecules, although some storage may occur as interstitial water (Tate and Meyer, 1983). In contrast, much of the POC moves episodically with storms, in alternating steps of storage and erosion (Cummins *et al.*, 1983). Storage of POC accounts for the very strong dependence of POC concentrations and export on discharge regime as described previously. As a consequence, flow conditions that occur during only a small fraction of the annual discharge cycle can account for a very large fraction of annual exports. In a small woodland stream in New England, high discharges (more than $40\,l\,s^{-1}$) representing just 1% of the annual discharge regime accounted for 20% of water export and fully 70% of annual export of FPOM (Bilby and Likens, 1980; Bilby, 1981). In this system, and in boreal forest streams of Quebec (Naiman, 1982), half or more of annual TOC export was associated with high runoff events, and in both studies POC export was more strongly pulsed than export of DOC.

DOC concentrations may be greatest during periods of high flows or during low flows, depending on the pathways by which DOC enters the stream as discussed earlier. However, even when DOC concentrations do not correlate with discharge, the annual export pattern is strongly influenced by hydrologic regime because seasonal variation in discharge usually exceeds seasonal variation in DOC concentrations. In pristine boreal forest streams in Quebec, Canada, DOC concentrations were relatively low and constant during the spring freshet, and higher during the productive summer and autumn (Naiman, 1982). Nonetheless, the two-month spring freshet accounted for roughly half of annual discharge and 50% of DOC export.

The influence of human disturbance also is seen most strongly for POC export. Webster *et al.* (1988) compared 17 streams ranging from undisturbed reference sites to watersheds that had been disturbed by logging and associated activities such as road-building 7–34 years before. Seston concentrations were highest within recently disturbed watersheds, but effects were detectable as much as 30 years after the logging occurred. When log and debris jams were removed from streams or prevented from forming, export of POC increased greatly, while increased export of DOC was small or negligible (Bilby and Likens, 1980). The mechanisms responsible for this are considered in a subsequent section.

12.2.2 Instream processes affecting DOC and POC

DOC is removed from streamwater by both abiotic and biotic processes (chapter 5). The principal biotic processes are uptake by microorganisms, especially bacteria, assimilation of the organic carbon into microbial biomass, consumption of this heterotrophic production by other microorganisms and some members of the macroinvertebrates (chapter 6), and eventual remineralization of the carbon to carbon dioxide by community respiration. POC is a substrate for microbial growth and is directly ingested by consumers (chapters 5 and 6), hence also is eventually converted to carbon dioxide.

Because these processes were described in detail in previous chapters, the focus here is on their integration into an ecosystem perspective. The extent to which the microbial loop (Figure 6.1) serves to link DOC and POC via microbial production to metazoans at higher trophic levels, or acts as a sink in which carbon is re-mineralized without any substantial energy transfer to larger organisms, remains uncertain.

Moreover, there are likely to be major differences associated with river size, hydrologic regime, and the mix of energy inputs. Suffice it to say that DOC and POC undergo transformations within the stream channel, by abiotic as well as biotic pathways, and so the river ecosystem exerts some control over the carbon that passes through its channel. However, because export is such a major term, especially under high-flow conditions, the opportunities for in-stream processing are critically dependent upon the capacity of the river channel to store organic carbon temporarily. We turn now to the natural features of streams that affect their retentiveness.

12.2.3 Organic matter retention

Physical features of a stream, such as boulders that trap accumulations of woody debris and a topography that permits flooding rivers to overflow their banks, slow the passage of water and materials downstream. By increasing the retentiveness of stream segments, such features should increase the amount of organic matter respired by the consumer community and decrease the amount exported downstream. As retention varies among locations, seasons or stream types, so should the relationship between processing and export.

Undisturbed watersheds can be highly retentive (Naiman, 1982), transporting most organic matter inputs downstream only after processing them to small particles and DOC. Most export occurs during storms and at seasonally high flows, demonstrating the importance of discharge. The amount of material exported annually correlates with watershed area for various biome types (Schlesinger and Melack, 1981; Mulholland and Watts, 1982). However, correlations between discharge and transport of organic carbon are weak, especially during non-storm periods (Cuffney and Wallace, 1989), indicating that particle supply and retention are important processes as well.

Factors influencing retention are many. Biological processes including uptake of DOC and consumption of POC retain organic carbon within the biological community, eventually metabolizing it to carbon dioxide. Physical–chemical processes such as adsorption and flocculation can remove DOC and also make it more available to organisms. However, the primary focus in studies of retention is on the influence of physical structure and hydraulics on the opportunity for these removal processes to act. Debris dams are particularly important retention devices in low-order streams of forested watersheds, while bars, alcoves and eddies are of greater importance in higher order streams. In large rivers the floodplain can be a principal site of POC deposition and storage. When beaver (*Castor canadensis*) were unexploited they must have contributed greatly to organic matter storage over large areas of the north temperate zone. Where beaver occur at natural densities today, their activities influence 2–40% of the length of second- to fifth-order streams, and increase the retention time of carbon roughly six-fold (Naiman, Melillo and Hobbie, 1986).

Organic debris dams are accumulations of organic matter that obstruct water flow. They are formed when large woody debris becomes lodged against obstructions, trapping smaller debris and leaves into a nearly watertight structure. Sediments and organic matter settle in the pools formed upstream of debris dams, forming hotspots of heterotrophic activity. In streams of the Hubbard Brook Experimental Forest, community respiration was nearly three times greater in sediments of organic debris dams compared with other sediments of the streambed (Hedin, 1990). Stream width and channel pattern are more variable when woody debris is present (Zimmerman, Goodlet and Comer, 1967), and the capacity of the stream to transport material is substantially altered (Bilby and Likens, 1980; Bilby, 1981).

Experimental removal of all organic debris dams from a 175 m stretch of a small New Hampshire stream demonstrated their importance to organic matter dynamics (Bilby and Likens, 1980; Bilby, 1981). Following this removal, total organic matter export increased by a factor of 2.5. Export of fine particulates increased greatly (about sevenfold), coarse material less so (about 2.5 times) and DOC export increased only slightly (18%). With dams intact, DOC comprised some 70% of TOC exports; after dam removal DOC accounted for only about one-third.

Debris dams tend to become scarcer as stream size increases. An inventory of organic matter standing stock by Bilby and Likens revealed that 75% was contained in debris dams in first-order streams, compared with 58% at second-order and 20% at third-order sites. However, the history of logging at the site and maturity of the forest will influence the amount and size of available large woody debris. This New Hampshire location was logged some 60 years previously, and debris dams rarely occurred at channel widths over 7 m, whereas much larger debris dams were found in streams in Oregon flowing through mature stands of Douglas Fir. There also is evidence that debris dams may be more abundant in coniferous compared with deciduous forests (Harmon *et al.*, 1986).

Debris dams may be especially important sites of organic matter accumulation in high-gradient streams with stony bottoms, compared with low-gradient streams with finer sediments. In first-order streams on Virginia's Coastal Plain, Smock, Metzler and Gladden (1989) found that half or more of organic matter storage was in the sediments rather than debris dams. Nonetheless, debris dams were important in this situation also. Macroinvertebrate biomass, organic matter content, and trapping of leaves in transport were enhanced by the presence of natural and experimental dams.

If invertebrate abundance is determined by the availability of CPOM, which in turn is under the influence of retentiveness, then hydraulic forces of flow and the presence of debris dams should strongly affect the numbers of leaf-shredding organisms. Smock, Metzler and Gladden (1989) confirmed this expectation in streams of the Coastal Plain of Virginia. Macroinvertebrate biomass was more than five times higher in debris dams *versus* sediments, and the percentage of shredders increased where debris dams were constructed. In streams of the Ashdown Forest in southern England, Hildrew *et al.* (1989) quantified differences in retentiveness among sites by measuring the downstream transport distance of leaves and by employing the standard hemispheres developed by Statzner (p. 58). Sites that differed in retentiveness also differed in the amount of leaf litter on the stream bottom and in the abundances of shredding macroinvertebrates and of copepods (a major component of the meiofauna). The addition of artificial retention devices made from coarse screening resulted in an increase in shredder abundance, except at the most naturally retentive site where the effect was negligible.

Debris dams are the best-studied aspect of retention, but are not the only agent. Substrate, gradient, stream geomorphology and hydraulic regime are likely to be important factors. Locale-specific hydraulic regime led to variation in retentiveness among sites in the Ashdown Forest streams, described above. Rainfall and watershed characteristics that determine the 'flashiness' of runoff also are likely to influence the ability of retention devices to hold material.

Not only does the relative importance of retention devices vary among low-, middle- and high-order streams, but mechanisms of retention are likely to differ for CPOM and FPOM, and under baseflow *versus* stormflow conditions. Experimental studies of transport and retention in 10 m × 30.5 cm channels by Webster *et al.* (1987) demonstrated that under normal flows, CPOM retention varied between leaf species, increased in the presence of obstacles, and decreased with higher discharge. FPOM retention was much more variable, but retention generally was highest for larger size fractions, coarser substrates and low flows.

In summary, riverine geomorphological features and debris dams act as important retention devices, counteracting the tendency for export to dominate the fate of organic matter in flowing water. These natural obstructions clearly play a significant role in ecosystem function, allowing organic matter to accumulate. This enhances ecosystem processing relative to downstream export, and perhaps favors the formation of localized hotspots of biological activity. In the absence of retention devices the stream functions more like a pipe, allowing inputs to be flushed from the system, including a higher fraction of particulates.

At this point we have described the main stores and transfers of a stream ecosystem (Figure 12.1), and also the underlying processes. A number of studies have attempted to estimate all of the inputs and outputs for some ecosystem segment, and we turn now to a consideration of this approach.

12.3 Organic matter budgets

Mass balance calculations, or budgets, are based on an accounting of all inputs to and outputs from some delimited area of an ecosystem. This can be a length of stream or river, or in the case of small headwater streams, the entire catchment. The object of study is organic matter, measured as ash-free dry mass or carbon; or elements such as phosphorus, nitrogen or calcium, considered in chapter 13. Budgets document rate of loss of materials from the land, the magnitude of various input and output pathways, and storage within the study section *versus* export to downstream ecosystems. They can reveal important transformations that occur within the study system (e.g. CPOM might dominate inputs while FPOM dominates outputs), thereby lending insight into the physical and biological processes that alter the quantity and quality of material within the stream. Coupled with measurement of internal fluxes and the underlying mechanisms, the budget approach can provide considerable insight into the flow of material through ecosystems.

Ideally, a budget is constructed from independent measurement of all inputs, outputs and changes in storage. Some terms are sufficiently difficult to measure that, in practice, they are ignored or considered to be in steady state. Budget contruction begins with a water balance, in which inputs from groundwater, sub-surface and overland flow, precipitation and upstream must equal outputs to downstream, evaporation and possibly groundwater recharge. Next, water and seston samples are used to estimate DOM and POM concentrations over a discharge range, allowing calculation of annual imports and exports from the product of discharge and the established relationship between concentration and discharge (e.g. Figures 12.2 and 12.4). If primary production is negligible, all organic matter enters as POM or DOM, and so imports will have been adequately assessed. If algal or macrophyte photosynthesis is a substantial source of carbon, then it must of course be quantified by appropriate measurements (chapter 4) and extrapolated to the entire area under study. If one assumes that POM is neither accruing to nor eroding from the system (i.e. that organic matter storage is in steady state) then the unquantified loss is respiratory conversion of organic carbon to carbon dioxide. Some studies estimated respiration by subtraction, but benthic respirometers now provide a means to measure respiration directly.

12.3.1 Examples of organic matter budgets

In their landmark study of a 1700 m reach of Bear Brook, a small woodland stream in New Hampshire, Fisher and Likens (1973) pioneered the use of organic matter budgets in running waters. Inputs from litter, throughfall, surface and sub-surface water were quantified. Because impermeable bedrock underlies this drainage basin, all outputs could be estimated from streamflow and concentrations measured at a weir. The amount of stored material in Bear

TABLE 12.2 The annual energy budget for Bear Brook, New Hampshire based on a mass balance approach to inputs and outputs. Width varied between 2.2 and 4 m, and bankfull stream area of the study segment was 5877 m^2. The estimate of micro-consumer respiration, which accounts for virtually all of the energy processing within the stream segment, was determined by difference between inputs and outputs. (From Fisher and Likens, 1973)

Organic energy category	*Entire stream segment (kg)*	*kcal m^{-2}*	*Percentage of kcal m^{-2}*
Inputs			
Litter fall			
Leaf	1990	1370	22.7
Branch	740	520	8.6
Miscellaneous	530	370	6.1
Wind transport			
Autumn	422	290	4.8
Spring	125	90	1.5
Throughfall	43	31	0.5
Fluvial transport			
CPOM	640	430	7.1
FPOM	155	128	2.1
DOM, surface	1580	1300	21.5
DOM, sub-surface	1800	1500	24.8
Moss production	13	10	0.2
Total inputs	8051[a]	6039	99.9
Outputs			
Fluvial transport			
CPOM	1370	930	15.0
FPOM	330	274	5.0
DOM	3380	2800	46.0
Respiration			
Macro-consumers	13	9	0.2
Micro-consumers	2930	2026	34.0
Total outputs	8020[a]	6039	100.2

[a] Budget in kg does not balance exactly because of different caloric equivalents of budgetary components.

Brook was assumed to be constant, and on this basis respiration was determined from the excess of imports over exports.

From the annual energy budget for Bear Brook (Table 12.2), it appears that over 99% of the energy input was due to allochthonous material (with particulates contributing slightly more than dissolved matter), and about 65% of this was exported downstream. Inputs of DOM equalled export, and the absence of net change indicates that microbial uptake removed leaf leachate as it was released. More POM left the study segment than entered it, and this difference was made up by inputs of litter fall. Virtually all internal processing was attributed to micro-organisms. Organic matter budgets have since been constructed for a number of river ecosystems spanning a range of conditions. A predominance of allochthonous inputs seems to be the rule wherever there is riparian vegetation, although perhaps not so overwhelmingly as in Bear Brook. The Fort River, Massachusetts, and Red Cedar River, Michigan, both flow through farmland as well as wooded areas, are wide enough to receive ample light, and support macrophytic plants and benthic algae (King and Ball, 1967; Fisher and Carpenter, 1976). Nonetheless, somewhat over 50% of the total energy budget originated with allochthonous material in each instance. Of the autochthonous production, considerably more resulted from the periphyton than from macrophytes in the Fort and Red Cedar rivers. Indeed, with the exception of highly productive Silver Springs (Odum, 1957), it appears that macrophytes rarely contribute greatly to energy budgets even in rivers where they are abundant. Some estimates of the contribution of macrophytes to total organic inputs include: Fort River 9%, Red Cedar River 13%, a British chalk stream 18% (Westlake *et al.*, 1972), and the Thames at Reading 1.2% (Mann *et al.*, 1972).

The New River, Virginia, should be expected to have substantial autochthonous production because it is wide (about 170 m), shallow (depths often less than 1 m), and has a largely bedrock substrate. Studies of periphyton, macrophytes and litter fall (Hill and Webster, 1982a, b, 1983) indicate that macrophytes contributed about 13–20% of inputs, periphyton another 20%, and allochthonous materials 60%. Thus the New River also derives the majority of its energy from beyond the streambanks.

The Kuparuk River, originating in the Brooks Range of Alaska and flowing northwards into the Arctic Ocean, illustrates how a carbon budget can be almost totally dominated by allochthonous inputs (Peterson *et al.*, 1986). In this tundra stream meandering through peatland, allochthonous inputs exceed periphyton primary production by roughly an order of magnitude. Cold temperatures and limitation by phosphorus apparently account for low periphyton levels in this unshaded stream.

It appears that *in situ* plant production truly dominates the organic matter budget mainly in aridland streams (Minshall, 1978; Fisher *et al.*, 1982). In the budget for Sycamore Creek, Arizona (Table 12.3), production substantially exceeded respiration. The excess was accounted for by biomass accrual by the community and by downstream export.

Not surprisingly, organic matter budgets have been constructed primarily for sections of small streams. Large rivers with extensive floodplains are highly productive, as their fish harvests attest (Welcomme, 1979), but the ultimate origin of the organic matter and the trophic network through which it passes are just beginning to be examined carefully. Bayley (1989) constructed a carbon budget for a 187 km stretch (maximum inundated area of 5330 km^2) of the Solimões River (Amazon above Manaus) that clearly is quite preliminary. Nonetheless, it indicates that only a small fraction of the total carbon supply originates with transport of material from upstream (less than 1%), or as primary production by river phytoplankton (5.4%) and periphyton attached to macrophytes (1.5%). Production by aquatic and terrestrial macrophytes in the littoral regions and floodplain, and litter inputs from the flooded forest, collectively account for about 90% of carbon production, and so river–floodplain interactions appear to be of far greater consequence than events within the channel. Meyer and Edwards (1990) reached a similar conclusion in their study of a sixth-order blackwater river in Georgia. River channel gross primary production accounted for only about one-fifth of total inputs, which were dominated by floodplain organic matter from extensive riparian swamps of up to 1–2 km in width (Figure 12.6).

12.3.2 Limitations of the mass balance approach

A complete accounting of all organic matter imports and exports is an enormous undertaking. Budget calculations are simplified by assuming a zero value for some terms and determining others by subtraction, with effects on accuracy that are difficult to specify. Some values that are particularly difficult to measure include losses by uptake processes such as flocculation, adsorption and microbial activity, inputs from primary production and exchanges via groundwater and sub-surface flows. Storage is often assumed to be in steady state, but this may rarely

TABLE 12.3 The organic matter budget for a 500 m section of Sycamore Creek, Arizona, during a 63 day recovery period between major flood events. (From Fisher *et al.*, 1982)

	kg	*g m^{-2} day^{-1}*	*Percentage of g m^{-2} day^{-1}*
Inputs			
P_G	1060	5.43	89.9
Net import	119	0.61	10.1
Total input	1179	6.04	100.0
Outputs			
Macroinvertebrate respiration	138	0.71	15.3
Algal and microorganism respiration	590	3.02	65.1
Export (drying on streambanks)	54	0.28	6.0
Export downstream	122	0.63	13.6
Total output	904	4.64	100.0
Community biomass accrual	278	1.40	–

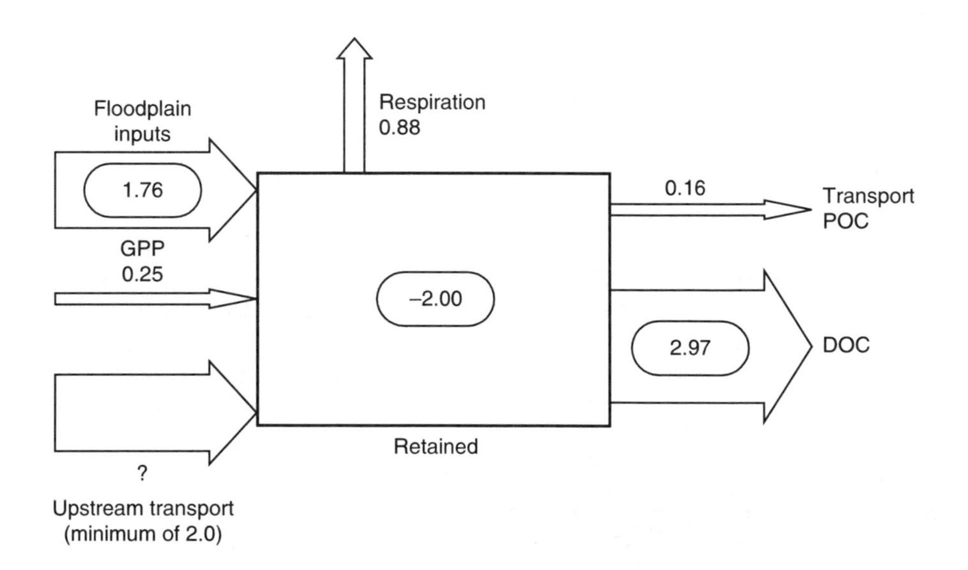

FIGURE 12.6 Annual carbon budget for one square meter of Ogeechee River channel, in kg C m^{-2} yr^{-1}. GPP, gross primary production; POC, particulate organic carbon; DOC, dissolved organic carbon. (From Edwards and Meyer, 1987.)

be correct. For these reasons, budget terms should be viewed with caution. Each budget is a unique study, and interpretation of any differences among estimates is to some extent speculative.

Ideally, any ecosystem budget should be placed in a historical context, for several reasons. Budget terms will be affected by hydrologic events during the year of study, and also by events during preceding years because of the influence of previous flood history on storage of organic matter. Long-term changes taking place in the terrestrial ecosystem, such as fire, succession and disturbance, can affect discharge and terrestrial productivity. Such episodic events of greater than annual frequency are important to ecosystem dynamics, and they are unlikely to be incorporated in a 1 year 'snap-shot'. This is illustrated nicely by a 2 year study of an old-growth conifer watershed in Oregon (Cummins *et al.*, 1983). Precipitation was well below average in the first year and well above average in the second, and import and export terms differed considerably between years. Because three of the 10 highest stormflows for the preceding 24 yr record occurred just prior to the study, stored POM likely was depleted at the time of both studies.

The amount of stored organic matter is difficult to measure, and exhibits high spatial variability. Storage occurs on or within the steambed in association with debris dams, within channel sediments, and on banks, bars and flood-plains. The latter sites are particularly important in low-gradient and larger rivers, while debris dams play a greater role in high-gradient, stony streams. The rate of decomposition of stored material will depend on many factors, including temperature, availability of oxygen, and whether the stored organic matter is exposed or buried. In addition, the amount of stored organic matter might be increasing or decreasing from year to year, depending on terrestrial productivity and recent history of discharge. In a comparison of 23 organic matter budgets from rivers of various sizes, located in different biomes, only one was in steady state (Cummins *et al.*, 1983). Substantial accrual of stored organic matter occurred in 14, while exports exceeded imports in the remaining eight.

How the physical area is delimited also affects the outcome of budget calculations. In practice, budgets usually are expressed per unit area, based on bankfull width; ideally, they should be based on the actual wetted area, which most of the time is much less. Thus, choice of area can change numbers substantially. In addition, budget calculations emphasize storms because that is when exports are greatest, with the consequence that hydrologic rather than biological processes dominate (Meyer and Likens, 1979).

12.3.3 Ecosystem analysis based on energy budgets

Analyses of energy budgets traditionally have emphasized the importance of the ratio of gross primary production to community respiration, the P/R ratio. An ecosystem that respires all the energy fixed by its primary producers has a P/R value of one. Export or accrual must be invoked to account for P/R values in excess of unity, while an ecosystem with P/R less than one is relying on energy imported from elsewhere, almost invariably as non-living organic matter. Most studies of stream ecosystems have found P/R to be well below one (Table 12.4).

All ecosystems have some energy flux across their boundaries, but imports and exports are especially important in energy budgets of running waters. By using the ratio of imports to exports (I/E) along with the P/R ratio, stream ecosystems can be characterized as either autotrophic or heterotrophic (Fisher and Likens, 1973). When instream primary production is low and organic matter inputs from outside the stream section are substantial, then P is less than R and I is greater than E. Such systems are referred to as heterotrophic, and Bear Brook (Table 12.2) is an example. Conversely, when high

TABLE 12.4 Gross primary production (GPP), respiration (R), and P/R values from various stream studies, in $g\ O_2\ m^{-2}\ day^{-1}$. (From Edwards and Meyer, 1987)

River	*Period of sampling*	*GPP*	*R*	*P/R*	*Source*
Silver Springs, FL	Winter	8–35	2.8–5.0	2.9–7.0	Odum, 1957
Blue River, OK	Annual	3.0–21.5	7.7–12.6	0.39–1.67	Duffer and Dorris, 1966
River Ivel, G.B.	Summer	9.6	8.5	1.1	Edwards and Owens, 1962a
Truckee River, NV	August	8.1–9.5	11.4	0.83	Thomas and O'Connell, 1966
Buffalo Creek, PA	August	5.6	2.2	2.6	McDiffet, Carr and Young, 1972
Sycamore Creek, AZ	Summer	8.5	5.1	1.7	Busch and Fisher, 1981
Madison River, WY	July–August	4.8	1.6	3.0	Wright and Mills, 1967
New Hope Creek, NC	Annual	0.8	1.3	0.7	Hall, 1972
Neuse River system, NC	May–October	0.29–9.8	1.68–21.5	0.2–0.7	Hoskin, 1959
Fort River, MA	Annual	1.8	3.7	0.5	Fisher and Carpenter, 1976
Bear Brook, NH	Annual	0.01	1.5	0.01	Fisher and Likens, 1973
McKenzie R system, OR	Annual	0.4–0.9	0.02–0.07	–[a]	Naiman and Sedell, 1980
Bayou Chevreuil, LA	Annual	2.1	2.7	0.7	Day, Butter and Conner, 1977
Salmon River, ID	Annual	0.54–2.53	0.34–1.73	–[a]	Bott *et al.*, 1985
Kalamazoo River, MI	Annual	0.13–6.39	0.55–5.79	–[a]	Bott *et al.*, 1985
White Clay Creek, PA	Annual	0.46–2.65	0.64–3.81	–[a]	Bott *et al.*, 1985
La Trobe River, Australia	Annual	0.15–1.90	2.97–4.61	0.05–0.50	Chessman, 1985
Ogeechee River, GA	Annual	2.2	6.7	0.3[b]	Edwards and Meyer, 1987

[a] For seasonal trends in P/R see Figure 12.8.
[b] Seasonal data show P/R < 0.5 throughout year.

plant production results in P greater than R and organic matter either accrues or is exported, as in Sycamore Creek (Table 12.3), the system is autotrophic.

Although the term autotrophy traditionally has been reserved for ecosystems with P/R greater than 1, this definition arguably is too stringent, as it is primarily an indicator of exporting ecosystems. Indeed, since P/R values greater than 0.5 indicate that over half of the respired energy is attributable to autochthonous primary production, this might be a more sensible dividing line between streams whose energy base lies primarily within its banks, rather than beyond.

Based on nine studies that gave values for autochthonous and allochthonous energy inputs into streams, Minshall (1978) argued that the role of within-stream primary production has been under-appreciated. Photosynthetic P_N exceeds litter inputs in a number of examples (Table 12.5), and there is a fairly obvious alternation in their relative importance depending on forest canopy development. *In situ* primary production is far greater in open compared with closed canopy locations, and highest values have been reported from grassland and desert streams.

The importance of various energy inputs to the support of community metabolism and

TABLE 12.5 Comparison of energy inputs from net primary production (P_N) and litter fall for a number of spring and running water studies. Additional inputs (e.g. groundwater, transport from upstream) are not considered here. (From Peterson *et al.*, 1986; after Minshall, 1978)

River	Energy input (g C m^{-2} yr^{-1})		Reference
	Autochthonous P_N	*Allochthonous Litter inputs*	
Bear Brook, NH	0.6	251	Fisher and Likens, 1973
Kuparuk River, AK	13	100–300	Peterson *et al.*, 1986
Root Spring, MA	73	261	Teal, 1957
New Hope Creek, NC	73	238	Hall, 1972
Fort River, MA	169	213	Fisher, 1977
Cone Spring, IA	119	70	Tilly, 1968
Deep Creek, ID 1	206	0.2	Minshall, 1978
Deep Creek, ID 2	368	7	Minshall, 1978
Deep Creek, ID 3	761	1.1	Minshall, 1978
Thames River, England	667	16	Mann *et al.*, 1970
Silver Springs, FL	981	54	Odum, 1957
Tecopa Bore, CA[a]	1229	0	Naiman, 1976

[a] Thermal spring.

production also merits consideration. Organic matter budgets account for the flux of carbon through a physical unit, such as a stream reach. They are not restricted to carbon that is utilizable by the biota. Budgets are based on measurements of quantities, not their biological turnover, and refractory material is quantitatively dominant. If budget terms were weighted by either their turnover rate or measurements of biological uptake and assimilation into new tissue, a very different view would emerge. As a speculative example, assigning assimilation efficiencies of 0.2% to DOM, 2% to POM and 20% to benthic primary production would dramatically change their importance and perhaps give an autotrophic complexion (from the perspective of support of the biological community) to an otherwise heterotrophic system. This sort of reasoning may help to explain a striking discrepancy between Bayley's (1989) analysis of carbon flux in the Rio Solimões, which showed most carbon originating as detritus from aquatic and floodplain macrophytes, and other investigations directed at the food chain itself. Analysis of the stable isotope of carbon, ^{13}C, in fish tissue and in various plants demonstrates that the food chain supporting an abundant group of detritivorous fishes, the Characiformes, begins with phytoplankton and not macrophyte detritus as might be expected (Araujo-Lima *et al.*, 1986). In a similar vein, Coffman, Cummins and Wuycheck's (1971) analysis of energy flow in a woodland stream in Pennsylvania indicates that benthic algae are important to the metazoan food chain in a setting where allochthonous inputs probably are of the same magnitude as in Bear Brook.

12.4 The river continuum concept

Virtually all statements concerning ecological processes in running waters have to be qualified by some reference to the physical location and specific circumstances of the site. Whether it is a headwater stream or a lowland river, shaded by forest canopy or exposed to full sunlight, modified by human activity or pristine, our interpretation of particular findings is incomplete without such information. The river continuum concept (RCC) is a bold attempt to

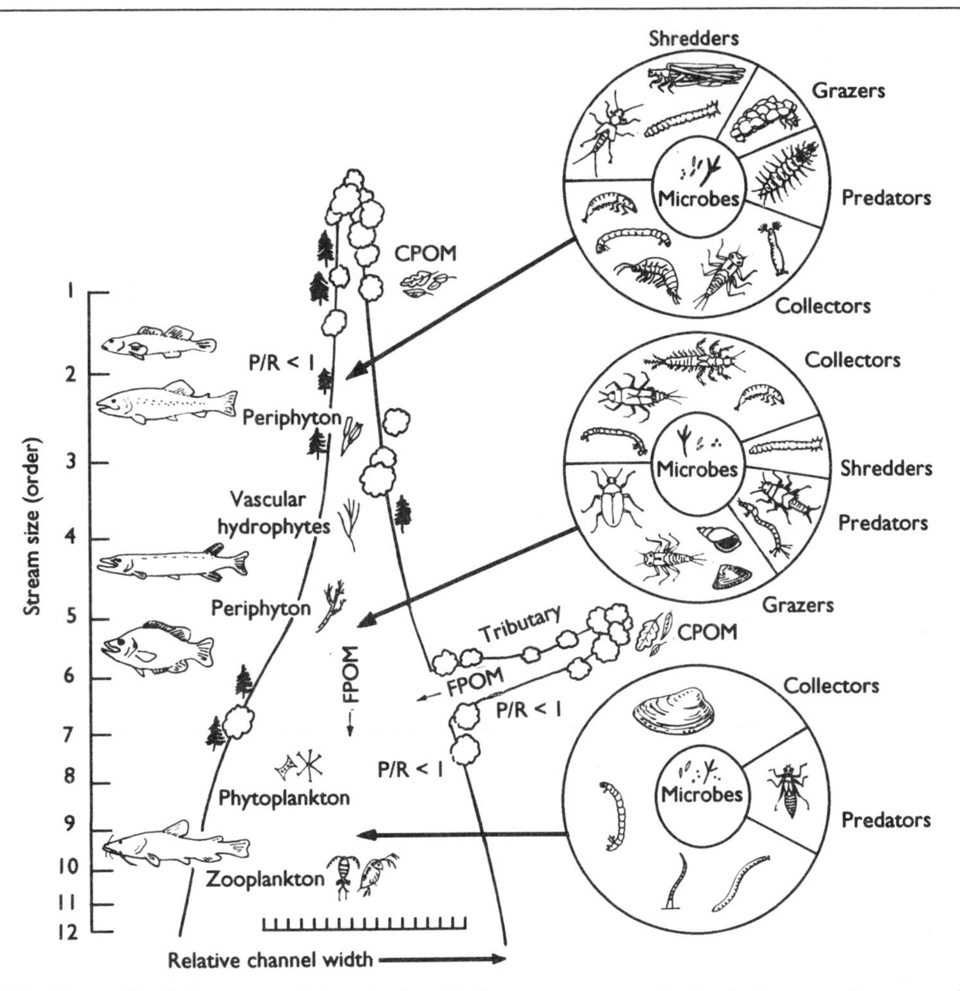

FIGURE 12.7 Generalized depiction of the relationship between stream size (order), energy inputs and ecosystem function expected under the river continuum concept. (From Vannote *et al.*, 1980.)

construct a single synthetic framework to describe the function of lotic ecosystems from source to mouth, and to accommodate the variation among sites that results from differences in their terrestrial setting (Vannote *et al.*, 1980). It is a useful way to think about the changes that occur along the length of a river, and it has provided the stimulus and conceptual context for a great deal of running water research over more than a decade. Just how well it portrays reality is uncertain, and perhaps the generality of the RCC is a handicap when it is applied to a multitude of specific situations. Nonetheless, the river continuum concept deserves careful review because it is a useful conceptual framework for describing how ecological functioning varies among riverine ecosystems.

The physical basis of the RCC is size and location along the gradient from tiny springbrook to mighty river. Along its length the stream swells in size, gathers tributaries and drains an increasingly larger catchment area. Stream order, discharge and watershed area have each been advocated as the physical measure of position along the river continuum, and all three are highly correlated (chapter 1). Stream order is the

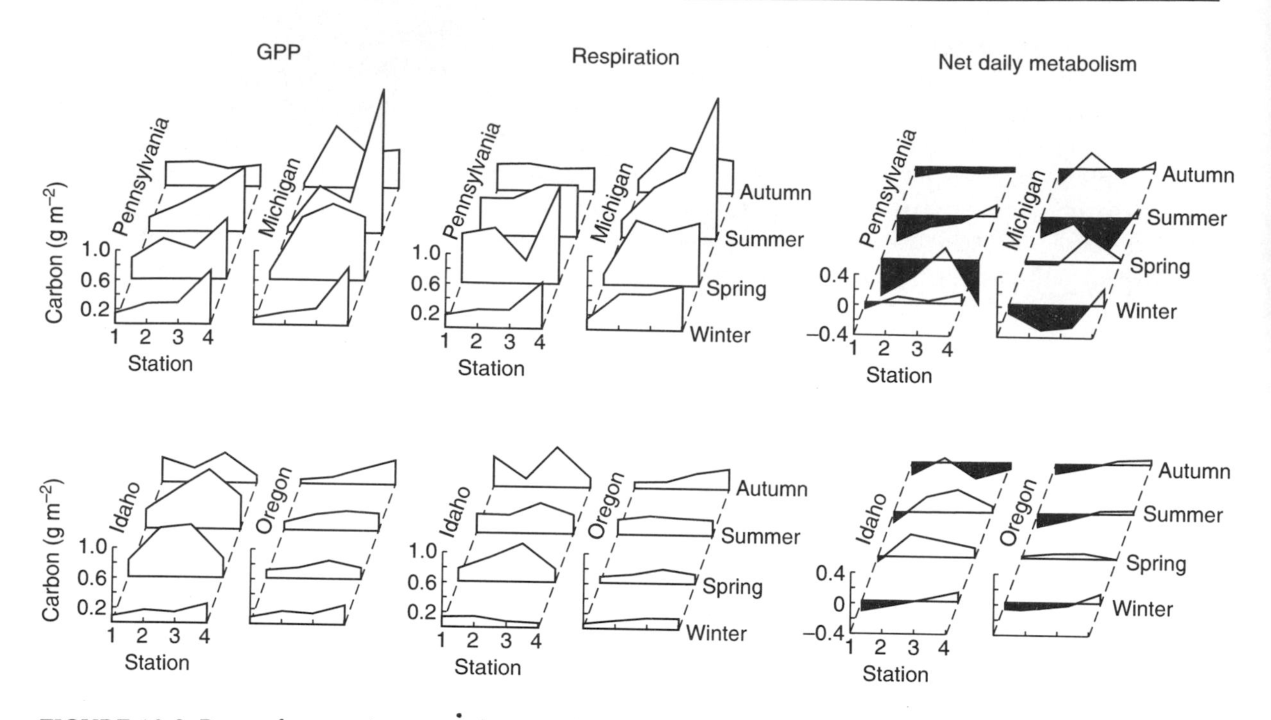

FIGURE 12.8 Rates of gross primary production (GPP), 24 h community respiration (R) and net daily metabolism (gross primary production – community respiration), determined by placing substrate material in enclosed chambers and measuring changes in oxygen concentration. Four sites (1, uppermost; 4, lowermost) in each region were studied in each season. Heterotrophy predominates in upstream sites, and the relative importance of autotrophy increases downstream. Note seasonal and regional variation in these trends, however. (From Minshall *et al.*, 1983.)

easiest to obtain and visualize, and so is the most widely used.

Biological changes along the river continuum are many. As initially conceived for a temperate woodland stream (Vannote *et al.*, 1980), low-order sites are envisioned as shaded headwater streams where inputs of CPOM provide a critical resource base for the consumer community (Figure 12.8). As the river broadens at mid-order sites, energy inputs are expected to change. Shading and CPOM inputs will be minimal, and ample sunlight should reach the stream bottom to support significant periphyton production. In addition, biological processing of CPOM inputs at upstream sites is expected to result in the transport of substantial amounts of FPOM to downstream ecosystems. Macrophytes become more abundant with increasing river size, particularly in lowland rivers where reduced gradient and finer sediments form suitable conditions for their establishment and growth. Extension of the river continuum concept to large rivers is problematic because they have received less study, and in most instances are highly modified by human engineering of the river and development of the landscape. Nonetheless, it generally is true that in high-order rivers the main channel is unsuitable for macrophytes or periphyton due to turbidity, swiftness of current, and scarcity of stable substrates. The only autochthonous production is by phytoplankton, and they are likely to be severely limited by turbidity and mixing (chapter 4). Allochthonous inputs of organic matter thus are expected to be the primary energy source in larger rivers. In their original formulation, Vannote *et al.* emphasized energy inputs in the form of FPOM imported from upstream systems; in later considerations, the role of lateral inputs from the floodplain have received more emphasis (Minshall *et al.*, 1985).

Such a pattern of longitudinal change in energy inputs should have profound consequences for the composition of consumer communities and the functioning of ecosystem segments along the river continuum. Functional group composition is expected to change in accord with resource inputs (chapter 6). Most obviously, shredders should prosper in low-order streams and grazers in mid-order. Low-order streams are expected to exhibit the lowest P/R ratio and highest CPOM:FPOM ratio. Proceeding downstream, one anticipates a steady decline in CPOM:FPOM, and a mid-order peak in P/R. Thus, heterotrophic inputs should especially dominate headwaters and large rivers, while autotrophy should play a greater role in mid-order streams. Lastly, in mid-order streams the variety of energy inputs appears to be greatest; as a consequence one might expect to find a peak in biological diversity as well.

12.4.1 Tests of the river continuum concept

The most ambitious effort to evaluate the RCC examined a range of stream sizes in four distinct regions of the USA (Minshall *et al.*, 1983). Each stream system was located in a relatively undisturbed watershed, had as its uppermost station a forested headwater site, and took for its lowermost station the largest stream site that was relatively undisturbed (seventh order). Terrestrial vegetation and rainfall varied considerably among biomes where the four river systems were located. Oregon sites receive abundant precipitation, mostly as winter rain, and support dense conifer forests. Idaho sites lie in a cold, arid region of the northern Rocky Mountains, where forest cover is less than other biomes and runoff is dominated by melt of the winter snowpack. Coniferous forest and scattered deciduous tree species are found at upper elevations, and sagebrush/grass vegetation at lower elevations. Michigan and Pennsylvania sites are in the eastern deciduous biome, with less pronounced seasonality in precipitation and runoff than either of the western sites.

Observed variation in the standing crop of periphyton among sites, seasons and locations was consistent with what is known of the various factors that limit algae. However, the patterns reported by Minshall *et al.* (1983) do not support a general claim that periphyton levels are highest in mid-order streams. Neither did they find the expected downstream increase in the ratio of FPOM:CPOM in transit, except for a single site in Idaho. However, rates of daily metabolism conformed reasonably well to expectations (Figure 12.8). Headwater reaches generally were heterotrophic, downstream systems were relatively more autotrophic, and heterotrophy usually predominated except in Idaho where most sites were unshaded. Differences between regions and seasons were substantial. In short, the longitudinal distribution of energy inputs was broadly consistent with predictions. However, sites along a particular river continuum were less distinctly different than the RCC suggests, and differences among regions were considerable.

Shifts in the relative abundance of invertebrate functional groups likewise provided mixed support. Shredders generally were most abundant in headwaters, as expected, but trends for the remaining functional groups did not fit neatly with expectations (Minshall *et al.*, 1983).

Because the Salmon River, Idaho, is a free-flowing river unimpounded over its entire length, Minshall *et al.* (1992) were able to extend their analysis up to an eighth-order site. Patterns of energy input corresponded well with expectations under the RCC. CPOM was prevalent only in the headwaters, primary production by periphyton was greatest at mid-order sites, and FPOM and heterotrophy were predominant at the farthest downstream site. However, functional group representation again was not a close match to predictions, which Minshall *et al.* attributed to lower than expected CPOM loading to headwaters, and high sediment inputs into mid-reaches from a major tributary.

It is a logical extension of the river continuum concept that alterations of the terrestrial vegetation should have predictable consequences for stream communities. Hawkins, Murphy and Anderson (1982) tested this by comparing stream sections flowing through old-growth conifer stands, recently clear-cut areas, and deciduous forest that had re-grown since logging. Although some predictions were met, opening the canopy did not cause the expected shifts in functional group composition. All guilds were more abundant in the open stream, and shredders were not more abundant under forest canopy, contrary to expectation. As one possible explanation, Hawkins and colleagues suggested that lack of correspondence between energy inputs and functional groups might be due to the investigators' inability to distinguish between low and high quality foods. Winterbourn, Rounick and Couie (1981) also found little correspondence between expectations based on the RCC and invertebrate populations in New Zealand streams. Food specialization was minimal, shredders were absent, and energy inputs associated with organic microlayers apparently were of greater importance than CPOM, FPOM or algae. Biogeographical isolation and the depauperate status of the New Zealand fauna may partly explain these differences. However, Winterbourn and colleagues make the more general point that a highly unpredictable hydrologic regime can override expectations from the RCC, and if this is true, it might be important in other regions than just New Zealand.

Some natural systems simply differ from the idealized scheme of Figure 12.7. Headwaters are not always forested, particularly in xeric or montane locations (Minshall, 1978; Minshall *et al.*, 1992), and so autotrophy rather than CPOM inputs can dominate headwater regions. Large tropical rivers with extensive floodplains and an annual cycle of inundation receive substantial energy inputs from floodplain detritus and floodplain lakes (Junk, Bayley and Sparks, 1989). The RCC very likely places too much emphasis on FPOM imported from up-river, and not enough on river–floodplain interactions. Large temperate rivers also can be strongly influenced by allochthonous inputs from extensive floodplains. Inputs of OM from extensive riparian forests resulted in an increase in heterotophy from second-order to sixth-order sites on the Ogeechee River, Georgia, contrary to expectations under the RCC (Meyer and Edwards, 1990). The Ogeechee is a blackwater river, as is the boreal river system studied by Naiman (1982), and such rivers exhibit a number of features that set them apart from clearwater rivers (Meyer, 1990). From the perspective of the RCC, their tea-colored waters and high DOC content act together to favor heterotrophy over autotrophy over their entire length.

Evidence from large temperate rivers is scarce because dams and levees prevent floodplain inundation in most instances. Indeed, the human modification of rivers and their valleys is sufficiently extensive to negate the expected longitudinal sequence of energy inputs at low- as well as high-order sites (Statzner and Higler, 1985).

The river continuum concept has met only limited success when specific predictions are put to the test. Its applicability to running waters world-wide has been debated. Moreover, the hypothesized pattern of biological events from headwaters to river mouth is plainly idealized. For these reasons it has been subjected to some criticisms (Winterbourn, Rounick and Cowie, 1981; Statzner and Higler, 1985; Lake *et al.*, 1986). On the other hand, the RCC has much in its favor as an attempt to search for general principles that are applicable to many streams across different biomes. Comparisons along the length of a river and among river systems have been made more explicit, and some standardization of approaches and methods has resulted from the framing of specific questions (Barmuta and Lake, 1982; Minshall *et al.*, 1992). The basic premise, that energy inputs vary among streams with predictable consequences for the biota and for ecosystem processes, is sensible and in many instances is supported by data. The reasons for the weak match of invertebrate functional groups to energy sources remains a puzzle that further examination of energy pathways leading to consumers might help to resolve.

12.5 Summary

The study of organic matter within lotic ecosystems requires an evaluation of all of its various states, where and how organic matter originates, the magnitude of different stores, and the processes that transfer organic matter from one state or one store to another. DOC is the largest pool of organic matter in rivers, exceeding POC by roughly 2:1 on a world average. However, absolute and relative amounts of DOC and POC vary greatly, depending on the mix of inputs, biological activity, and hydrologic events. Because fine particles tend to accumulate during low flows and be flushed by high flows, organic seston varies much more with discharge than does the dissolved material. Ultimately, it appears that of all the organic carbon entering river systems, about a quarter is processed within the river, about a quarter is stored as sediment POC, and half is exported to the world's oceans.

Transformations of DOC and POC within lotic ecosystems involve numerous abiotic and biotic processes, discussed more fully in previous chapters. Because rivers are such open systems with large export terms, any factor that acts to retard the rate of passage of organic matter downstream increases the likelihood that biological processes will re-mineralize a greater fraction of total organic material. Debris dams in small streams, floodplains of large rivers, and variation in channel structure and hydrologic regime are important mechanisms influencing organic matter retention.

The mass balance or budget, constructed from an accounting of organic matter inputs and outputs for a river segment or sub-watershed, provides insight into the flux and transformation of organic matter through stream ecosystems. Allochthonous energy sources invariably are substantial, although their importance is reduced in open canopy streams during seasons of high instream primary productivity. In aridland streams, autochthonous primary production can contribute most of the energy. All ecosystems import energy from 'upstream' and export energy 'downstream', but these terms are of particular importance in running waters because of the predominance of heterotrophy and the influence of hydrology. An adequate characterization for open ecosystems such as streams and rivers requires knowledge not only of the production to respiration (P/R) ratio, but also of the import to export (I/E) ratio. Strongly heterotrophic streams are characterized by P less than R and I greater than E; strongly autotrophic systems have P greater than R and either store or export biomass.

Everything about lotic ecosystems is highly context dependent. The kinds and amounts of energy inputs vary with the size of the stream or river, the surrounding vegetation, and general position along the gradient from headwaters to river mouth. The river continuum concept provides a general framework to describe how changes in energy pathways are likely to affect the balance between heterotrophy and autotrophy, which feeding guilds should predominate, and other functional aspects of running waters. Broad trends in the relative importance of autotrophic *versus* heterotrophic inputs accord well with expectations, and functional group representation is supportive in some studies but not all. Differences among sites due to local and regional conditions appear often to override RCC predictions. However, the underlying premise, that the biota of a stream reflects the nature of organic matter inputs, continues to serve as a useful conceptual framework.

Chapter thirteen

Nutrient dynamics

Solutes are materials chemically dissolved in water. Their dynamics depend on transport in the water column, and on all the transfer processes linking the water column to the streambed, the streambanks, and the land. Solute dynamics are closely coupled with the physical movement of water, and the net flux is downhill. Studies of stream solutes can be viewed from two perspectives. To geochemists and geophysicists, the transport of materials from the land to the sea is of central interest: rivers are the "gutters down which flow the ruins of continents" (Leopold, Wolman and Miller, 1964). The primary concerns of biologists, not surprisingly, lie with those elements essential for life and whose supply potentially limits metabolic processes in streams. These are called nutrients, and their uptake, transformation, and eventual release are the principal topics studied by stream biologists.

Not all solutes are readily utilized by the biota or otherwise transformed from state to state by physical and chemical processes. Such materials are referred to as conservative, and their passage downstream is presumably of little relevance to stream ecosystems. Reactive solutes, especially those that regulate metabolic processes, have their downstream passage retarded by uptake and temporary storage. Their dynamics may be significantly influenced by both abiotic and biotic processes. This distinction between conservative and reactive solutes, while useful, is not absolute, and may depend on the relationship between supply and demand for a particular element at a particular time and place.

Traditionally, the transfer of metabolically important elements between abiotic and biotic compartments is thought of as taking place within the fixed spatial bounds of a particular ecosystem. Some transfer of elements across boundaries of course occurs, and the net flux is always downhill on a geological time scale (Leopold, 1949). Clearly, however, strong unidirectional flow must impart unusual properties to the cycling of nutrients in running waters. Nutrients generated at one location typically will be transported some distance before subsequent re-utilization. Thus, although a given nutrient atom might be re-used many times, each cycle is displaced downstream from the previous cycle. The term 'nutrient spiraling' describes the interdependent processes of nutrient cycling and downstream transport (Webster and Patten, 1979).

The principal features of solute dynamics are summarized in Figure 13.1. Most of the materials under consideration are transported as solutes in the water column, although phosphorus, hydrophobic organics and many trace

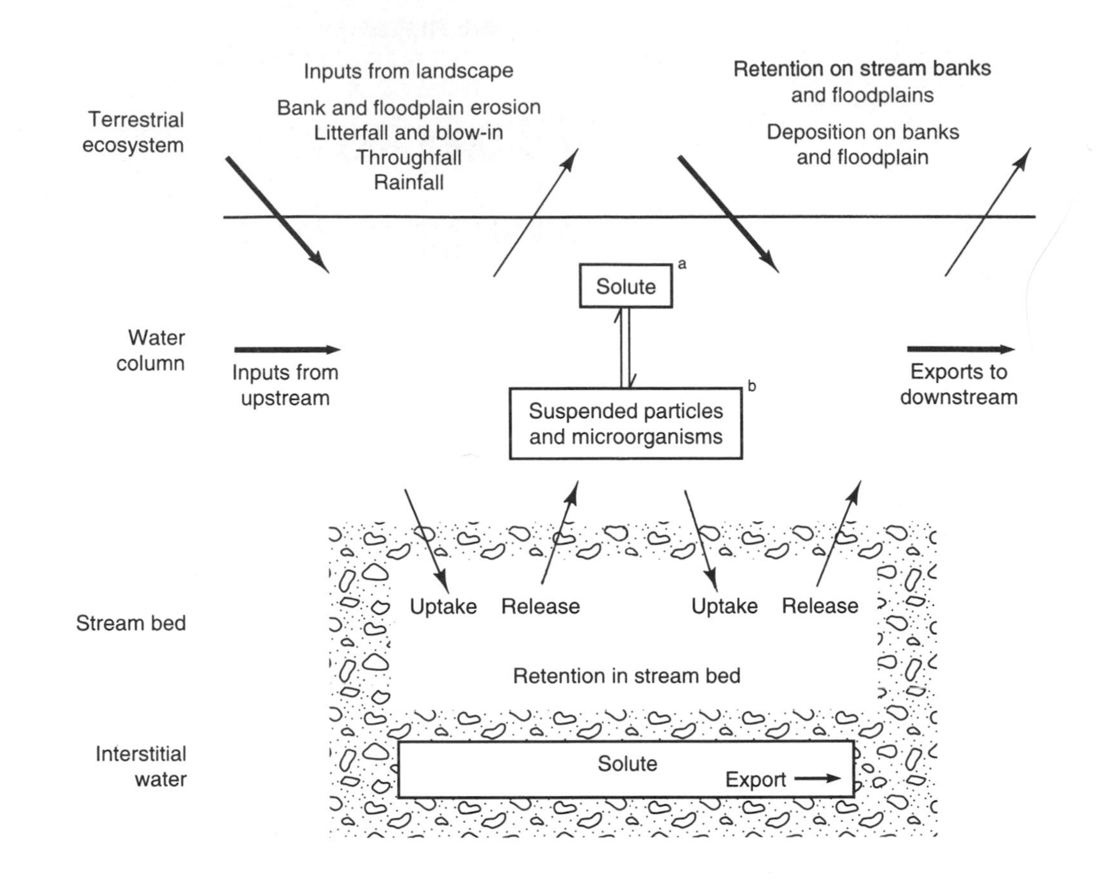

FIGURE 13.1 Conceptual diagram of solute processes in streams. Arrow widths are intended to approximate magnitude of process. Most materials are transported in dissolved form (a), but phosphorus, trace metals, and hydrophobic organics are transported mainly as particulates (b). (Modified from Stream Solute Workshop, 1990.)

metals are transported mainly as particulates, and nitrogen often is as well. A number of processes affect solutes, including transformations from one chemical species to another, physical changes such as adsorption, desorption and chemical precipitation, and biological uptake and release. These transformations, which are associated largely with streambed processes, as well as with deposition on streambanks and the floodplain, are principal mechanisms of retention. Transport of solutes also occurs within the streambed in sub-surface water, and exchange between channel and interstitial water can greatly influence solute dynamics.

Early experiments with the release of radioactive isotopes of phosphorus demonstrated that added nutrient was rapidly removed from the water column (Ball and Hooper, 1963). Moreover, while current would be expected to displace the tracer downstream, some ^{32}P was found to disperse upstream (Ball, Woitalik and Hooper, 1963), indicating that the biological community was somehow influencing phosphorus dynamics.

In this chapter we will examine the often complex relationships between availability of inorganic nutrient and its utilization by the biological community. There are, broadly speaking, two perspectives: how nutrient supply affects biological productivity, and how the stream ecosystem in its totality influences the supply of nutrients being transported downstream. These are of course intertwined. The limiting effects of nutrients on instream production were discussed in chapters 4 and 5, and here we will pay particular attention to nutrient dynamics in ecosystems. Before proceeding to the details of nutrient cycling in streams, it is useful to review briefly the central issues concerning nutrient cycling as elucidated from studies of estuaries (Day *et al.*, 1989) and lakes (Wetzel, 1983).

13.1 Basic principles of nutrient cycling

Energy enters the biosphere principally as sunlight and is transformed into energy-rich organic substrates by the autotrophic processes of photosynthesis and chemosynthesis. This organic material provides the fuel for most biotic metabolism and eventually is entirely transformed into heat. Various chemical compounds of life must be acquired by plants and microbes in order for synthesis of new organic matter to take place, and by animals to sustain their growth and metabolism. Although energy thus exhibits a one-way flow from synthesis to dissipation via ecosystem metabolism, it is a fundamental rule of ecology that the chemical constituents of living organisms are continuously re-utilized as they cycle between the biota and the environment.

Various solutes and gases occur dissolved in water, and a number of these may be incorporated into living tissue. Heterotrophs obtain most of their chemical constituents the same way they obtain their energy, by consuming other organisms, and also by ingesting or absorbing water. Usually it is energy, not the availability of materials, that limits most animals, but microbes sometimes are nutrient limited. In contrast, autotrophs commonly are limited by the availability of certain chemical constituents, including C, N, P, Si, S, K, Mg, Na, Ca, Fe, Mn, Zn, Cu, B, Mo, V and the vitamins thiamin, cyanocobalamin and biotin (Hutchinson, 1967). Carbon, nitrogen, phosphorus and silicon are the most heavily utilized, and since carbon typically is abundant as dissolved carbon dioxide, the usual presumption is that nitrogen, phosphorus and silicon are of primary importance. These are referred to as the macronutrients. More generally, the term nutrient refers to any inorganic material that is necessary for life.

Nutrients occur in various chemical forms as ions or dissolved gases in solution in water (Table 13.1). They are transformed into the particulate phase by physical and chemical processes as well as by metabolic activities. Adsorption describes the attachment of ions onto inorganic surfaces. It is especially important for highly charged anions such as phosphate and is reversible (desorption) depending on ion concentration and environmental conditions. Other important physical–chemical processes include the coalescing of colloidal-sized particles into larger aggregates, termed flocculation; and chemical precipitation, which occurs under oxidized conditions and reverses under anaerobic conditions. The principal biotic processes are assimilation and excretion, although nitrogen cycling is further complicated by additional transformations carried out by various bacteria.

External inputs of solutes originate from two sources (chapter 2). Materials released by the weathering of rocks and soils are transported in solution by runoff. Wind-eroded terrestrial dust, and salts originating in sea spray, enter aquatic systems via wet and dry precipitation. In many systems the inputs are much too low to support observed productivity, underscoring the importance of internal recycling as an additional source. Whether recycling plays as critical a role in running water ecosystems as it does in lakes and estuaries is questionable and constitutes one of the major issues to be addressed in this subject area. The principal pathways of cycling of course differ among nutrients, as can be illustrated by a brief description of transformations

TABLE 13.1 Major forms of nitrogen and phosphorus found in natural waters (After Meybeck, 1982). Nitrogen is also present as dissolved N_2 gas (not shown)

Form		
Nitrogen		
Dissolved inorganic nitrogen (DIN)	Total dissolved nitrogen (TDN)	Total nitrogen
NO_3^- nitrate		
NO_2^- nitrite		
NH_4^+ ammonium		
Dissolved organic nitrogen (DON)		
Particulate organic nitrogen (PON)		
Phosphorus		
Dissolved inorganic phosphorus (DIP)	Total dissolved phosphorus (TDP)	Total phosphorus
PO_4^{-3} orthophosphate		
Dissolved organic phosphorus (DOP)		
Particulate organic phosphate (POP)		
Particulate inorganic phosphorus (PIP)		

influencing three macronutrients, phosphorus, silicon and nitrogen.

(a) Phosphorus

Phosphorus frequently limits the productivity of plants and other autotrophs in aquatic environments. Dissolved inorganic phosphorus (DIP) is assimilated by plants and microbes into cellular constituents, thereby being transformed into particulate organic phosphorus (POP). It may be excreted or released during cell lysis directly as DIP, or released as dissolved organic phosphorus (DOP), which subsequently is broken down to DIP by bacterial activity. In addition to these biological processes, phosphorus availability is influenced by physical–chemical transformations. Sorption of phosphate ions onto charged clays and charged organic particles occurs at relatively high DIP concentrations, while desorption is favored by low concentrations. Sorption–desorption reactions thus act as a buffer on DIP. In addition, under aerobic conditions both DIP and DOP may complex with metal oxides and hydroxides to form insoluble precipitates. This phosphate is released under anaerobic conditions, and since the extent of the anaerobic zone tends to vary seasonally with organic matter loading, the availability of dissolved phosphate varies accordingly.

(b) Silicon

Silicon is limiting only to diatoms, but since these are important primary producers in aquatic ecosystems, especially in streams and smaller rivers, the availability of silicon is of some importance. The silicon cycle is extremely simple. The principal dissolved form is silicic acid, originating from the weathering of rocks and from anthropogenic inputs, mainly sewage discharge (chapter 2). It is assimilated by diatoms for the formation of their frustules, and released by chemical dissolution (although it is possible that microbial enzymes may augment the process).

(c) Nitrogen

The nitrogen cycle is complex (Figure 13.2), due to the many chemical states in which nitrogen is found and the central role of bacteria in its transformation from one form to another. Urea, uric acid and amino acids are principal forms of dissolved organic nitrogen (DON), while dissolved inorganic nitrogen (DIN) includes NH_4^+,

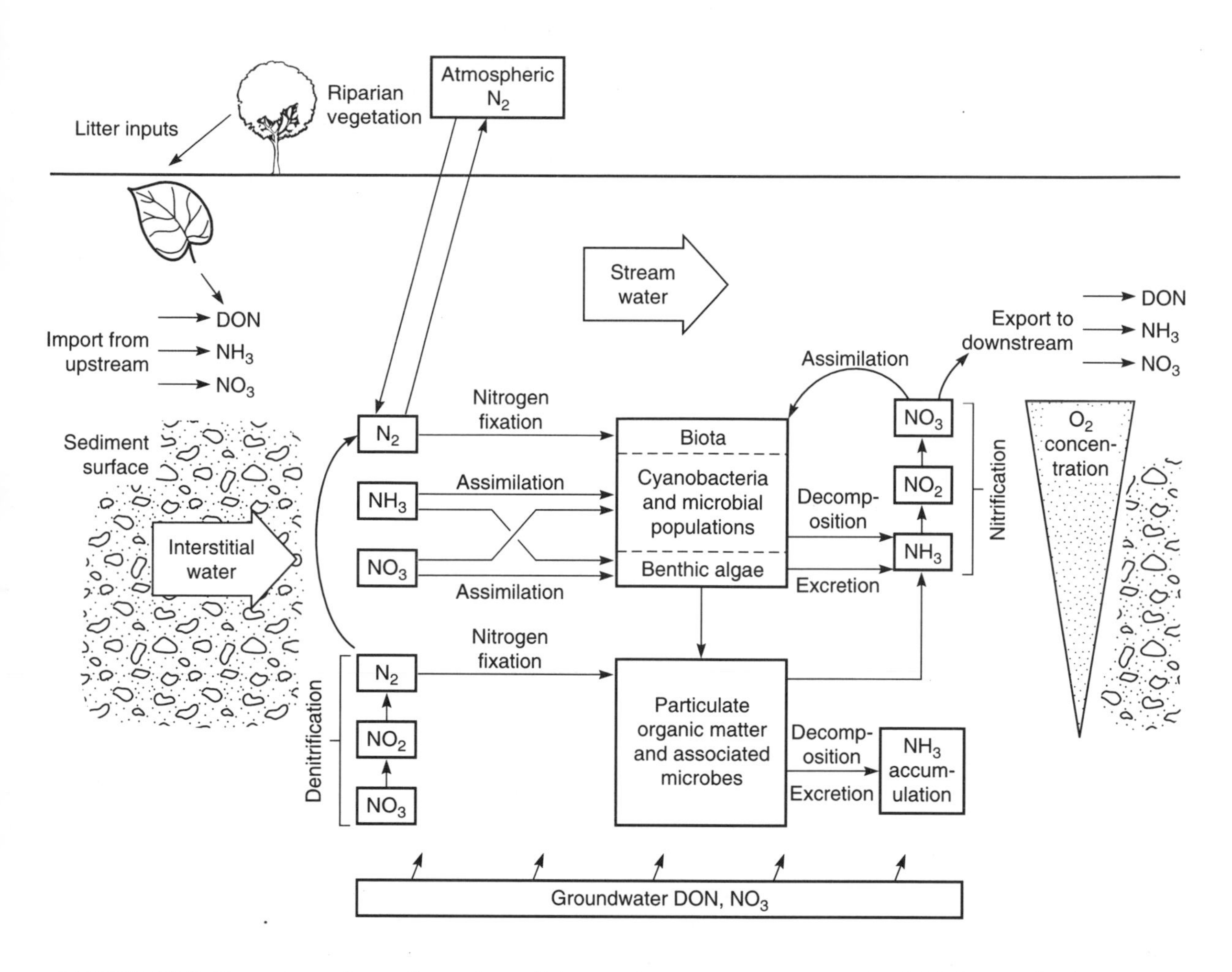

FIGURE 13.2 Dynamics and transformations of nitrogen in a stream ecosystem. Available inorganic nitrogen consists mainly of nitrate and ammonia, which the biota can assimilate directly. Excretion, decomposition and production of exudates are the principal pathways by which elements are recycled to an inorganic state. Various transformations by bacteria add to the complexity of the nitrogen cycle. Cyanobacteria and other microorganisms capable of nitrogen fixation transform N_2 gas into ammonia. Nitrification, which takes place under aerobic conditions, and denitrification, which takes place under anaerobic conditions, further influence the quantities and availability of dissolved inorganic nitrogen.

NO_3^-, and NO_2^-. Nitrogen also is available as the gases N_2 and N_2O. The sources of these various forms of nitrogen include atmospheric diffusion, runoff and anthropogenic inputs from sewage discharge and agricultural fertilizers. In addition, atmospheric nitrous oxides resulting from industrial pollution are a significant source of nitrogen enrichment in many regions receiving acid precipitation.

To understand the complexities of the nitrogen cycle, it helps to recognize that some transformations are to obtain nitrogen for structural synthesis, while others are energy-yielding reactions. Nitrogen fixation and assimilation of DIN are in the former category, whereas nitrification and denitrification are reactions where bacteria obtain energy by using ammonia as a fuel or nitrate as an oxidizing agent (Day *et al.*, 1989).

Autotrophs, bacteria and fungi assimilate nitrogen as DIN, with NH_4^+ generally used preferentially over nitrate and nitrite. Since ammonium-N usually is a small fraction of DIN compared with nitrate, processes influencing the availability of NH_4^+ may be of considerable importance. Thus nitrogen fixation, in which bacteria and cyanobacteria convert nitrogen gas to NH_4^+ and incorporate this NH_4^+ into bacterial biomass, may be favored under nitrogen limitation. Ammonium-N is regenerated by excretion and decomposition, and its concentrations are buffered to some extent by sorption–desorption reactions with clays and humic materials. Lastly, nitrification–denitrification pathways result in its loss, as ammonium is first converted to nitrate, then to nitrogen gas. The oxidation of NH_4^+ to nitrate by nitrifying bacteria occurs under aerobic conditions, whereas denitrifying bacteria use nitrate as an electron acceptor to oxidize organic matter anaerobically. In the well-oxygenated sediments of gravel-bed streams, denitrification should be of little consequence. However, in deeper sediments where oxygen levels are low, denitrification may be of greater importance, and as these same conditions do not favor nitrification, ammonia may accumulate (Figure 13.2).

Although phosphorus, nitrogen and silicon are only three of a long list of potentially limiting nutrients, we discussed the major transformations of each because they are widely viewed as critical to ecosystem metabolism. In addition, a comparison of their cycles makes several important points. Transformations occur by abiotic processes as well as by biological uptake and release, and some ions such as phosphate are strongly influenced by physical–chemical processes. These cycles vary considerably in their complexity and in specific details. It should be kept in mind that phosphorus and nitrogen potentially are limiting to heterotrophic production associated with the decomposition of organic matter, as well as to photosynthetic primary production. Lastly, under circumstances where a significant amount of biological productivity is due to the anaerobic decomposition of imported organic matter, a common finding in estuaries (Day *et al.*, 1989) and perhaps also in large or slow-moving rivers, we would need to consider the supply of sulfate and carbonate. These act as electron acceptors during reduction of organic matter in the absence of oxygen. Hence their supply is critical to heterotrophic metabolism, and the presence of their end products (the gases hydrogen sulfide and methane, respectively) reveal the extent of this activity.

13.2 Nutrient concentrations in running waters

For a number of reasons it is difficult to infer the importance of nutrients solely from knowledge of their concentrations, and later sections will examine evidence from the study of processes that relate nutrient availability to biological activity. However, nutrient concentrations predict phytoplankton abundance and trophic status of lakes with considerable accuracy (Wetzel, 1983), and they may prove useful in characterizing the productivity of riverine ecosystems as well. The chemistry of river water is highly variable, especially in streams and small rivers. The amount of dissolved material in freshwaters, and its chemical composition, varies greatly with location and season, due to local geology and rainfall, position along the river continuum, and the extent of human influence (chapter 2). We now consider the availability of important nutrients in a variety of running water situations.

13.2.1 Nutrient chemistry of large rivers

Large rivers, especially large lowland rivers, often have their nutrient concentrations modified by industrial emissions, sewage and agricultural effluents, and they also are likely to have their flow regime regulated by dams. Because their catchments include a substantial land area, each large river has its individual climate and geology. For these reasons it is difficult to generalize, and interested readers should consult studies of specific rivers (e.g. Whitton, 1975b; Davies and Walker, 1986; Dodge, 1989; and detailed descriptions of individual river systems in the Monographiae Biologicae series by W. Junk, The Hague). Even to summarize data for a given river is problematic, because values change greatly along a river's length, seasonally, and between cleaner and polluted sections.

Meybeck (1982) provides a comprehensive review of the various forms (dissolved and particulate, organic and inorganic) of carbon, nitrogen and phosphorus in world rivers. Natural levels are estimated from small streams in the temperate zone and major rivers of the tropics and subarctic, where anthropogenic inputs are thought to be minimal. Nutrient concentrations from these systems are very similar to the average content of rain, supporting the view that they represent an unpolluted state. Natural levels of dissolved phosphorus are very low, around $0.01\,mg\,l^{-1}$ for PO_4^{-3} and $0.025\,mg\,l^{-1}$ for total dissolved phosphate, which includes the organic form. Natural levels of DIN (including ammonium, nitrate and nitrite) are also low compared with rivers affected by human activities, about $0.12\,mg\,l^{-1}$, and nitrate is the major fraction. Ammonia averages 15%, and nitrite about 1%, of DIN. Because NH_4^+ is usually the preferred form of inorganic nitrogen, data on nitrate alone or total inorganic nitrogen may be of limited usefulness.

Human activities have profoundly altered nutrient levels in many of the world's surface waters. On a global basis, total dissolved phosphorus and nitrogen have doubled, and in areas of Western Europe and North America they have increased by factors of 10 to 50. Nutrient concentrations from selected rivers vary greatly because of differences in their natural state and extent of local human influence (Table 13.2). The Glama is a large Scandinavian river draining a region of predominantly hard crystalline rocks, within a largely forested watershed. The highest values in the Glama occur in its agricultural, sparsely settled lower reaches, and are still very low. Tropical rivers in general have low concentrations of chemical constituents compared with temperate rivers. This is because tropical rivers are for the most part precipitation-dominated, although local geology can also be important (chapter 2).

Almost invariably, higher nutrient concentrations are found at the lowermost sites along large rivers, due to a variety of human activities. Nitrate values typically are elevated by runoff of agricultural fertilizer, and phosphorus values by sewage effluent. However, changing geological conditions along a river's length may also contribute to longitudinal trends. Lowlands are more heavily farmed and settled than upland regions because they are more suitable for agriculture, and so the relative importance of human influence *versus* underlying conditions is difficult to separate. The strong influence of human presence is unmistakable, however. Long-term data from the Mississippi River establish that nitrate concentrations changed little from the turn of the century until the 1950s and then roughly doubled in the following 35 years, coincident with a steady rise in fertilizer application of the same time period (Turner and Rabalais, 1991). Human population density in the watersheds of 42 major rivers of the world was found to explain 76% of the variation in average annual nitrate concentration (Peierls *et al.*, 1991).

Nutrient concentrations often vary seasonally due to influences of hydrology, the growing season and changes in anthropogenic inputs. Whenever the supply of an element or nutrient is relatively constant, or at least varies less than discharge, high flows will dilute and low flows will concentrate the material. In the Wye River of Wales, where the major source of phosphorus is sewage effluent and the supply is therefore

TABLE 13.2 Phosphorus and nitrogen concentrations representative of various large river systems from different regions. Values indicate range across sites or seasons measured in $mg\,l^{-1}$. (Sources: Golterman, 1975; Whitton, 1975b; Sioli, 1984)

	PO_4-P	*NH_4-N*	*NO_3-N*
Temperate rivers, receiving various amounts of anthropogenic inputs			
Glama (Norway)	0.002–0.008	–[a]	0.065–0.33
Lot (France)	0.001–0.013	–	0.32–0.52
Meuse (Belgium)	<0.005–0.6	<0.002–0.66	<0.1–6.3
Tees (UK)	0.025–1.27	0.017–0.280	0.166–1.05
Upper Volga	0.018–0.26	0.53–1.4	0.58–0.93
Wye (Wales)	near 0–0.13	–	0.1–2
Tropical rivers, generally less influenced by human activities			
South America			
Amazon			
Mainstem	0.012	–	0.40
Andean streams	–	–	very low
Whitewaters	0.015	–	very low
Clearwaters	<0.001	–	very low
Blackwaters	0.006	–	very low
Upper Parana	0.02–0.18	0.02–0.62	0.03–0.86
Africa			
Blue Nile	0.002–0.12	–	0.001–0.10
White Nile	0.005–0.10	–	0.010–0.09
Orange	0.003–0.10	–	0.030–1.40
Asia			
Gombak	0.010–0.41	0.05–0.08	0.40–0.88
Sumutra	0.0070	–	0.175

[a] Data not available.

relatively constant, concentrations are greatest during the summer and maximal during droughts (Edwards and Brooker, 1975). Where groundwater is a major input, as is true for nitrate in a small tributary of the Wye, the same pattern is found. On the other hand, if heavy rainfall flushes materials from the surrounding terrestrial soils, including agricultural fertilizers, as apparently is true for nitrate in the main Wye catchment, concentrations increase at high flows.

Biological uptake provides another mechanism for seasonal variation, and in large rivers with a well developed potamoplankton, nutrient concentrations are highest in winter when algal growth is minimal. The Volga demonstrates such seasonal variation at some locations (Golterman, 1975). Nutrient concentrations in the Lot are quite low, but not for a lack of supply (Décamps, Capblanq and Turenq, 1984). This highly regulated river in southern France receives considerable domestic, industrial and agricultural waste which, combined with lentic conditions due to the water's long residence time, allows substantial phytoplankton blooms to develop. Dense algal populations strip nutrients from the water, and so conditions typical of a highly eutrophic lake occur in the presence of nutrient concentrations that are, for many rivers, in the low range.

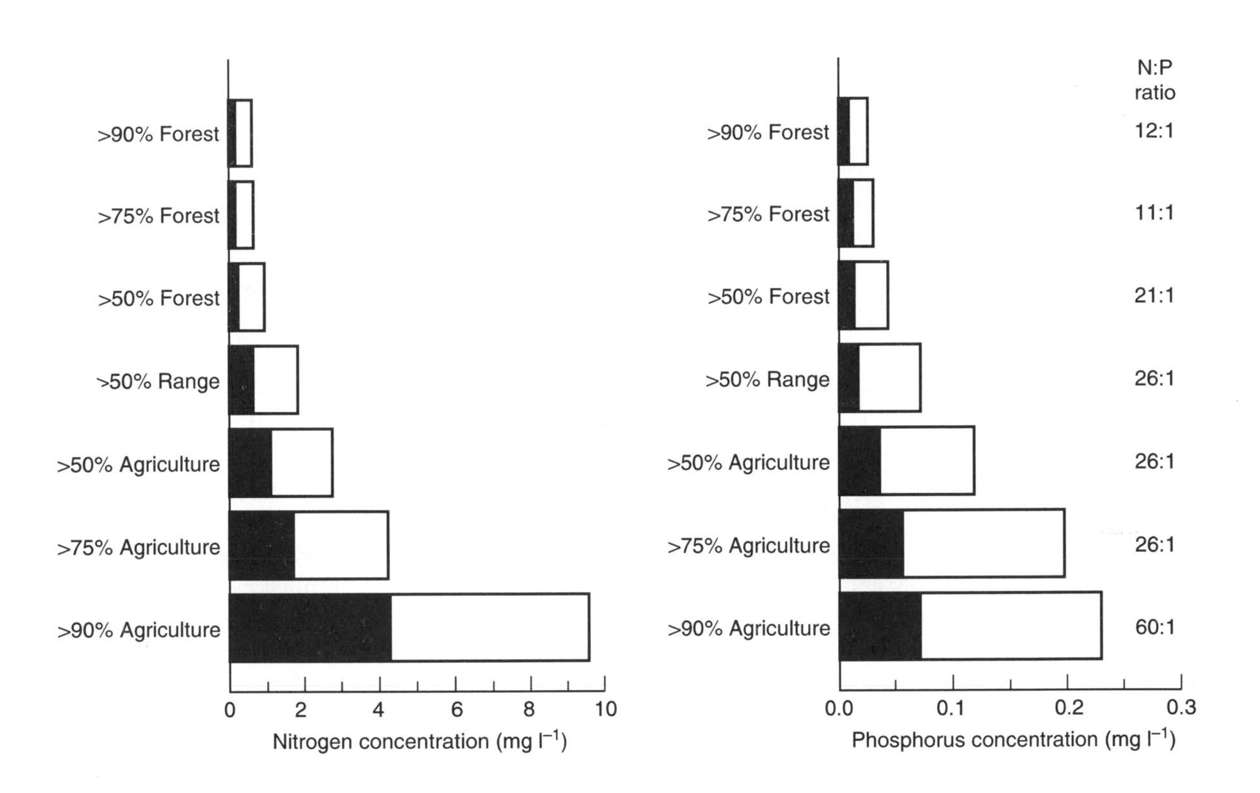

FIGURE 13.3 Concentrations of nitrogen and phosphorus, and N:P ratios, for small, relatively unpolluted streams of the USA. The inorganic fraction of the total is shaded for each nutrient. (Based on Omernik, 1977.)

13.2.2 Nutrient chemistry of streams and small rivers

A eutrophication survey of 928 streams throughout the USA provides a comprehensive description of phosphorus and nitrogen levels in small to mid-sized streams (Omernik, 1977). Sites were selected to be representative of a region and uncontaminated by such things as municipal and industrial discharge and animal feedlots, although fertilizer runoff from agricultural land was an unknown contribution to streamwater values. The reported concentrations of nitrate, phosphate, total dissolved nitrogen and total dissolved phosphorus exhibited a clear relationship with land use, and surprisingly little dependency on underlying geology. Streams draining agricultural land had higher nutrient concentrations than those draining forested land. The concentrations were proportional to percentage of land in agriculture and inversely proportional to percentage of land in forest (Figure 13.3). The inorganic fraction of total phosphorus was between 40 and 50% across all land-use types, whereas the inorganic fraction of total nitrogen varied widely, due at least partly to the influence of agriculture. DIN increased from about 18% of total nitrogen in streams draining forested watersheds to nearly 80% in streams draining agricultural watersheds (Figure 13.3), presumably due to the use of nitrogen fertilizers.

It should again be noted that the differences between agricultural and forested watersheds cannot be attributed solely to human influences. Forested land typically remains as forest because it lacks the topography, soil and climatic conditions to be suitable for agriculture, and so differences in nutrient concentration must reflect different physical environments in addition to land-use practices.

Other studies of nutrients in streams have indicated that geology might play as great or even a greater role than land use in determining the nutrient concentrations of unpolluted waters (Likens and Borman, 1974). Dillon and Kirchner (1975) reported greater phosphorus export from Ontario streams draining watersheds of sedimentary origin compared with those of igneous origin. Sedimentary watersheds with phosphate-bearing limestone contained substantially more phosphate and nitrate in streamwater in comparison to watersheds where sandstone and shales were the main geologic types present (Thomas and Crutchfield, 1974). On the other hand, a study of one watershed with limestone and one without limestone revealed that stream phosphate levels were unrelated to geology but were related to anthropogenic activities (Stone, 1974). Omernik (1977) was unable to find a significant relationship between geological classification and nutrient concentrations in his study of over 900 watersheds.

13.2.3 Nitrogen:phosphorus ratio

Algal growth in lakes is usually limited by the supply of inorganic phosphorus, and so lake productivity is primarily a function of phosphorus availability. This has the important practical application that reduction of phosphorus concentrations can prevent lake eutrophication. Nitrogen is considered to be secondary, although in some regions nitrogen acts as the primary nutrient, and sometimes nitrogen and phosphorus together influence algal growth (Elser, Marzolf and Goldman, 1990). Oligotrophic lakes are those in which the average concentrations of total phosphorus are 0.01 $mg\,l^{-1}$ or less, while lakes with more than 0.030 $mg\,l^{-1}$ usually are eutrophic (Wetzel, 1983). As can be seen from Figure 13.3, phosphorus concentrations in running waters are often above the level that induces a eutrophic state in lakes. This is especially true for larger rivers influenced by human activity and for streams flowing through agricultural land.

The ratio of N:P indicates which nutrient is likely to limit algal growth in lakes. Carbon, nitrogen and phosphorus occur in algal tissue in a remarkably consistent ratio of atomic weights of 106:16:1, termed the Redfield ratio (Redfield, Ketchum and Richards, 1963). Based on this natural ratio and numerous bioassay experiments with lake algae, it is generally accepted that when N:P ratios fall below 16:1, algae will have less nitrogen available per unit of phosphorus and thus experience nitrogen limitation, whereas ratios above 16:1 indicate phosphorus limitation. At N:P ratios between 10:1 and 20:1, joint limitation by both nutrients is likely.

Most estimates of the ratio of nitrogen to phosphorus in freshwaters are above 16:1, implying that nitrogen is in surplus supply. Atomic ratios of N:P for a number of watersheds, mainly in the northeastern USA, almost invariably were well above the Redfield ratio during the non-growing season (Meyer, Likens and Sloane, 1981). Interestingly, N:P ratios during the growing season were lower, although still above 16:1 in most instances. Omernik's (1977) survey also revealed generally high N:P ratios calculated from TDN:TDP (Figure 13.3). Although such data do not indicate which form of inorganic N (NH_4^+, NO_3^-, or NO_2^-) is present, these data nonetheless suggest that phosphorus is more likely than nitrogen to limit algal growth, particularly in watersheds with some agricultural land use.

Regional variation in the relative availability of nitrogen and phosphorus is evident in Omernik's (1977) survey and from individual studies. Comparing forested watersheds of the USA, Omernik reported mean annual phosphorus concentrations from streams in the west to be roughly twice as high as in the east. In contrast, DIN is two to three times higher in streams of eastern forested watersheds, compared with the central and western states. Anthropogenic inputs must be partly responsible for higher inorganic nitrogen levels in the eastern USA due to the nitric acid component of acid rain, and in the corn belt as a consequence of agriculture (Omernik, 1977).

Low N:P ratios have been observed in some instances, indicative of nitrogen limitation, and algal response to nutrient enrichment provides supporting evidence. Based on a survey of 92 streams of the arid and semi-arid southwestern USA, the mean N:P ratio under low-flow conditions was 11.4, and 80% of sites had ratios less than 10 (Grimm and Fisher, 1986). When nutrients were added to one such stream, Sycamore Creek, Arizona, periphyton growth responded strongly to nitrogen enrichment, whereas phosphorus alone had no effect and nitrogen and phosphorus together produced no greater response than nitrogen alone. Streams of the northwestern USA also generally have low nitrogen concentrations. Periphyton growth is stimulated by nitrogen addition in unshaded streams in Oregon and in northern California (Gregory, 1980; Triska *et al.*, 1984).

Year-to-year variation in atomic ratios of N:P can result in different algal responses to nutrient enrichment between studies carried out in the same stream. In Walker Branch, a small woodland stream in Tennessee, Newbold *et al.* (1983a) found no increase in algal biomass to enrichment with ammonium-N when the average N:P ratio was 34:1, whereas Rosemond, Mulholland and Elwood (1993) found co-limitation of periphyton by nitrogen and phosphorus in two years when the N:P ratios averaged 23:1 and 12:1, respectively.

The response of periphyton to nutrient enrichment in areas where N:P ratios are high establishes that phosphorus typically is the limiting factor under these conditions. Phosphorus addition has been shown to stimulate algal growth in a British Columbia rainforest stream (Stockner and Shortreed, 1978), an Alaskan tundra stream (Peterson, Hobbie and Corliss, 1983), and in a woodland stream in Michigan (Pringle and Bowers, 1984) and in Walker Branch, Tennessee (Elwood *et al.*, 1981). Nitrogen enrichment by itself had no effect in the latter two studies, although the combined effect of nitrogen and phosphorus produced a greater response than phosphorus alone in Pringle and Bower's study and in other years in Walker Branch (Rosemond, Mulholland and Elwood, 1993). Because these studies are from diverse regions of North America, the implication that phosphorus limitation is widespread seems to be borne out.

It should be noted that the terms oligotrophic and eutrophic do not have the same utility for running waters that they do for lakes, nor is there evidence for a natural process of eutrophication corresponding to lake succession (Hynes, 1969). Nonetheless, nutrient limitation of metabolic activity is the underlying basis for the study of the dynamics of elements, and it is apparent that variation in nutrient concentrations and N:P ratios have predictable consequences for stream metabolism. We now turn to an examination of nutrient cycling in running waters, where downstream transport and a variety of abiotic and biotic factors play prominent roles.

13.3 Transport and transformation of nutrients

13.3.1 Physical transport models

Engineers and hydrologists have devoted considerable effort to the development of models that describe solute transport in flowing water (e.g. Fischer *et al.*, 1979). If a tracer or contaminant is released at some point, and measurements are made at another location downstream, the concentration will be observed to rise, reach a plateau, and then decline as the pulse passes the point of monitoring (Figure 13.4). A first

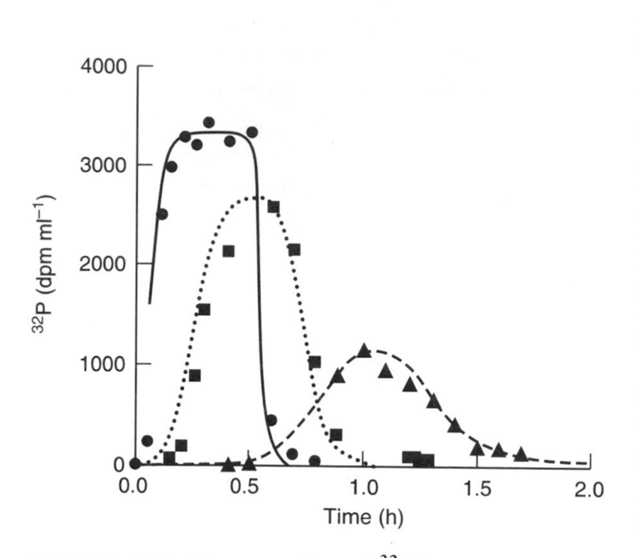

FIGURE 13.4 Concentration of ^{32}P in streamwater at 15 (●), 47 (■) and 120 m (▲) locations downstream of a 30 min release of radioactive phosphorus in a small woodland stream. (From Newbold *et al.*, 1983a.)

approximation to this curve can be achieved with a basic equation describing advection and dispersion, taking into account stream dimensions and water velocity. The solute disperses from its point of release due to the unidirectional force of current (advection), and diffusion and turbulent mixing throughout the stream. Under certain simple conditions, which include a uniform channel, constant discharge, and minimal turbulence and sub-surface flow, the change in solute concentration (C) over time is described as follows.

$$\frac{\partial C}{\partial t} = -U\frac{\partial C}{\partial x} + D'\frac{\partial^2 C}{\partial x^2} \qquad (13.1)$$

The first term describes downstream advection and is proportional to water velocity, u. The second term describes mixing of the solute randomly throughout the water mass according to a dispersion coefficient, D'.

More complicated models are needed to account for additional variables such as groundwater and tributary inputs, channel storage and sub-surface flow (Stream Solute Workshop, 1990). Solute dynamics are less complex in large rivers compared with small streams because large rivers generally have low slopes, are deeper than the roughest bed feature, and have relatively uniform and perhaps regulated flows. Small streams tend toward the opposite characteristics and in addition are highly variable (Bencala and Walters, 1983). Inclusion of terms for transient storage is particularly important to the description of solute dynamics in small streams and so merits some elaboration here. Transient storage refers to the temporary retention of solutes in slowly moving or even stationary water, and the eventual movement of solutes and water back into the stream channel. Adding this term permits the model to account for significant features of the observed passage of a solute pulse that equation (13.1) is unable to mimic. Specifically, measured passage of a tracer pulse usually shows the rising shoulder of the actual pulse to be more gradual and the descending tail to be prolonged relative to the symmetrical curve generated by equation (13.1).

'Storage zones' (also called 'dead zones') are included in models to account for the more gradual rise and extended decline of the passing pulse. These zones are stagnant regions that gain solute during the initial phase of the passage of the pulse and release this material back into the stream as the pulse passes and stream concentrations decline.

The equations to describe passage of a conservative solute, including transient storage, are:

$$\frac{\partial C}{\partial t} = -U\frac{\partial C}{\partial x} + D'\frac{\partial^2 C}{\partial x^2} + a(C_S - C)$$

$$\frac{\partial C_S}{\partial t} = -a\frac{A}{A_S}(C_S - C) \qquad (13.2)$$

where A_S is the cross-sectional area of a hypothetical storage zone. The rate of dispersion of solute in or out of this zone is proportional to the difference between solute concentration in the storage zone (C_S) and the water column (C), and an exchange coefficient (a).

It should be recognized that these models are empirically useful descriptions of observed dynamics, in which transient storage clearly takes place. However, the storage zone component of the model is an abstraction. In contrast to cross-sectional area (A) of the stream channel, which can be measured directly, storage zone area (A_S) is determined by fitting the model to observed solute dynamics. Nonetheless, storage zones exist and are numerous. Bencala and Walters (1983) recognized five in their study of solute transport in a small mountain stream: (1) turbulent eddies generated by large-scale bottom irregularities; (2) large but slowly moving recirculating zones along the sides of pools; (3) small but rapidly recirculating zones behind flow obstructions, especially in riffles; (4) side pockets; and (5) flow in and out of beds of coarse substrate.

The effectiveness of obstructions and roughness features in creating storage zones and the magnitude of channel flow *versus* flow through the streambed are both expected to vary widely. Model estimates of A_S provide a useful index of the size of the transient storage effect. In a small riffle-pool mountain stream, Bencala and Walters (1983) estimated A_S to be $0.7\,m^2$, compared with an A of $0.4\,m^2$. Estimates of $A_S{:}A$ for first- to fifth-order streams in the Appalachian and Cascade mountains ranged from 1.2 to 0.1 (D'Angelo *et al.*, 1993). Largest values were found in headwater streams where in-stream channel complexity was greatest. $A_S{:}A$ decreased downstream and with increasing velocity. The lowest ratio was recorded in the fifth-order stream (Lookout Creek, OR) at sites where the active channel width was constrained by channel geomorphology. Thus, although A_S can not be measured directly, estimates from equation (13.2) vary in accord with stream features as one might expect. In general, transient storage is likely to be of greatest importance in small streams with rough or porous beds, and be inversely related to stream size and discharge (Bencala, 1984; Bencala and Walters, 1983; Munn and Meyer, 1990; D'Angelo *et al.*, 1993).

These processes of advection, dispersion and transient storage describe only the hydrologic influences over transport of a solute downstream. Such models adequately describe the passage of a conservative tracer such as a dye, chloride or labeled water, but reactive solutes are subject to additional processes that biologists wish to study. Models of nutrient transport, discussed in the next section, have these hydrologic models as their basis. Because uptake and exchange of reactive solutes occurs mainly at substrate surfaces, the extent of transient storage and its importance in the trapping and utilization of solutes will prove to be of considerable interest.

13.3.2 The nutrient spiraling concept

In any ecosystem, nutrient cycling describes the passage of an atom or element from a phase where it exists as dissolved available nutrient, through its incorporation into living tissue and passage through perhaps several links in the food chain, to its eventual release by excretion and decomposition and re-entry into the pool of dissolved available nutrients. In most ecosystems one thinks of nutrients as cycling largely in place, with minimal transport. In running waters, however, transport must be incorporated into our conceptual framework. An atom or element occurring in the water column as dissolved available nutrient is transported some distance as a solute, then becomes incorporated into the biota, and eventually is returned to the water column in dissolved form. Since the cycle involves downstream transport, it is best described as a spiral (Webster and Patten, 1979).

Application of the spiraling concept requires a method to quantify the distance traveled by an atom in completing a cycle. Because the uptake of dissolved available nutrient from the water column is easier to quantify than its subsequent release, most studies measure uptake rate and distance. A theoretical framework for nutrient spiraling (Newbold *et al.* 1981, 1982) provides a useful measure of spiraling distance and thus allows investigation of the factors that influence this quantity.

In its simplest form, the model considers spiraling length (S, in meters) to be the sum of transport in two compartments: the water column, where the element occurs as inorganic solute; and the biota, where the element occurs in particulate form (Figure 13.5). Thus,

$$S = S_W + S_B \qquad (13.3)$$

S_W is a measure of distance traveled by an atom as inorganic solute, from when it becomes available in the water column until its uptake and incorporation into the biota. S_B is a measure of distance traveled by the atom within the biota until its eventual release back into the water column, thus completing one spiral passage downstream. The biota generally are associated with the streambed, as attached microorganisms, periphyton and benthic invertebrates. Typically an atom will travel the greatest distance in the water column, and so one expects S_W to be greater than S_B. Low values (short distances) for S_W reflect greater demand for the nutrient in question and more frequent cycling of each atom over a given length of stream.

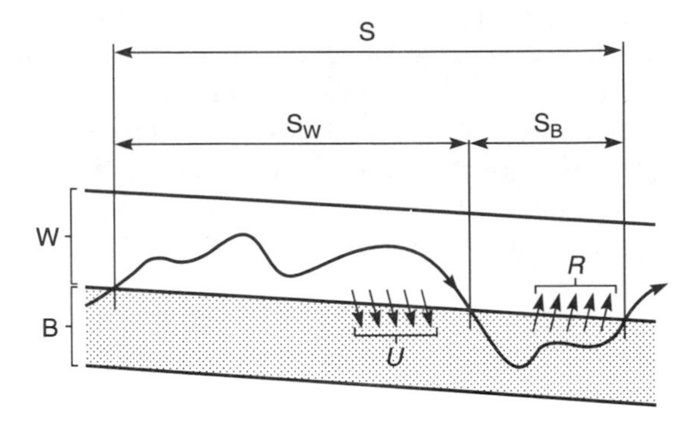

FIGURE 13.5 Two-compartment nutrient spiraling model. The spiraling length S is the average distance a nutrient atom, such as phosphorus, travels downstream during one cycle. A cycle begins with the availability of the nutrient atom in the water column, includes its distance of transport in the water (S_w) until its uptake (U) and assimilation by the biota, and whatever additional distance the atom travels downstream within the biota (S_B) until that atom is eventually re-mineralized and released. (Modified from Newbold, 1992)

The model can be made more complex by subdividing the aggregated variable S_B. For example, Newbold *et al.* (1981) separated the biota into a particulate compartment S_P (microbes and attached algae) and the consumers S_C (mainly macroinvertebrates). In another study, Newbold *et al.* (1983a) constructed a 12 compartment model that included the water column, CPOM, benthic and suspended FPOM, *Aufwuchs* and various consumer compartments. To obtain estimates of these terms in a stream is complicated, and the papers of Newbold *et al.* (1981, 1983) should be consulted for specific details. Study of S_B poses a number of challenges, requires laboratory measurements of flux through the particulate compartment versus the water column, and is greatly facilitated by the use of radioisotopes.

Unlike S_B, S_W is straightforward to estimate, and this ease of measurement is the appeal of the method. While a full modeling of nutrient spiraling can become quite complex, S_W can be determined from plateau measurements of solute concentration at successive points downstream from its release (Figure 13.4; Stream Solute Workshop, 1990). Certain conditions must be met and are evidenced by the simultaneous arrival of a definite plateau of both conservative and reactive solute. Plateau concentrations of the reactive solute, corrected for dilution by dividing through by the conservative tracer, will form a straight line plot against distance on a logarithmic scale. The slope is $1/S_W$.

Newbold *et al.* (1981) tested the spiraling model in a small woodland stream, Walker Branch, Tennessee, with an experimental release of $^{32}P\text{-}PO_4$ along with tritiated water as a conservative tracer. Phosphorus concentration declined exponentially, indicating that uptake of ^{32}P was proportional to the concentration remaining in the water column. Spiraling length in the water column then was estimated as the reciprocal of the exponent of the decline in water column concentration, and a value of 167 m was obtained. From laboratory studies, Newbold *et al.* derived a value of 26 m for S_B. Note that a small value of S_B relative to S_W does not imply that the role of the biota is negligible, but only that downstream movement of the biota does not significantly limit nutrient availability.

Although estimates of S_B and S_W do not of themselves reveal underlying processes, they are useful in evaluating how various factors influence nutrient spiraling. Nutrient uptake by autotrophs and microbes will reduce S_W by incorporating nutrients into benthic biomass. Consumers can stimulate nutrient uptake and release in several ways (Fenchel and Harrison, 1976). By cropping algal and microbial populations and associated organic microlayers, grazers should prevent self-limitation and senescence of *Aufwuchs*, thereby increasing nutrient demand. Animal exudates can stimulate growth of periphyton and microbes. Consumers also may increase spiraling distance, either by diminishing benthic populations responsible for uptake, or by fragmenting large into small particles, which then are more likely to be transported downstream (chapter 5). Environmental variables also will strongly influence spiraling distance, especially via fluctuations in discharge regime. S_W is likely to increase at high flows because of greater water velocity, and S_B may increase because of particle erosion. We shall first examine the abiotic factors and then turn to the biological processes that govern the spiraling of nutrients in lotic ecosystems.

13.3.3 Abiotic controls of nutrient spiraling

(a) Physical–chemical processes

Precipitation and sorption onto sediments are physical–chemical processes that influence the concentrations of some ions. These abiotic transformations can have a strong influence upon phosphate and a lesser influence upon ammonium-N, whereas nitrate-N apparently is little affected by physical–chemical removal. These differences are important to keep in mind when comparing estimates of S_W for phosphate *versus* various forms of nitrogen, since it may be misleading to attribute shorter spiraling distances for phosphate to biotic uptake alone.

Precipitation of phosphorus as iron and aluminum phosphates is important only at relatively high levels of dissolved P (about 25–100 $\mu g\,l^{-1}$, Stumm and Morgan, 1981) and hence is unlikely to be substantial in relatively unpolluted streams. At low levels of dissolved P, sorption of organic and inorganic phosphorus compounds by sediments occurs rapidly, particularly in fine-grained sediments, and may effectively regulate streamwater concentrations of dissolved phosphorus (Meyer, 1979). In Bear Brook, New Hampshire, Meyer concluded that microbial uptake played a relatively minor role in removing DIP from the water column, and that adsorption largely accounted for its disappearance following experimental enrichment. In a study of nitrogen dynamics in Bear Brook, Richey, McDowell and Likens (1985) concluded that adsorption of ammonium to sediments was a significant store in summer and autumn. Concentrations of silica are generally under strong physical–chemical control as well, and in unpolluted waters range between 8 and 12 $mg\,l^{-1}$ due to the solubility of SiO_2 and adsorption–desorption reactions (chapter 2).

(b) Hydrologic influences

Variation in discharge on both seasonal and annual time scales strongly influences whether nutrients are stored or exported. The importance of high discharge periods is evident from estimates of inputs, outputs and storage of nutrients in stream sections. From a phosphorus budget for Bear Brook, New Hampshire, Meyer and Likens (1979) estimated that 48% of annual inputs and 67% of exports occurred during 10 days of the water year. In contrast, phosphorus

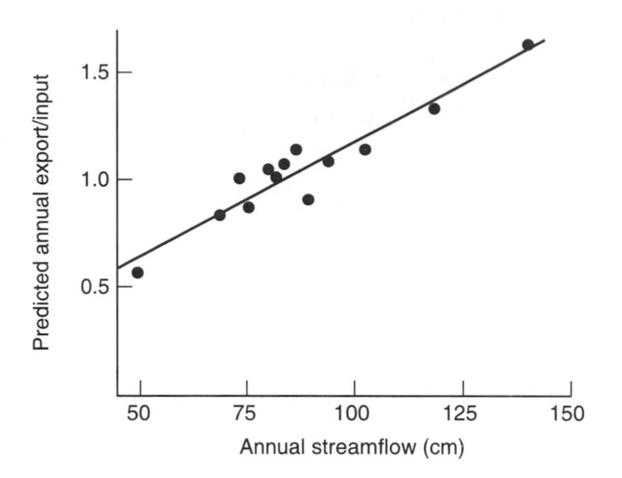

FIGURE 13.6 The ratio of annual exports to imports of phosphorus in Bear Brook, New Hampshire, calculated from a phosphorus mass balance for a single year and hydrologic data for a 13 year period. Although the study section experienced net accrual in some years and net loss in others, over the entire period exports and imports were essentially equal. (From Meyer and Likens, 1979.)

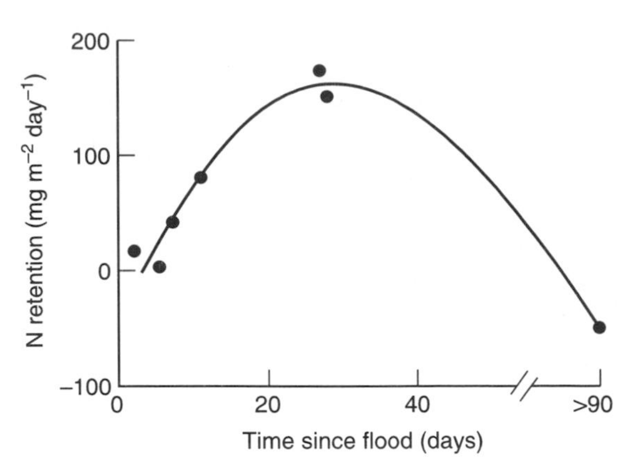

FIGURE 13.7 Retention rate of inorganic nitrogen determined from seven diel nitrogen budgets as a function of time since a flood scoured the biological community. Biomass of the algal mat increased rapidly at first, then leveled off after 20 days. (From Grimm, 1987.)

accumulated for 319 days of the year of study, and this storage presumably increased the opportunity for instream processing. Nutrients accumulate during low flows because they occur mainly in association with fine particulates, whose transport is dependent on discharge. Indeed, the dichotomy between high-flow and low-flow periods is so pronounced that one can speak of a stream as being in 'throughput' mode or 'processing mode', respectively.

Differences in streamflow between years also have a substantial effect on nutrient dynamics. In years of overall low flow, Meyer and Likens estimated exports of phosphorus to be less than imports, whereas the reverse occurred in high-flow years (Figure 13.6). Over the 13 year record available, however, total inputs and exports roughly balanced.

Hydrologic regime also influences nutrient uptake by its effect on the standing stock biomass and productivity of the biological community, and thus the amount of material stored in biological tissue. The amount of nitrogen retained in the biological community of a desert stream increases in proportion to biomass accrual (Figure 13.7). As described earlier, flash floods in Sycamore Creek 'reset' the biomass essentially to zero, and community development is highly dependent on time since the last flood (Fisher *et al.*, 1982). Similarly, storms can influence nutrient uptake by affecting the standing stock of leaf litter and organic matter. In a study of seasonal variation in S_W in Walker Branch (Mulholland *et al.*, 1985b), the shortest distance for S_W was obtained in the autumn. Because fallen leaf material usually is still abundant into early winter, one might expect short uptake distances in January as well. However, a storm some two weeks prior to the January tracer release depleted standing stocks of CPOM, thus eliminating the organic matter substrate that was the presumed site of nutrient uptake.

As was true for organic matter, hydrology interacts with instream retention devices and geomorphological features of river channels to determine the relative importance of transport and storage to nutrient dynamics. Retention and uptake are favored by low flow, a high ratio of streambed area to channel volume, retention

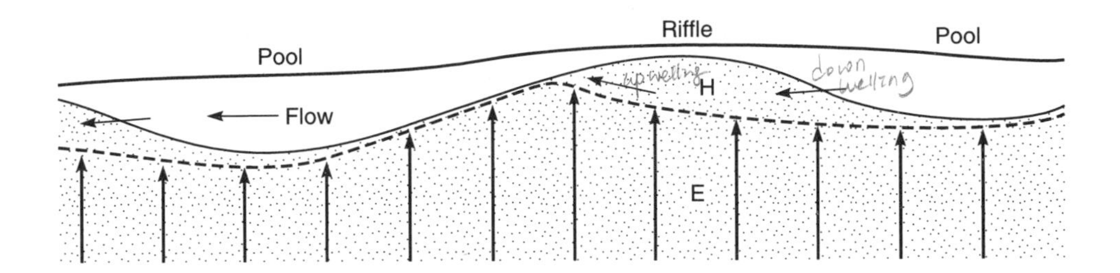

FIGURE 13.8 Postulated distribution of hyporheal zones (H) and groundwater zones (E) beneath a pool–riffle–pool sequence in a Michigan river, as inferred from temperature profiles. (From White, Elzinga and Hendricks, 1987.)

devices such as debris dams and beaver ponds, and permeable substrates that allow substantial interstitial flow. Throughflow of inputs and export of stored materials are favored by the opposite conditions. As a consequence, we expect decreasing retentiveness along a continuum from small stream to large river. Estimates of spiraling distance as a function of stream order appear to support this expectation (Minshall *et al.*, 1983; Naiman *et al.*, 1987), although the method used to estimate S_W in these studies did not adequately correct for the simple effect of physical dimensions and greater average mean velocities. Moreover, large rivers are likely to be affected by additional processes, particularly their association with side channels and floodplains, that have not received sufficient study to be incorporated into this scheme.

Sub-surface flow and storage zone mechanisms constitute a second important aspect of hydrologic influence over nutrient dynamics. Because abiotic and biotic retention mechanisms operate principally at or within the streambed, especially in smaller rivers and streams, hydrologic processes that favor exchange between surface and interstitial waters will enhance the retention and recycling of nutrients. Greater contact of solutes with the streambed provides greater opportunity for sorption to occur. Most importantly from an ecosystem perspective, periphyton, microbes and their consumers typically occur in intimate association with the substrate, and only in large rivers are biological transformations in the water column likely to be significant (chapter 5). Thus the opportunities for biological uptake and recycling of nutrients will be strongly influenced by hydrologic processes that increase exchange of water between the stream channel and streambed.

Depending on the proximity of the water table to the streambed–streamwater interface, both upward percolation of groundwater into the stream channel (discharge), and downward movement of streamwater into the water table (recharge), may occur (Figure 1.3). The amount of sub-surface flow is likely to vary from place to place along the streambed, depending on such variables as discharge, gradient, bed permeability and longitudinal bed surface profile (Vaux, 1968; White, Elzinga and Hendricks, 1987). In general, downwelling of streamwater should occur where the longitudinal bed profile is convex or streambed elevation increases, for instance where the downstream end of a pool empties into the head of a riffle. The opposite conditions exist at the downstream end of riffles and favor upwelling.

The extent of sub-surface flow is measureable in various ways. White, Elzinga and Hendricks (1987) exploited the considerable temperature difference between lake water (27°C) and groundwater (8°C) inputs to a small stream in northern Michigan to map downwelling and upwelling of streamwater between channel and substratum. Cool water was much nearer to the streambed–streamwater interface at the downstream end of riffles, while the influence of warm streamwater was detectable to depths of 50 cm at the head of riffles (Figure 13.8). Collections of invertebrates from deep in the streambed (Hynes *et al.*, 1976; Danielpol, 1989) also point to considerable interchange between channel and interstitial water, since organic matter and oxygen must be continually replenished to support the hyporheic fauna. Grimm and Fisher (1984) found rates of community metabolism at 30 cm depth in sandy sediments of Sycamore Creek to be comparable to surface values, which implies adequate sub-surface flow. Likewise, in an Appalachian headwater stream with up to 10 cm of gravel sediments over bedrock, chloride tracer dispersed rapidly, especially in coarser sediments, reaching the deepest sediments within minutes and equilibrating within hours (Munn and Meyer, 1988). As bacterial densities and benthic organic matter content were similar at 2 and 8 cm, high exchange between underflow and channel appears to be reflected in biotic functioning. Clearly, underflow and sediment characteristics acting together may potentially exert strong influence over solute retention and thus over transport distance and spiraling length.

13.3.4 Biotic controls of nutrient spiraling

In addition to the abiotic influences just described, a number of biological processes influence the ability of stream ecosystems to utilize, retain and recycle particulate and dissolved nutrients. Rapid uptake and short transport distance are expected under strong nutrient limitation, whereas at the other extreme all solutes behave conservatively and simply are in transit downstream. Thus the 'tightness' of the spiral should serve as an index of the stream's ability to retain and use nutrients.

(a) Uptake and assimilation

The biological removal of nutrients from streamwater is accomplished primarily by autotrophs and microbes. Wherever their biomass is large and populations are metabolically active, removal rates will be highest. The following examples illustrate this for periphyton, heterotrophic microbes associated with detritus and macrophytes.

The influence of the periphyton community over nutrient uptake is nicely illustrated by Grimm's (1987) study of successional events in a desert stream following a flood that eliminated virtually all of the biota. Biomass first accumulated rapidly and then more slowly as the system acquired a thick periphyton mat and high densities of invertebrates. Retention of inorganic nitrogen increased similarly, due to nitrogen storage in living tissue. As community biomass approached steady state the system no longer stored nitrogen, and system outputs became equal to system inputs (Figure 13.7).

A similar effect can be caused by microbial populations at the time of maximum leaf litter availability. Mulholland *et al.* (1985b) compared uptake rates from four 1 h releases of ^{32}P, spaced over the seasons of the year. The shortest distance for S_W occurred in the autumn, indicating that abundant CPOM as an organic carbon substrate favors rapid uptake (Figure 13.9).

Aquatic macrophytes are capable of removing substantial amounts of nutrients from flowing water. Meyer (1979) recorded significant removal of phosphorus as a pulse passed over a bryophyte bed in a forested stream. One might expect the highly productive chalk streams of southern England, because of their dense macrophyte stands, to remove great quantities of nutrients. In fact they do (Westlake, 1968; Ladle and Casey, 1971; Casey and Newton, 1972), but comparisons of removal to throughflow indicate that temporary storage in plant biomass is not very large. Westlake estimated that macrophyte production would require only 2% of the throughflow of nitrate-N. At another site, growth of *Ranunculus* during April and May was estimated to remove 94 kg of nitrate-N, compared with

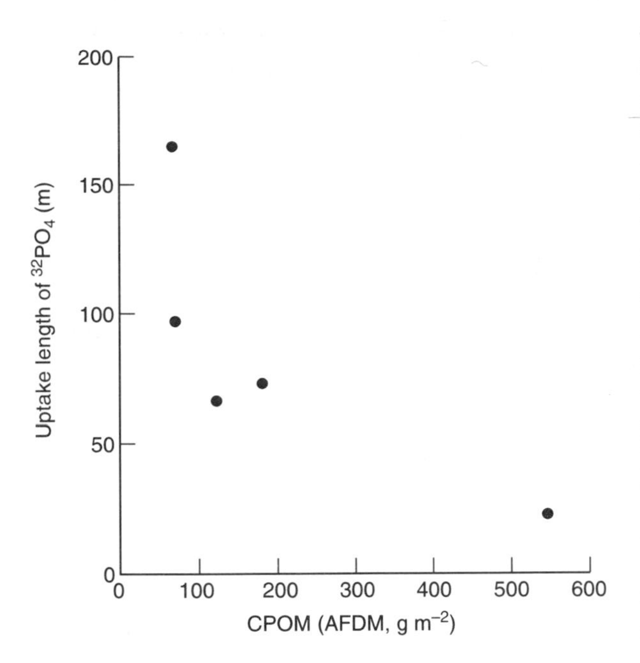

FIGURE 13.9 Uptake length of $^{32}PO_4$ compared with the amount of benthic CPOM from studies conducted in different seasons in a small woodland stream. (From Mulholland *et al.*, 1985b.)

a throughflow of 47 420 kg. This should effectively make the point that significant uptake from the standpoint of the biotic compartment may not be equivalent to significant removal from the standpoint of the stream-water.

(b) Influence of the animal community

The animal community influences nutrient cycling in many ways. Direct consumption of algal and microbial populations can either reduce or stimulate productivity and nutrient uptake. Animals also affect nutrient dynamics by returning nutrients to the water by excretion and egestion, by changing particle size, by storage of nutrients in consumer biomass, and by their movements.

Direct consumption of periphyton and microbial populations reduces standing stocks and presumably reduces uptake rates although, in theory, modest grazing pressure may stimulate productivity (Figure 8.6) and thus uptake rates as well. However, the limited experimental evidence now available indicates that consumers reduce uptake rates and lengthen spiraling distance. Mulholland *et al.* (1983, 1985a) conducted experiments in artificial channels to determine the effect of grazing and shredding activity on spiraling length, measured with ^{32}P. The snail, *Elimia clavaeformes*, consumed both periphyton and autumn-shed leaves, and so was used as the consumer in each study. Although weight-specific metabolic rates of microbial populations were enhanced by consumer activity, as expected, periphyton and microbial biomass were so reduced that overall biotic uptake of phosphorus also was reduced. Thus, spiraling distance was shortest in the absence of snails in both studies.

Consumers can also enhance the rate of regeneration of nutrients. Excreted and egested materials are most likely to contribute significantly to nutrient dynamics in highly productive systems where nutrients are scarce. Sycamore Creek in the Sonoran Desert of Arizona is strongly nitrogen limited (Grimm and Fisher, 1986) yet sustains very high rates of secondary production, which probably requires reingestion of feces (Fisher and Gray, 1983). From laboratory estimates of mass-specific excretion and egestion rates, Grimm (1988) estimated that up to a third of ingested nitrogen was converted to NH_3, a form of DIN readily used by autotrophs. A larger fraction (42–64%) was egested as fecal material. Presumably this is recycled via reingestion by consumers, or following leaching or microbial breakdown. Grimm was unable to determine exactly the amount of recycled N, which was estimated to fall between 15 and 70% of whole stream nitrogen retention. However, even the lower value implies a significant role for animals in nutrient regeneration in this highly productive, N-limited system.

Movements and migrations by animal populations can result either in inputs or outputs. Emergence of the adult stages of aquatic insects is one such process, but various authors agree that it is a small fraction overall (less than 1%;

Meyer, Likens and Sloane, 1981; Naiman and Melillo, 1984; Triska *et al.*, 1984; Grimm, 1987). In contrast, spawning runs of fish may import substantial amounts of nutrients to streams and lakes by excretion, release of gametes, and their own mortality, especially if many or all die after reproducing. Richey, Perkins and Goldman (1975) reported a phosphate peak following the die-off of Kokanee salmon (*Onchorhynchus nerka*) in a small tributary of Lake Tahoe, California. Periphyton visibly increased, as did carbon fixation by water column phytoplankton and heterotrophic activity in the benthos. The alewife, a member of the herring family, spawns in streams and ponds along the Atlantic coast of North America in sufficient numbers to demonstrably stimulate pond phytoplankton blooms and enhance rates of leaf decomposition (Durbin, Nixon and Oviatt, 1979). In Sashin Creek, Alaska, isotope analysis showed that nitrogen and carbon derived from a spawning run of Pacific salmon were incorporated into periphyton, macroinvertebrates and fish (Kline *et al.*, 1990).

Where death by reproductive exhaustion is not the natural fate of spawners, the death toll due to predators and their subsequent excretion of nutrients may accomplish the same end. In some tropical waters, nutrients are so scarce that virtually all are tied up in living tissue, just as is true of the soils and vegetation of tropical forests. Under these circumstances, it appears that egestion and excretion by top consumers recycle the nutrients that are critical to sustained productivity. Fittkau (1970) speculates that caiman played this role in the nutrient-poor Amazonian waterways. At one time caiman were so abundant that the naturalist Bates (1863) could write: "It is scarcely exaggerating to say that the waters of the Solimões are as well stocked with large alligators in the dry season as a ditch in England is in summer with tadpoles." Fish harvests declined following the near-extirpation of caiman by hunting. Fittkau suggests that the once-productive side channels of the Amazon River depended on the consumption of fishes by caiman and other top carnivores, whose excretion replenished the nutrients needed for continued productivity.

(c) Other influences of the biota

Assimilation and regeneration are central processes in the cycling of all nutrients, but other transformations also may play a role. This is particularly true for the nitrogen cycle, because of the importance of bacteria to the various transformations of N.

Nitrogen fixation by cyanobacteria and heterotrophic microbes is favored by calm water, high temperatures, moderately high DON and low DIN. These conditions are more typical of standing than flowing waters. However, under some conditions N fixation by cyanobacteria in streams appears to be important. *Nostoc* occurs in stony streams and can make a small local contribution to a system's nitrogen income (Home and Carmiggelt, 1975). Preliminary estimates from desert streams by Grimm, Fisher and Petrone (1994) indicate that nitrogen fixation by cyanobacteria in late-successional periphyton assemblages is substantial. Nitrogen fixation also has been estimated for microbes associated with woody debris and sediments. Estimated rates are low, although the contribution to budgets may be substantial in locations with abundant organic sediments, such as in beaver ponds (Naiman and Melillo, 1984).

Denitrification of nitrate-N to nitrogen gas by bacterial action is a possible pathway of nitrate loss from running waters. It is most likely to occur in sites rich in organic matter and under anaerobic conditions. Debris dams and pools in high-gradient streams, backwaters and side channels of larger rivers, beaver ponds, and other areas of organic matter deposition and fine sediments thus are likely sites for denitrification. This expectation was borne out by a survey of denitrifying activity in the River Dorn, which drains an agricultural area near Oxford, England. Highest activity was found in accumulations of fine-grained sediments at meander bends, and lowest activity in cores of mixed sand–gravel (Cooke and White, 1987). By their rough estimate, denitrification reduced the nitrate load of this stream by 15% under summer baseflow conditions.

Other situations where significant denitrifying activity has been demonstrated include within senescent mats of *Cladophora*, which are rich in organic matter and low in oxygen (Triska and Oremland, 1981); in sediment-rich streams in agricultural watersheds (Kaushik and Robinson, 1976; Sain *et al.*, 1977), and in forested watersheds after clear-cutting caused substantial sedimentation (Swank and Caskey, 1982). The latter study established a positive correlation of denitrifying activity with organic nitrogen and organic matter content. Denitrification also has been shown to vary with sediment type, being greatest in silt-rich deposits and less where the sand and gravel content is high (Hill, 1979). Denitrification was negligible in Bear Brook, New Hampshire, evidently because sediments were shallow and contained relatively little organic matter (Richey, McDowell and Likens, 1985).

13.4 Summary

Nutrients are inorganic materials necessary for life, the supply of which is potentially limiting to biological activity within stream ecosystems. Their uptake, transformation and release are influenced by a number of abiotic and biotic processes. Important metabolic processes likely to affect and be affected by the supply of nutrients include primary production and the microbial decomposition of organic matter. Both phosphorus and nitrogen have been shown to stimulate primary production, and silica likely is important to diatoms because of the high silica content of their cell walls. In addition to these major nutrients, many other chemical constituents of freshwater are potentially important as well.

DIP and several forms of DIN are present in river and stream water in a wide range of concentrations. Levels are very low in unpolluted waters, but are greatly elevated in many areas including most large temperate rivers due to human inputs of agricultural fertilizers, sewage, and industrial pollution. Nitrogen and phosphorus concentrations in rivers often greatly exceed the levels that cause eutrophication in standing water. At ratios of N:P above 16, nitrogen is expected to be in surplus and phosphorus to be limiting. This seems to be true in most streams, although some are N-limited and in some instances additions of both nutrients stimulate growth.

Nutrient cycling describes the passage of an atom or element from dissolved inorganic nutrient through its incorporation into living tissue to its eventual re-mineralization by excretion or decomposition. In most ecosystems nutrients largely cycle in place, but in lotic ecosystems downstream flow stretches cycles into spirals. The uptake length S_W, or distance traveled by an atom as inorganic solute until its uptake by the biota, is the simplest measure of nutrient spiraling. Total spiral length equals S_W plus S_B, where the latter term refers to distance traveled by the atom within the biotic compartment until its eventual release back into the water column.

A number of abiotic and biotic processes influence nutrient spiraling. Some uptake, especially of phosphorus, is by physical–chemical sorption of sediments. High flows reduce the opportunity for biological uptake and increase downstream transport. Low flows, stream channel retentiveness, and interchange between sub-surface and surface flows increase opportunities for uptake, thus shortening S_W. In addition to direct uptake by autotrophs and microbes, the biological community affects nutrient dynamics through consumption and egestion, and by a number of microbial transformations. Although rivers and streams unquestionably transport large quantities of materials downstream, biologically reactive elements cycle repeatedly during their passage, governed by a complex and interdependent set of biological, hydrological and physical–chemical processes.

Chapter fourteen

Modification of running waters by humankind

As Benjamin Franklin noted in 1772, 'Rivers are ungovernable things, especially in hilly countries. Canals are quiet and very manageable.' Perhaps no other ecosystems have been as significantly modified by human activity as have rivers and streams. From the dawn of agriculture we may presume that landscape changes due to farming, grazing and deforestation have influenced watershed characteristics directly and through effects on climate. Attempts to control the flow of rivers likewise date far back in time, and over human history there has been a continuous increase in the variety of ways and intensity with which humankind has modified the physical, chemical and biological nature of running waters. With our growing understanding of the tight coupling of the stream and its valley (Hynes, 1975), we now must recognize that our knowledge of pristine rivers is sparse, and most studies are of systems subject to varying degrees of modification from their ancestral state.

14.1 Brief history of river modification

Although the first efforts to tame the flow of water for human benefit left no enduring physical record, various accounts document that great waterworks accompanied great civilizations. Large-scale river regulation was underway by 5000 years ago along the Nile, the Tigris–Euphrates, and Indus rivers. The earliest known irrigation ditches, in Egypt, date to 3200 BC and the first dam at Sadd el Kafara to 2759 BC (Petts, 1989). When the Greek historian Herodotus visited Egypt in the fifth century BC, he reported the Lake of Moeris, created and fed by canals, to have a circumference of over 650 km with two pyramids rising in its center. The Code of Hammurabi, named for the King of Babylon nearly 4000 years ago, included warnings against opening sluice gates such that neighbors' fields would be flooded.

The full effect of ancient river regulation is nearly impossible to specify, as the earliest dams and canals were vulnerable to the passage of time and vanished without a trace. Even the Erie canal exists today as a patchwork of isolated ditches, yet this engineering marvel of the early 1800s fell into disuse little more than 100 years ago (Payne, 1959). In Roman times, however, engineering of water flow produced structures the remnants of which can be marveled at today.

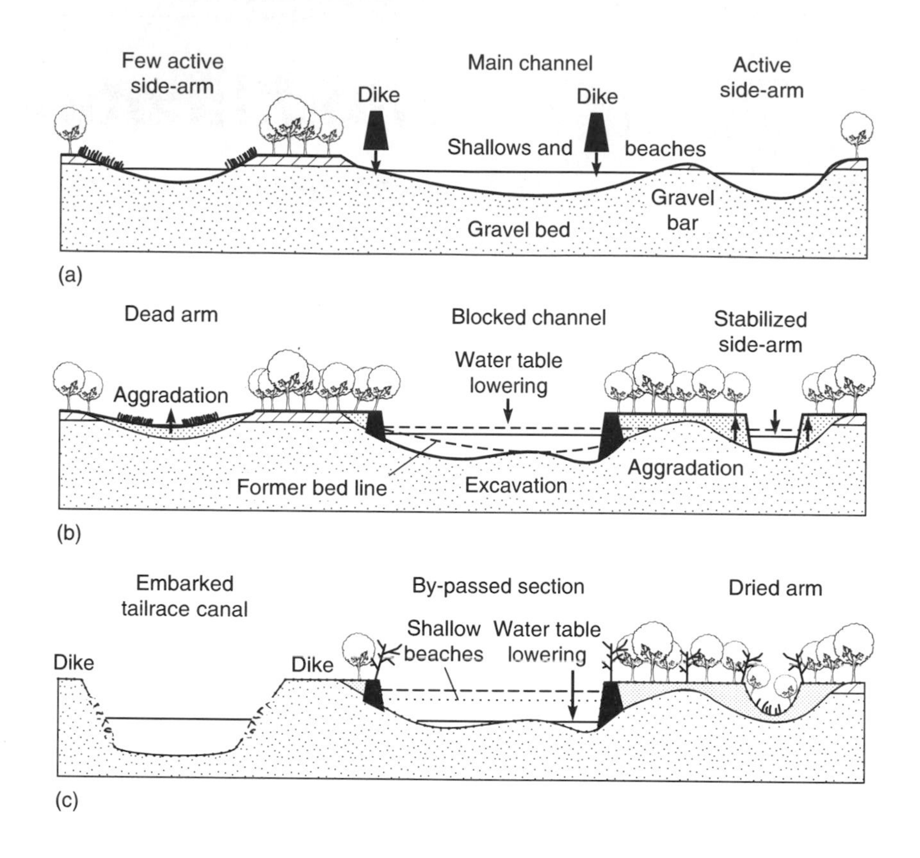

FIGURE 14.1 Changes in the lower River Rhône from 1870 to the present. (a) In 1870, the natural river was braided with a tendency to meander. Construction of levees beginning in the mid-nineteenth century and excavation confined the river to a single, deep channel, improving navigation and protecting riverside residents from flooding. (b) The channelized braided riverbed in 1971. Hydroelectric developments and channelization since the 1960s have largely dewatered the old main channel and furthered the drying of old side channels. (c) The regulated braided riverbed in 1980. The stippled area represents gravel, the unshaded area represents silt. (From Fruget, 1992.)

Great aqueducts are perhaps the most vivid reminder of this era. The aqueduct at Metz crossed the Moselle on a bridge over 1 km in length and over 30 m high. The Romans also constructed great drainage tunnels such as the Cloaca Maxima that carried away Rome's wastes in the 6th century BC, and exists today in modern form.

Not all areas of the world were subject to great water works, of course. Even in Europe the technology and scale of river modification was modest until perhaps 1750, when the industrial revolution and scientific advances set the stage for the beginning of the current era. Petts (1989) recognizes four phases to the modern era of river modification. Phase 1, from about 1750 to 1900, saw the implementation of ambitious regulation schemes for most of the large European rivers for navigation, flood control and utilization of floodplain land. The Tisza lost an estimated 12.5 million ha of floodplain marshland and some 340 km in length. Changes in the channel of the Rhône (Figure 14.1) are typical. The USA had a particularly compressed history of canal construction during this era, as little of substance was constructed before 1800, and the railroads largely outcompeted barge traffic within 60 or so years. In between, however, some 7000 km of artificial waterways were hand-dug, with only blasting powder as an aid.

Phase II, between 1900 and 1940, marks the development of the technology to build great dams and their first proliferation in North America, Europe and southeast Asia (Petts, 1989). Hoover Dam, the first great dam on the Colorado River, was completed in 1936. Today it still ranks fourteenth in height and twenty-third in volume among world dams. Phase III, from 1950 to 1980, marks the peak activity in dam-building world-wide. Such was the pace during the late 1970s that dams over 15 m in height were being completed in North America at the rate of over 200 per year, and world-wide at over 700 per year.

In phase IV, from 1980 to the present, the pace of dam-building has slowed to about 500 per year world-wide. The preferred sites have been utilized in many areas, forcing dam construction to move toward headwater streams. In the USA excluding Alaska, only 51 rivers over 100 km in length remain free-flowing from headwaters to major confluence (Stanford and Ward, 1979). Based on a Nationwide Rivers Inventory completed in 1982, Benke (1991) reports that only 42 high-quality, free-flowing rivers greater than 200 km in length remain in the 48 contiguous states. However, very large dams continue to be constructed in areas that previously were little affected, including great rivers of the far north, of South America, and in parts of Asia. Hydropower generation in Latin America and the Caribbean increased roughly fivefold between 1970 and 1990, a trend that is likely to continue. Hydropower provides over two-thirds of the total electrical power generation of this region, and it is estimated that only 14% of the technically usable potential has been tapped (Sanchez-Sierra, 1993). A number of rivers of high northern latitudes remain unaffected by dams at present, largely in Alaska, Canada and the Federal Republic of Russia (Table 14.1). Great development projects such as those in the James Bay region of Canada represent the next wave of regulation of the largest remaining free-flowing rivers in North America (Rosenberg and Bodaly, 1994). It is estimated that, by the year 2000, over 60% of the total streamflow in the world will be regulated (Petts, 1989).

Although dams are perhaps the most conspicuous human influence, river systems are subject to a wide array of additional threats. Water removal, ranging from irrigation canals and other local withdrawals to transbasin diversions, reduces river flow and alters instream habitat. Pollution from organic and industrial wastes has been somewhat reduced over the past 20 years, but poses a continuing threat to water quality in the heavily used rivers of North America and Europe and is a growing threat in countries undergoing development. Intensive land use affects streams by altering the landscape, through deforestation, the spread of agriculture, and growth of settlements. Moreover, changes that previously were of minor importance threaten to take on greater seriousness in the future. Water needs of the twenty-first century will only intensify the pressure for water withdrawal and interbasin transfer schemes (Gleick, 1993). The spread of alien species, which has been intensifying throughout the twentieth century, promises to alter species composition in many areas. Finally, global climate change in the twenty-first century, including not just airborne pollution from great distances, but altered temperature and rainfall regimes, is likely to change the physical state of many river systems. Clearly, the rivers of the world continue to be subject to diverse and changing forces that impact upon their structure and function. To manage and where possible restore and preserve the biological integrity of rivers effectively, we must explicitly address their function in the context of human influence.

In concluding this historical perspective, it is

TABLE 14.1 Number of large northern rivers (mean annual discharge > 350 $m^3 s^{-1}$) subject to differing degrees of human impoundment and diversions. Circumstances that designate rivers as strongly affected include regulation of the main channel as opposed to tributaries, of upstream rather than downstream sites, and a high ratio of total storage capacity relative to total annual discharge (From Dynesius and Nilsson, 1993)

	Unaffected	*Moderately affected*	*Strongly affected*	*Total*
River systems in Europe and the former Soviet Union				
Receiving sea				
Pacific Ocean	5	1	0	6
Arctic Ocean	13	3	3	19
Atlantic Ocean	0	3	10	13
Baltic Sea	1	2	10	13
Mediterranean	0	0	5	5
Black Sea	0	1	5	6
Interior basins (Caspian, Aral, Balkhash)	0	0	7	7
Total	19	10	40	69
River systems in Canada and USA				
Nation/drainage				
USA (except Alaska)	0	2	13	15
Canada/Atlantic Ocean	3	0	9	12
Canada/Ungava + Hudson Bay	16	1	7	24
Canada/Arctic Ocean + Alaska and British Columbia	15	3	0	18
Total	34	6	29	69

useful to place the timespan of human influence within the even longer history of natural change. The age of a watershed is approximately the age of the landform it drains, often late Cenozoic or more recent. This is old in comparison to all lakes but those of tectonic origin. However, fluvial geomorphology holds that a river constantly seeks a quasi-dynamic equilibrium with its grade, and maintains its floodplain by regularly overflowing its banks and by lateral shifts over the inundated region (chapter 1). Thus river systems are ever-changing as well. When one considers that a dozen or more advances and retreats of glaciation over a time span of hundreds of thousands of years have repeatedly lowered and raised the sea level, thus steepening and then reducing the gradient, a recent geological history of continuous change becomes apparent. Moreover, at glaciated latitudes, much of what we see of at least the lower order networks of rivers has been strongly influenced by events over the past tens of thousands of years. On a more recent timeframe corresponding to the rise of human culture, the vegetation of river valleys almost certainly has undergone marked change on the time scale of thousands of years, due to climate change in formerly glaciated areas and changes in rainfall at low as well as higher latitudes. The time frame of geological and climatic change thus intergrades with the time when human modification of rivers begins, and it may be misleading to imagine that river ecosystems have ever experienced much constancy.

14.2 Dams and impoundments

The damming of so many rivers, largely a phenomenon of the twentieth century, has occurred at such a furious pace that scientific and public concern for the effects of dams is fairly recent (e.g. Hynes, 1970; Baxter, 1977; Petts, 1984). The first symposium on regulated rivers (Ward and Stanford, 1979) marks a shift from studies focused mainly on reservoir limnology to studies of the effects of dams on the river itself. Subsequent symposia (Lillehammer and Saltveit, 1984; Craig and Kemper, 1987; Gore and Petts, 1989) attest to the growth of studies documenting the effects of what has come to be called river regulation. The numerous adverse effects of impoundments and other forms of river regulation are now well documented, but a better understanding is needed of cause-and-effect relationships and of best management alternatives to minimize negative impacts.

Dams vary widely in their size, purpose and operation, and these differences influence their impact on river ecosystems (Petts, 1984). Dams also differ in whether water is released from the surface of the dam, near the bottom, or both. Water supply impoundments require a large storage volume to meet projected needs and outlast droughts. Dams constructed for irrigation must store as much water as possible during the rainy season for release during the growing season. Flood control reservoirs maintain only a small permanent pool in order to maximize storage capacity, and draw down as soon as possible after a flood event to restore their capacity. Navigation requires water storage in upper reaches to offset seasonal low-flow conditions, and may be complemented by a system of locks and dams. Hydroelectric dams store water for release to meet regional energy demands, which can vary seasonally or over the course of 24 h. Hydroelectric dams can differ substantially in their operation. 'Run-of-the-river' dams release water at the rate it enters the reservoir, usually are of low height, and are thought to have relatively small adverse effects. 'Peaking' hydropower dams meet daily fluctuations in energy demand by allowing water to flow through turbines only at certain times, usually from mid-morning through early evening, and are considered to affect aquatic life seriously. Finally, reservoirs may also serve recreational purposes including fisheries, but typically this is a secondary function of a multiple purpose facility.

Dams have many negative effects on rivers. Unquestionably they cause fundamental changes in community structure and ecosystem function as a naturally free-flowing and continuous river course is transformed into river segments interrupted by impoundments. The human and economic benefits can be considerable, which is the argument for construction of dams. At least in the view of some, the aquatic system is improved by the recreational opportunities within the reservoir and in the regulated river section downstream of the dam.

14.2.1 Physical effects of dams

The effects of impoundments include a series of changes in the physical conditions downstream of the dam, especially modification of the flow and temperature regimes, and usually greater water clarity. Water quality changes may be slight or considerable, depending on water residence time in the reservoir and whether surface or deep water is released. The modified physical and chemical conditions result in changes in the plant and animal life of the river. In addition, reservoirs impede the downstream passage of young migratory fishes, while dams are a barrier to upstream migrations. More indirectly, dams, and especially a series of dams, break the upstream–downstream connectivity that is a natural feature of rivers. Which of these effects is most damaging is to a large degree situational, depending upon the location, type and operating procedures of the particular dam.

The presence of an impoundment obviously changes the discharge and current for some distance downstream, and a series of large dams will completely alter a river's natural periodicity of flow. In addition to this dampening of flow

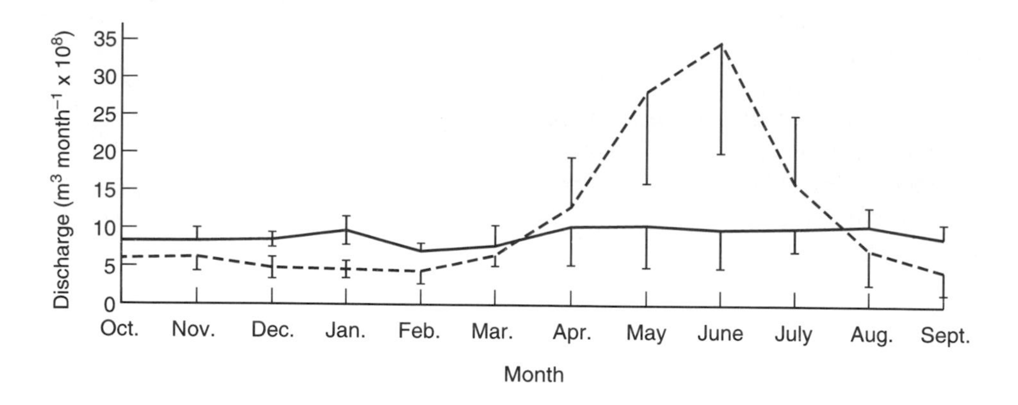

FIGURE 14.2 Monthly means and ranges of discharge in the Colorado River at Lees Ferry before (1944–1962; ----) and after (1963–1977; ——) impoundment of Lake Powell. (From Paulson and Baker, 1981.)

variability, transbasin diversions and evaporative losses can result in an overall reduction in discharge. In its natural state, the Colorado River at Lees Ferry experienced peak flows of 2400–5600 $m^3 s^{-1}$ in June, and very low flows in late summer (Stanford and Ward, 1986). As a consequence of the impoundment of Lake Powell in 1963 and the export of water via numerous transbasin diversions, flows now range between 130 and 764 $m^3 s^{-1}$ year-round, and seasonal extremes have been eliminated (Figure 14.2). In some circumstances dams virtually de-water riverbeds, as has occurred in the lower Gila River of the Colorado Basin (Stanford and Ward, 1986). Peaking hydropower results in extreme daily fluctuations in discharge, velocity and available habitat. In the Nelson River, part of the large Churchill–Nelson hydroelectric project in Manitoba, Canada, daily discharge variation over the period 1979–1988 was in the range of 2000 to 3000 $m^3 s^{-1}$, at a location where the natural average discharge was 2170 $m^3 s^{-1}$ (Figure 14.3; Rosenberg and Bodaly, 1994).

Transport of suspended particles and the amount of fine sediments on the streambed are affected by the presence of the reservoir and the river's altered flow regime (Figure 14.4). Inflowing sediments settle out of suspension under reduced current velocities within the reservoir, sometimes leading to dramatic loss of water storage capacity. The Cali Dam in Colombia is reported to have lost 80% of its storage within 12 years, despite expensive dredging operations (Barrow, 1987).

Dams that release very high discharges cause scouring of fine materials and armoring of the streambed, a process in which the surface substrate becomes tightly compacted. Because channel morphometry is flow-dependent (chapter 1), channel form is likely to adjust to the altered flow regime. Channel and bank erosion, and down-cutting of the streambed, are possible consequences of extremely high flows, and sediments accumulate whenever flushing flows are completely eliminated.

A river's temperature regime is altered to varying degrees by impoundments, strongly so in the case of large reservoirs with deep release dams located on temperate rivers. Because the large volume of a reservoir has considerable thermal inertia, diel temperature fluctuations

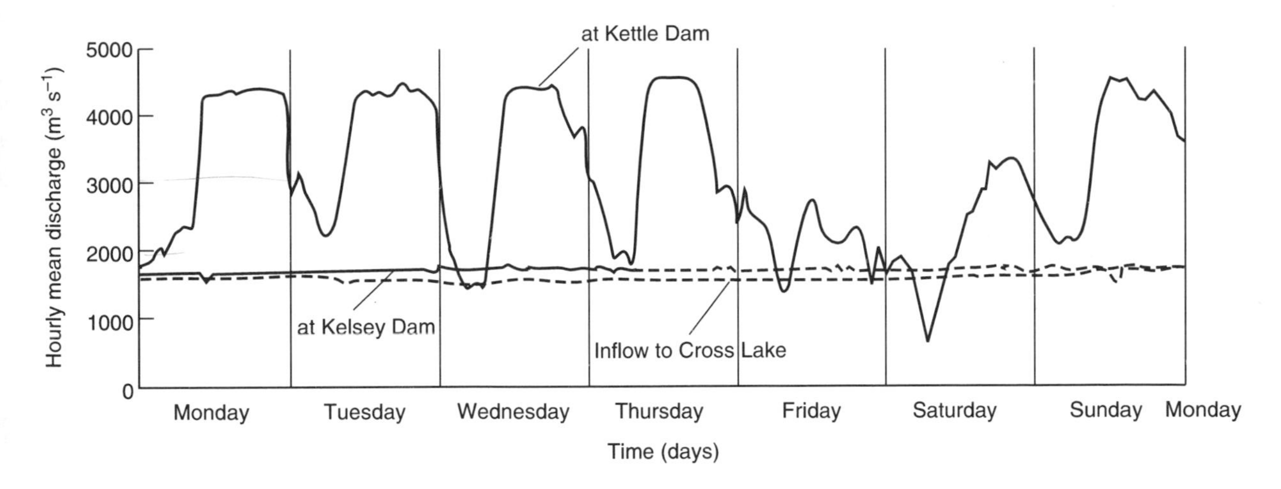

FIGURE 14.3 Hourly mean discharge for the Nelson River below Kettle Dam during one week in July, 1984. (From Environment Canada and Department of Fisheries and Oceans, 'Federal Ecological Monitoring Program', Volume One, 1992.)

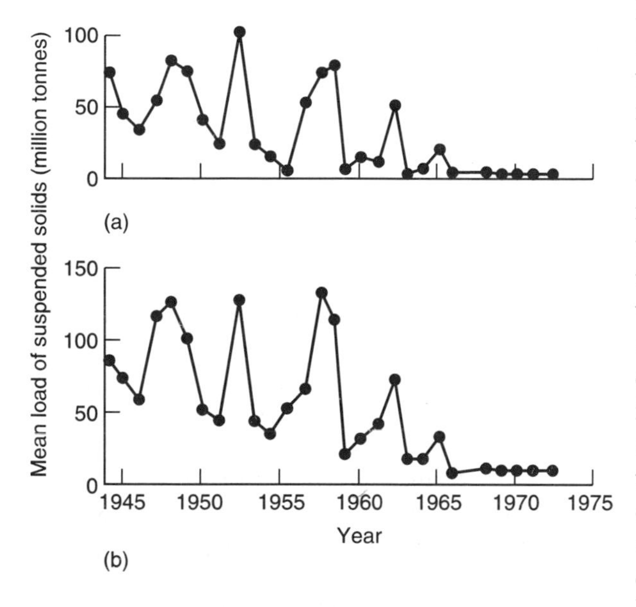

FIGURE 14.4 Mean load of suspended solids (a) during spring runoff and (b) annually in the Colorado River at Lees Ferry. (From Paulson and Baker, 1981.)

are reduced or eliminated. Deep release reservoirs also may reduce the extent of seasonal temperature variation. Provided the reservoir has sufficient depth and residence time, thermal stratification will occur during warm months, in which the epilimnion, or surface waters, are much warmer than the hypolimnion, or deep waters. Water released from the hypolimnion during the summer thus is much cooler than normal, and this has allowed the development of tailwater salmonid fisheries very popular with the fishing public. Water released during the winter is near 4°C, because that is the temperature at which water reaches its maximum density and thus the temperature of deep water in a lake or large reservoir. As a consequence, the seasonal pattern is for downstream river water to be warmer than normal during winter, cooler during summer, and of reduced seasonal amplitude overall (Figure 14.5). Total annual degree days may be quite similar to an unregulated river, but the seasonal pattern of degree day accumulation can change substantially (Munn and Brusven, 1991).

Downstream changes in water quality depend

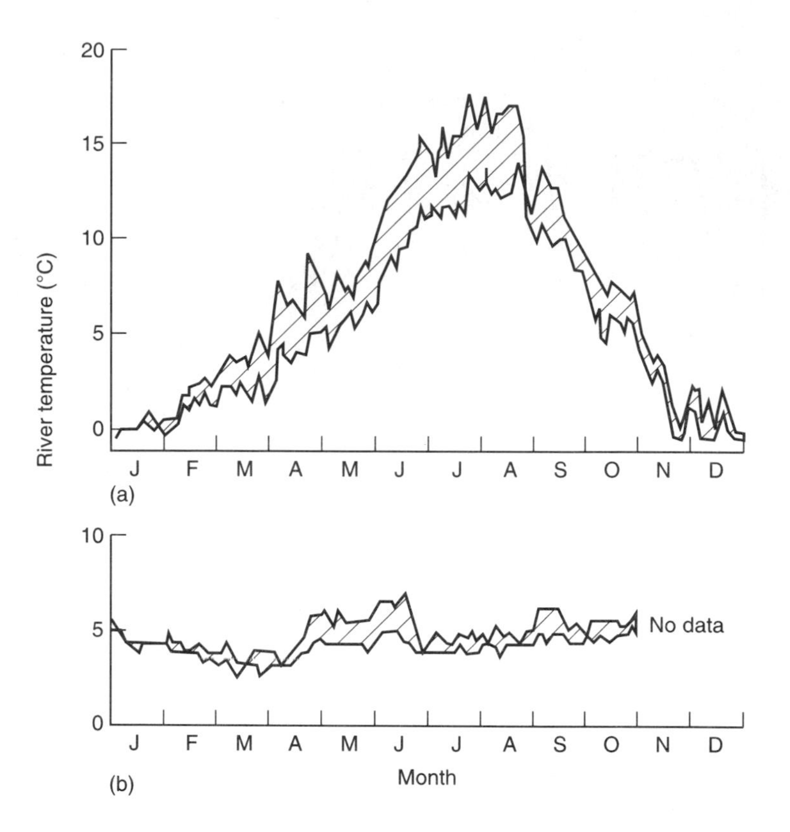

FIGURE 14.5 Thermal regimes of the (a) unregulated Middle Fork and (b) regulated South Fork of the Flathead River, Montana, during 1977. The shaded area approximates the daily temperature range (From Ward and Stanford, 1979b.)

on limnological processes within the reservoir and depth of water release. Deep release dams generally cause the most adverse effects. Water clarity typically increases because of the reduced sediment in transport, often with significant effects on plant life. Nutrient concentrations may be elevated, especially below deep release dams, and H_2S is occasionally released from reservoir sediments. By far the most serious change in water quality is due to the release of oxygen-depleted water from the hypolimnion of a deep reservoir. However, turbulence usually re-oxygenates the water within a short distance, and artificial aeration is a relatively simple solution.

14.2.2 Biological effects of dams

Biological changes downstream of impoundments are substantial and well documented (Ward and Stanford, 1979, 1987; Armitage, 1984; Petts, 1984). The type of dam and its mode of operation again are important determinants of the kind and magnitude of effects. In some circumstances dams cause permanent biological change, as when a species of migratory fish is eliminated. Often the effects of a dam cease to be evident at some distance downstream of the impoundment. This can be as short as a few kilometers, in the case of a dam on a tributary river just above its confluence with another tributary (Armitage, 1978) or the mainstem (Munn and Brusven, 1991), and as long as 80 km for a deep release dam on the mainstem of a river (Stanford and Ward, 1989).

Because of reduced and altered river flow, dams help to sever the river's historic connection with its floodplain, leading to reduced productivity in both habitats. Especially in the large rivers of the tropics, many of which still flood seasonally, freshwater fish production is strongly dependent on regular inundation of the floodplain (Junk, Bayley and Sparks, 1989). Riverine fisheries are an important resource, and are more productive than the reservoirs that supplant them (Petrere, 1989). In Brazil, freshwater fishes contributed 24% of the total commercial catch in 1984 (Petrere, 1989). Based on the argument that a high-amplitude flood of long duration results in the greatest fish yield, Bayley (1991) suggests restoration of the ancestral flood pulse in temperate rivers, a goal that might take a century or more to achieve.

Whenever dams cause enhanced water clarity and reduced variability of streamflow, there is usually a greater abundance of periphyton or higher plants than is found elsewhere in the river (Stanford and Ward, 1979). For example, a dense growth of the aquatic moss, *Fontinalis neomexicana*, developed in riffle habitats of a regulated reach of an Idaho river (Munn and Brusven, 1991), and this in turn appeared to be responsible for substantial changes in the benthic fauna. However, dams that release high enough flows to scour the streambed, and especially dams operated to meet peak power needs, are much more likely to eliminate plants along with much of the fauna. Finally, on large rivers with many dams, such as the Loire (Decámps, 1984), extensive phytoplankton blooms often develop as a consequence of the slowed downstream passage of water (chapter 4).

The benthic invertebrate community immediately below dams often shows a reduction in species richness but an increase in overall abundance of invertebrates. In the Gunnison River, Colorado, the greatest number of species of Trichoptera (Hauer, Stanford and Ward, 1989) and Plecoptera (Stanford and Ward, 1989) are found in unregulated stretches, and the lowest species richness in the tailwaters below deep release dams. A comparison of the Chironomidae between a regulated section of the River Tees and unregulated Maize Brook, in northern England, reported the lowest species richness from the site nearest the dam, which was dominated by a common and widely distributed species of *Orthocladius* (Armitage and Blackburn, 1990).

These changes in the invertebrate community are due to the altered physical and chemical environment below impoundments. An overall reduction in habitat heterogeneity likely accounts for a reduction in species diversity and a greater abundance of those species favored by the altered conditions. Below Cow Green Dam on the Tees, for example, reduced flow variability permits dense algal and moss cover, which in turn favors an abundance of orthoclad midges and oligochaetes of the genus *Nais* (Armitage, 1977). Silt deposition benefits larvae of the Tanytarsini, and the absence of extreme flows favors the snail, *Lymnaea peregra*. The reservoir provides a rich outflow of zooplankton, and this nourishes a large *Hydra* population in the Tees. This last observation is unusual, as it is more common to find net-spinning caddis larvae and suspension-feeding black fly larvae below dams that release surface water.

Often the faunal changes are fairly predictable from a knowledge of habitat requirements. From a survey of the effects of dams on mayflies, Brittain and Saltveit (1989) report that low flows usually cause a shift from lotic to lentic species such as *Cloeon, Paraleptophlebia*, and *Siphlonurus*. In contrast, high flows favor mayflies of torrential conditions, including *Baetis*, *Rhithrogena* and *Epeorus*.

Changes in temperature regime of the magnitude found below deep release dams (Figure 14.5) have a significant impact on the benthic fauna. A reduction in species richness is likely for several reasons. Warmer than normal winter temperatures eliminate the thermal cues needed by many species to break egg diapause (Figure 3.14). Cool summer temperature can also have an adverse effect, because there are too few degree days to complete development, life cycles lose their synchrony, or because of changes in the temporal pattern of growth and development. The mechanisms by which altered temperature influences invertebrates clearly warrants further study, but unquestionably the total effect can be profound. Only one family of insects, the Chironomidae, persisted in a section of the Saskatchewan River impacted by a deep release dam, compared with 30 families representing 12 orders in the unaltered river (Lehmkuhl, 1974).

The effect of dams on populations of migratory fishes is well known and of serious concern, not only because of the economic value of these fishes, but also for their contribution to regional biodiversity and their significance to native and regional cultures. Dams block the upstream passage of anadromous fishes such as salmon, American shad, and sea-run populations of trout; of catadromous fishes such as eels; and some fishes that are stream residents but migrate to spawn, including a number of popular sports fishes in North America and important food fishes in South America such as *Prochilodus* and some large pimelodid catfish (Lowe-McConnell, 1987). Young salmon migrating downstream to the sea are damaged by water pressures encountered in turbines and spillways. In addition, currents that once speeded their seaward journey are reduced within impoundments, weakening juvenile fishes which are now forced to expend significant energy in swimming.

The salmon of northwestern North America, and unquestionably many other species as well, are now recognized to consist of many sub-populations. Known as stocks, these sub-populations originate from specific watersheds and possess unique life history and/or migratory patterns that ensure their reproductive distinctiveness. In the Fraser River of British Columbia, for example, the sockeye salmon (*Oncorhynchus nerka*) once comprised about 40 separate stocks (Ricker, 1972). The existence of multiple stocks is important both to the total fish production and the genetic diversity of the species. Tragically, however, some 214 native stocks of Pacific salmon, steelhead, and sea-run cutthroat trout of California, Oregon, Idaho and Washington presently are threatened or endangered, and nearly half are at high risk of extinction (Nehlsen, Williams and Lichatowich, 1991). The number of extirpated populations is less certain, but of a similar magnitude.

The decline and extirpation of salmonid stocks has many causes. Habitat damage in headwater reaches required for spawning and recruitment, genetic introgression from hatchery-reared fish, and over-harvesting have all played a role. Nonetheless, the contribution of dams has been considerable. Dams that lack fish passage facilities have completely eliminated a number of upstream salmon populations (Nehlsen, Williams and Lichatowich, 1991). Iron Gate Dam constructed in 1916 on the Klamath River in northen California caused failure of chinook salmon (*O. tshawytscha*) stocks in a number of Oregon rivers. Completed as recently as 1974, the Dworshak Dam on the Clearwater River of Idaho totally blocked chinook and steelhead (sea-run rainbow trout, *O. mykiss*) from this tributary of the Snake River. One-third of the salmon and steelhead habitat of the Columbia River basin is believed to have been lost as a result of impassable dams.

The regulation of free-flowing rivers clearly brings about fundamental change in their structure and function. Unaltered lotic ecosystems form a continuous strand from headwaters to mouth (Figure 12.7), in which processes taking place upstream strongly influence downstream dynamics, and to some extent the reverse occurs as well (Vannote *et al.*, 1980). A substantial fraction of the energy base of streams is allochthonous (chapter 12), by both surface and subsurface input pathways, and this is just one of the ways that the surrounding valley determines much of what happens within the stream itself. Dams disrupt the natural connections, both longitudinal and lateral, that strongly influence lotic ecosystems.

14.2.3 Other aspects of river regulation

From small drainage ditches that return water to a river a short distance downstream, to interbasin diversions that connect historically distinct river systems, to massive plumbing schemes that alter drainage patterns of large regions, canals and water transfers probably have had an effect on rivers as great as that of dams. Indeed, diversion projects typically are combined with impoundments and dams, either to bring water to dams for power, or provide water to canals for irrigation and navigation.

As was true for dams, these other forms of river regulation have a long history. The ancient Egyptians constructed a navigation canal to bypass a waterfall on the Nile more than 4000 years ago. In Szechwan, China, a canal network dating to 250 BC diverts the spring floods of the Min River through over 1000 km of canals to irrigate some 200 000 ha of land (Postel, 1992). In countless rivers, removal of snags and deepening of the main channel has occurred to improve navigability for trade and transport. As Sedell and Froggatt (1984) document for the Willamette River in Oregon, the resulting loss of riverine channel structure has been great (Figure 14.6). The snags themselves provide important habitats for invertebrates, and in rivers with soft bottoms snags are major sites of secondary production (Table 3.5; Benke *et al.*, 1985).

The extent of canal construction and channelization is considerable. Within the USA, some 26 550 km of channelization work had been completed by 1977, and a further 16 090 km were proposed (Leopold, 1977). Surveys conducted in the UK and Denmark also document extensive channelization (Brookes, 1989). Quinn (1987) counts 54 interbasin diversions in Canada transferring mean annual flow (MAF) over 25 cubic feet per second (approximately $1\ m^3\ s^{-1}$) scattered across nine provinces. The total volume of diverted flow, if consolidated, would be the third largest Canadian river, after the St. Lawrence and McKenzie. Nearly all of this is attributable to hydro-development (Bocking, 1987).

Although diversion projects may be physically less spectacular than great dams, large diversions are great engineering accomplishments that profoundly alter the physical conditions of rivers and their basins. Probably no example illustrates this more forcefully than the Aral Sea of central Asia. Once the size of Ireland and the world's fourth largest freshwater lake, the Aral Sea has been steadily shrinking due to diversions for irrigation. Since 1960, the lake level has dropped 15 m, its surface area has shrunk by 40% and its volume by 60%, and the salinity level has tripled. The Aral's fish yield was 44 000 metric tons in the 1950s, but its fishery has since collapsed and all 24 native species of fish have disappeared. Salinization of the lake and desertification of catchment is underway, with serious consequences for the human and economic health of the region (Postel, 1992). The former Soviet Union had plans to reverse the flow of three north-flowing rivers (Ob, Irtysh and Yenisei) and send $120\ km^3$ of water annually some 2200 km to central Asia, partly to offset changes to the Aral Sea.

While plans to divert and re-direct river flow at the scale of subcontinents hopefully will not come to pass, one has only to look to the Colorado River and the waterscape of California to see that water diversion at what we might call an 'intermediate scale' is widespread. The sprawling metropolis of Los Angeles was made possible by completion in 1913 of an aqueduct system nearly 400 km in length, bringing snowmelt from the Sierra Nevada mountain range to the city of Los Angeles and ruin to the ranching communities of the mountain valleys. Today approximately 90% of water use in California is

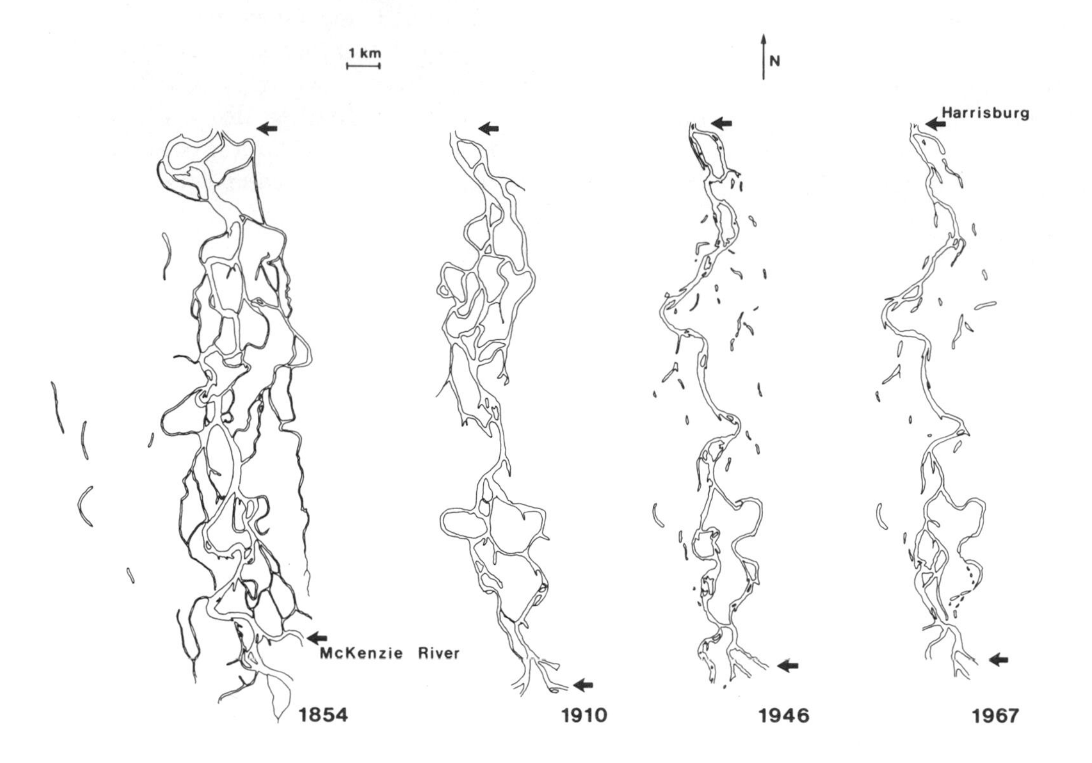

FIGURE 14.6 Historic change of the Willamette River, Oregon. Early settlers commented on the extensive floodplain 1.5–3 km in width, the thick underbrush and multiple shifting channels filled with snags. Snag removal for steamboat travel beginning in 1868, followed by expansion of agriculture on the old floodplain and construction of 11 major dams since 1946, have transformed the Willamette into a very different river than it was less than 150 years ago. The estimated length of shoreline in a 25 km stretch was 250 km in 1854, 120 km in 1910, 82 km in 1946, and 64 km in 1967. (From Sedell and Froggat, 1984.)

for irrigated agriculture, and roughly half of all water used in southern California comes from outside the state, in aqueducts and canals diverted from the Colorado River. The Imperial Valley, now the largest expanse of irrigated agriculture in the Western Hemisphere, was aptly named the Colorado Desert until it was developed with diversions from the Colorado River and a network of irrigation canals (Kahrl, 1979).

Irrigation at today's scale was impractical so long as the flow of the Colorado River varied with the seasons and spring floods were allowed to escape to the sea. Regulation of the Colorado's flow became imperative not just for hydropower, but to ensure a year-round water supply for diversion. Nine major storage reservoirs, anchored by Lake Mead in the Lower Basin and Lake Powell in the Upper Basin, have a collective storage capacity of roughly four times the long-term average virgin flow of $18.5 \times 10^9 m^3$ (15 million acre-feet). Practically no flow has entered the Gulf of California since 1961, and today the mighty Colorado ends in a pipe in Tiajuana, 225 km north of its natural terminus, according to an agreement that allots Mexico a final share of the river.

14.2.4 Biological effects of water diversions

The biological effects of water diversions are not as well documented as those of dams. Extensive diversion projects typically require impoundments, and their effects were described previously. Canal systems facilitate invasion by non-native species, surely one of their most serious consequences. And water transfers obviously must decrease the flow in some places and augment it in others, causing a host of physical and chemical changes in the affected river systems.

Studies of the fish fauna of the Phoenix Metropolitan Area clearly demonstrate the role of canal systems as aquatic habitat. Development of water resources has lowered groundwater to the extent that formerly perennial rivers now flow only during periods of flooding. As a consequence canal systems now comprise a significant portion of the flowing-water habitat at low elevation in Arizona, and may soon constitute the major lotic habitat in the state (Marsh and Minckley, 1982). The native fish fauna of the Salt River and its laterals originally comprised 15 species. By 1981, nine of these were sufficiently rare to warrant listing under the US Endangered Species Act. The fishes of the Phoenix area include just four of the original native species and 19 exotics. Although canals are now significant lotic environments in such areas, they are hardly ideal habitats. Their physical uniformity and routine dewatering for maintenance are probable causes of the loss of native species and the almost continual change in the composition of the exotic species.

The ready colonization of canals underscores their significance as a dispersal pathway for species. Transbasin diversions of large scope are especially likely to permit biotic interchange, a topic that is discussed more fully below.

The physical and chemical changes resulting from diversions constitute the other major class of adverse effects. Riverbeds may be entirely dewatered, as was mentioned earlier for the Gila River; but there also are examples of augmented flows. Prior to an interbasin transfer from the Orange River, completed in 1977, the Great Fish River of South Africa was characterized by irregular seasonal flow of water of high mineral content. Inflow from the Orange River converted the Fish River to perennial flow and reduced the concentrations of sodium, magnesium, chloride, and sulphate, but not calcium or total alkalinity (O'Keeffe and De Moor, 1988). Invertebrate communities of riffles changed substantially, although overall densities apparently were not altered. The dominant simuliids were replaced by a blood-feeding pest of livestock (*Simulium chutteri*), and the dominant species of hydropsychids and chironomids also changed. More permanent flow and increased area of erosional habitats in the Fish River were the likely causes of these changes.

Channel structure and instream habitat are also likely to change in relation to flow alteration, as was discussed previously for dams. Loss or reduction of flushing flows is expected to be a general consequence of water management aimed at capturing 'surplus' flows. Cross and Moss (1987) describe how aquatic habitats and the fish assemblages of plains streams of Kansas have changed due to a variety of human influences, including diversions and impoundments for agriculture. In the Kansas high plains, the distinctive local fish fauna was adapted to shallow streams subject to fluctuating flows and shifting sand beds. By eliminating flood peaks, diversions and impoundments have caused channels to become narrower, more uniform in depth, and firmer in substrate. The absence of flood peaks has eliminated fishes dependent on floods to trigger spawning, while increased water clarity has favored a different species assemblage, including introduced piscivorous game fish.

14.2.5 Minimum flow reguirements

Water management including dams, diversions and withdrawals collectively results in reduction of streamflows below their natural levels in certain seasons or throughout the year. While there clearly must be some minimum flow needed to maintain a healthy, functioning river community, methods to establish minimum flows have proved controversial. Although by no means restricted to arid areas, the effects of reduced flow may be particularly severe in regions where water is scarce relative to demand, and at times of low flow when conflicting needs are most apparent. Most efforts have been directed at minimum acceptable flows, but the topic also includes maximum flows and rate and frequency of change of flow (Morehardt, 1986).

Some methods specify minimum flows based on easily obtained measures such as discharge, basin area or wetted perimeter (distance from streambank to streambank along the bottom). A particular approach known as instream flow incremental methodology (IFIM) has been developed in the western USA, initially for small salmonid streams (Bovee, 1982). Although recent in inception and still debated, the IFIM approach is rapidly becoming a legal requirement in various states of the USA and in other countries as well. Thus this is an important area of applied river ecology and deserves some review. The IFIM generates a prediction of the amount of usable habitat for fish as a function of discharge. It does this by coupling two models, one which simulates physical habitat preferences of the fishes of the system under study, and a second which estimates how available habitat space varies with discharge.

Habitat suitability curves are derived from fish abundance and distribution data for each life stage of each target species over a range of habitat conditions. Depth, velocity and substrate are the habitat variables measured, although substrate is sometimes omitted. The hydraulic model simulates changes in the availability of habitat based on changes in depth and velocity in cells in a cross-sectional profile of the channel, over incremental changes in flow. By combining habitat suitability curves with the hydraulic model, the instream flow incremental methodology produces estimates of the relative amounts of fish habitat as weighted usable area (WUA) for a given stream reach over a range of discharge conditions (Figure 14.7).

The IFIM offers the benefit of a standardized methodology that requires only physical measurements (once habitat suitability curves have been established for enough fishes), although depth and velocity must be surveyed quite precisely over a range of flows. However, a closer look at the details reveals serious limitations. The hydraulic simulation has received little testing and has often proved difficult to calibrate (Mathur *et al.*, 1985; Morehardt, 1986). Evaluation of one hydraulic model in Illinois streams (Osborne, Wiley and Larrimore, 1988) revealed several problems. The calibration guidelines for the model could not be met almost half the time, particularly at low flows. In those instances where the model was successfully calibrated, 33–50% of model output values failed to correspond with observed depth and velocity measurements.

The habitat component of the IFIM has received the most evaluation and also the most criticism. Its use in the model assumes that fish populations are limited by the three physical variables measured, and that these variables can be treated as if they were independent. Other variables such as cover, temperature, instream habitat structure, resource levels and presence of other species may be of great importance (e.g. Binns and Eiserman, 1979; Wesche, 1980; Baltz, Moyle and Knight, 1982), but these typically are not included. An implicit assumption of the calculation of available fish habitat (WUA) is that fish abundance is quite precisely regulated by habitat availability. Yet, as Moyle and Baltz (1985) report for California streams, year-to-year variation in fish abundance may be 5–12-fold for biomass, 7–21-fold for numbers. At times of high abundance, fish are found in apparently marginal habitats from which they otherwise are missing. In addition, present

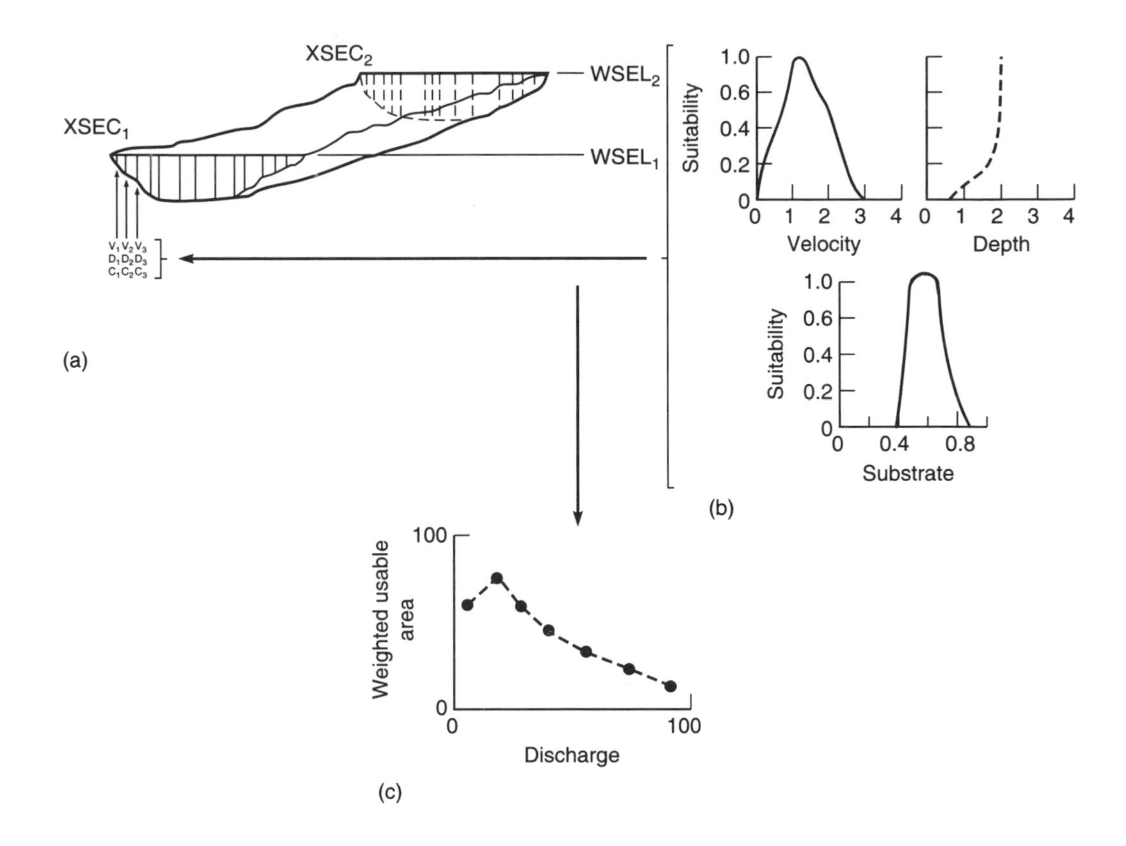

FIGURE 14.7 The instream flow incremental methodology (IFIM). (a) Velocity, depth and substrate from stream cross-sections (XSEC) are combined with water surface elevation (WSEL) at low to high discharges to construct a hydraulic model that predicts habitat availability as a function of stream discharge. (b) Habitat suitability is determined for target species or life stages. (c) Weighted usable area as a function of discharge is generated as model output. (From Gore and Nestler, 1988.)

evidence suggests that short-term changes in flow, either natural or experimental, cause changes in the distribution rather than the abundance of fish.

Most importantly, for the IFIM approach to be valid, one should expect to find a positive linear relationship between WUA and fish biomass over increments in flow. While some examples appear convincing, critics of the IFIM (Mathur *et al.*, 1985; Morehardt, 1986; Scott and Shirvell, 1987) all emphasize the lack of substantiating evidence. Scott and Shirvell surveyed 11 studies where the linear relationship between fish biomass and WUA was examined. Of a total of 444 relationships, 74% showed no trend, and the remainder were both positive and negative. It appeared that the best relationships were found for older age class fish in cold-water trout streams, when several streams were compared. There was no example demonstrating a response to streamflow perturbation, which is the intended use.

Thus, despite the evident value of being able to predict the response of stream biota to changes in flow regime, existing methods cannot do so reliably at the present time. Clearly there is great need for development of more reliable methods that are rigorously tested against the biological

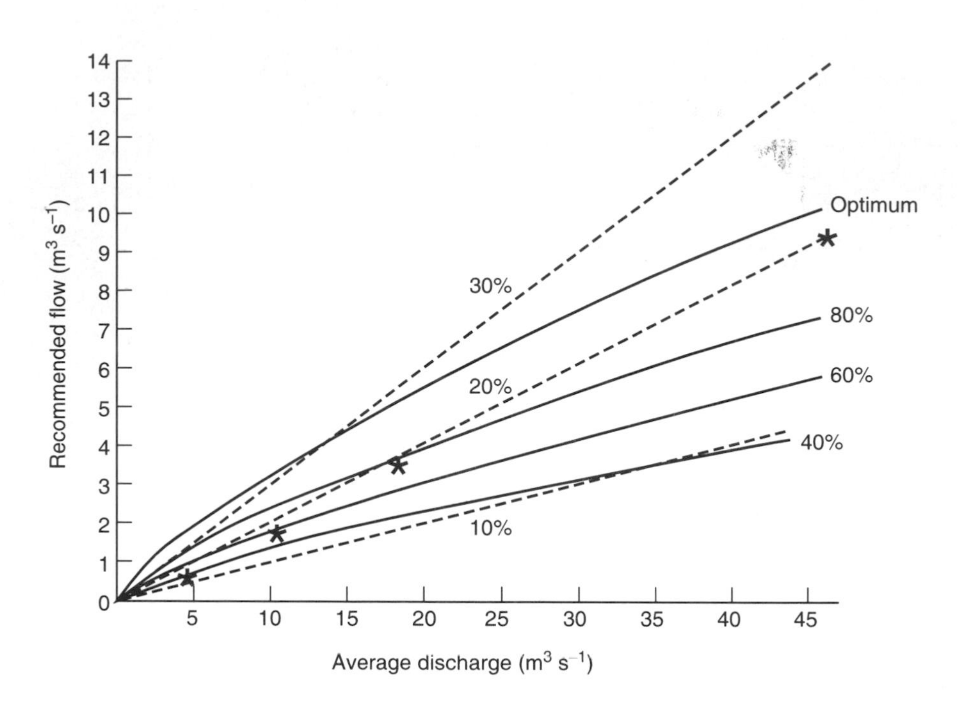

FIGURE 14.8 Recommended minimum flows to maintain fish habitat in four Virginia streams according to IFIM (solid lines), the Montana method (dashed lines) and aquatic base flow (ABF; asterisks). The Montana method considers 10% of mean annual flow (MAF) to provide minimal habitat and 30% of MAF to provide good habitat for fishes (Tennant, 1976). The ABF method uses the median daily flow for the low-flow month of the year (US Fish and Wildlife Service, 1981). (From Orth and Leonard, 1990.)

variables they are intended to predict. Simplicity also is a virtue needed in this area. Orth and Leonard (1990) compared predictions from the IFIM approach to simpler, discharge-based measures and found reasonable correspondence. For example, recommended flows under the Montana method, which considers 10% of MAF to be the minimum acceptable flow and 30% of MAF to maintain good quality habitat for aquatic life (Tennant, 1976), bracketed the range of values from the more labor-intensive IFIM (Figure 14.8). Future efforts are likely to include both simple methods for ease of application and more sophisticated approaches for better understanding. Additional biological detail (Orth, 1987), inclusion of invertebrates (Gore and Judy, 1981) and non-game fishes (Moyle and Baltz, 1985), and perhaps other modeling approaches (Morehardt, 1986) are needed in order fully to explain how lotic populations respond to changes in streamflow.

14.3 Transformation of the land

Although major water projects constitute an important set of forces causing modification of rivers and streams, various transformations of the landscape probably are responsible for the most widespread and serious damage. Draining of floodplains, timber harvest, road building, spread of human settlement and the intensification of agriculture are some principal forces behind changes in land use, with attendant consequences for hydrology, vegetation cover and terrestrial–aquatic linkages. Collectively these landscape changes have led to degradation and fragmentation of aquatic habitat at a level so broad that it appears commonplace. Of all these changes, agriculture probably has had the widest impact, although localized effects of urban and industrial development may cause the most intensive change within restricted areas (Karr, Toth and Dudley, 1985).

14.3.1 Agriculture and settlement

Degradation of the ecological health of running waters is commonplace wherever significant human settlement has occurred. Typically, channel morphologies are made straighter, wider and deeper to promote drainage of low-lying areas. Channelization results in an increased amount of tillable land, a reduction in flooding of riverfront towns, and substantial loss of aquatic habitat. Historic connections between rivers and their floodplains, already affected by dams, are further reduced by levees and dikes with a resultant loss of natural ecosystem function and biological production. Loss or reduction of nearstream vegetation is another common consequence of maximizing the amount of tillable land. Together, these habitat alterations profoundly affect key qualities of the stream ecosystem (Karr and Schlosser, 1978). Streams with natural channel morphologies suffer less bank erosion and export fewer sediments. Streamside vegetation reduces both sediment and nutrient transport, which tend to be related because substantial nutrient loss from agricultural watersheds occurs in association with sediments. Shading by riparian forest canopy ameliorates temperature extremes, resulting in lower maximum values in summer and higher minimum values in winter.

Conversion of native vegetation to agricultural cropland has transformed large land areas of many countries. Estimates for two agricultural watersheds of east-central Illinois indicate that deforestation since European settlement has reduced total riparian forested area by over 70% (Wiley, Osborne and Larimore, 1990). The broad alluvial floodplain of the Mississippi River historically supported the largest expanse of forested wetlands within the USA. Its conversion to production crops such as cotton, corn and soybeans is extensive and relatively recent. Some 55% of the stream edges of the Tensas Basin in Louisiana were forested in 1957, compared with 15% in 1987 (Gosselink *et al.*, 1990).

Agricultural runoff is a major source of pollutants to aquatic habitats. In the USA, some 80% of the 4.9 billion metric tons of soil eroded from non-federal rural land occurred on cropland. Agricultural sources are responsible for 46% of the sediment, 47% of total phosphorus and 52% of total nitrogen discharged into waterways within the USA (Gianessi *et al.*, 1986). Because these pollutants originate over a large area rather than from a single point, this is referred to as non-point source pollution.

A historical review of two major rivers of the midwestern USA, the Maumee and the Illinois, documents the cumulative effects of agricultural and other impacts on river health (Karr, Toth and Dudley, 1985). Land use in both rivers is dominated by agriculture, and Trautman's (1981) description of historical changes to Ohio landscapes and streams probably applies well to these midwestern USA sites. Despite the limitations of such a historical analysis, this study leaves no question that valuable aquatic resources have been lost. River systems that once yielded individual fish large enough to feed an exploring party and supported a productive commercial fishery at the turn of the century have had their fish populations decimated. Of the original 98 species of the Maumee River, 17 have been extirpated and an additional 26 have become less abundant, for an overall species decline of 44%. Loss of spawning habitat in headwaters is considered to be a critical factor leading to the decline of many midwestern fish species. The imperiled status of California's freshwater fishes similarly is attributed largely to modifications of habitat and water flow associated with agriculture (Moyle and Williams, 1990).

(a) Effects on the biota

The replacement of native vegetation by agricultural crops causes numerous changes in the physical condition and energy base of streams. Some effects are broadly predictable from basic principles of stream ecology (Figure 12.7), from which we would predict an overall shift from heterotrophy to autotrophy. Removal of streamside vegetation results in a number of changes, including higher temperatures, altered channel structure due to reduced input of woody debris, fewer inputs of leaf litter and less retentiveness. These changes may be less dramatic when the riparian is left unaltered (Gregory *et al.*, 1991) or when crops replace prairie rather than forest, but detailed evidence is disappointingly scarce. Rivers in agricultural and urban landscapes suffer from additional problems, of course. Their altered hydrology makes them more likely to experience droughts and floods. High rates of soil erosion and runoff of fertilizers, pesticides and herbicides are common problems of agriculturally impacted rivers, while municipal and industrial wastes are important pollutants of urbanized rivers.

Agricultural activity modifies the hydrology of running waters through a combination of impoundments, channelization and drain tiles to enhance runoff following storms. Floodwaters that normally recharge soils and aquifers are rapidly exported, or stored behind impoundments for later use. As a consequence, water tables are lowered and summer baseflows are reduced. Usable habitat likewise is lessened, and perennial streams may become intermittent (Trautman, 1981). Water stress debilitates riparian vegetation that requires access to a permanent water supply, especially under the high evaporative demand of hot summer days. Shrinkage of the riparian corridor and shifts in the composition of streamside vegetation are likely consequences (Smith *et al.*, 1991).

Floodplains are a natural feature of large lowland rivers, a fact that is easy to forget in North America and Europe due to construction of dams, dikes and levees to control flooding and permit agricultural use and human settlement. In unmodified large river systems, floodplains are sustained by flooding that exhibits considerable regularity in timing of onset and duration. Biological productivity is highly dependent on lateral exchanges between river and floodplain, as discussed previously. Because the number of species in river systems increases as one proceeds downstream (Figure 11.5), it is apparent that intact floodplains are essential to the maintenance of biodiversity (Welcomme, 1979; Lowe-McConnell, 1987). Unfortunately, these areas are often ideal for agriculture and settlement. Although some medium-size rivers in the USA have their floodplains largely intact, those of most large rivers are highly modified, and only in a handful of major river systems of the tropics are floodplains essentially unaltered.

An increased load of silt and sediments is typical of rivers draining agricultural and urbanized landscapes. The turbidity of the Tensas River has increased substantially over the past 30 years, which correlates with the decline in riparian vegetation described above. Sedimentation affects the distribution of fish species, which vary widely in their tolerance for silty conditions. The diversity and abundance of species associated with riffles, pools and runs changed significantly in Ozarks streams as accumulations of fine sediments reduced the distinction among these habitat types (Berkman and Rabeni, 1987). As one might expect, less tolerant species disappear as intensive land use brings about habitat degradation. In an agricultural river system in southeastern Michigan, the percentage of species classified as intolerant was directly proportional to overall habitat quality (Figure 14.9), as measured by a standard index used by management agencies (habitat metrics are described by Platts, Megahan and Minshall (1983), Plafkin *et al.* (1989) and Petersen (1992)).

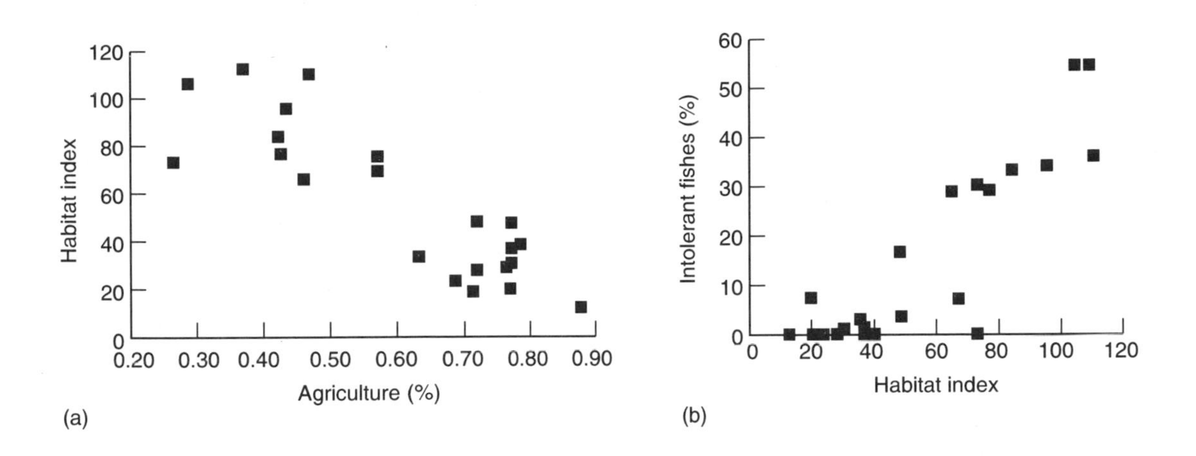

FIGURE 14.9 Influence of agricultural land use at 23 stream sites in southeastern Michigan. (a) Instream habitat quality declined with increasing percentage of land upstream of site that is in agriculture. Habitat quality is a composite of 10 variables that include habitat heterogeneity and evidence of degradation. (b) The percentage of fish species at a site that are intolerant of silty and degraded habitat conditions varies directly with instream habitat quality. (From Roth, 1994.)

Moreover, the habitat index declined as agriculture occupied an increasing fraction of land use in the sub-basin associated with each collecting site.

Increased nutrient concentrations are a serious and well-known consequence of a greater human presence within a watershed. Agriculture increases nutrient levels due to fertilizers and animal wastes, and also by increasing soil erosion, which particularly affects the transport of phosphorus. Municipal wastes and fertilizers are significant nutrient sources from urban areas. An association of high nutrient levels with amount of land in agriculture was established in Omernik's (1977) eutrophication survey of 928 watersheds within the USA (Figure 13.3). However, such comparisons are always complicated by the fact that lands left in forest are often less suitable for farming. Nutrient studies in the Salt Fork River, an agricultural watershed in east-central Illinois that includes two substantial urban areas, indicated that urbanization was at least as important as agriculture in controlling instream concentrations (Osborne and Wiley, 1988). This was true for soluble reactive phosphorus throughout the year, and for nitrate-N during half of the year. During the winter and spring, nitrogen fertilization of agricultural fields was the main determinant of instream concentrations. Nitrate concentrations in the Mississippi River doubled between the 1950s and 1980s, coincident with steady growth in the application of fertilizers over the same time period (Turner and Rabalais, 1991). Deterioration of the Danube River takes many forms, including intense algal blooms in the Danube Delta due to increased nutrient loading over the past several decades (Pringle *et al.*, 1993).

(b) Changes in the energy base

Collectively, the physical and chemical changes that result from agriculture and human settlement change the ecology of rivers in many ways. Typically, the energy base becomes less heterotrophic and more autotrophic, at least in small streams. Habitat quality usually declines, and

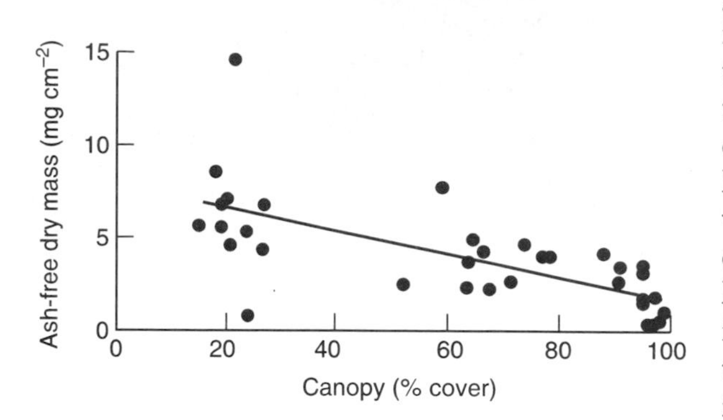

FIGURE 14.10 Relationship between riparian canopy cover and periphyton biomass on tiles elevated on platforms in a coastal California stream. (From Feminella, Power and Resh, 1989.)

biological communities become dominated by a smaller number of species that tolerate these degraded conditions.

A shift from heterotrophy to autotrophy is likely to occur in streams affected by intensive agricultural activity because of reduced shading and enhanced nutrient levels. Feminella, Power and Resh (1989) documented an inverse relationship between riparian canopy cover and periphyton biomass in coastal streams of northern Califomia (Figure 14.10). In Lapwai Creek, an agricultural non-point-source-polluted stream in northern Idaho, the amount of periphyton chlorophyll *a* was two to ten times higher than reported for comparable undisturbed streams (Delong and Brusven, 1992). Nutrient levels were also high for the region, especially following rain events. Enhanced exposure to direct solar radiation due to removal of riparian vegetation did not appear to be a major factor in this study.

Measurements of the amount of organic matter occuring on and within the streambed document a reduction in allochthonous energy inputs in agriculturally impacted streams. Delong and Brusven (1993) found low amounts of benthic organic matter in Lapwai Creek relative to comparable, undisturbed streams. Biomass was greatest at sites receiving the most litterfall inputs, resulting in site-specific patterns in benthic organic matter rather than the longtitudinal trends predicted by the river continuum concept for an unaltered river. Reduced inputs of litter have also been documented for logged watersheds (Webster *et al.*, 1990). Moreover, changes in riparian vegetation composition are likely to affect the timing and quality of litter inputs. Because litter from herbaceous vegetation generally decays more rapidly than the litter from woody plants (Figure 5.1), the year-round availability of organic matter may be altered.

14.3.2 Timber harvest

Unless carefully managed, timber harvest substantially alters the physical environment of streams. Changes in streamflow and increased sediment production are among the most serious consequences of logging activities because they have long-term effects on channel and habitat features. Logging also exposes the streambed to increased solar radiation, resulting in warmer temperatures and greater autotrophic production. The biological consequences of canopy removal may be of relatively short duration, due to regrowth of the vegetation.

Timber removal affects streamflow via multiple pathways of the hydrologic cycle (Figures 1.1 and 1.2). With less standing vegetation, the importance of interception and evapotranspiration is reduced, resulting in higher soil moisture levels relative to unlogged areas and higher streamflows at some times of the year. In addition, roads, landings and skid trails cause soil compaction and thus increase surface runoff. Road building also can cause marginally stable slopes to fail, increasing the amount of debris avalanche erosion, and roads may capture surface runoff and channel it more directly into streams.

The magnitude of these effects depends on many variables, including precipitation regime, slope steepness and soils, as well as timber management practices. Summer streamflows are often higher in logged catchments, especially in areas receiving some summer rainfall. Summer streamflow increased markedly in a small stream in New Hampshire following timber harvest and suppression of re-growth by herbicide treatment (Likens, 1984). In the northwestern USA, summers usually are dry and the greatest effects of logging on streamflow occur with small, early winter storms immediately after timber harvest. However, comparison of paired watersheds subject to different harvest practices indicates that larger stormflows generally are unaffected. In Caspar Creek, a coastal watershed in northern California, road building and selective tractor harvest resulted in greater storm response in very small storms. Storm volumes and peaks increased, and the lag between rainfall and hydrograph peak decreased (Wright *et al.*, 1990). However, runoff following large storms (occurring less frequently than eight times per year) showed no change. Timber harvest on two small watersheds in western Oregon resulted in some increase in water yield and summer low flows, but no change in size or timing of peak flow events (Harr, Levno and Mersereau, 1982). There is some evidence that large stormflow events are affected when the total area occupied by roads, skid trails and landings exceeds 12% of watershed area (Harr, Harper and Krygier, 1975). In regions subject to occasional warm fronts during winter, rain falling on existing snow can result in high streamflows, and clear-cutting often increases the size of peak flows during 'rain-on-snow' events (Harr, 1986).

Sediment delivery to streams is an important long-term consequence of logging and other intensive land-use activities. In two experimental watersheds in coastal Oregon, sediment production increased following timber harvest and road construction, particularly in the more extensively logged area (Beschta, 1978). Major sources of sediments include landslides on deforested slopes, surface scour from logging roads, and erosion of sediments stored on stream banks or with the streambed itself due to greater flooding (Scrivener and Brownlee, 1989). Erosion of gravel roads is an important source of fine sediments, and these cause the greatest harm to fish and water quality. A heavily used gravel road contributes more than 100 times as much sediment as an abandoned road, or a paved road along which ditches and cut slopes are the only sources of sediment (Reid and Dunne, 1984). The amount of fine sediments correlated with area of roads in the Clearwater watershed of Washington State (Cederholm, Reid and Salo, 1981). However, in Carnation Creek, British Columbia, neither roads not landslides appeared to be a major sediment source, indicating that scouring of the stream channel and banks, and reduced effectiveness of debris dams, resulted in the majority of sediments transported.

(a) Effects on the biota

There is ample evidence that poorly regulated forest harvest has resulted in substantial degradation in habitat and fish populations (Bisson *et al.*, 1992). Simplification of channel structure and reduction in habitat complexity are common features in forests managed for timber harvest. As more of the basin is logged, the frequency and size of pools declines (Figure 14.11) due to filling of pools with sediments and loss of pool-forming large woody debris. Complex channel margins that provide important edge habitat also become simplified in the absence of large flow obstructions such as logs and boulders. Catastrophic events including floods and landslides are more likely, especially where slopes are steep and unstable. Mass soil movements are much more frequent in areas with roads compared with roadless areas (Swanson *et al.*, 1987), and also in recent clear-cuts because the root system of the logged forest decomposes before there is full replacement by the regenerating forest (Franklin, 1992).

These changes in stream habitat result in numerous and often complex changes in the aquatic biota. In general one observes a reduction

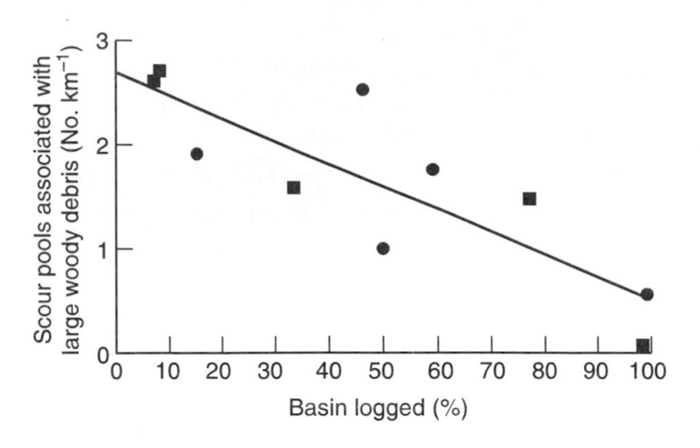

FIGURE 14.11 Frequency of pools associated with large woody debris in ten Oregon coastal streams with different logging histories and differing geology. ■, Basalt; ●, sandstone. (From Bisson *et al.*, 1992.)

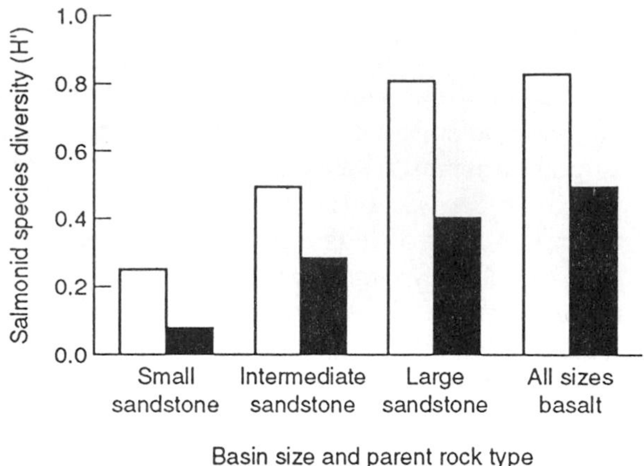

FIGURE 14.12 Diversity of salmonid fishes in logged (solid areas) and unlogged (clear areas) Oregon coastal streams with different parent rock types. Diversity is expressed as H′ (equation 11.2). (From Bisson *et al.*, 1992.)

in species diversity, attributed to habitat simplification, and an increase in standing crop biomass, attributed to greater light penetration and autotrophic production. In the Pacific Northwest, conversion of pool to riffle habitat favors juvenile coho salmon and older, larger trout (Bisson *et al.*, 1992). Surveys of streams in logged and unlogged watersheds reveal that salmonid species diversity is lower in logged areas regardless of underlying geology (Figure 14.12). Aquatic invertebrates are also adversely affected by clear-cut timber harvest. In streams in northern California, the diversity of the macrobenthos was lower although the density was higher in logged as compared with unlogged streams (Figure 14.13).

Riparian vegetation directly influences instream water quality through its moderating effect on temperature. Taxa adapted to cool waters are likely to be eliminated by temperature increases following canopy removal, as was shown by a study of the suitability of 38 streams in southern Ontario for maintenance of trout populations (Barton, Taylor and Biette, 1985). Streams that maintained adequate trout populations were characterized by summer maximum temperatures less than 22°C. Sites experiencing higher temperatures at best supported marginal trout populations. The fraction of forested streambank within 2.5 km upstream of a site was an effective predictor of weekly maximum water temperature (Figure 14.14).

An increased amount of fine sediments within the streambed reduces its permeability to water movement, affecting the delivery and removal of gases, nutrients and metabolites, and potentially restricting movements by animals. These changes have particularly serious consequences for the successful spawning of salmonid fishes, and undoubtedly influence other animals as well. Reduced quality of spawning gravels owing to sedimentation has been documented in the Alsea watershed, Oregon (Beschta, 1978), Clearwater River, Washington (Cederholm, Reid and Salo, 1981), and in Alaskan rivers (Everest *et al.*, 1987). In Carnation Creek, British Columbia, fine sediments increased by about 5% following logging, and survival

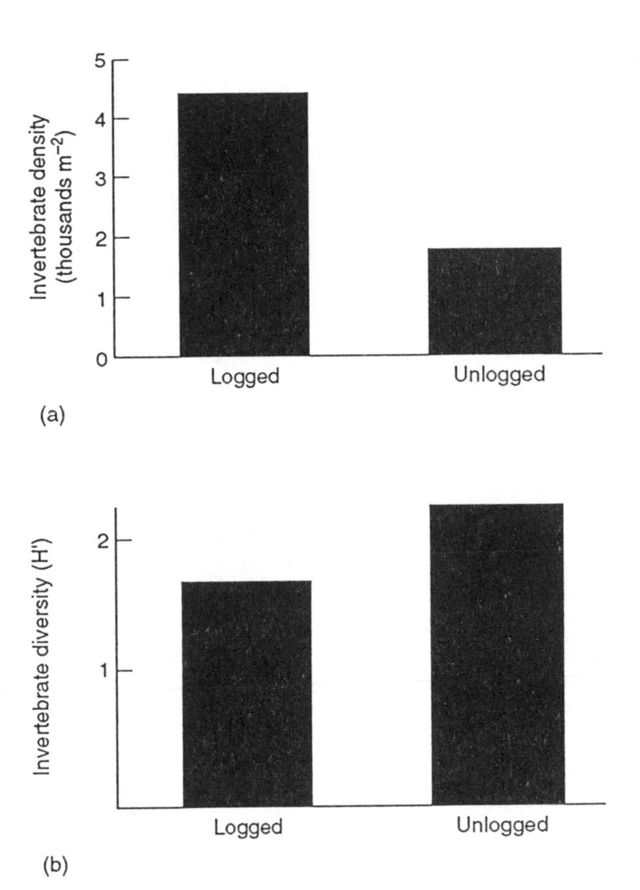

FIGURE 14.13 Density (a) and species diversity H′ (b) of aquatic invertebrates in logged and unlogged streams in northern California. (From Bisson *et al.*, 1992.)

to emergence of coho (*O. kisutch*) and chum (*O. keta*) salmon was approximately halved (Scrivener and Brownlee, 1989). Corn and Bury (1989) compared amphibian abundances in headwater streams in uncut forests to streams in second growth forests that had been logged between 14 and 40 years previously. Species richness was lower (Figure 14.15) and the percentage of fine sediments was greater (Figure 14.16) in previously logged streams, demonstrating the long-term consequences of timber harvest.

As noted above, removal of overhanging canopy vegetation provides one example where the effects of logging appear favorable, at least by some measures. Immediately after clear-cutting, increased solar insolation and higher water temperatures often result in increased biomass at all trophic levels. Invertebrate abundances may be greater (Hawkins, Murphy and Anderson, 1982; Figure 14.13) and mayfly growth rates higher (Hawkins, 1986). Salmonid biomass and densities generally are greater in recent clear-cuts compared with forested sections (Murphy, Hawkins and Anderson, 1981), perhaps because of greater availability of prey and improved foraging in unshaded streams (Figure 7.4), or because warmer temperatures promote faster growth (Holtby, 1988). Effects of altered environmental conditions can be offsetting and thus difficult to predict. In Carnation Creek, British Columbia, warmer stream temperatures following clear-cut logging resulted in increased size and improved overwinter survival of coho salmon fry. However, warm spring temperatures initiated earlier seaward migration of smolts, and this probably resulted in decreased survival of smolts to adulthood (Holtby, 1988). In contrast to the adverse effects of logging on channel structure, which must require a very long recovery time, recovery of shading due to forest re-growth will occur in approximately 10–30 years, perhaps even more rapidly in warm climates. Thus, overall, the effects of an open canopy do not appear to be as serious as changes to the channel and streambed.

14.3.3 Addressing the adverse effects of habitat degradation

Fortunately, much can be done to address the adverse effects of agricultural practices and timber harvest. Switching from conventional to minimum tillage and use of grass or forested buffer strips can greatly reduce average erosion rate, and also the supply of nutrients to streams

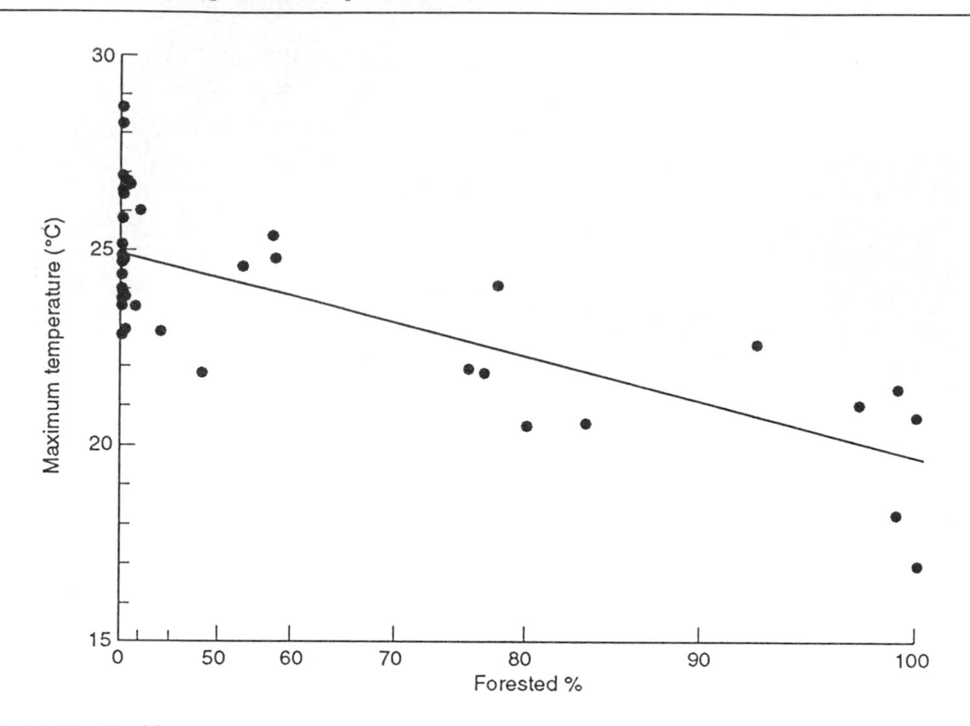

FIGURE 14.14 Weekly maximum stream temperature compared with the percentage of forested streambank upstream of the study site. Note that the horizontal axis is non-linear. (From Barton, Taylor and Bietta, 1985.)

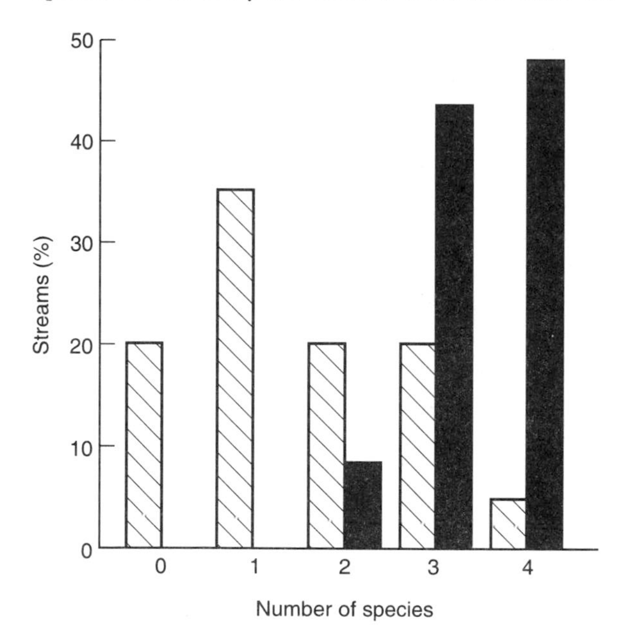

FIGURE 14.15 Numbers of amphibian species present in 23 streams in uncut forests (solid areas) and 20 streams in logged stands (shaded areas) in western Oregon. (From Corn and Bury, 1989.)

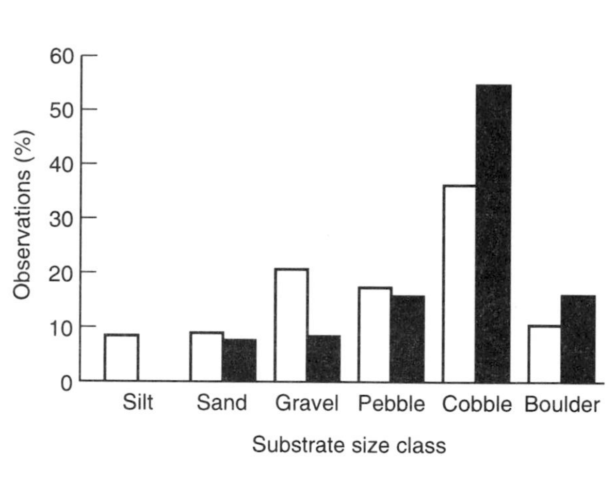

FIGURE 14.16 Distribution of substrates in different stream reaches for 23 streams in uncut forests (solid areas) and 20 streams in logged stands (clear areas). (From Corn and Bury, 1989.)

(Prato *et al.*, 1989). Riparian vegetation in buffer strips as little as 10–30 m wide significantly ameliorate the transport of nutrients and sediments into streams (Karr and Schlosser, 1978). However, the long-term effectiveness of these vegetated buffer strips (VBS) is questionable (Osborne and Kovacic, 1993). It is probable that accumulation of sediments will eventually limit the effectiveness of VBS as sediment traps. Likewise, their ability to retain nutrients depends on continued plant uptake, which differs between growing and mature vegetation (Omernik, Abernathy and Male, 1981) and will benefit from harvest and removal of plant biomass (Lowrance *et al.*, 1984). An especially serious challenge to the effectiveness of VBS are drain tiles beneath croplands which carry sub-surface water directly into stream channels, bypassing the riparian zone completely. This makes some additional mechanism of nutrient trapping necessary; Osborne and Kovacic (1993) suggest artificial wetlands lateral to the streambank and separated by a berm from the main channel.

There appears to be little information concerning how far a VBS must extend upstream of a site to be effective, and clearly this also is of considerable importance. From the standpoint of temperature amelioration (Barton, Taylor and Biette, 1985), the distance of vegetated riparian zone upstream of a site may be more critical than riparian width.

The implementation of better methods for the management of streamside and instream habitat requires effective means to document adverse effects. Unfortunately, the widespread perception that running waters are biologically degraded has not resulted in uniform implementation of biomonitoring. Instead, the relative ease of measurement and standardization of physical and chemical variables had led to the use of biological oxygen demand and other chemical indicators under the assumption that they are useful surrogates (Karr, 1991). At present there is growing interest in a number of biological measures, including invertebrate (Plafkin *et al.*, 1989) and fish-based (Karr, 1981) indices. The Index of Biotic Integrity (IBI) proposed by Karr is a composite of 10–12 individual measures (Table 14.2) including species richness and composition, local indicator species, trophic composition, fish abundance and fish condition. Biotic integrity has been shown to vary within a region in relation to land-use measures and other indicators of environmental condition.

The IBI proved to be a reasonable indicator of stream condition at 209 stream locations on 10 watersheds near Toronto, Canada (Steedman, 1988). Previously forested watersheds have undergone substantial urban development, especially in the area to the south of Toronto along Lake Ontario. Overall, variation in the proportion of urban *versus* forested land cover explained 68% of the variance in the IBI, and suggested a threshold for water quality degradation ranging from 75% removal of riparian forest at 0% urbanization to 0% removal of riparian forest at 55% urbanization (Figure 14.17). Roth (1994) likewise found a significant relationship between the IBI and land use in an agricultural watershed in southeastern Michigan. The IBI declined with increasing agricultural land use, which explained roughly half of the variation in water quality as measured by biotic integrity.

Application of the IBI requires a good deal of local calibration. Measures of species richness must be based on expected values for streams of a given size and zoogeographic region, and require suitable undisturbed locations to serve as reference sites. Some studies use 10 rather than 12 metrics, and cold-water streams may lack some of the species groups of metrics 2–6 (Table 14.2). Nevertheless, the IBI is a valuable tool because of its ability to convert the relative abundance data of a species assemblage into a single index of biotic integrity which demonstrably varies with environmental degradation.

TABLE 14.2 Metrics included in the Index of Biotic Integrity, used to assess the biological integrity of fish assemblages. Ratings of 5, 3 and 1 are assigned to each metric according to whether its value approximates to, deviates somewhat from, or deviates greatly from the expected value based on a reference site of high quality (From Karr, 1991)

Metric	Rating of metric		
	5	*3*	*1*
Species richness and composition			
1 Total number of native fish species	Expectations for metrics 1–5 vary with stream size and region		
2 Number and identity of darter species (benthic species)			
3 Number and identity of sunfish species (water column species)			
4 Number and identity of sucker species (long-lived species)			
5 Number and identity of intolerant species			
6 Percentage of individuals as tolerant species	< 5	5–20	>20
Trophic composition			
7 Percentage of individuals as omnivores	<20	20–45	>45
8 Percentage of individuals as insectivorous cyprinids	>45	45–20	<20
9 Percentage of individuals as piscivores (top carnivores)	> 5	5–1	< 1
Fish abundance and condition			
10 Number of individuals in a sample	Expectations vary with stream size and location		
11 Percentage of individuals as hybrids or exotics	0	0–1	> 1
12 Percentage of individuals with disease or deformities	0–2	2–5	> 5

14.4 Alien species

Biological invasions include the introduction of species exotic to a region, and the transplantation of indigenous species into locales where they presently do not occur. Introductions are well documented for a wide variety of aquatic plants and fishes, less so for aquatic invertebrates. Introduced species are of special concern for several reasons. Once alien species are established in a new environment, they are usually there to stay and often are capable of reproducing and dispersing far beyond the point of origin. In contrast to chemical pollutants that can be eliminated at their source, or habitats that might potentially be restored, species introductions are usually impossible to undo. Natural enemies are often lacking, and the impacts of introduced species in new habitats are highly unpredictable because of differences in the nature of species interactions under novel ecological conditions.

Over 160 species of exotic fishes from some 120 countries are listed on the international register kept by the Food and Agriculture Organization (FAO) of the United Nations, attesting to the global level of this problem (Courtenay and Stauffer, 1984; Welcomme 1984; Bruton and van As, 1986). The common carp (*Cyprinus carpio*) was probably the earliest fish species to be transferred into new habitats, and its story illustrates how good intentions can lead to unwanted outcomes. Although its original distribution is not known with certainty, the common carp is believed to have originated in central Asia and been introduced into China and Japan in ancient times. The Romans brought carp to Italy, and it subsequently dispersed through Europe with the spread of Christianity, as fish escaped from monastery ponds. Carp were later introduced into southern Africa in the early 1700s (Bruton and van As, 1986), and were brought to the New World in the mid 1800s (Courtenay *et al.*, 1984). Indeed, the

FIGURE 14.17 Contour plot of Index of Biotic Integrity (IBI) ratings as a function of urban land use and riparian forest in 10 watersheds near Toronto, Canada. Fair water quality is indicated by an IBI between 23 and 30, and good water quality by values between 31 and 41. (From Steedman, 1988.)

lowly carp was brought to North America with such enthusiasm that according to Madson (1985) "states anxiously awaited their quotas, and when a modest shipment of young carp arrived it might be greeted at the railway station by a brass band and paraded through town on its way to a river or pond."

Thus, the common carp has been transformed from a fish of modest distribution in central Asia, to a species of global status.

14.4.1 Causes of species invasions

Reasons for purposeful and accidental introductions of aquatic organisms are many and have varied historically. During the nineteenth century and lasting until about World War II, fish introductions were largely an outgrowth of colonialism and nostalgia by settlers in their newly adopted homelands for the familiar species and surroundings left behind (Welcomme, 1984). For example, acclimatization societies proliferated in New Zealand in the late 1800s, with the aim of 'enhancing' the fauna with fishes from Europe and North American (McDowall, 1990). Such was the zeal to augment the fauna that of the 47 fish species now found in New Zealand, some 20 species are exotic.

The rate of species introductions is itself increasing sharply (Figure 14.18). In 1920 only six fish species of exotic origin were established in US waters; by 1945 just three more had been added. The big boom in fish introductions occurred after 1950. By 1980 some 35 exotics had become established and about 50 other exotic species had been recorded (Courtenay and Hensley, 1980). Government agencies are responsible for 12 of these introductions, as sport fishes, for fish culture, and as agents of biological control. The aquarium fish-culture industry accounts for the release of the remaining 23 species.

Some of the most widely promoted exotics include popular sports fishes such as salmonids and largemouth bass (*Micropterus salmoides*). The rainbow trout (*Onchorhynchus mykiss*), once confined to the Pacific Northwest of North America, now has a world-wide distribution thanks to ease of breeding, appeal to the angler and culinary graces (Figure 14.19; also MacCrimmon, 1971). Indeed, stocking of non-native fish remains a cornerstone of management efforts to provide the most desirable species for sport or commercial fisheries. That alien species comprise more than 25% of the recreational fisheries' catch of freshwater fish in the continental USA (Moyle, Li and Barton, 1986), attests to the pervasiveness of this practice. Programs aimed at eradicating non-sport species are well known from many parts of the USA. In some cases, efforts to purge native fishes in order to 'improve' recreational stocks have been so extensive that extinctions of native species have occurred. The state of Oregon intentionally eradicated the endemic Miller Lake lamprey using ichthyocides because it was a predator of stocked trout (Miller, Williams and Williams, 1989).

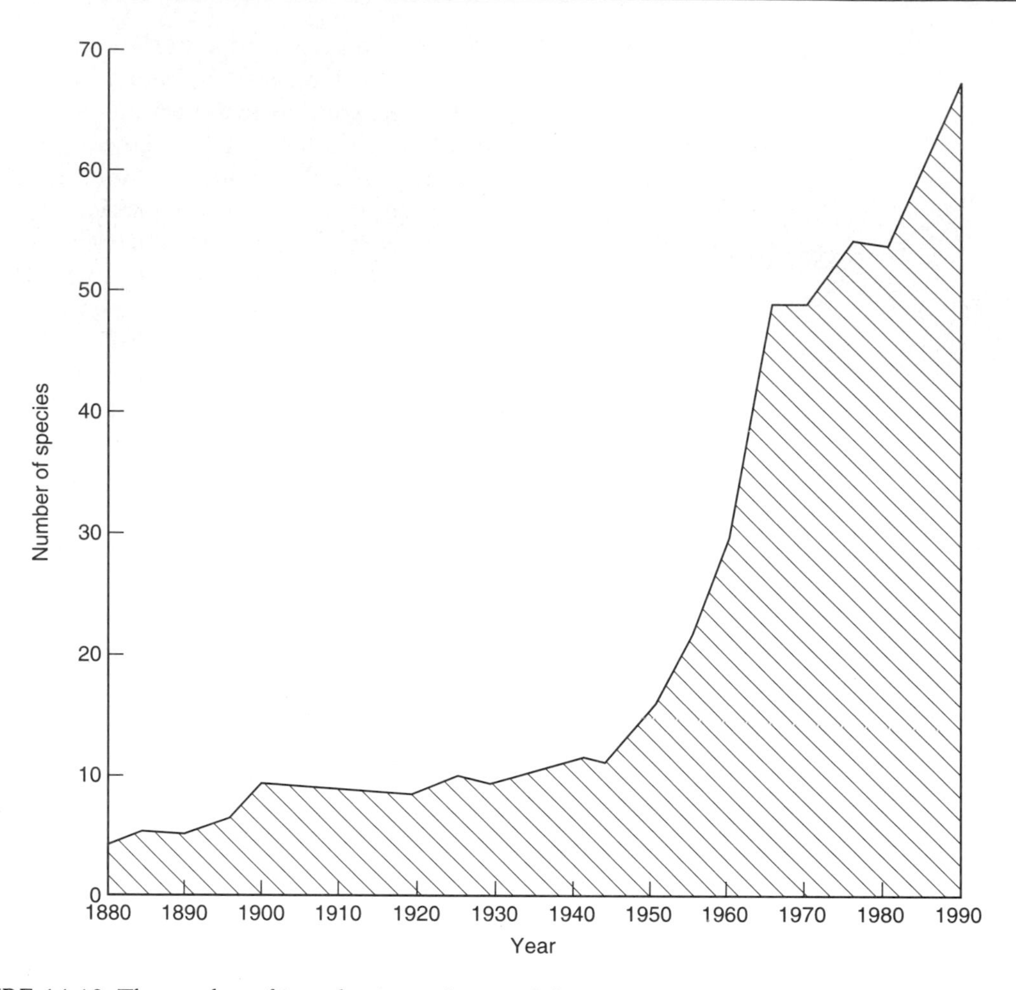

FIGURE 14.18 The number of introductions of exotic fish species to US freshwaters as of 1990. Translocations of fishes within the USA are not included. (From Allan and Flecker, 1993.)

In recent years, introductions have often occurred for other purposes, such as aquaculture and biological control. Although relatively few species are extensively used in aquaculture (Welcomme, 1984), they are among the most widely transferred fishes. Introductions associated with fish culture expanded considerably in the 1970s, as international development agencies promoted aquaculture to provide protein for rapidly expanding human populations, and for commerce. Fishes also have been introduced as biological control agents to combat disease vectors and noxious aquatic weeds (Welcomme, 1984). Mosquito control has been a frequent objective, using species such as mosquitofish (*Gambusia affinis*) and guppies (*Poecilia reticulata*), which now enjoy world-wide distributional status. The mollusk-eating cichlid, *Astatoreochromis alluaudi*, has been used as a biological control agent of the snail host of bilharzia. Control of aquatic weeds, many of which themselves are exotic, is another frequent objective (Shireman, 1984), with tilapia and carp the most commonly promoted species.

In addition to purposeful introductions, generally for sport fishing, aquaculture and

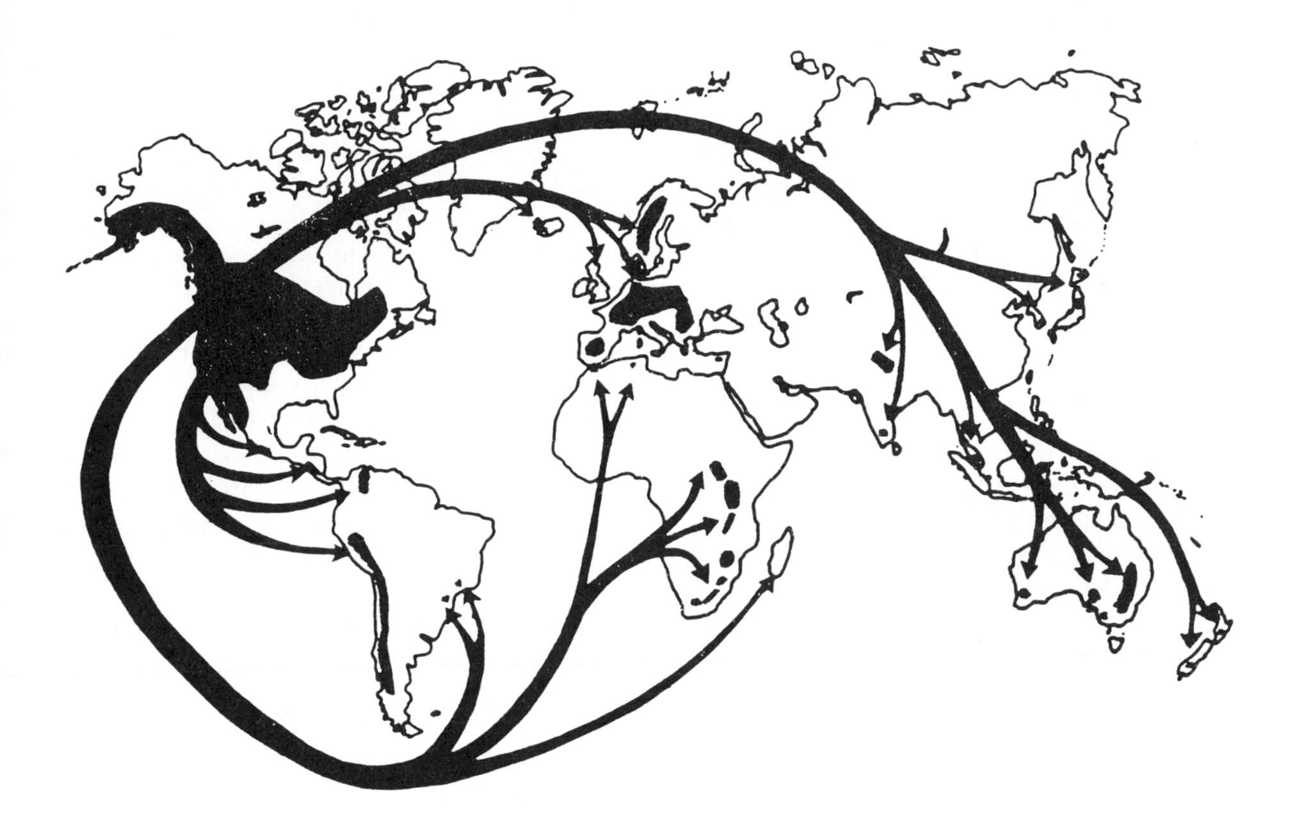

FIGURE 14.19 Transfer of rainbow trout (*Oncorhynchus mykiss*) from its original range in western North America (shaded area) to every continent but Antarctica. (From Petersen *et al.*, 1987.)

biological control, there are several sources of accidental introductions. Chief among these are the release or escape of aquarium pets, and various means of unintentional hitch-hiking. Over 1200 tropical aquarium species are shipped to various parts of the world, and as many as 6000 species may ultimately be of interest to the pet trade overall (Welcomme, 1984). Most tropical fish introductions have occurred since 1960, when techniques greatly improved for the live transport of fish.

About 10% of international transfers of exotic fishes have been the result of truly non-purposeful introductions (Welcomme, 1984). Many cases of accidental transfers have involved small cyprinids included with shipments of juvenile carp species. Recently, ballast water introductions have been a focus of concern, illustrated by invasions into the Laurentian Great Lakes of the ruffe (*Gymnocephalus cernuus*), the zebra mussel (*Dreissena polymorpha*) and the mitten crab (*Eriocheir sinensis*) (Yount, 1990).

Inter-catchment water transfers may also contribute to species translocations. This has been a widespread problem in southern Africa, because almost all of that region's major river systems are connected by tunnels, pipes and canals (Bruton and van As, 1986). At least five fish species have invaded the Orange River from the Great Fish River, which prior to their connection had distinct faunas with high numbers of endemics. In Venezuela, a tunnel connecting the Uribante and Caparo Rivers was recently completed, but effects on the local aquatic fauna have not been described. On the positive side of the ledger, high risk plans in North America that would link the Peace and Fraser drainages of the Pacific Northwest, and the Missouri–Mississippi with Hudson Bay drainages, have thus far remained unimplemented because of the concern over mixing of faunas.

14.4.2 Effects of invading species

Negative impacts of exotics on native stream fauna have been implicated in a variety of geographical settings, most often from correlational analyses (e.g. Taylor, Courtenay and McCann, 1984; Ross, 1991). An analysis of 31 case studies of fish introductions to stream communities found that 77% of the cases documented a subsequent decline in the native species (Ross, 1991). Examples included the decline of native species in the southwest USA following the introduction of mosquitofish (*Gambusia affinis*), and declines of the native brook trout (*Salvelinus fontinalis*) following the introductions of brown and rainbow trout. Trout and galaxiids in New Zealand apparently are incompatible; formerly widespread populations of galaxiids are now fragmented into remnant populations restricted to regions above barrier waterfalls inaccessible to trout (Townsend and Crowl, 1991). In all likelihood the present day galaxiid distribution is a mid-course snapshot of the fate that has already befallen other native species in New Zealand. The New Zealand grayling (*Prototroctes oxyrhynchus*), an endemic fish once so abundant that it was taken by the cartload (McDowall, 1990), precipitously declined following the introduction of brown trout and is now considered extinct.

Although declines in native species following fish introductions are well known, surprisingly little is known of the mechanisms by which exotics affect native species in stream systems. Potential impacts of introduced species include habitat alterations, introductions of diseases or parasites, trophic alterations, hybridization and spatial alterations (Taylor, Courtenay and McCann, 1984). Predation appears to be a common cause of the replacement of native species by exotics (Taylor, Courtenay and McCann, 1984; Ross, 1991). Perhaps the most spectacular and catastrophic effects of an introduced freshwater fish are now taking place in Lake Victoria, Africa, where the Nile perch threatens the existence of literally hundreds of species of endemic cichlids (Kaufman, 1992).

Changes in habitat use by the native fauna are a non-lethal effect of introduced species. Of the studies reviewed by Ross (1991) where resource use was examined (10 studies), one-half found habitat shifts following fish introductions. For example, a variety of native species including the Sacramento sucker, rainbow trout, California roach and threespine stickleback shifted patterns of habitat use in the presence of the Sacramento squawfish, a predatory cyprinid introduced into the Eel River of California (Brown and Moyle, 1991).

Invading species also affect native species by hybridization, which Miller, Williams and Williams (1989) found to be a factor in 38% of the recorded extinctions of North American fish species. In most instances some other factor apparently resulted in the initial decline, and hybridization was the final blow. Examples include at least two subspecies of native cutthroat trout (*O. clarki*) which have become extinct because of interbreeding with stocked rainbow trout (Miller, Williams and Williams, 1989), the Snake River sucker (*Chasmites muriei*) which hybridized with the Utah sucker, and the blue pike (*Stizostedion vitreum glaucum*) which hybridized with the walleye. Many native stocks of Pacific salmon have been influenced to an unknown extent by interbreeding with their hatchery-reared counterparts of different genetic make-up.

A host of diseases and parasites are associated with alien species (Hoffman and Schubert, 1984; Bruton and van As, 1986), posing yet another threat to the invaded community. A fungal parasite causing crayfish plague affected native crayfish throughout Europe following the introduction of resistant crayfish species from North America (Reynolds, 1988). Likewise in fish, a number of examples are known of the introduction of associated diseases (Hoffmann and Schubert, 1984). Of particular concern in fish are parasites that affect a wide variety of species such as the cestode, *Bothriocephalus acheilognathi*, which was introduced from Germany into southern Africa by grass carp (Bruton and van As, 1986).

Introduced species may cause a variety of indirect effects via food web interactions. Native fish and crustacean species now are absent from river sites in Hawaii where smallmouth bass has been introduced (Maciolek, 1984). In New Zealand, introduced trout exert greater top-down control over invertebrates than do the native galaxiids, resulting in a reduction in benthic grazing and an increase in algal biomass (Flecker and Townsend, 1994).

The future promises a continuing spread of exotic species and mixing of faunas. As a consequence, managers of aquatic ecosystems will increasingly be confronted with a shifting mix of native and non-native species. The Laurentian Great Lakes serve as one object lesson, and in this example managers have been comparatively successful, albeit at considerable cost and with the continual necessity of researching new problems. At the other extreme, by all accounts Lake Victoria presently is undergoing disastrous change. What mix of successes and disasters will emerge from the continuing mixing of the biota of running waters will reveal itself over the next years and decades, but an overall loss of biological diversity seems certain.

14.5 Climate change

Running waters face a future threat of unknown severity in the projected climate change of the twenty-first century. Atmospheric carbon dioxide is expected to double from its pre-industrial concentration of 270 ppm by some time late in the next century. In concert with other greenhouse gases, this elevated carbon dioxide level is expected to bring about significant climatic change. Global mean temperature is forecast to increase by roughly 3°C, and the extent of warming likely will be most pronounced at high latitudes and in winter (Levine, 1992). Global average precipitation is predicted to increase by 3–15%, but regional variation is likely to be substantial (Gleick, 1993).

Should these predictions prove true, biological systems will experience rapid and possibly unprecedented climate change. Some likely results of the expected changes in temperature, flow regime, and the composition of riparian and catchment vegetation are described below. It should be noted, however, that a great deal of uncertainty surrounds these predicted changes as well as the anticipated consequences. The ability of global climate models to forecast precipitation change is limited, and temperature cannot be resolved at a regional scale. Prediction is further complicated by feedback among variables. For example, warming is expected to increase evaporation, which should reduce surface runoff, but elevated carbon dioxide causes the stomata of plants to close down, decreasing plant transpiration and making more water available as runoff. Because these results are offsetting, the effect of warming on surface runoff in uncertain. Modeling of this interaction suggests that small changes in precipitation potentially may cause large changes in surface runoff, especially in aridland streams (Wigley and Jones, 1985).

14.5.1 Effects of climate change

An increase in temperature will have numerous direct and indirect effects on the biota of rivers. Species ranges are likely to shift toward higher latitudes, and locales that previously supported species assemblages characteristic of cold waters will experience replacement by warm-water species (Regier and Meisner, 1990). Because the headwaters of mid-latitude streams generally are cooler than downstream sections, climate warming may permit species from warmer waters further downriver to expand into headwaters, with a corresponding reduction in diversity within watersheds.

Although the local extirpation of some taxa is likely, such as the cool-adapted species of head-water streams, species with broad geographic ranges or able to disperse to cooler regions should persist, albeit within an altered range. However, streams of the southern Great Plains of the USA are likely to experience species loss if significant warming occurs (Matthews and Zimmerman, 1990). Summer temperatures in these streams are already near the thermal tolerances of their inhabitants, with late summer maxima as high as 38 and 39°C. Only a small increase would suffice to cause heat death. Equally important, virtually all Great Plains river systems drain from west to east or southeast (Figure 14.20). Only by an eastward dispersal of about 1000 km would residents be able to move northward, and some rivers drain directly into the Gulf of Mexico, providing no escape at all. At least 20 species of the Great Plains and southwest have ranges such that total extinction could result from a several-degree warming (Matthews and Zimmerman, 1990).

Higher temperatures will affect the growth, metabolism and life histories of the biota. An increase in total annual degree days effectively places the organism at a lower latitude (Figure 3.12), influencing growth, survival and overall fitness (Sweeney *et al.*, 1992). In the short term, species may benefit from greater opportunities for growth, provided that other resources are available to meet their higher metabolic demands, but over time they are likely to be replaced by taxa whose life cycles (Figure 3.19) best suit the new temperature regime.

Temperature change will affect ecosystem processes via changes in riparian and catchment vegetation, and higher metabolic rates of microbial populations (Meyer and Pulliam, 1992). An average temperature difference of 3°C corresponds to a shift in latitude of 250 km, and so this amount of warming potentially will bring about considerable vegetation change (Davis, 1989). The most serious effects may be during the transition, if tree mortality occurs prior to the establishment of a new vegetation cover, perhaps allowing exotic riparian species to increase in importance (Sweeney *et al.*, 1992). Warmer temperatures will increase the rate at which organic matter is processed and nutrient transformations take place, which will change ecosystem dynamics in unknown ways. At high latitudes, where winter warming constitutes a substantial climate change, ecosystem effects are likely to be many and complex. One likely effect, an increased release of carbon dioxide, potentially will exacerbate climate warming in a positive feedback loop (Oswood, Milner and Irons, 1992).

Future climate change will also affect river systems by altering the amount, timing and variability of flow. It presently is not possible to predict how the flow regime of rivers will change in response to greenhouse warming. However, changes in the frequency of floods and droughts are likely to be of particular importance, and the rivers of certain regions may therefore be more vulnerable. Perennial streams in arid areas may become intermittent, or a snowmelt-driven hydrograph (Figure 11.6c) may change to a winter-rain pattern (Figure 11.6d). These alterations in flow regime are likely to influence ecosystem processes. The concentration and transport of elements (chapters 2 and 13) and organic matter (chapter 12) are typically flow-

FIGURE 14.20 Major drainages of the Great Plains of the central USA. Fishes of this region experience summer maximum temperatures near their natural limits, and lack northward dispersal corridors should greenhouse warming affect their environments. (From Matthews and Zimmerman, 1990.)

dependent. Reduced flow favors storage and processing of materials, whereas high flow increases transport; hence mass balances and input/output ratios should change in response to hydrology (Meyer and Pulliam, 1992). Changes in flow may also result in concentration or dilution of pollutants. Because river hydrology and geomorphology are strongly interrelated (chapter 1), changes in flow regime will have consequences for many aspects of physical structure and habitat. Reductions in flow may be expected to bring about change similar to dams and diversions, including reduced connectivity of the river with its floodplain and riparian zone, and a gradual narrowing of the channel.

In summary, the impact of global climate

TABLE 14.3 Status of selected animal groups in North America, based on species (not subspecies), as of 15 April 1990. Ranks as developed by Heritage Network of The Nature Conservancy; data are for global (G) occurrences (From Master, 1990)

Number of US species ranked as:	*Mammals*	*Birds*	*Reptiles*	*Amphibians*	*Fishes*	*Crayfish*	*Unionid mussels*
GX (extinct)	1	20	0	3	18	1	12
GH (historical; possibly extinct)	0	2	0	1	1	2	17
G1 (critically imperiled)	8	25	6	23	78	62	88
G2 (imperiled)	23	9	10	17	72	49	49
G3 (rare, not imperiled)	19	23	25	26	110	84	35
G4-G5 (widespread and abundant)	330	628	251	153	549	106	73
G? (not yet ranked)	62	55	9	3	24	9	26
Total	443	762	301	226	852	313	300
GX-G3 as percentage of total (%)	13	11	14	28	34	65	73

change on the biota of rivers and streams is difficult to predict, due to uncertainty regarding future climate scenarios and the difficulty of anticipating the ecological consequences. The effects of altered precipitation and discharge are potentially the most serious consequences of future climate change for rivers, because of the many ways that flow regime influences individual species, ecosystem processes and land–water linkages. However, higher temperatures unquestionably will bring about changes in species distributions, and at least local extirpation for those species unable to disperse to higher latitudes. Changes in ecosystem metabolism at high latitudes are particularly worrying because of their potential to influence the supply of greenhouse gases.

14.6 Imperilment of the biota

The cumulative impact of anthropogenic change upon rivers is alarmingly evident in the imperilment of river-associated species (Allan and Flecker, 1993). A substantial fraction of the rare and threatened species of North America are aquatic, and primarily freshwater, taxa (Table 14.3). Within North American freshwaters, fish, mollusks and crayfish have received the most attention, and much less is known concerning the imperilment of many other groups. An accurate description of the status of river-inhabiting organisms is difficult for several reasons. Not only are our listings far from complete, but they are strongly biased toward vertebrates. In addition, the distinction between running waters, ponds and lakes is complicated by the refusal of many species to recognize such boundaries.

The extent of the endangerment of North American freshwater fishes is well summarized by Williams *et al.* (1989), who categorize 103 species as endangered, 114 as threatened and 147 as deserving of special concern. This represents roughly one out of three species and subspecies of North American fishes. As mentioned previously, some 214 native stocks of salmonids of northwestern North America are considered threatened or endangered (Nehlsen, Williams and Lichatowich, 1991).

The molluskan fauna of North American rivers, which historically harbored an exceedingly high diversity of snails (family Pleuroceridae) and mussels and clams (family Unionidae), now face a catastrophic level of extinctions. Of 297 species and subspecies of mussels, 13 are extinct, 40 are endangered and 74 are candidates for listing under the USA Endangered Species Act, for a total of 43% of the taxa (Neves, 1991).

Many factors contribute to the extinction of a species, and often it is difficult to identify a single cause, due either to inadequacy of data or, often, because multiple factors play a role. According to Miller, Williams and Williams (1989), a total of three genera, 27 species and 13 subspecies of North American fishes became extinct over the past century. Of these, some 15 were primarily from flowing water habitats and two ascended streams to spawn, so about 40% can be viewed as losses to lotic systems. Miller and colleagues concluded that more than one factor was responsible in most of these extinctions. Habitat loss and species introductions were about equally common (73% and 68% of cases, respectively), followed by chemical pollution (38%) and hybridization (also 38%, and often related to species invasions). Over-harvesting contributed substantially to the demise of 15% of the losses, but these were exclusively large fishes of large lakes.

14.7 Recovery and restoration of running waters

Streams and rivers are unusual in one regard. Although all ecosystems have permeable boundaries, importing and exporting matter, energy, and organisms, the throughput in running water ecosystems is unusually high. This property provides a natural cleansing ability (Hynes, 1970) and also a regular supply of propagules delivered from upstream to downstream locations except in the most drastically degraded situations. As a consequence rivers have some natural recovery capability which facilitates the restoration of river ecosystems (Gore, 1985).

Rehabilitation of physical habitat and improvement in water quality are the principal categories of stream restoration measures. Regrettably, we can do little to address large-scale land transformation and the invasion of alien species, two of the most critical forces affecting the health of rivers. Improvement in water quality, particularly when a manageable number of point source inputs can be identified, is readily achievable. When the degradation of water quality is due to a diversity of causes, including non-point sources, mitigation is more difficult. However, gradual improvement in water quality of the Rhine over the past two decades (Lelek and Köhler, 1990) establishes that progress can be made even in difficult cases.

The Thames River and estuary is an interesting example because of its long and relatively well documented history (Gameson and Wheeler, 1977). Comments as early as 1620 by the Bishop of London brought attention to the foulness of Thames water. Nonetheless, it was a good fishing river, providing a living for fishermen and food for the inhabitants of the city. By 1850, this era was ended by pollution from human and animal wastes, and the river became so foul that 1858 was known as the 'Year of the Great Stink'.

Advances in sewage treatment led to improved conditions into the first decades of the twentieth century, and some recovery of fish populations occurred. But by 1955 the increased volume of sewage had reduced water quality to an all-time low, hydrogen sulphide gas released from anaerobic sections of the Thames was again a public nuisance, and a 70 km section around London was devoid of fish life except eels. These conditions spurred further improvements in the treatment of organic wastes, and within a decade the region around London again had sufficient oxygen for fish to survive passage. By 1973, some 62 species of fish were represented by more than a single capture within the river and estuary. Isolated captures raised the total to 80 species. This startling recovery of fish diversity in a decade was paralleled by a similarly dramatic return of waterfowl, and, although the evidence is scanty, it appears that the invertebrate fauna has likewise recovered. The Thames thus provides a heartening example of the rapidity with which grossly polluted rivers can recover.

A recent review of more than 150 case studies of recovery in freshwater systems establishes that resilience varies with the type of disturbance, with biological attributes of the community, and with degree of isolation from a source of colonists (Niemi *et al.*, 1990). The authors distinguish two types of disturbances: pulse events, which are of limited and definable duration; and press events, which are longer in duration. Typical examples of pulse events are chemical spills that are rapidly diluted, compared with press events such as habitat alteration by channelization.

Unsurprisingly, recovery generally was more rapid from pulse than from press disturbances. Especially for pulse disturbances, the recovery process depended on the opportunities for colonization and subsequent population growth, so the rapidity of organisms' life cycles, their dispersal capabilities, and the availability of refugia become significant rate controllers. Press disturbances generally were synonymous with habitat alteration. Recovery times of years to decades were common when habitat mitigation was accomplished; without intervention the recovery could be longer yet.

Habitat restoration measures are well developed mainly for game fish such as trout (Hunter, 1991), and within this restricted focus the efficacy of such measures is well established. Wesche (1985) describes a number of physical structures designed to improve habitat for fish or the insects they feed on. The guiding principle of all such efforts is to modify the instream habitat to resemble a natural stream as much as possible, and the efforts usually result in greater numbers of invertebrates and fish. An even greater challenge is provided by the many rivers that have had their channels directly modified by dredging, removal of obstructions, and other channelization procedures, and indirectly affected by alteration of flow. In these situations we need river engineering to be guided by geomorphic and ecological principles so that channel design is as natural as possible (Brookes, 1988, 1989).

Restoration of ecosystem function in running waters clearly requires that we look beyond the banks, to the qualities of the riparian and perhaps the entire landscape. Protection of streambanks via vegetated buffer strips is attractive because there is ample evidence of their beneficial functions. Moreover, protection of riparian regions undoubtedly will benefit wildlife directly as habitat and by serving as connecting corridors between isolated fragments of high quality terrestrial habitat. Determining the best design and management practices for riparian areas poses numerous challenges for further research (Petersen and Petersen, 1992; Osborne and Kovacic, 1993).

It is uncertain what can be done to remedy large-scale human modification of river systems. Altered landscapes, regulated flows and engineered channels bring about extreme change in the structure and function of rivers. Their effects are not easily offset, and clearly pose a significant challenge in terms of the application of ecological and geomorphic principles to river management.

Clearly there is need for a network of protected rivers, just as parks and wilderness areas protect valued terrestrial ecosystems. Some protection is already afforded wherever a stream section flows through a protected landscape, or is specifically recognized under legislation such as the USA Wild and Scenic Rivers Act. However, the extent to which rivers are influenced by events in their valleys and events upstream (Figure 12.8) dictates that small segments are not of themselves sufficient. An effective strategy should include the protection of headwater areas, of the riparian zone, and of habitats that meet special needs of individual species. Refuges of high quality are of special importance (Sedell *et al.*, 1990), particularly in altered landscapes, and unfortunately we do not yet understand enough about their required spatial arrangement to plan with confidence. Preservation of biodiversity and ecosystem function likely will require a protective system that includes a few large reserves of high diversity, combined with many smaller preserves that protect specific habitat. Criteria for preserve design and integration of individual preserves into an effective network have scarcely been considered for running waters, and clearly this is an urgent need (Moyle and Sato, 1991).

14.8 Summary

Running waters are ever-changing systems, ecologically and geomorphologically. The Earth's changing geology and climate naturally affect the biological functioning of rivers by influencing slope, basin and channel characteristics, as well as valley vegetation. However, human-induced change increasingly surpasses such natural change in both rate and magnitude. The extent of the problem is well documented by a 1982 Rivers Inventory which found that barely 2% (less than 100 000 km) of the 5.2 million km of stream in the contiguous 48 states of the USA had sufficient high quality features to warrant federal protection status (Benke, 1990). By the year 2000, over 60% of the total streamflow of the world will be regulated. Historic connections between rivers and their floodplains have been altered in many rivers of large and moderate size. A significant fraction of the fishes and mollusks of North America are imperiled. Clearly, running waters are in urgent need of both restoration and preservation.

River regulation by dams, diversions, channelization and other physical controls over natural flow regimes has substantially affected the majority of rivers in developed countries and is rapidly spreading to other areas. The physical, chemical and biological characteristics of rivers below dams differ greatly from those of a free-flowing river. A reduction in natural variability in discharge and temperature are among the most critical changes. Shifts in species composition occur in response to altered habitat conditions; typically, overall abundance increases but species richness declines. Dams and impoundments are barriers to the upstream and downstream movements of migratory fishes, causing significant decline of species of economic importance. Diversions cause substantial flow reduction and provide dispersal corridors for alien species wherever they connect regions with distinctive faunas. As pressures to regulate and divert flow continue to grow, there is increasing need to specify minimum flow requirements and natural flow variation for healthy, functioning river systems.

In addition to directly altering river flow, human activities have transformed the landscape through which the river flows. Principal changes that accompany the spread of human settlement include draining of floodplains, the intensification of agriculture, timber harvest and many others that may be less obvious but are cumulatively important. Through a combination of influences over flow regime, water tables are often lowered and drainage is increased, making both droughts and floods more likely. Erosion delivers more sediments to the stream channel, with detrimental effects on instream habitat. Removal of riparian vegetation leads to many changes, including higher temperatures, a shift from heterotrophy to autotrophy, reduced bank stability, and loss of a natural capacity to prevent sediments and nutrients from reaching stream channels. Human activities also increase the chemical wastes that enter rivers, from both agricultural and urban sources. Collectively these changes in land use comprise the most broad and serious class of threats affecting rivers. Because the causes are broad, no single remedy will suffice. However, protection of headwater and riparian areas clearly is warranted, along with a greater reliance on biological monitoring of the integrity of river systems.

The spread of alien species both by planned and accidental means constitutes an especially serious threat because, once established, exotics are extremely difficult to control. Native species may decline in the face of new predation pressure or diseases, and the impact of species replacements can cascade through several trophic levels. Rapid shifts in temperature and rainfall regime brought on by global climate change constitute an uncertain but potentially serious threat with effects on both community structure and ecosystem function, especially at higher latitudes where climate change is expected to be greatest.

Protection and restoration of running waters is an urgent need. Rivers have considerable ability to recovery from 'pulse' events, such as chemical inputs from an accidental spill or a point source. Unfortunately, the most serious threats are essentially continuous, or 'press' events, such as channelization and other forms of river regulation. These require much more effort to address, and one hopes that the principles of ecology and geomorphology will play an increasingly important role in the design and maintenance of structural changes to rivers. The diffuse and widespread effects of human activities on the land, termed non-point source pollution, can potentially be met through appropriate protection of riparian and headwater areas in a management network designed around the watershed as a unit. Finally, there is need for a system of parks and preserves for running waters that have as their goal the protection of the species and functions of these special ecosystems.

References

Adams, J. (1980) The feeding behaviour of *Bdellocephala punctata*, a freshwater triclad. *Oikos*, **35**, 2–7.

Alexander, R. McN. (1970) *Functional Design in Fishes*, Hutchinson University Library, London.

Alexander, G.R. (1979) Predators of fish in coldwater streams, in *Predator-prey Systems in Fisheries Management*, (ed. H. Clepper), Sports Fishing Institute, Washington DC, pp. 153–70.

Allan, J.D. (1975) The distributional ecology and diversity of benthic insects in Cement Creek, Colorado. *Ecology*, **56**, 1040–53.

Allan, J.D. (1976) Life history patterns in zooplankton. *Am. Nat.*, **110**, 165–80.

Allan, J.D. (1978) Trout predation and the size composition of stream drift. *Limnol. Oceanogr.*, **23**, 1231–7.

Allan, J.D. (1981) Determinants of diet of brook trout (*Salvelinus fontinalis*) in a mountain stream. *Can. J. Fish. Aquat. Sci.*, **38**, 184–92.

Allan, J.D. (1982a) Feeding habits and prey consumption of three setipalpian stoneflies (Plecoptera) in a mountain stream. *Ecology*, **63**, 26–34.

Allan, J.D. (1982b) The effects of reduction in trout density on the invertebrate community of a mountain stream. *Ecology*, **63**, 1444–55.

Allan, J.D. (1984) Hypothesis testing in ecological studies of aquatic insects, in *The Ecology of Aquatic Insects*, (eds V.H. Resh and D.M. Rosenberg), Praeger, New York, pp. 484–507.

Allan, J.D. (1985) The production ecology of Ephemeroptera in a Rocky Mountain stream. *Verh. Int. Ver. Theor. Ang. Limnol.*, **22**, 3233–7.

Allan, J.D. (1987) Macroinvertebrate drift in a Rocky Mountain stream. *Hydrobiologia*, **144**, 261–8.

Allan, J.D. and Flecker, A.S. (1988) Prey preference in stoneflies: A comparative analysis of prey vulnerability. *Oecologia*, **76**, 495–503.

Allan, J.D. and Flecker, A.S. (1993) Biodiversity conservation in running waters. *BioScience*, **43**, 32–43.

Allan, J.D., Flecker, A.S. and Kohler, S.L. (1990) Diel changes in epibenthic activity and gut fullness of some mayfly nymphs. *Verh. Int. Ver. Theor. Ang. Limnol.*, **24**, 2882–5.

Allan, J.D., Flecker, A.S. and McClintock, N.C. (1986) Diel epibenthic activity of mayfly nymphs, and its nonconcordance with behavioral drift. *Limnol. Oceanogr.*, **31**, 1057–65.

Allan, J.D., Flecker, A.S. and McClintock, N.C. (1987a) Prey size selection by carnivorous stoneflies. *Limnol. Oceanogr.*, **31**, 864–72.

Allan, J.D., Flecker, A.S. and McClintock, N.C. (1987b) Prey preference of stoneflies: sedentary vs. mobile prey. *Oikos*, **49**, 323–32.

Allan, J.D. and Russek, E. (1985) The quantification of stream drift. *Can. J. Fish. Aquat. Sci.*, **42**, 210–15.

Allen, K.R. (1941) Studies on the biology of the early stages of the salmon (*Salmo salar*). *J. Anim. Ecol.*, **10**, 47–76.

Allen, K.R. (1951) The Horokiwi stream: A study of a trout population. *N. Z. Mar. Dep. Fish. Bull.*, **10**.

Allen, K.R. (1969) Distinctive aspects of the ecology of stream fishes: A review. *J. Fish. Res. Board Can.*, **26**, 1429–38.

Ambühl, H. (1959) Die Bedeutung der Strömung als ökologischer Faktor. *Schweiz Z. Hydrol.*, **21**, 133–264.

American Public Health Association (1989) *Standard Methods for the Examination of Water and Wastewater*, 17th edn. American Public Health Association (APHA), Washington DC, 1550 pp.

Anderson, N.H. (1966) Depressant effect of moonlight on activity of aquatic insects. *Nature* (*London*), **209**, 319–20.

Anderson, N.H. (1976) Carnivory by an aquatic detritivore, *Clistoronia magnifica* (Trichoptera: Limnephilidae). *Ecology*, **57**, 1081–5.

Anderson, N.H. and Cummins, K.W. (1979) The influences of diet on the life histories of aquatic insects. *J. Fish. Res. Board Can.*, **36**, 335–42.

Anderson, N.H. and Sedell, J.R. (1979) Detritus processing by macroinvertebrates in stream ecosystems. *Ann. Rev. Entomol.*, **24**, 351–77.

Anderson, N.H., Sedell, J.R., Roberts, L.M. and Triska, F.J. (1978) The role of aquatic invertebrates in processing of wood debris in coniferous forest streams. *Am. Mid. Nat.*, **100**, 64–82.

Andersson, K.G., Brönmark C. and Herrmann, J. *et al.* (1986) Presence of sculpins (*Cottus gobio*) reduces drift and activity of *Gammarus pulex* (Amphipoda). *Hydrobiologia*, **133**, 209–15.

Angermeier, P.L. and Karr, J.R. (1983) Fish communities along environmental gradients in a system of tropical streams. *Environ. Biol. Fish.*, **9**, 117–35.

Araujo-Lima, C., Forsberg, B., Victoria, R. and Martinelli, L. (1986) Energy sources for detritivorous fishes in the Amazon. *Science*, **234**, 1256–8.

Armitage, P.D. (1977) Invertebrate drift in the regulated River Tees and an unregulated tributary, Maize Beck, below Cow Green dam. *Freshwater Biol.*, **7**, 167–84.

Armitage, P.D. (1978) Downstream changes in the composition, numbers and biomass of bottom fauna in the Tees below Cow Green Reservoir and in an unregulated tributary Maize Beck, in the first five years after impoundment. *Hydrobiologia*, **58**, 145–56.

Armitage, P.D. (1984) Environmental changes induced by stream regulation and their effect on lotic macroinvertebrate communities, in *Regulated Rivers*, (eds A. Lillehammer and S.J. Saltveit), Universitetforlaget AS, Oslo, pp. 139–66.

Armitage, P.D. and Blackburn, J.H. (1990) Environmental stability and communities of Chironomidae (Diptera) in a regulated river. *Regulated Rivers: Res. Manage.*, **5**, 319–28.

Arsuffi, T.L. and Suberkropp, K. (1984) Leaf processing capabilities of aquatic hyphomycetes: interspecific differences and influence on shredder feeding preference. *Oikos*, **42**, 144–54.

Arsuffi, T.L. and Suberkropp, K. (1986) Growth of two stream caddisflies (Trichoptera) on leaves colonized by different fungal species. *J. N. Am. Benthol. Soc.*, **5**, 297–305.

Bagnold, R.A. (1966) An approach to the sediment transport problem of general physics. *U. S. Geol. Surv. Prof. Pap.*, 422–I.

Bailey, R.G. (1966) Observations on the nature and importance of organic drift in a Devon river. *Hydrobiologia*, **27**, 353–67.

Baker, A.L. and Baker, K.K. (1979) Effects of temperature and current discharge on the concentration and photosynthetic activity of the phytoplankton in the upper Mississippi River. *Freshwater Biol.*, **9**, 191–8.

Baker, J.A. and Ross, S.T. (1981) Spatial and temporal resource utilization by southeastern cyprinids. *Copeia*, 1981, 178–89.

Baker, R.L. and Dixon, S.M. (1986) Wounding as an index of aggressive interactions in larval Zygoptera (Odonata). *Can. J. Zool.*, **64**, 893–7.

Ball, R.C. and Hooper, F.F. (1963) Translocation of phosphorus in a trout stream ecosystem, in *Radioecology*, (eds V. Schultz and A.W. Klement, Jr.) Reinhold Publishing, New York, pp. 217–28.

Ball, R.C., Woitalik, T.A. and Hooper, F.F. (1963) Upstream dispersion of radiophosphorus in a Michigan trout stream. *Michigan Acad. Sci. Arts Lett.*, **48**, 57–64.

Baltz, D.M., Moyle, P.B. and Knight, N.J. (1982) Competitive interactions between benthic stream fishes, riffle sculpin, *Cottus gulosus*, and speckled dace, *Rhinichthys osculus*. *Can. J. Fish. Aquat. Sci.*, **39**, 1502–11.

Bannon, E. and Ringler, N.H. (1986) Optimal prey size for stream resident brown trout (*Salmo trutta*): tests of predictive models. *Can. J. Zool.*, **64**, 704–13.

Barbour, C.D. and Brown, J.H. (1974) Fish species diversity in lakes. *Am. Nat.*, **108**, 473–89.

Bardach, J.E., Todd, J.H. and Crickmer, R. (1967) Orientation by taste in fish of the genus *Ictalurus*. *Science*, **155**, 1276–8.

Barlöcher, F. (1982a) The contribution of fungal enzymes to the digestion of leaves by *Gammarus fossarum* Koch (Amphipoda). *Oecologia*, **52**, 1–4.

Barlöcher, F. (1982b) On the ecology of Ingoldian fungi. *Bioscience*, **32**, 581–6.

Barlöcher, F. (1983) Seasonal variation of standing crop and digestibility of CPOM in a Swiss Jura stream. *Ecology*, **64**, 1266–72.

Barlöcher, F. (1985) The role of fungi in the nutrition of stream invertebrates. *Bot. J. Linn. Soc.*, **91**, 83–94.

Barlöcher, F. and Kendrick, B. (1973a) Fungi in the diet of *Gammarus pseudolimnaeus* (Amphipoda). *Oikos*, **24**, 295–300.

Barlöcher, F. and Kendrick, B. (1975) Leaf-conditioning by micro-organisms. *Oecologia*, **20**, 359–62.

Barlöcher, F., Kendrick, B. and Michaelides, J. (1978) Colonization and conditioning of *Pinus resinosa* needles by aquatic hyphomycetes. *Arch. Hydrobiol.*, **81**, 462–74.

Barmuta, L.A. and Lake, P.S. (1982) On the value of the river continuum concept. *N. Z. J. Mar. Freshwater Res.*, **16**, 227–31.

Barns, H.H. (1967) Roughness characteristics of natural channels. *U.S. Geol. Surv. Water-Supply Pap.*, **1849**, 213 pp.

Barrett, S.C.H. (1989) Waterweed invasions. *Sci. Am.*, October, 90–7.

Barrow, C.J. (1987) The environmental impacts of the Tucurí Dam on the middle and lower Tocantins River Basin, Brazil. *Regulated Rivers: Res. Manage.* **1**, 49–60.

Barton, D.R., Taylor, W.D. and Biette, R.M. (1985) Dimensions of riparian buffer strips required to maintain trout habitat in southern Ontario streams. *N. Am. J. Fish. Manage.*, **5**, 364–78.

Bates, H.W. (1863) *The Naturalist on the River Amazon*. I. Murray, London.

Baxter, R.M. (1977) Environmental effects of dams and impoundments. *Ann. Rev. Ecol. Syst.*, **8**, 255–84.

Bayley, P.B. (1989) Aquatic environments in the Amazon Basin, with an analysis of carbon sources, fish production, and yield, in *Proceedings of the International Large River Symposium*, (ed. D.P. Dodge),

Canadian Special Publication of Fisheries and Aquatic Sciences, **106**, pp. 399–408.

Bayley, P.B. (1981) The flood pulse advantage and the restoration of river-floodplain systems. *Regulated Rivers: Res. Manage.*, **6**, 75–86.

Beauchamp, R.S.A. and Ullyot, P. (1932) Competitive relationships between certain species of freshwater triclads. *J. Ecol.*, **20**, 200–8.

Bellamy, L.S. and Reynoldson, T.B. (1974) Behaviour in competition for food amongst lake-dwelling triclads. *Oikos*, **25**, 356–64.

Bencala, K.E. (1984) Interactions of solutes and streambed sediment 2. A dynamic analysis of coupled hydrologic and chemical processes that determine solute transport. *Water Resour. Res.*, **20**, 1804–14.

Bencala, K.E. and Walters, R.A. (1983) Simulation of solute transport in a mountain pool-and-riffle stream: a transient storage model. *Water Resour. Res.* **19**, 718–24.

Benfield, E.F., Jones, D.R. and Patterson, M.F. (1977) Leaf pack processing in a pastureland stream. *Oikos*, **29**, 99–103.

Benfield, E.F., Paul, R.W. and Webster, J.R. (1979) Influence of exposure technique on leaf breakdown rates in streams. *Oikos*, **33**, 386–91.

Benfield, E.F. and Webster, J.R. (1985) Shredder abundance and leaf breakdown in an Appalachian mountain stream. *Freshwater Biol.*, **15**, 113–20.

Benke, A.C. (1990) A perspective on America's vanishing streams. *J. N. Am. Benthol. Soc.*, **9**, 77–88.

Benke, A.C. (1993) Concepts and patterns of invertebrate production in running waters. *Verh. Int. Ver. Theor. Ang. Limnol.*, **25**, 15–38.

Benke, A.C., Henry, R.L., III, Gillespie D.M. and Hunter, R.J. (1985) Importance of snag habitat for animal production in southeastern streams. *Fisheries*, **10**, 8–13.

Benke, A.C., Van Arsdall, T.C., Jr., Gillespie, D.M. and Parrish, F.K. (1984) Invertebrate productivity in a subtropical blackwater river: the importance of habitat and life history. *Ecol. Monogr.*, **54**(1), 25–63.

Benke, A.C. and Wallace, J.B. (1980) Trophic basis of production among net-spinning caddisflies in a southern Appalachian stream. *Ecology*, **61**, 108–18.

Bennett, J.P., Woodward, J.W. and Shultz, D.J. (1986) Effect of discharge on the chlorophyll *a* distribution in the tidally-influenced Potomac River. *Estuaries*, **9**, 250–60.

Berg, M.B. (1994) Larval food and feeding behaviour, in *Chironomidae: Biology and Ecology of Non-biting Midges*, (eds P.D. Armitage, P.S. Cranston and L.C.V. Pinder), Chapman & Hall, London, 136–68.

Berkman, H.E. and Rabeni, C.F. (1987) Effect of siltation on stream fish communities. *Environ. Biol. Fish.*, **18**, 285–94.

Berner, E.K. and Berner, R.A. (1987) *The Global Water Cycle*, Prentice-Hall, Englewood Cliffs, New Jersey.

Berner, L.M. (1951) Limnology of the lower Missouri River. *Ecology*, **32**, 1–12.

Berrie, A.D. (1972a) The occurrence and composition of seston in the River Thames and the role of detritus as an energy source for secondary production in the river. *Mem. Ist. Ital. Idrobiol.*, **29** (supplement), 475–83.

Berrie, A.D. (1972b) Productivity of the Thames at Reading, in *Conservation and Productivity in Natural Waters. Symposium of the Zoological Society of London*. (eds R.W. Edwards and D.J. Garrod). pp. 69–86.

Beschta, R.L. (1978) Long-term patterns of sediment production following road construction and logging in the Oregon coast range. *Water Resour. Res.*, **14**, 1011–16.

Beukema, J.J. (1968) Predation by threespine stickleback (*Gasterosteus aculeatus* L.): The influence of hunger and experience. *Behavior*, **31**, 1–126.

Bilby, R.E. (1981) Role of organic debris dams in regulating the export of dissolved and particulate matter from a forested watershed. *Ecology*, **62**, 1234–43.

Bilby, R.E. and Liken, G.E. (1979) Effect of hydrologic fluctuations on the transport of fine particulate organic carbon in a small stream. *Limnol. Oceanogr*, **24**, 69–75.

Bilby, R.E. and Liken, G.E. (1980) Importance of organic debris dams in the structure and function of stream ecosystems. *Ecology*, **61**, 1107–13.

Binns, N.A. and Eiserman, F.M. (1979) Quantification of fluvial trout habitat in Wyoming. *Trans. Am. Fish. Soc.*, **108**, 215–28.

Birch, L.C. (1957) The meanings of competition. *Am. Nat.*, **91**, 5–18.

Bird, G.A. and Hynes, H.B.N. (1981) Movements of immature insects in a lotic habitat. *Hydrobiologia*, **77**, 103–12.

Bishop, J.E. (1969) Light control of aquatic insect activity and drift. *Ecology*, **50**, 371–80.

Bishop J.E. and Hynes, H.B.N. (1969a) Downstream drift of the invertebrate fauna in a stream ecosystem. *Arch. Hydrobiol.*, **66**, 56–90.

Bishop, J.E. and Hynes, H.B.N. (1969b) Upstream movements of the benthic invertebrates in the Speed River, Ontario. *J. Fish. Res. Board Can.*, **26**, 279–98.

Bisson, P.A., Quinn, T.P., Reeves, G.H. and Gregory, S.V. (1992) Best management practices, cumulative effects, and long-term trends in fish abundance in Pacific Northwest river systems, in *Watershed Management: Balancing Sustainability and Environmental Change*, (ed. R.J. Naiman), Springer-Verlag, New York, pp. 189–233.

Bjarnov, N. (1972) Carbohydrases in *Chironomus*, *Gammarus*, and some Trichoptera larvae. *Oikos*, **23**, 261–3.

Blum, J.L. (1960) Algal populations of flowing waters. *Spec. Publ. Pymatuning Lab Field Biol.*, **2**, 11–21.
Bocking, R.C. (1987) Canadian water: a commodity for export? in *Canadian Aquatic Resources. Canadian Bulletin of Fisheries and Aquatic Sciences* 215, (eds M.C. Healey and R.R. Wallace), pp. 105–35.
Bohle, H.W. (1978) Beziehungen zwischen dem Nahrungsangebot, der Drift und der raumlichen Verteilung bei Larven von *Baetis rhodani* (PICTET) (Ephemeroptera: Baetidae). *Arch. Hydrobiol.*, **84**, 500–25.
Borgh, T. (1927) Sitzungsprotokoll de 'Zoologiska Seminaret' der Universitat Uppsala vom 22.12.1927.
Bormann, F.H., Likens, G.E., Siccama, T.G., Pierce, R.S. and Eaton, J.S. (1974) The export of nutrients and recovery of stable conditions following deforestation at Hubbard Brook. *Ecol. Monogr.*, **44**, 255–77.
Bothar, A. (1987) The estimation of production and mortality of *Bosmina longirostris* (O.F. Müller) in the River Danube (Daubialia Hungarica, CIX). *Hydrobiologia*, **145**, 285–91.
Bothwell, M.L. (1989) Phosphorus-limited growth dynamics of lotic periphyton diatom communities: areal biomass and cellular growth rate responses. *Can. J. Fish. Aquat. Sci.*, **46**, 1293–301.
Bott, T.L. (1983) Primary productivity in streams, in *Stream Ecology*, (eds J.R. Barnes and G.W. Minshall), Plenum, New York, pp. 29–53.
Bott, T.L., Brock, J.T., Cushing, C.E., Gregory, S.V., King, D. and Petersen, R.C. (1978) A comparison of methods for measuring primary productivity and community respiration in streams. *Hydrobiologia*, **60**, 3–12.
Bott, T.L., Brock, J.T., Dunn, C.S., Naiman, R.J., Ovink, R.W. and Petersen, R.C. (1985) Benthic community metabolism in four temperate stream systems: an inter-biome comparison and evaluation of the river continuum concept. *Hydrobiologia*, **123**, 3–45.
Bott, T.L. and Kaplan, L.A. (1990) Potential for protozoan grazing of bacteria in streambed sediments. *J. N. Am. Benthol. Soc.*, **9**, 336–45.
Bott, T.L., Kaplan, L.A. and Kuserk, F.T. (1984) Benthic bacterial biomass supported by streamwater dissolved organic matter. *Microbial Ecol.*, **10**, 335–44.
Boulton, A.J. and Boon, P.I. (1991) A review of methodology used to measure leaf litter decomposition in lotic environments: Time to turn over an old leaf? *Aust. J. Mar. Freshwater Res.*, **42**, 1–43.
Bournard, M. and Thibault, M. (1973) La derive des organismes dans eaux courantes. *Ann. Limnol.*, **4**, 11–49.
Bovbjerg, R.V. (1970) Ecological isolation and competitive exclusion in two crayfish (*Orconectes virilis* and *Orconectes immunis*). *Ecology*, **51**, 225–36.
Bovee, K.D. (1982) A guide to steam habitat analysis using the instream flow incremental methodology. Instream flow information paper 12. FWS/OBS-82/86. United States Department of the Interior, Cooperative Instream Flow Service Group, Fort Collins Colorado.
Bowen, S.H. (1983) Detritivory in neotropical fish communities. *Environ. Biol. Fish.*, **9**, 137–44.
Bowlby, J.N. and Roff, J.C. (1986) Trophic structure in southern Ontario streams. *Ecology*, **67**, 1670–90.
Brett, J.R. (1971) Energetic responses of salmon to temperature. A study of some thermal relations in the physiology and freshwater ecology of sockeye salmon (*Oncorhynchus nerka*). *Am. Zool.*, **11**, 99–113.
Briand, F. and Cohen, J.E. (1987) Environment correlates of food chain length. *Science*, **238**, 956–60.
Briggs, J.C. (1986) Introduction to the zoogeography of North American fishes, in *Zoogeography of North American Freshwater Fishes*, (eds C.H. Hocutt and E.O. Wiley), Wiley Interscience, New York, pp. 1–16.
Britt, N.W. (1962) Biology of two species of Lake Erie mayflies, *Ephoron album* (Say) and *Ephemera simulans* Walker. *Bull. Ohio Biol. Surv.*, **5**, 70.
Brittain, J.E. and Saltveit, S.J. (1989) A review of the effect of river regulation on mayflies (Ephemeroptera). *Regulated Rivers: Res. Manage.*, **3**, 191–204.
Brönmark, C. (1985) Interactions between macrophytes and herbivores: an experimental approach. *Oikos*, **45**, 26–30.
Brönmark, C. and Malmqvist, B. (1986) Interactions between the leech *Glossiphonia complanata* and its gastropod prey. *Oecologia*, **69**, 268–76.
Brookes, A. (1988) *River Channelization: Perspective for Environmental Management*, John Wiley, Chichester, UK.
Brookes, A. (1989) Alternative channelization procedures, in *Alternatives in Regulated River Management*, (eds J.A. Gore and G.E. Petts), CRC Press, Boca Raton, Florida, pp. 139–62.
Brown, A.V. and Armstrong, M.L. (1985) Propensity to drift downstream among various species of fish. *J. Freshwater Ecol.*, **3**, 3–17.
Brown, L.R. and Moyle, P.B. (1991) Changes in habitat and microhabitat partitioning within an assemblage of stream fishes in response to predation by Sacramento squawfish (*Ptychocheilus grandis*). *Can. J. Fish. Aquat. Sci.*, **48**, 849–56.
Bruce, R.C. (1985) Larval periods, population structure, and the effects of stream drift in larvae of the salamanders *Desmognathus quadramaculatus* and *Leurognathus marmoratus* in a southern Appalachian stream. *Copeia*, **1985**, 847–54.
Brusven, M.A. and Rose, S.T. (1981) Influence of substrate composition and suspended sediment on insect predation by the torrent sculpin, *Cottus rhotheus*. *Can. J. Fish. Aquat. Sci.*, **38**, 1444–8.
Bruton, M.N. and Van As, J. (1986) Faunal invasions of aquatic ecosystems in southern Africa, with sugges-

tions for their management, in *The Ecology and Management of Biological Invasions in Southern Africa*, (eds I.A.W. MacDonald, F.J. Kruger and A.A. Ferrar), Oxford University Press, Cape Town, South Africa, pp. 47–61.

Bryan, J.E. and Larkin, P.A. (1972) Food specialization by individual trout. *J. Fish. Res. Board Can.*, **29**, 1615–24.

Burton, T.M., Standford, R.M. and Allan, J.W. (1985) Acidification effects on stream biota and organic matter processing. *Can. J. Fish. Aquat. Sci.*, **42**, 669–75.

Busch, D.E. and Fisher, S.G. (1981) Metabolism of a desert stream. *Freshwater Biol.*, **11**, 301–7.

Butcher, R.W. (1933) Studies on the ecology of rivers. I. On the distribution of macrophytic vegetation in the rivers of Britain. *J. Ecol.*, **21**, 58–91.

Cadwallader, P.L. (1975) The food of the New Zealand common river galaxias, *Galaxias vulgaris* Stokel (Pisces: Salmoniformes). *Aust. J. Mar. Freshwater Res.*, **26**, 15–30.

Calow, P. (1970) Studies of the natural diet of *Lymnaea pereger obtusa* (Kobelt) and its possible ecological implications. *Proc. Malacol. Soc. London*, **39**, 203–15.

Calow, P. (1975a) The feeding strategies of two freshwater gastropods, *Ancylus fluviatilis* Mull. and *Planorbis contortus* Linn. (Pulmonata), in terms of ingestion rates and absorption efficiencies. *Oceologia*, **20**, 33–49.

Calow, P. (1975b) Defecation strategies of two freshwater gastropods, *Ancylus fluviatilis* Mull. and *Planorbis contorta* Linn. (Pulmonata) with a comparison of field and laboratory estimates of food absorption rate. *Oceologia*, **20**, 51–63.

Calow, P. and Calow, L.J. (1975) Cellulase activity and niche separation in freshwater gastropods. *Nature*, **255**, 478–80.

Campbell, I. (1980) Diurnal variations in the activity of *Mirawara purpurea* Riek (Ephemeroptera, Siphlonuridae) in the Aberfeldy River, Victoria, in *Advances in Ephemeroptera Biology*, (eds J.F. Flannagan and K.E. Marshall), Plenum, New York, pp. 297–308.

Campbell, R.N.B. (1985) Comparison of the drift of live and dead *Baetis* nymphs in a weakening water current. *Hydrobiologia*, **126**, 229–36.

Campbell, H.W. and Irvine, A.B. (1977) Feeding ecology of the West Indian manatee *Trichechus manatus* Linnaeus. *Aquaculture*, **12**, 249–51.

Cargill, A.S., Cummins, K.W., Hanson, B.J. and Lowry, R.R. (1985) The role of lipids, fungi, and temperature on the nutrition of a shredder caddisfly, *Clistoronia magnifica* (Trichoptera: Limnephilidae). *Freshwater Invert. Biol.*, **4**, 64–78.

Carling, P.A. (1992) The nature of the fluid boundary layer and the selection of parameters for benthic ecology. *Freshwater Biol.*, **28**, 273–84.

Carlough, L.A. and Meyer, J.L. (1990) Rates of protozoan bacterivory in three habitats of a southeastern blackwater river. *J. N. Am. Benthol. Soc.*, **9**, 45–53.

Carlsson, M., Nilsson, L.M., Svensson, Bj. *et al.* (1977) Lacustrine seston and other factors influencing the black flies (Diptera: Simuliidae) inhabiting lake outlets in Swedish Lapland. *Oikos*, **29**, 229–38.

Carrick, T.R. (1979) The effect of acid water on the hatching of salmonid eggs. *J. Fish Biol.*, **14**, 165–72.

Casey, H. and Newton, P.V.R. (1972) The chemical composition and flow of the South Winterbourne in Dorset. *Freshwater Biol.*, **2**, 229–34.

Cederholm, C.J., Reid, L.M. and Salo, E.O. (1981) Cumulative effects of logging road sediment on salmonid populations in the Clearwater River, Jefferson County, Washington, in Proceedings of a conference 'Salmon spawning gravel: a renewable resource in the Pacific Northwest?' State of Washington Water Resource Center, Washington State University, Pullman, Washington USA, Report 39, pp. 38–74.

Cerri, R.D. and Fraser, D.F. (1983) Predation and risk in foraging minnows: balancing conflicting demands. *Am. Nat.*, **121**, 552–61.

Chambers, P.A., Prepas, E.E., Bothwell, M.L. and Hamilton, H.R. (1989) Roots versus shoots in nutrient uptake by aquatic macrophytes in flowing waters. *Can. J. Fish. Aquat. Sci.*, **46**, 435–9.

Chance, M.M. (1970) The functional morphology of the mouthparts of black fly larvae (Diptera: Simuliidae). *Quaestiones Entomologicae*, **6**, 245–84.

Chance, M.M. and Craig, D.A. (1986) Hydrodynamics and behaviour of Simuliidae larvae (Diptera) *Can. J. Zool.*, **64**, 1295–309.

Chandler, C.M. (1966) Environmental factors affecting the local distribution and abundance of four species of stream-dwelling triclads. *Invest. Indiana Lakes Streams*, **7**, 1–56.

Chapman, D.W. and Demory, R. (1963) Seasonal changes in the food ingested by aquatic insect larvae and nymphs in two Oregon streams. *Ecology*, **44**, 140–6.

Chaston, I. (1969) The light threshold controlling the periodicity of invertebrate drift. *J. Anim. Ecol.*, **38**, 171–80.

Chaston, I. (1972) Non-catastrophic invertebrate drift in lotic systems, in *Essays in Hydrobiology*, (eds R.B. Clark and R.J. Wootton), University of Exeter, pp. 33–51.

Chessman, B.C. (1985) Estimates of ecosystem metabolism in the La Trove River, Victoria. *Aust. J. Mar. Freshwater Res.*, **36**, 873–80.

Chesson, J. (1983) The estimation and analysis of preference and its relationship to foraging models. *Ecology*, **64**, 1297–304.

Chow, V.T. (1981) *Open-channel Hydraulics*, McGraw-Hill, New York.
Chudyba, H. (1965) *Cladophora glomerata* and accompanying algae in the Skawa River. *Acta Hydrobiol*, 7/Suppl 1, 93–126.
Chutter, F.M. and Noble, R.G. (1966) The reliability of method of sampling stream invertebrates. *Arch. Hydrobiol.*, **62**, 95–103.
Coffman, W.P., Cummins, K.W. and Wuycheck, J.C. (1971) Energy flow in a woodland stream ecosystem: I. Tissue support structure of the autumnal community. *Arch. Hydrobiol.*, **68**, 232–76.
Cohen, R.H., Dressler, P.V., Phillips, E.J.P. and Cory, R.L. (1984) The effects of the Asiatic clam, *Corbicula fluminea*, on phytoplankton of the Potomac River, Maryland. *Limnol. Oceanogr.*, **29**, 170–80.
Colbo, M.H. and Porter, G.N. (1979) Effects of food supply on the life history of Simuliidae (Diptera). *Can. J. Zool.*, **57**, 301–6.
Cole, G.A. (1979) *Limnology*, Mosby, St. Louis.
Cole, J.J., Caraco, N.F. and Peierls, B. (1991) Phytoplankton primary production in the tidal, freshwater Hudson River, New York (USA). *Verh. Int. Ver. Theor. Ang. Limnol.*, **24**, 171–9.
Cole, J.J., Caraco, N.F. and Peierls, B. (1992) Can phytoplankton maintain a positive carbon balance in a turbid, freshwater tidal estuary? *Limnol. Oceanogr.*, **37**, 1608–17.
Cole, J.J., Findlay, S. and Pace, M.L. (1988) Bacterial production in fresh and saltwater ecosystems: a cross-system overview. *Mar. Ecol. Prog. Ser.*, **43**, 1–10.
Colletti, P.J., Blinn, D.W., Pickart, A. and Wagner, V.T. (1987) Influence of different densities on the mayfly grazer *Heptagenia criddlei* on lotic diatom communities. *J. N. Am. Benthol. Soc.*, **6**, 270–80.
Connell, J.H. (1975) Some mechanisms producing structure in natural communities, in *Ecology and Evolution of Communities*, (eds M.L. Cody and J.M. Diamond), Harvard University Press, Cambridge, pp. 460–90.
Connell, J.H. (1978) Diversity in tropical rain forests and coral reefs. *Science*, **199**, 1302–10.
Connell, J.H. (1980) Diversity and the coevolution of competitors, or the ghost of competition past. *Oikos*, **35**, 131–8.
Cooke, J.G. and White, R.E. (1987) Spatial distribution of denitrifying activity in a stream draining an agricultural catchment. *Freshwater Biol.*, **18**, 509–19.
Cooper, S.D. (1984) The effects of trout on water striders in stream pools. *Oecologia*, **63**, 376–9.
Cooper, S.D., Smith, D.W. and Bence, J.R. (1985) Prey selection by freshwater predators with different foraging strategies. *Can. J. Fish. Aquat. Sci.*, **42**, 1720–32.
Corbet, P.S. (1959) Notes on the insect food of the Nile crocodile in Uganda. *Proc. R. Entomol. Soc. London*, **A34**, 17–22.
Corbet, P.S. (1960) The food of a sample of crocodiles from Lake Victoria. *Proc. Zool. Soc. London*, **133**, 561–72.
Corbet, P.S. (1980) Biology of Odonata. *Annu. Rev. Entomol.*, **25**, 189–217.
Corkum, L.D. and Clifford, H.F. (1980) The importance of species associations and substrate types of behavioural drift, in *Advances in Ephemeroptera Biology*, (eds J.F. Flannagan and K.E. Marshall), Plenum, New York, pp. 331–41.
Corn, P.S. and Bury, R.B. (1989) Logging in Western Oregon: Responses of headwater habitats and stream amphibians. *Forest Ecol. Manage.*, **29**, 39–57.
Costa, J.E. (1978) Holocene stratigraphy in flood frequency analysis. *Water Resour. Res.*, **14**, 626–32.
Costello, M.J., McCarthy, T.K. and O'Farrell, M.M. (1984) The stoneflies (Plecoptera) of the Corrib catchment area, Ireland. *Ann. Limnol.*, **20**, 25–34.
Courtenay, W.R. and Hensley, D.A. (1980) Special problems associated with monitoring exotic species, in *Biological Monitoring of Fish*, (eds C.H. Hocutt and J.R. Stauffer, Jr.), Lexington Books, Lexington, MA, pp. 281–307.
Courtenay, W.R., Hensley, D.A., Taylor, J.N. and McCann, J.A. (1984) Distribution of exotic fishes in the continental United States, in *Distribution, Biology, and Management of Exotic Fishes*, (eds W.R. Courtenay and J.R. Stauffer, Jr.), The Johns Hopkins University Press, Baltimore, MD, pp. 41–77.
Courtenay, W.R. and Stauffer, J.R., Jr. (eds) (1984) *Distribution, Biology, and Management of Exotic Fishes*, The Johns Hopkins University Press, Baltimore, MD.
Cowell, B.C. and Carew, W.C. (1976) Seasonal and diel periodicity in the drift of aquatic insects in a subtropical Florida stream. *Freshwater Biol.*, **6**, 587–94.
Craig, J.F. and Kemper, J.B. (1987) *Regulated Streams–Advances in Ecology*, Plenum, New York.
Cressey, G.B. (1963) *Asia's Lands and Peoples*, 3rd edn, McGraw-Hill, New York.
Crisp, D.T. and Howson, G. (1982) Effect of air temperature upon mean water temperature in streams of the North Pennines and English Lake District. *Freshwater Biol.*, **12**, 359–67.
Crocker, M.T. and Meyer, J.L. (1987) Interstitial dissolved organic carbon in sediments of a southern Appalachian headwater stream. *J. N. Am. Benthol. Soc.*, **6**, 159–62.
Cross, F.B. and Moss, R.E. (1987) Historic changes in fish communities and aquatic habitats in plains streams of Kansas, in *Community and Evolutionary Ecology of North American Stream Fishes*, (eds W.J. Matthews and D.C. Heins), University of Oklahoma Press, Norman, OK, pp. 155–65.

Crosskey, R.W. (1990) *The Natural History of Blackflies*, John Wiley, New York.

Crowl, T.A. and Covich, A.P. (1990) Predator-induced life-history shifts in a freshwater snail. *Science*, **247**, 949–51.

Crowther, R.A. and Hynes, H.B.N. (1977) The effect of road de-icing salt on the drift of stream benthos. *Environ. Poll.*, **14**, 113–26.

Cudney, M.D. and Wallace, J.B. (1980) Life cycles, microdistribution, and production dynamics of six species of net-spinning caddisflies in a large south-eastern (USA) river. *Holarctic Ecol.*, **3**, 169–82.

Cuffney, T.F. and Wallace, J.B. (1989) Discharge-export relationships in headwater streams: the influence of invertebrate manipulations and drought. *J. N. Am. Benthol. Soc.*, **8**, 331–41.

Cummins, K.W. (1962) An evaluation of some techniques for the collection and analysis of benthic samples with special emphasis on lotic waters. *Am. Midl. Nat.*, **67**, 477–504.

Cummins, K.W. (1973) Trophic relations of aquatic insects. *Ann. Rev. Entomol.*, **18**, 183–206.

Cummins, K.W. (1974) Structure and function of stream ecosystems. *Bioscience*, **24**, 631–41.

Cummins, K.W., Wilzbach, M.A., Gates, D.M. *et al.* (1989) Shredders and riparian vegetation. *Bioscience*, **39**, 24–30.

Cummins, K.W., Coffman, W.P. and Roff, P.A. (1966) Trophic relationship in a small woodland stream. *Verh. Int. Ver. Theor. Ang. Limnol.*, **16**, 627–38.

Cummins, K.W. and Klug, M.J. (1979) Feeding ecology of stream invertebrates. *Annu. Rev. Ecol. Systematics*, **10**, 147–72.

Cummins, K.W. and Lauff, G.H. (1969) The influence of substrate particle size on the microdistribution of the stream macrobenthos. *Hydrobiologia*, **34**, 145–81.

Cummins, K.W., Petersen, R.C. and Howard, F.O. *et al.* (1973) The utilization of leaf litter by stream detritivores. *Ecology*, **54**, 336–45.

Cummins, K.W., Sedell, J.R., Swanson, F.J. *et al.* (1983) Organic matter budgets for stream ecosystems: problems in their evaluation, in *Stream Ecology – Application and Testing of General Ecological Theory*, (eds J.R. Barnes and G.W. Minshall), Plenum Press, New York, pp. 299–353.

Currie, D.C. and Craig, D.A. (1987) Feeding strategies of larval black flies, in *Black Flies: Ecology, Population Management and Annotated World List*, (eds K.C. Kim and R.W. Merritt), Pennsylvania State University, University Park, pp. 155–170.

Cushing, C.E. (1964) Plankton and water chemistry in the Montreal River lake-stream system, Saskatchewan. *Ecology*, **45**, 306–13.

Cushing, C.E., Minshall, G.W. and Newbold, J.D. (1993) Transport dynamics of fine particulate organic matter in two Idaho streams. *Limnol. Oceanogr.*, **38**, 1101–15.

D'Agostino, A.S. and Provasoli, L. (1970) Dixenic culture of *Daphnia magna* Straus. *Biol. Bull.*, **139**, 485–94.

D'Angelo, D.J., Webster, J.R., Gregory, S.V. and Meyer, J.L. (1993) Transient storage in Appalachian and Cascade mountain streams as related to hydraulic parameters. *J. N. Am. Benthol. Soc.*, **12**, 223–35.

Dahm, C.N. (1981) Pathways and mechanisms for removal of dissolved organic organic carbon from leaf leachate in streams. *Can. J. Fish. Aquat. Sci.*, **38**, 68–76.

Danielpol, D.L. (1989) Groundwater fauna associated with riverine aquifers. *J. N. Am. Benthol. Soc.*, **8**, 18–35.

Davies, B.R. and Walker, K.F. (1986) *The Ecology of River Systems*. Monographiae Biologicae, Vol. 60. Dr. W. Junk, The Hague.

Davis, J.A. (1986) Boundary layers, flow micro-environments and stream benthos, in *Limnology in Australia*, (eds P. DeDeckker and W.D. Williams), Commonwealth Scientific and Industrial Research Organization Australia, Melbourne, pp. 293–312.

Davis, J.A. and Barmuta, L.A. (1989) An ecologically useful classifiction of mean and near-bed flows in streams and rivers. *Freshwater Biol.*, **21**, 271–82.

Davis, M.B. (1989) Insights from paleoecology on global change. *Bull. Ecol. Soc. Am.*, **70**, 222–8.

Day, J.W., Jr., Butler, T.J. and Conner, W.H. (1977) Productivity and nutrient export studies in a cypress swamp and lake system in Louisiana, in *Estuarine Processes, Volume II*, (ed. M. Wiley), Academic Press, New York, pp. 225–70.

Day, J.W., Hall, C.A.S., Jr., Kemp, W.M. and Yanez-Arancibia, A. (1989) *Estuarine Ecology*, John Wiley, New York.

de March, B.G.E. (1976) Spatial and temporal patterns in macrobenthic stream diversity. *J. Fish. Res. Board Can.*, **33**, 463–76.

Deacon, J.E. and Minkley, W.L. (1974) Desert fishes, in *Desert Biology*, Vol. 2, (ed. R.W. Brown), Academic Press, New York, pp. 385–488.

Décamps, H. (1967) Écologie des Trichoptères de la vallée d'Aure (Hautes-Pyrénées). *Ann. Limnol.*, **3**, 399–577.

Decámps, H. (1984) Biology of regulated rivers in France, in *Regulated Rivers*, (eds A. Lillehammer and S.J. Saltveit), Universitetsforlaget AS, Oslo, pp. 495–514.

Décamps, H., Capblanq, J. and Turenq, J.N. (1984) Lot, in *Ecology of European Rivers*, (ed. B.A. Whitton), Blackwell Scientific Publications, Oxford, pp. 207–36.

Delong, M.D. and Brusven, M.A. (1992) Patterns of chlorophyll *a* in an agricultural non-point source impacted stream. *Water Resour. Bull.*, **28**, 731–41.

Delong, M.D. and Brusven, M.A. (1993) Storage and decomposition of particulate organic matter along the longitudinal gradient of an agriculturally-impacted stream. *Hydrobiologia*, **262**, 77–88.

Dendy, J.S. (1944) The fate of animals in stream drift when carried into lakes. *Ecol. Monogr.*, **14**, 333–57.

Denny, M.W. (1988) *Biology and the Mechanics of the Wave-swept Environment*, Princeton University Press, Princeton, New Jersey.

Devonport, B.F. and Winterbourn, M.J. (1976) The feeding relationships of two invertebrate predators in a New Zealand river. *Freshwater Biol.*, **6**, 167–76.

DeWald, L. and Wilzbach, M.A. (1992) Interactions between native brook trout and hatchery brown trout: effects on habitat use, feeding and growth. *Trans. Am. Fish. Soc.*, **121**, 287–96.

Diamond, J. (1986) Overview: Laboratory experiments, field experiments, and natural experiments, in *Community Ecology*, (eds J. Diamond and T.J. Case), Harper & Row, New York, pp. 3–22.

Diamond, J. and Case, T.J. (eds) (1986) *Community Ecology*, Harper & Row, New York.

Dill, L.M. (1983) Adaptive flexibility in the foraging of fishes. *Can. J. Fish. Aquat. Sci.*, **40**, 398–408.

Dill, L.M. and Fraser, A.H.G. (1984) Risk of predation and the feeding behavior of juvenile coho salmon (*Oncorhynchus kisutch*). *Behav. Ecol. Sociobiol.*, **16**, 65–71.

Dill, L.M., Ydenberg, R.C. and Fraser, A.H.G. (1981) Food abundance and territory size in juvenile coho salmon. *Can. J. Zool.*, **59**, 1801–9.

Dillon, P.J. and Kirchner, W.B. (1975) The effects of geology and land use on the export of phosphorus from watersheds. *Water Res.*, **9**, 135–48.

Dimond, J.B. (1967) Evidence that drift of stream benthos is density related. *Ecology*, **48**, 855–7.

Dodge, D.P. (1989) Proceedings of the International Large River Symposium. *Can. Spec. Publ. Fish. Aquat. Sci.*, 106.

Dorris, T.C., Copeland, B.J. and Lauer, G.L. (1963) Limnology of the middle Mississippi River. IV. Physical and chemical limniology of river and chute. *Limnol. Oceanogr.*, **8**, 79–88.

Douglas, B. (1958) The ecology of the attached diatoms and other algae in a small stoney stream. *J. Ecol.*, **46**, 295–322.

Dudgeon, O. (1992) Aspects of the micro-distribution of insect macrobenthos in a forest stream in Hong Kong. *Arch. Hydrobiol. suppl.*, **64**, 221–39.

Dudley, T. and Anderson, N.H. (1982) A survey of invertebrates associated with wood debris in aquatic habitats. *Melanderia*, **39**, 1–21.

Dudley, T.L., Cooper, S.D. and Hemphill, N. (1986) Effects of macroalgae on a stream invertebrate community. *J. N. Am. Benthol. Soc.*, **5**, 93–106.

Dudley, T.L., D'Antonio, C.M. and Cooper, S.D. (1990) Mechanisms and consequences of interspecific competition between two stream insects. *J. Anim. Ecol.*, **59**, 849–66.

Duffer, W.R. and Dorris, T.C. (1966) Primary productivity in a southern Great Plains stream. *Limnol. Oceanogr.*, **11**, 143–51.

Dumont, H.J. (1986) Zooplankton of the Nile system, in *The Ecology of River Systems*, (eds B.R. Davies and K.F. Walker), Dr. W. Junk, Dordrecht, The Netherlands, pp. 75–88.

Dunne, T. and Leopold, L.B. (1978) *Water in Environmental Planning*, W.H. Freeman, San Francisco.

Durbin, A.G., Nixon, S.W. and Oviatt, C.A. (1979) Effects of the spawning migration of the Alewife, *Alosa pseudoharengus*, on freshwater ecosystems. *Ecology*, **60**, 8–17.

Dussault, G.V. and Kramer, D.L. (1981) Food and feeding behavior of the guppy, *Poecilia reticula* (Pisces: Poecilidae). *Can. J. Zool.*, **59**, 684–701.

Dynesius, M. and Nilsson, C. (1993) *The Status of Northern River Systems: A Basis for Conservation Management of Watersheds*, World Wildlife Fund, Sweden, 24 pp.

Edington, J.M. (1968) Habitat preferences in net-spinning caddis larvae with special reference to the influence of water velocity. *J. Anim. Ecol.*, **37**, 675–92.

Edmunds, M. (1974) *Defense in Animals – a Survey of Antipredator Defenses*, Longman, New York, NY.

Edwards, R.T. and Meyer, J.L. (1987) Metabolism of a sub-tropical low gradient blackwater river. *Freshwater Biol.*, **17**, 251–63.

Edwards, R.T., Meyer, J.L. and Findlay, S.E.G. (1990) The relative contribution of benthic and suspended bacteria to system biomass, production, and metabolism in a low-gradient blackwater river. *J. N. Am. Benthol. Soc.*, **9**, 216–28.

Edwards, R.W. and Owens, M. (1962a) The effects of plants on river conditions. I. Summer crops and estimates of productivity of macrophytes in a chalk system. *J. Ecol.*, **48**, 151–60.

Edwards, R.W. and Owens, M. (1962b) The effects of plants on river conditions. IV. The oxygen balance of a chalk stream. *J. Ecol.*, **50**, 207–20.

Edwards, R.W. and Brooker, M.P. (1975) *The Ecology of the Wye. Monographiae Biologicae*, Vol. 50, Dr. W. Junk, The Hague.

Egglishaw, H.J. (1964) The distributional relationship between the bottom fauna and plant detritus in streams. *J. Anim. Ecol.*, **38**, 19–33.

Egglishaw, H.J. and Morgan, N.C. (1965) A survey of the bottom fauna of streams in the Scottish highlands. Part II. The relationship of the fauna to the chemical and geological conditions. *Hydrobiologia* **26**, 173–83.

Eichenberger, E. and Schlatter, A. (1978) Effect of

herbivorous insects on the production of benthic algal vegetation in outdoor channels. *Verh. Int. Ver. Theor. Ang. Limnol.*, **20**, 1806–10.

Elliott, J.M. (1967) Invertebrate drift in a Dartmoor stream. *Arch. Hydrobiol.*, **63**, 202–37.

Elliott, J.M. (1968) The daily activity patterns of mayfly nymphs (Ephemeroptera). *Can. J. Zool.*, **155**, 201–21.

Elliott, J.M. (1969) Diel periodicity in invertebrate drift and the effect of different sampling periods. *Oikos*, **20**, 524–8.

Elliott, J.M. (1970a) The diel activity patterns of caddis larvae. *Can. J. Zool.*, **160**, 279–90.

Elliott, J.M. (1970b) Methods of sampling invertebrate drift in running water. *Ann. Limnol.*, **6**, 133–59.

Elliott, J.M. (1971a) Upstream movements of benthic invertebrates in a Lake District stream. *J. Anim Ecol.*, **40**, 235–52.

Elliott, J.M. (1971b) The distances travelled by drifting invertebrates in a Lake District stream. *Oceologia*, **6**, 191–220.

Elliott, J.M. (1975) Effect of temperature on the hatching time of eggs of *Ephemerella ignita* (Poda) (Ephemeroptera: Ephemerellidae). *Freshwater Biol.*, **8**, 51–8.

Elliott, J.M. (1976) The energetics of feeding, metabolism and growth of the brown trout (*Salmo trutta* L.) in relation to body weight, water temperature, and ration size. *J. Anim. Ecol.*, **45**, 923–48.

Elliott, J.M. (1978) The growth rate of brown trout (*Salmo trutta* L.) fed on maximum rations. *J. Anim. Ecol.*, **44**, 805–21.

Elliott, J.M. (1987a) Egg hatching and resource partitioning in stoneflies: the six British *Leuctra* spp. (Plecoptera: Leuctridae). *J. Anim. Ecol.*, **56**, 415–26.

Elliott, J.M. (1987b) Population regulation in contrasting populations of trout *Salmo trutta* in two lake district streams. *J. Anim. Ecol.*, **56**, 83–98.

Elliott, J.M. (1988) Egg hatching and resource partitioning in stoneflies (Plecoptera): ten British species in the family Nemouridae. *J. Anim. Ecol.*, **57**, 201–16.

Elliott, J.M. and Minshall, G.W. (1968) The invertebrate drift in the river Duddon, English Lake district. *Oikos*, **19**, 39–52.

Elser, J.J., Marzolf, E.R. and Goldman, C.R. (1990) Phosphorus and nitrogen limitation of phytoplankton growth in the freshwaters of North America: a review and critique of experimental enrichments. *Can. J. Fish. Aquat. Sci.*, **47**, 1468–77.

Elwood, J.W. and Nelson, D.J. (1972) Periphyton production and grazing rates in a stream measured with a ^{32}P material balance method. *Oikos*, **23**, 295–303.

Elwood, J.W., Newbold, J.D., Trimble, A.F. and Stark, R.W. (1981) The limiting role of phosphorus in a woodland stream ecosystem: effects of P enrichment on leaf decomposition and primary producers. *Ecology*, **62**, 146–58.

Endler, J.A. (1983) Natural and sexual selection on color patterns in poeciliid fishes. *Environ. Biol. Fish.*, **9**, 173–90.

Englund, G. (1991) Assymetric resource competition in a filter-feeding stream insect (*Hydropsyche siltalai*; Trichoptera). *Freshwater Biol.*, **26**, 425–32.

Eriksen, C.H. (1963) The relation of oxygen consumption to substrate particle size in two burrowing mayflies. *J. Exp. Biol.*, **40**, 447–53.

Eriksen, C.H. (1964) The influence of respiration and substrate upon the distribution of burrowing mayfly naiads. *Verh. Int. Ver. Theor. Ang. Limnol.*, **15**, 903–11.

Erman, D.C. and Erman, N.A. (1984) The response of stream macroinvertebrates to substrate size and heterogeneity. *Hydrobiologia*, **108**, 75–82.

Ertel, J.R., Hedges, J.I., Devol, A.H., Richey, J.E. and Ribeiro, M. (1986) Dissolved humic substances of the Amazon River system. *Limnol. Oceanogr.*, **31**, 739–54.

Ertl, M. and Tomajka, J. (1973) Primary production of the periphyton in the littoral of the Danube. *Hydrobiologia*, **42**, 429–44.

Everest, F.H., Beschta, R.L., Scrivener, J.C., Koski, K.V., Sedell, J.R. and Cederholm, C.J. (1987) Fine sediment and salmonid production: a paradox, in *Streamside Management: Forestry and Fishery Interactions*, (eds E.O. Salo and T.W. Cundy), Contribution 57, Institute of Forest Resources, University of Washington, Seattle, Washington, pp. 98–142.

Fairchild, G.W. and Lowe, R.L. (1984) Artificial substrates which release nutrients: effects on periphyton and invertebrate succession. *Hydrobiologia*, **114**, 29–37.

Fausch, K.D., Karr, J.R. and Yant, P.R. (1984) Regional application of an index of biotic integrity based on stream fish communities. *Trans. Am. Fish. Soc.*, **113**, 39–55.

Fausch, K.D. and White, R.J. (1981) Competition between brook trout (*Salvelinus fontinalis*) and brown trout (*Salmo trutta*) for positions in a Michigan stream. *Can. J. Fish. Aquat. Sci.*, **38**, 1220–7.

Feeny, P. (1976) Plant apparency and chemical defense. *Recent Adv. Phytochem.*, **10**, 1–40.

Feminella, J.W., Power, M.E. and Resh, V.H. (1989) Periphyton responses to invertebrate grazing and riparian canopy in three northern California coastal streams. *Freshwater Biol.*, **22**, 445–57.

Feminella, J.W. and Resh, V.H. (1989) Submersed macrophytes and grazing crayfish: an experimental study of herbivory in a California freshwater marsh. *Holarctic Ecol.*, **12**, 1–8.

Fenchel, T. and Harrison, P. (1976) The significance of bacterial grazing and mineral cycling for the decomposition of particulate detritus, in *The Role of Terrestrial and Aquatic Organisms in Decomposition Processes*, (eds J.M. Anderson and A. Macfadyen), Blackwell, Oxford, pp. 285–99.

Feth, J.H. (1971) Mechanisms controlling world water chemistry: Evaporation-crystallization process. *Science*, **172**, 870–1.

Fetter, C.W. (1988) *Applied Hydrogeology*, 2nd edn, Merrill Publishing Company, Columbus, Ohio.

Filardo, M.J. and Dunstan, W.M. (1985) Hydrodynamic control of phytoplankton in low salinity waters of the James River Estuary, Virginia, U.S.A. *Estuar. Coastal Shelf Sci.*, **21**, 653–67.

Findlay, S., Meyer, J.L. and Risley, R. (1986a) Benthic bacterial biomass and production in two blackwater rivers. *Can. J. Fish. Aquat. Sci.*, **43**, 1271–6.

Findlay, S., Meyer, J.L. and Smith, P.J. (1984) Significance of bacterial biomass in the nutrition of a freshwater isopod (*Lirceus* sp.). *Oecologia*, **63**, 38–42.

Findlay, S., Meyer, J.L. and Smith, P.J. (1986b) Contribution of fungal biomass to the diet of a freshwater isopod (*Lirceus* sp). *Freshwater Biol.*, **16**, 377–85.

Finger, T.R. (1982) Interactive segregation among three species of sculpin (*Cottus*). *Copeia*, **1982**, 680–94.

Fischer, H.B., List, E.J., Koh, R.C.Y. *et al.* (1979) *Mixing in Inland and Coastal Waters*, Academic Press, New York.

Fisher, S.G. (1977) Organic matter processing by a stream-segment ecosystem: Fort River, Massachusetts, U.S.A. *Int. Rev. Gesamt. Hydrobiol.*, **62**, 701–27.

Fisher, S.G. and Carpenter, S.R. (1976) Ecosystem and macrophyte primary production of the Fort River, Massachusetts. *Hydrobiologia*, **47**, 175–87.

Fisher, S.G., Gray, L.J., Grimm, N.B. and Busch, D.E. (1982) Temporal succession in a desert stream ecosystem following flash flooding. *Ecol. Monogr.*, **52**, 93–110.

Fisher, S.G. and Likens, G.E. (1973) Energy flow in Bear Brook, New Hampshire: an integrative approach to stream ecosystem metabolism. *Ecol. Monogr.*, **43**, 421–39.

Fisher, S.G. and Gray, L.G. (1983) Secondary production and organic matter processing by collector macroinvertebrates in a desert stream. *Ecology*, **64**, 1217–24.

Fittkau, E.J. (1970) Role of caimans in the nutrient regime of mouth-lakes of Amazon effluents (an hypothesis). *Biotropica*, **2**, 138–42.

Flecker, A.S. (1984) The effects of predation and detritus on the structure of a stream insect community: a field test. *Oecologia*, **64**, 300–5.

Flecker, A.S. (1992a) Fish predation and the evolution of invertebrate drift periodicity: Evidence from neotropical streams. *Ecology*, **73**, 438–48.

Flecker, A.S. (1992b) Fish trophic guilds and the structure of a tropical stream: weak direct vs. strong indirect effects. *Ecology*, **73**, 927–40.

Flecker, A.S. and Feifarek, B. (1994) Disturbance and the temporal variability of invertebrate assemblages in two Andean streams. *Freshwater Biol.*, **31**, 131–42.

Flecker, A.S. and Townsend, C.R. (1994) Community-wide consequences of trout introduction in New Zealand streams. *Ecol. Appl.*, **4**.

Floyd, K.B., Hoyt, R.D. and Timbrook, S. (1984) Chronology of appearance and habitat partitioning by stream larval fishes. *Trans. Am. Fish. Soc.*, **113**, 217–23.

Fox, L.R. and Murdoch, W.W. (1978) Effects of feeding history on short-term and long-term fuctional responses in *Notonecta hoffmanni*. *J. Anim. Ecol.*, **47**, 945–59.

Franklin, J.F. (1992) Scientific basis for new perspectives in forests and streams, in *Watershed Management: Balancing Sustainability and Environmental Change*, (ed. R.J. Naiman), Springer-Verlag, New York.

Fraser, D.F. and Cerri, R.D. (1992) Experimental evaluation of predator-prey relationships in a patchy environment: Consequences for habitat use patterns in minnows. *Ecology*, **63**, 307–13.

Fredeen, F.J.H. (1964) Bacteria as a food for blackfly larvae (Diptera: Simuliidae) in laboratory cultures and in natural streams. *Can. J. Zool.*, **42**, 527–48.

Free, G.R., Browning, G.M. and Musgrave, G.W. (1940) *Relative Infiltration and Related Physical Characteristics of Certain Soils*, USDA Technical Bulletin 729, U.S. Government Printing Office, Washington, DC.

Frissell, C.A., Liss, W.L., Warren, C.E. and Hurley, M.D. (1986) A hierarchical framework for stream habitat classification: Viewing streams in a watershed context. *Environ. Manage.*, **10**, 199–214.

Fruget, J.F. (1992) Ecology of the lower Rhône after 200 years of human influence: a review. *Reg. Riv: Res. Manage.*, **7**, 233–46.

Fryer, G. (1959) The trophic interrelationship and ecology of some littoral communities of of Lake Nyasa with especial reference to the fishes, and a discussion of the evolution of a group of rock-frequenting Cichlidae. *Proc. Zool. Soc. London*, **132**, 153–281.

Fuller, R.L., Roelofs, J.L. and Fry, T.J. (1986) The importance of algae to stream invertebrates. *J. N. Am. Benthol. Soc.*, **5**, 290–6.

Gameson, A.L.H. and Wheeler, A. (1977) Restoration and recovery of the Thames Estuary, in *Recovery and Restoration of Damaged Ecosystems*, (eds J. Cairns, Jr., K.L. Dickson and E.E. Herricks), University of Virginia Press, Charlottesville, VA, pp. 72–101.

Gatz, A.J., Jr. (1973) Speed, stamina, and muscles in fish. *J. Fish. Res. Board Can.*, **30**, 325–8.

Gatz, A.J., Jr. (1979) Ecological morphology of freshwater stream fishes. *Tulane Studies Zool. Bot.*, **21**, 91–124.

Gaufin, A.R. (1959) Production of bottom fauna in the Provo River, Utah. *Iowa St. Coll. J. Sci.*, **33**, 395–419.

Gaufin, A.R., Harris, E.K. and Walter, J. (1956) A statistical evaluation of stream bottom sampling data obtained from samplers. *Ecology*, **37**, 643–8.

Gee, J.H. (1983) Ecological implications of buoyancy control in fish, in *Fish Biomechanics*, (eds P.W. Webb and D. Weihs), Praeger Scientific, New York, pp. 140–76.

Gee, J.H. and Northcote, T.G. (1963) Comparative ecology of two sympatric species of dace (*Rhinichthys*) in the Fraser River system, British Columbia. *J. Fish. Res. Board Can.*, **20**, 105–18.

Geesey, C.G., Mutch, R. and Costerton, J.W. (1978) Sessile bacteria: an important component of the microbial population in small mountain streams. *Limnol. Oceanogr.*, **23**, 1214–23.

Georgian, T. and Thorp, J.H. (1992) Effects of microhabitat selection on feeding rates of net-spinning caddisflies. *Ecology*, **73**, 229–40.

Georgian, T.J. and Wallace, J.B. (1981) A model of seston capture by net-spinning caddisflies. *Oikos*, **36**, 147–57.

Georgian, T.J. and Wallace J.B. (1983) Seasonal production dynamics in a guild of periphyton-grazing insects in a southern Appalachian stream. *Ecology*, **64**, 1236–48.

Gessner, F. (1955) Hydrobotanik I. Energiehaushalt. *Veb. Deutsch. Ver. Wissensch.*, Berlin.

Gessner, F. (1959) Hydrobotanik II. Stoffhaushalt. *Veb. Deutsch. Ver. Wissensch.*, Berlin.

Gianessi, L.P., Peskin, H.M., Crosson, P. and Puffer, C. (1986) Nonpoint source pollution controls: are cropland controls the answer? *Resources for the Future*, Washington, DC.

Gibbs, R.J. (1970) Mechanisms controlling world water chemistry. *Science*, **170**, 1088–90.

Gilliam, J.F., Fraser, D.F. and Sabat, A.M. (1989) Strong effects of foraging minnow on a stream benthic invertebrate community. *Ecology*, **70**, 445–52.

Ginitz, R.M. and Larkin, P.A. (1976) Factors affecting rainbow trout (*Salmo gairdneri*) predation on migrant fry of sockeye salmon (*Onchorhynchus nerka*). *J. Fish. Res. Board Can.*, **33**, 19–24.

Gleick, P. (1993) *Water in Crisis. A guide to the World's Fresh Water Resources*, Oxford University Press, New York.

Golladay, S.W. and Sinsabaugh, R.L. (1991) Biofilm development on leaf and wood surfaces in a boreal river. *Freshwater Biol.*, **25**, 437–50.

Golladay, S.W., Webster, J.R. and Benfield, E.F. (1983) Factors affecting food utilization by a leaf shredding aquatic insect: leaf species and conditioning time. *Holarctic Ecol.*, **6**, 157–62.

Golladay, S.W., Webster, J.R. and Benfield, E.F. (1987) Changes in stream morphology and storm transport of seston following watershed disturbance. *J. N. Am. Benthol. Soc.*, **6**, 1–11.

Golterman, H.L. (1975) Chemistry, in *River Ecology*, (ed. B.A. Whitton), University of California Press, Berkeley, pp. 39–80.

Golterman, H.L., Clymo, R.S. and Ohnstad, M.A.M. (1978) *Methods for Physical and Chemical Analysis of Fresh Waters*, Blackwell Scientific Publications, Oxford.

Gordon, N.D., McMahon, T.A. and Finlayson, B.L. (1992) *Stream hydrology. An Introduction for Ecologists*. John Wiley, Chichester.

Gore, J.A. (1978) A technique for predicting in-stream flow requirements of benthic macro-invertebrates. *Freshwater Biol.*, **8**, 141–51.

Gore, J.A. (1985) *Restoration of Rivers and Streams*, Butterworth Publishers, Boston, MA.

Gore, J.A. and Judy, R.D., Jr. (1981) Predictive models of benthic macroinvertebrate density for use in instream flow studies and regulated flow management. *Can. J. Fish. Aquat. Sci.*, **38**, 1363–70.

Gore, J.A. and Nestler, J.M. (1988) Instream flow studies in perspective. *Regulated Rivers: Res. Manage.*, **2**, 93–101.

Gore, J.A. and Petts, G.E. (1989) *Alternatives in Regulated River Management*, CRC Press, Boca Raton, FL.

Gorman, O.T. and Karr, J.R. (1978) Habitat structure and stream fish communities. *Ecology*, **59**, 507–15.

Gosselink, J.G., Shaffer, G.P., Lee, L.C. *et al.* (1990) Landscape conservation in a forested wetland watershed. *Bioscience*, **40**, 588–600.

Goulding, M. (1980) *The Fishes and the Forest*, University of California Press.

Graesser, A. and Lake, P.S. (1984) Diel changes in the benthos of stones and of drift in a southern Australian upland stream. *Hydrobiologia*, **111**, 153–60.

Grant, P.R. and Mackay, R.J. (1969) Ecological segregation of systematically related stream insects. *Can. J. Zool.*, **47**, 691–4.

Gray, J.R.A. and Edington, J.M. (1969) Effect of woodland clearance on stream temperature. *J. Fish. Res. Board Can.*, **26**, 399–403.

Gray, L.G. and Fisher, S.G. (1981) Postflood recolonization pathways of macroinvertebrates in a lowland Sonoran Desert stream. *Am. Midl. Nat.*, **106**, 249–57.

Greenberg, A.E. (1964) Plankton of the Sacramento River. *Ecology*, **45**, 40–9.

Gregg, W.W. and Rose, F.L. (1982) The effects of aquatic macrophytes on the stream microenvironment. *Aquat. Bot.*, **14**, 309–24.

Gregg, W.W. and Rose, F.L. (1985) Influences of aquatic macrophytes on invertebrate community structure, guild structure, and microdistribution in streams. *Hydrobiologia*, **128**, 45–56.

Gregory, S.V. (1980) Effects of light, nutrients, and grazing on periphyton communities in streams. Ph.D. Dissertation. Oregon State University, Corvallis, Oregon. 154 pp.

Gregory, S.V. (1983) Plant-herbivore interactions in stream systems, in *Stream Ecology*, (eds J.R. Barnes and G.W. Minshall, Plenum Press, New York, pp. 157–90.

Gregory, S.V., Swanson, F.J., McKee, W.A. and Cummins, K.W. (1991) An ecosystem perspective of riparian zones. *Bioscience*, **41**, 540–51.

Grimm, N.B. (1987) Nitrogen dynamics during succession in a desert stream. *Ecology*, **68**, 1157–70.

Grimm, N.B. (1988) Role of macroinvertebrates in nitrogen dynamics of a desert stream. *Ecology*, **69**, 1884–93.

Grimm, N.B. and Fisher, S.G. (1984) Exchange between interstitial and surface water: Implications for stream metabolism and nutrient cycling. *Hydrobiologia*, **111**, 219–28.

Grimm, N.B. and Fisher, S.G. (1986) Nitrogen limitation in a Sonoran Desert stream. *J. N. Am. Benthol. Soc.*, **5**, 2–15.

Grimm, N.B., Fisher, S.G. and Petrone, K.C. (1994) Nitrogen fization rates in a desert stream ecosystem: rates and spatial patterns. *Verh. Int. Ver. Theor. Ang. Limnol.*, **25**, 343.

Grossman, G.D., Moyle, P.B. and Whitaker, J.O., Jr., (1982) Stochasticity in structural and fuctional characteristics of an Indiana stream fish assemblage: A test of community theory. *Am. Nat.*, **120**, 423–54.

Grove, A.T. (1972) The dissolved and solid load carried by some West African rivers: Senegal, Niger, Benue and Shari. *J. Hydrol.*, **16**, 277–300.

Gunn, J.M., Qadri, S.U. and Mortimer, D.C. (1977) Filamentous algae as a food source for the brown bullhead (*Ictalurus nebulosus*). *J. Fish. Res. Board Can.*, **34**, 396–401.

Gurtz, M.E., Webster, J.R. and Wallace, J.B. (1980) Seston dynamics in southern Appalachian streams: effects of clear-cutting. *Can. J. Fish. Aquat. Sci.*, **37**, 624–31.

Hall, C.A.S. (1972) Migration and metabolism in a temperate stream ecosystem. *Ecology*, **53**, 585–604.

Hall, R.J., Driscoll, C.T. and Likens, G.E. (1987) Importance of hydrogen ions and aluminium in regulating the structure and function of stream ecosystems: an experimental test. *Freshwater Biol.*, **18**, 17–43.

Hall, R.J., Likens, G.E., Fiance, S.B. and Hendrey, G.R. (1980) Experimental acidification of a stream in the Hubbard Brook Experimental Forest, New Hampshire. *Ecology*, **61**, 976–89.

Hamilton, H.R. and Clifford, H.F. (1983) The seasonal food habits of mayfly (Ephemeroptera) nymphs from three Alberta, Canada, streams, with special reference to absolute volume and size of particles ingested. *Arch. Hydrobiol. Suppl.*, **65**, 197–234.

Hamilton, W.D. (1971) Geometry for the selfish herd. *J. Theor. Biol.*, **31**, 295–311.

Hansmann, E.W., Lane, C.B. and Hall, J.D. (1971) A direct method of measuring benthic primary production in streams. *Limnol. Oceanogr.*, **16**, 822–6.

Hargrave, B.T. (1976) The central role of invertebrate feces in sediment decomposition, in *The Role of Terrestrial and Aquatic Organisms in Decomposition Processes*, (eds J.M. Anderson and A. Macfadyen), Blackwell, Oxford, pp. 301–21.

Harmon, M.E., Franklin, J.F., Swanson, F.J. *et al.* (1986) Ecology of coarse woody debris in temperate ecosystems. *Adv. Ecol. Res.*, **15**, 133–302.

Harr, R.D. (1986) Effects of clear-cutting on rain-on-snow runoff in western Oregon: a new look at old studies. *Water Resour. Res.*, **22**, 1095–100.

Harr, R.D., Harper, W.C. and Krygier, J.T. (1975) Changes in storm hydrographs after road building and clear-cutting in the Oregon coast range. *Water Resour. Res.*, **11**, 436–44.

Harr, R.D., Levno, A. and Mersereau, R. (1982) Streamflow changes after logging 130-year-old Douglas Fir in two small watersheds. *Water Resour. Res.*, **18**, 637–44.

Harrison, A.D., Williams, N.V. and Greig, G. (1970) Studies on the effects of calcium bicarbonate concentrations on the biology of *Biomphalaria pfeifferi* (Krauss) (Gastropoda: Pulmonata). *Hydrobiologia*, **36**, 317–27.

Hart, D.D. (1978) Diversity in stream insects: regulation by rock size and microspatial complexity. *Verh. Int. Ver. Theor. Ang. Limnol.*, **20**, 1376–81.

Hart, D.D. (1981) Foraging and resource patchiness: field experiments with a grazing stream insect. *Oikos*, **37**, 46–52.

Hart, D.D. (1985a) Causes and consequences of territoriality in a grazing steam insect. *Ecology*, **66**, 404–13.

Hart, D.D. (1985b) Grazing insects mediate algal interactions in a stream benthic community. *Oikos*, **44**, 40–6.

Hartman, G.F. (1965) The role of behaviour in the ecology and steelhead trout (*Salmo gairdneri*). *J. Fish. Res. Board Can.*, **22**, 1035–81.

Haslam, S.M. (1988) *River Plants of Western Europe*, Cambridge University Press, Cambridge.

Hasler, A.D. (1957) Olfaction and gustatory senses of fish, in *The Physiology of Fishes*, (ed. M.E. Brown) Academic Press, New York, pp. 187–209.

Hauer, F.R., Stanford, J.A. and Ward, J.V. (1989) Serial discontinuities in a Rocky Mountain river. II. Distribution and abundance of Trichoptera. *Regulated Rivers: Res. Manage.*, **3**, 177–82.

Hawkins, C.P. (1986) Variation in individual growth

rates and population densities of ephemerellid mayflies. *Ecology*, **67**, 1384–95.
Hawkins, C.P. and Furnish, J.K. (1987) Are snails important competitors in stream ecosystems? *Oikos*, **49**, 209–20.
Hawkins, C.P., Murphy, M.L. and Anderson, N.H. (1982) Effects of canopy, substrate composition, and gradient on the structure of macroinvertebrate communities in Cascade Range streams of Oregon. *Ecology*, **63**, 1840–56.
Hearn, W.E. (1987) Interspecific competition and habitat segregation among stream-dwelling trout and salmon: a review. *Fisheries*, **12**, 24–31.
Hedges, J.I., Clark, W.I., Quay, P.D., Richey, J.E., Devol, A.H. and de M. Santos, U. (1986) Composition and fluxes of particulate organic matter in the Amazon River. *Limnol. Oceanogr.*, **31**, 717–38.
Hedin, L.O. (1990) Factors controlling sediment community respiration in woodland streams. *Oikos*, **57**, 94–105.
Hemphill, N. (1988) Competition between two stream dwelling filter-feeders, *Hydrospyche oslari* and *Simulium virgatum*. *Oecologia*, **77**, 73–80.
Hemphill, N. and Cooper, S.D. (1983) The effect of physical disturbance on the relative abundances of two filter-feeding insects in a small stream. *Oecologia*, **58**, 378–82.
Herbst, G.N. (1982) Efects of leaf type on the consumption rates of aquatic detritivores. *Hydrobiologia*, **89**, 77–87.
Hildebrand, S.G. (1974) The relation of drift to benthos density and food level in an artificial stream. *Limnol. Oceanogr.*, **19**, 951–7.
Hildrew, A.G., Dobson, M.K., Groom, A. *et al.* (1987) Flow and retention in the ecology of stream invertebrates. *Verh. Int. Ver. Theor. Ang. Limnol.*, **24**, 1742–7.
Hildrew, A.G. and Edington, J.M. (1979) Factors facilitating the coexistence of hydropsychid caddis larve (Trichoptera) in the same river system. *J. Anim. Ecol.*, **48**, 557–76.
Hildrew, A.G. and Townsend, C.R. (1976) The distribution of two predators and their prey in an iron rich stream. *J. Anim. Ecol.*, **45**, 41–57.
Hildrew, A.G. and Townsend, C.R. (1980) Aggregation, interference and foraging by larvae of *Plectrocnemia conspersa* (Trichoptera: Polycentropodidae). *Anim. Behav.*, **28**, 553–60.
Hildrew, A.G. and Townsend, C.R. (1982) Predators and prey in a patchy environment: a freshwater study. *J. Anim. Ecol.*, **51**, 797–816.
Hildrew, A.G., Townsend, C.R. and Francis, J. (1984a) Community structure in some southern English streams: the influence of species interactions. *Freshwater Biol.*, **14**, 297–310.
Hildrew, A.G., Townsend, C.R., Francis, J. and Finch, K. (1984b) Cellulolytic decomposition in streams of contrasting pH and its relationship with invertebrate community structure. *Freshwater Biol.*, **14**, 323–8.
Hildrew, A.G., Townsend, C.R. and Hasham, A. (1985) The predatory Chironomidae of an iron-rich stream: feeding ecology and food web structure. *Ecol. Entomol.*, **10**, 403–13.
Hill, A.R. (1979) Denitrification in the nitrogen budget of a river system. *Nature*, **281**, 291–3.
Hill, B.H. and Webster, J.R. (1982a) Periphyton production in an Appalachian river. *Hydrobiologia*, **97**, 275–80.
Hill, B.H. and Webster, J.R. (1982b) Aquatic macrophyte breakdown in an Appalachian river. *Hydrobiologia*, **89**, 275–80.
Hill, B.H. and Webster, J.R. (1983) Aquatic macrophyte contribution to the New River organic matter budget, in *Dynamics of Lotic Ecosystems*, (eds T. Fontaine and S. Bartell), Ann Arbor Science, Michigan, pp. 273–82.
Hill, W.R. (1992) Food limitation and interspecific competition in snail-dominated streams. *Can. J. Fish. Aquat. Sci.*, **49**, 1257–67.
Hill, W.R. and Knight, A.W. (1987) Experimental analysis of the grazing interaction between a mayfly and stream algae. *Ecology*, **68**, 1955–65.
Hill, W.R. and Knight, A.W. (1988) Concurrent grazing effects of two stream insects on periphyton. *Limnol. Oceanogr.*, **33**, 15–26.
Hixon, M.A. (1986) Fish predation and local prey diversity, in *Contemporary Studies on Fish Feeding*, (eds C.A. Simenstad and G.M. Cailliet), Dr. W. Junk, The Netherlands, pp. 235–57.
Hobbie, J.E. and Likens, G.E. (1973) Output of phosphorus, dissolved organic carbon, and fine particulate carbon from Hubbard Brook watersheds. *Limnol. Oceanogr.*, **18**, 734–42.
Hocutt, C.H. and Wiley, E.O. (eds) (1986) *Zoogeography of North American Freshwater fishes*, Wiley Interscience, New York.
Hoffman, G.L. and Schubert, G. (1984) Some parasites of exotic fishes, in *Distribution, Biology, and Management of Exotic Fishes*, (eds W.W. Courtenay and J.R. Stauffer), The Johns Hopkins University Press, Baltimore, MD, pp. 233–61.
Holeman, J.N. (1968) The sediment yield of major rivers of the world. *Water Resour. Res.*, **4**, 737–47.
Holt, C.S. and Waters, T.F. (1967) Effect of light intensity on the drift of stream invertebrates. *Ecology*, **48**, 225–34.
Holtby, L.B. (1988) Effects of logging on stream temperatures in Carnation Creek, British Columbia, and associated impacts on the coho salmon (*Oncorhynchus kisutch*). *Can. J. Fish. Aquat. Sci.*, **45**, 502–15.
Hora, S.L. (1928) Animal life in torrential streams. *J. Bombay Nat. Hist. Soc.*, **32**, 111–26.

Horne, A.J. and Carmiggelt, C.J.W. (1975) Algal nitrogen fixation in California streams: seasonal cycles. *Freshwater Biol.*, **5**, 461–70.

Hornick, L.E., Webster, J.R. and Benfield, E.F. (1981) Periphyton production in an Appalachian mountain stream. *Am. Midl. Nat.*, **106**, 22–36.

Horton, R.E. (1945) Erosional development of streams and their drainage basins: hydrophysical approach to quantitative morphology. *Bull. Geol. Soc. Am.*, **56**, 275–350.

Horwitz, R.J. (1978) Temporal variability patterns and the distributional patterns of stream fishes. *Ecol. Monogr.*, **48**, 307–21.

Hoskin, C.M. (1959) Studies of oxygen metabolism of streams of North Carolina. *Publ. Inst. Mar. Sci., Univ. Texas*, **6**, 186–92.

Huet, M. (1949) Apercu des relations entre la pente et les populations piscicoles deseaux courantes. *Schweiz. Z. Hydrol.*, **11**, 333–51.

Hughes, D.A. (1966) The roles of responses to light in the selection and maintenance of microhabitat by the nymphs of two species of mayfly. *Anim. Behav.*, **14**, 17–33.

Hughes, D.A. (1970) Some factors affecting drift and upstream movements of *Gammarus pulex*. *Ecology*, **51**, 301–5.

Hughes, R.M. and Omernik, J.M. (1983) An alternative for characterizing stream size, in *Dynamics of Lotic Ecosystems*, (eds T.D. Fontaine and S.M. Bartell), Ann Arbor Science, pp. 87–102.

Hultin, L. (1968) A method of trapping freshwater Amphipoda migrating upstream. *Oikos*, **19**, 400–2.

Hultin, L., Svensson, B.W. and Ulfstrand, S. (1969) Upstream movements of insects in a south Swedish small stream. *Oikos*, **20**, 553–57.

Hunter, C.J. (1991) Better Trout Habitat, Island Press, Washington, DC.

Hurlbert, S.H. (1984) Pseudoreplication and the design of ecological field experiments. *Ecol. Monogr.*, **54**, 187–211.

Hutchinson, G.E. (1957) *A Treatise on Limnology. Volume 1. Geography, Physics, and Chemistry*, John Wiley, New York.

Hutchinson, G.E. (1967) *A Treatise on Limnology. Volume 2. Introduction to Lake Biology and the Limnoplankton*, John Wiley & Sons, New York.

Hutchinson, G.E. (1975) *A Treatise on Limnology. Volume 3. Limnological Botany*. Wiley-Interscience, New York.

Hutchinson, G.E. (1981) Thoughts on aquatic insects. *Bioscience*, **31**, 495–500.

Hyatt, K.D. (1979) Feeding strategy, in *Fish Physiology, Vol. 8*, (eds W.S. Hoar, D.J. Randall and J.R. Brett), Academic Press, New York, pp. 71–119.

Hynes, H.B.N. (1941) The taxonomy and ecology of the nymphs of British Plecoptera, with notes on the adults and eggs. *Trans. R. Entomol. Soc. London*, **91**, 459–557.

Hynes, H.B.N. (1955) Distribution of some freshwater Amphipoda in Britain. *Verh. Int. Ver. Theor. Ang. Limnol.*, **12**, 620–8.

Hynes, H.B.N. (1969) The enrichment of streams, in *Eutrophication: Causes, Consequences, Correctives*, National Academy of Sciences, Washington, DC, pp. 188–96.

Hynes, H.B.N. (1970) *The Ecology of Running Waters*, University of Toronto Press.

Hynes, H.B.N. (1975) The stream and its valley. *Verh. Int. Ver. Theor. Ang. Limnol.*, **19**, 1–15.

Hynes, H.B.N. (1988) Biogeography and origins of the North American stoneflies (Plecoptera). *Mem. Ent. Soc. Can.*, **144**, 31–7.

Hynes, H.B.N., Williams, D.D. and Williams, N.E. (1976) Distribution of the benthos within the substratum of a Welsh mountain stream. *Oikos*, **27**, 307–10.

Hyslop, E.J. (1980) Stomach contents analysis – a review of methods and their application. *J. Fish. Biol.*, **17**, 411–29.

Ide, F.P. (1942) Availability of aquatic insects as food of the speckled trout, *Salvelinus fontinalis*. *Trans. N. Am. Wildl. Conf.*, **7**, 442–50.

Illies, J. (1969) Biogeography and ecology of neotropical freshwater insects, especially those from running waters, in *Biogeography and Ecology in South America*, (eds E.J. Fittkau *et al.*) Dr. W. Junk, The Hague, pp. 685–708.

Illies, J. (1971) Emergenz 1969 im Breitenbach. *Arch. Hydrobiol.*, **69**, 14–59.

Illies, J. and Botosaneanu, L. (1963) Problèmes et méthodes de la classification et de la zonation écologique des eaux courantes, considerées surtout du point de vue faunistique. *Mitt. Int. Ver. Theor. Ang. Limnol.*, **12**, 57 pp.

Interagency Advisory Committee on Water Data. 1982. *Guideleines for Determining Flood Frequency*, Hydrology Subcommittee Bulletin 17B, U.S. Geological Survey, Reston VA.

Ittekkot, V. (1988) Global trends in the nature of organic matter in river suspensions. *Nature*, **332**, 436–8.

Iverson, T.M. (1973) Decomposition of autumn-shed leaves in a springbrook and its significance for the fauna. *Arch. Hydrobiol.*, **72**, 305–12.

Iverson, T.M., Thorup, J., Hansen, T., Lodal, J. and Olsen, J. (1985) Quantitative estimates and community structure of invertebrates in a macrophyte rich stream. *Arch. Hydrobiol.*, **102**, 291–301.

Ivlev, V.S. (1961) *Experimental Ecology of the Feeding of Fishes*, Yale University Press, New Haven CT.

Jaag, O. and Ambühl, H. (1964) The effect of the current on the composition of biocoenoses in flowing water streams. *Adv. Water Pollut. Res.*, **1**, 31–49.

Jacobi, G.Z. (1979) *Ecological Comparisons of Stream Sections with and without Fish Populations, Yellowstone National Park*. National Park Service Research Center, University of Wyoming, Laramie.

Jansson, A. and Vuoristo, T. (1979) Significance of stridulation in larval Hydropsychidae (Trichoptera). *Behaviour*, **171**, 167–86.

Jazdzewski, K. (1980) Range extensions of some gammaridian species on European inland waters caused by human activity. *Crustaceana*, Supplement b, **84**, 107.

Jeffries, M.J. and Lawton, J.H. (1984) Enemy free space and the structure of ecological communities. *Biol. Linnean Soc.*, **23**, 269–86.

Jenkins, T.M. (1969) Night feeding of brown and rainbow trout in an experimental stream channel. *J. Fish. Res. Board Can.*, **26**, 3275–8.

Johnson, D.M. and Crowley, P.H. (1980) Odonate 'hide-and-seek': habitat-specific rules? in *Evolution and Ecology of Zooplankton Communities*, (eds W.C. Kerfoot), New England University Press, Hanover NH, pp. 569–79.

Johnson, C.M. and Needham, P.R. (1966) Ionic composition of Sagehen Creek, California, following an adjacent fire. *Ecology*, **47**, 636–9.

Johnson, P.L. and Swank, W.T. (1973) Studies of cation budgets in the southern Appalachians on four experimental watersheds with contrasting vegetation. *Ecology*, **54**, 70–80.

Jones, J.B. and Smock, L.A. (1991) Transport and retention of particulate organic matter in two low-gradient headwater streams. *J. N. Am. Benthol. Soc.*, **10**, 115–26.

Jones, J.G. (1974) A method for observation and enumeration of epilithic algae directly on the surface of stones. *Oecologia*, **16**, 1–8.

Jones, J.G. (1978) Spatial variation in epilithic algae in a stony stream (Wilfin Beck) with particular reference to *Cocconeis placentula*. *Freshwater Biol.*, **8**, 539–46.

Jones, J.R.E. (1949) A further ecological study of calcareous streams in the 'Black Mountain' District, South Wales. *J. Anim. Ecol.*, **18**, 142–59.

Jones, J.R.E. (1950) A further ecological study of the River Rheidol: the food of the common insects of the main-stream. *J. Anim. Ecol.*, **19**, 159–74.

Junk, W.J., Bayley, P.B. and Sparks, R.E. (1989) The flood pulse concept in river-floodplain systems, in *Proceedings of the International Large River Symposium*, (ed. D.P. Dodge), Canadian Special Publication of Fisheries and Aquatic Sciences, 106, pp. 110–27.

Kahrl, W.L. (ed.) (1979) *California Water Atlas*, William Kaufman, Los Altos, CA.

Kaplan, L.A. and Bott, T.L. (1982) Diel fluctuations of DOC generated by algae in a Piedmont stream. *Limnol. Oceanogr.*, **27**, 1091–100.

Kaplan, L.A. and Bott, T.L. (1989) Diel fluctuations in bacterial activity on streambed substrata during vernal algal blooms: Effects of temperature, water chemistry, and habitat. *Limnol. Oceanogr.*, **34**, 718–33.

Karr, J.R. (1981) Assessment of biological integrity using fish communities. *Fisheries*, **6**, 21–7.

Karr, J.R. (1991) Biological integrity: a long-neglected aspect of water resource management. *Ecol. Appl.*, **1**, 66–84.

Karr, J.R. and Schlosser, I.J. (1978) Water resources and the land water interface. *Science*, **201**, 229–34.

Karr, J.R., Toth, L.A. and Dudley, D.R. (1985) Fish communities of midwestern rivers: A history of degradation. *Bioscience*, **35**, 90–5.

Kaufman, L. (1992) Catastrophic change in species-rich freshwater ecosystems. *Bioscience*, **42**, 846–58.

Kaushik, N.K. and Hynes, H.B.N. (1971) The fate of autumn-shed leaves that fall into streams. *Arch. Hydrobiol.*, **68**, 465–515.

Kaushik, N.K. and Robinson, J.B. (1976) Preliminary observations of nitrogen transport during summer in a small spring-fed Ontario stream. *Hydrobiologia*, **49**, 59–63.

Kazmierczak, R.F., Jr., Webster, J.R. and Benfield, E.F. (1987) Characteristics of seston in a regulated Appalachian mountain river, U.S.A. *Regulated Rivers: Res. Manage.*, **1**, 287–300.

Kehew, A.E. and Lord, M.L. (1987) Glacial-lake outbursts along the mid-continent margins of the Laurentide ice-sheet, in *Catastrophic Flooding: Binghamton Symposia in Geomorphology* (Number 18), (eds L. Mayer and D. Nash), Allen & Unwin, Boston, Massachusetts, pp. 95–120.

Keithan, E.D. and Lowe, R.L. (1985) Primary productivity and spatial structure of phytolithic growth in streams in the Great Smokey Mountains National Park, Tennessee. *Hydrobiologia*, **123**, 59–67.

Kelly, P.M. and Cory, J.S. (1987) Operculum closing as a defense against predatory leeches in four British freshwater prosobranch snails. *Hydrobiologia*, **144**, 121–4.

Kempe, S., Fettine, M. and Cauwet, G. (1991) Biogeochemistry of European rivers, in *Biogeochemistry of Major World Rivers*, (eds E.T. Degens, S. Kempe and J.E. Richey), John Wiley, New York, pp. 169–212.

Kerfoot, W.C. and Sih, A. (1987) *Predation: Direct and Indirect Impacts on Aquatic Communities*, University Press of New England, Hanover, NH.

Kilham, P. (1990) Mechanisms controlling the chemical composition of lakes and rivers: Data from Africa. *Limnol. Oceanogr.*, **35**, 80–3.

King, D.L. and Ball, R.C. (1966) A qualitative and quantitative measure of *Aufwuchs* production. *Trans. Am. Microsc. Soc.*, **85**, 232–40.

King, D.L. and Ball, R.C. (1967) Comparative energetics of a polluted stream. *Limnol. Oceanogr.*, **12**, 27–33.

Kline, T.C., Goering, J.J., Mathisen, O.A. and Hoe, P.H. (1990) Recycling of elements transported upstream by runs of pacific salmon: 1. ^{15}N and ^{13}C evidence in Sashin Creek, Southeastern Alaska. *Can. J. Fish. Aquat. Sci.*, **47**, 136–44.

Klotz, R.L., Cain, J.R. and Trainor, F.R. (1976) Algal competition in a epilithic flora. *J. Phycol.*, **12**, 363–8.

Knight, A.W. and Gaufin, A.R. (1963) The effect of water flow, temperature, and oxygen concentration on the Plecoptera nymph, *Acroneuria pacifica* Banks. *Proc. Utah Acad. Sci., Arts, Lett.*, **40**, 175–84.

Knight, A.W., Simmons, M.A., and Simmons, C.S. (1976) A phenomenological approach to the growth of the winter stonefly, *Taenyopteryx nivalis* (Fitch) (Plecoptera: Taeniopterygidae). *Growth*, **40**, 343–67.

Kohler, S.L. (1983) Positioning on substrates, positioning changes, and diel drift periodicities in mayflies. *Can. J. Zool.*, **61**, 1362–8.

Kohler, S.L. (1984) Search mechanism of a stream grazer in patchy environments: the role of food abundance. *Oecologia*, **62**, 209–18.

Kohler, S.L. (1985) Identification of stream drift mechanisms: An experimental and observational approach. *Ecology*, **66**, 1749–61.

Kohler, S.L. (1992) Competition and the structure of a benthic stream community. *Ecol. Monogr.*, **62**, 165–88.

Koslucher, D.G. and Minshall, G.W. (1973) Food habits of some benthic invertebrates in a northern cool-desert stream (Deep Creek, Curlew Valley, Idaho-Utah). *Trans. Am. Microsc. Soc.*, **92**, 441–52.

Kovalak, W.P. (1976) Seasonal and diel changes in the positioning of *Glossosoma nigrior* Banks (Trichoptera: Glossosomatidae) on artifical substrates. *Can. J. Zool.*, **54**, 1585–94.

Kovalak, W.P. (1978) Diel changes in stream benthos density on stones and artificial substrates. *Hydrobiologia*, **58**, 7–16.

Kuehne, R.A. (1962) A classification of streams illustrated by fish distribution in an eastern Kentucky creek. *Ecology*, **43**, 608–14.

Kuusela, K. (1979) Early summer ecology and community structure of the macrozoobenthos on stones in the Javajankoski Rapids on the River Lestijoki, Finland. *Acta Univ. Ouluensis Ser. A*, Number 87.

Lack, T.J. (1971) Quantitative studies on the phytoplankton of the Rivers Thames and Kennet at Reading. *Freshwater Biol.*, **1**, 213–24.

Lacoursiere, J.O. and Craig, D.A. (1993) Fluid transmission and filtration efficiency of the labral fans of black fly larvae (Dipetra: Simuliidae): hydrodynamic, morphological, and behavioural aspects. *Can. J. Zool.*, **71**, 148–62.

Ladle, M. and Casey, H. (1971) Growth and nutrient relationships of *Ranunculus penicillatus* var. *calcareus* in a small chalk stream. Proceedings of European Weed Research Council, 3rd. International Symposium on Aquatic Weeds, pp. 53–62.

Ladle, M., Cooling, D.A., Welton, J.S. and Bass, J.A.B. (1985) Studies on Chironomidae in experimental recirculating stream systems. II. The growth, development and production of a spring generation of *Orthocladius* (*Euorthocladius*) *calvus* Pinder. *Freshwater Biol.*, **15**, 243–55.

Ladle, M. and Griffiths, B.S. (1980) A study on the faeces of some chalk stream invertebrates. *Hydrobiologia*, **74**, 161–71.

Lake, P.S., Barmuta, L.A., Boulton, A.J., Campbell, I.C. and St. Clair, R.M. (1986) Australian streams and northern hemisphere stream ecology: Comparisons and problems. *Proc. Ecol. Soc. Aust.*, **14**, 61–82.

Lamberti, G.A., Ashkenas, L.R., Gregory, S.V. and Steinman, A.D. (1987) Effects of three herbivores on periphyton communities in laboratory streams. *J. N. Am. Benthol. Soc.*, **6**, 92–104.

Lamberti, G.A., Feminella, J.W. and Resh, V.H. (1987) Herbivory and intraspecific competition in a stream caddisfly population. *Oecologia*, **73**, 75–81.

Lamberti, G.A., Gregory, S.V., Ashkenas, L.R., Steinman, A.D. and McIntire, C.D. (1989) Productive capacity of periphyton as a determinant of plant-herbivore interactions in streams. *Ecology*, **70**, 1840–56.

Lamberti, G.A. and Moore, J.W. (1984) Aquatic insects as primary consumers, in *The Ecology of Aquatic Insects*, (eds V.H. Resh and D.M. Rosenberg), Praeger Scientific, New York, pp. 164–95.

Lamberti, G.A. and Resh, V.H. (1983) Stream periphyton and insect herbivores: an experimental study of grazing by a caddisfly population. *Ecology*, **64**, 1124–35.

Lamp, W.O. and Britt, N.W. (1981) Resource partitioning by two species of stream mayflies (Ephemeroptera: Heptageniidae). *Great Lakes Entomol.*, **14**, 151–7.

Lang, H.H. (1980) Surface wave discrimination between prey and nonprey by the backswimmer *Notonecta glauca* L. (Hemiptera, Heteroptera). *Behav. Ecol. Sociobiol.*, **6**, 233–46.

Langbein, W.B. and Dawdy, D.R. (1964) Occurrence of dissolved solids in surface waters in the United States. *U.S. Geol. Surv. Prof. Pap.*, **501**, D115–17.

Larkin, P.A. and McKone, D.W. (1985) An evaluation by field experiments of the McLay model of stream drift. *Can. J. Fish. Aquat. Sci.*, **42**, 909–18.

Larson, R.A. (1978) Dissolved organic matter of a low-coloured stream. *Freshwater Biol.*, **8**, 91–104.

Lawson, D.L., Klug, M.J. and Merritt, R.W. (1984) The influence of physical, chemical, and microbiological characteristics of decomposing leaves on the growth of the detritivore *Tipula abdominalis* (Diptera: Tipulidae). *Can. J. Zool.*, **62**, 2339–43.

Ledger, D.C. (1981) The velocity of the River Tweed and its tributaries. *Freshwater Biol.*, **11**, 1–10.

Lee, D.S., Gilbert, C.R., Hocutt, C.H., Jenkins, R.E., McAllister, D.E. and Stauffer, J.R., Jr. (1980) *Atlas of North American Freshwater Fish*, North Carolina State Museum of Natural History.

Lehmann, U. (1967) Drift and Populationsdynamik von *Gammarus pulex fossarum* Koch. *Z. Morphol. Okol. Tiere*, **60**, 2227–74.

Lehmann, U. (1972) Tagesperiodisches Verhalten und Habitatwechsel der Larven von *Potamophylax luctuosus* (Trichoptera). *Oecologia*, **9**, 265–78.

Lehmkuhl, D.M. (1974) Thermal regime alterations and vital environmental physiological signals in aquatic systems, in *Thermal Ecology*, (eds J.W. Gibbons and R.R. Sharitz), AEC Symposium Series (CONF-730505), pp. 216–22.

Lelek, A. and Köhler, C. (1990) Restoration of fish communities of the Rhine River two years after a heavy pollution wave. *Regulated Rivers: Res. Manage.*, **5**, 57–66.

Leopold, A. (1949) *A Sand County Almanac*, Oxford University Press, New York.

Leopold, L.B. (1962) Rivers. *Am. Sci.*, **50**, 511–37.

Leopold, L.B. (1977) A reverence for rivers. *Geology*, **5**, 429–30.

Leopold, L.B. and Maddock, T., Jr. (1953) The hydraulic geometry of stream channels and some physiographic implications. *U.S. Geol. Surv. Prof. Pap.*, **242**, 57 pp.

Leopold, L.B., Wolman, M.G. and Miller, J.P. (1964) *Fluvial Processes in Geomorphology*, W.H. Freeman, San Francisco.

Levine, J.S. (1992) Global climate change, in *Global Climate Change and Freshwater Ecosystems*, (eds P. Firth and S.G. Fisher), Springer-Verlag, New York, pp. 1–25.

Levine, J.S. and MacNichol, M.F., Jr. (1979) Visual pigments in teleost fishes: effects of habitat, microhabitat, and behavior on visual system evolution. *Sensory Processes*, **3**, 95–131.

Lewis, W.M. (1988) Primary production in the Orinoco River. *Ecology*, **69**, 679–92.

Likens, G.E. (1984) Beyond the shoreline: A watershed-ecosystem approach. *Verh. Int. Ver. Theor. Ang. Limnol.*, **22**, 1–22.

Likens, G.E. and Bormann, F.H. (1974) Linkages between terrestrial and aquatic ecosystems. *Bioscience*, **24**, 447–56.

Likens, G.E., Bormann, F.H., Johnson, N.M., Fisher, D.W. and Pierce, R.S. (1970) Effects of forest cutting and herbicide treatment on nutrient budgets in the Hubbard Brook watershed-ecosystem. *Ecol. Monogr.*, **40**, 23–47.

Likens, G.E., Bormann, F.H., Johnson, N.M. and Pierce, R.S. (1967) The calcium, magnesium, potassium and sodium budgets for a small forested ecosystem. *Ecology*, **48**, 772–84.

Likens, G.E., Bormann, F.H., Pierce, R.S., Eaton, J.S. and Johnson, N.M. (1977) *Biogeochemistry of a Forested Ecosystem*, Springer-Verlag, New York.

Lillehammer, A. and Saltveit, S.J. (1984) *Regulated Rivers*, Universitetsforlaget AS, Oslo.

Lingdell, P.E. and Müller, K. (1979) Migrations of *Leptophlebia vespertina* L. and *L. marginata* L. (Ins.: Ephemeroptera) in the estuary of a coastal stream. *Aquat. Insects*, **1**, 137–42.

Linsley, R.K., Kohler, M.A. and Paulhus, J.L.H. (1958) *Hydrology for Engineers*, McGraw-Hill, New York.

Livingstone, D.A. (1963) Chemical composition of rivers and lakes. *Geol. Surv. Pap.*, 440-G (pages G1-G64).

Lock, M.A. (1981) River epilithon – a light and organic energy transducer, in *Perspectives in Running Water Ecology*, (eds M.A. Lock and D.D. Williams), Plenum Press, New York, pp. 3–40.

Lock, M.A. and Hynes, H.B.N. (1976) The fate of 'dissolved' organic carbon derived from autumn-shed maple leaves (*Acer saccharum*) in a temperate hard-water stream. *Limnol. Oceanogr.*, **21**, 436–43.

Lock, M.A., Wallace, R.R., Costerton, J.W., Ventullo, R.M. and Charlton S.E. (1984) River epilithon: toward a structural-functional model. *Oikos*, **42**, 10–22.

Lock, M.A., Wallis, P.M. and Hynes, H.B.N. (1977) Colloidal organic carbon in running waters. *Oikos*, **29**, 1–4.

Lodge, D.M. (1991) Herbivory on freshwater macrophytes. *Aquat. Bot.*, **41**, 195–224.

Lodge, D.M. and Lorman, J.G. (1987) Reductions in submersed macrophyte biomass and species richness by the crayfish *Orconectes rusticus*. *Can. J. Fish. Aquat. Sci.*, **44**, 591–7.

Lopez, G.R. and Levinton, J.S. (1987) Ecology of deposit-feeding animals in marine sediments. *Quart. Rev. Biol.*, **62**, 235–60.

Lowe, R.L., Golladay, S.W. and Webster, J.R. (1986) Periphyton response to nutrient manipulation in streams draining clearcut and forested watersheds. *J. N. Benthol. Soc.*, **5**, 221–9.

Lowe-McConnell, R.H. (1964) The fishes of the Rupununi savanna district of British Guiana, Part 1. Groupings of fish species and effects of seasonal cycles on the fish. *J. Linnean Soc.* (*Zool.*), **45**, 103–44.

Lowe-McConnell, R.H. (1987) *Ecological Studies in Tropical Fish Communities*, Cambridge University Press, Cambridge, UK.

Lowrance, R., Todd, R., Fail, J., Jr., Hendrickson, O., Jr., Leonard, R. and Asmussen, L. (1984) Riparian forests as nutrient filters in agricultural watersheds. *Bioscience*, **34**, 374–7.

Lubchenco, (1978) Plant species diversity in a marine intertidal community: importance of herbivore food preference and algal competitive abilities. *Am. Nat.*, **112**, 23–9.

Lush, D.L. and Hynes, H.B. (1973) The formation of particles in freshwater leachates of dead leaves. *Limnol. Oceanogr.*, **18**, 968–77.

Lush, D.L. and Hynes, H.B.N. (1978) The uptake of dissolved organic matter by a small spring stream. *Hydrobiologia*, **60**, 271–5.

Macan, T.T. (1974) *Freshwater Ecology*, 2nd edn, John Wiley, New York.

Macan, T.T. (1977) The influence of predation on the composition of freshwater animal communities. *Biol. Rev.*, **52**, 45–70.

MacArthur, R.H. and Wilson, E.O. (1967) *The Theory of Island Biogeography*, Princeton University Press, Princeton, NJ.

MacCrimmon, H.R. (1971) World distribution of rainbow trout (*Salmo gairdneri*). *J. Fish. Res. Board Can.*, **25**, 2527–48.

Maciolek, J.A. (1984) Exotic fishes in Hawaii and other islands of Oceania, in *Distribution, Biology, and Management of Exotic Fishes*, (eds W.W. Courtenay and Stauffer, J.R., Jr.), The Johns Hopkins University Press, Baltimore, MD, pp. 131–61.

Mackay, R.J. (1969) Aquatic insect communities of a small stream on Mont St. Hilaire, Quebec. *J. Fish. Res. Board Can.*, **26**, 1157–83.

Mackay, R.J. (1977) Behavior of *Pycnopsyche* (Trichoptera: Leptophlebiidae) on mineral substrates in laboratory streams. *Ecology*, **58**, 191–5.

Mackay, R.J. and Kalff, J. (1969) Seasonal variation in standing crop and species diversity of insect communities in a small Quebec stream. *Ecology*, **50**, 101–9.

Mackay, R.J. and Kalff, J. (1973) Ecology of two related species of caddisfly larvae in the organic substrates of a woodland stream. *Ecology*, **54**, 499–511.

Madsen, B.L., Bengtson, J. and Butz, I. (1973) Observations on upstream migrations by imagines of some Plecoptera and Ephemeroptera. *Limnol. Oceanogr.*, **18**, 678–81.

Madsen, B.L. and Butz, I. (1976) Population movements of adult *Brachyptera risi* (Plecoptera). *Oikos*, **27**, 273–80.

Madson, J. (1985) *Up on the River*, Penguin Books, New York.

Mahon, R. (1984) Divergent structure in fish taxocenes of north temperate streams. *Can. J. Fish. Aquat. Sci.*, **41**, 330–50.

Malas, D. and Wallace, J.B. (1977) Strategies for coexistence in three species of net-spinning caddisflies (Trichoptera) in second-order southern Appalachian streams. *Can. J. Zool.*, **55**, 1829–40.

Malde, H.E. (1968) The catastrophic late Pleistocene Bonneville flood in the Snake River Plain, Idaho. *U.S. Geol. Surv. Prof. Pap.*, **595**, 52 pp.

Malmqvist, B., Nilsson, L.M. and Svensson, B.S. (1978) Dynamics of detritus in a small stream in southern Sweden and its influence on the distribution of the bottom animal communities. *Oikos*, **31**, 3–16.

Malmqvist, B. and Sjöström, P. (1980) Prey size and feeding patterns in *Dinocras cephalotes* (Perlidae). *Oikos*, **35**, 311–16.

Malmqvist, B. and Sjöström, P. (1984) The microdistribution of some lotic insect predators in relation to their prey and to abiotic factors. *Freshwater Biol.*, **14**, 649–56.

Mann, K.H., Britton, R.H., Kowalczewski, A., Lack, T.J., Mathews, C.P. and McDonald, I. (1972) Productivity and energy flow at all trophic levels in the River Thames, England, in *Productivity Problems of Freshwaters*, (eds Z. Kajak and A. Hillbricht-Illkowska), Polish Scientific Publishers, Warszawa, pp. 579–96.

Margalef, R. (1960) Ideas for a synthetic approach to the ecology of running waters. *Int. Rev. Gesamt. Hydrobiol.*, **45**, 133–53.

Marker, A.F.H. (1976) The benthic algae of some streams in southern England. I. Biomass of the epilithon in some small streams. *J. Ecol.*, **64**, 343–58.

Marker, A.F.H., Nusch, E.A., Rai, H. and Reimann, B. (1980) The measurement of photosynthetic pigment in freshwaters and standardization of methods: Conclusions and recommendations. *Arch. Hydrobiol. Beih. (Ergebn. Limnol.)*, **14**, 91–106.

Marksen, J. (1988) Evaluation of the importance of bacteria in the carbon flow of an open grassland stream, The Breitenbach. *Arch. Hydrobiol.*, **111**, 339–50.

Marsh, P.C. and Minckley, W.L. (1982) Fishes of the Phoenix Metropolitan Area in Central Arizona. *N. Am. J. Fish. Manage.*, **4**, 395–402.

Martin, M. and Meybeck, M. (1979) Elemental mass-balance of material carried by world major rivers. *Mar. Chem.*, **7**, 173–206.

Martin, M.M. and Kukor, J.J. (1984) Role of mycophagy and bacteriophagy in invertebrate nutrition, in *Current Perspectives in Microbial Ecology*, (eds M.J. Klug and C.A. Reddy), American Society for Microbiology, Washington DC, pp. 257–63.

Martin, M.M., Martin, J.S., Kukor, J.J. and Merritt, R.W. (1981a) The digestive enzymes of detritus-feeding stonefly nymphs (Plecoptera: Pteronarcyidae). *Can. J. Zool.*, **59**, 1947–51.

Martin, M.M., Kukor, J.J., Martin, J.S., Lawson, D.L. and Merritt, R.W. (1981b) Digestive enzymes of larvae of three species of caddisflies (Trichoptera). *Insect Biochem.*, **11**, 505–5.

Martin, M.M., Kukor, J.J., Martin, J.S. and Merritt,

R.W. (1985) The digestive enzymes of larvae of the black fly, *Prosimulium fuscum* (Diptera: Simuliidae). *Comp. Biochem. Physiol.*, **82B**, 37–9.

Martin, M.M., Martin, J.S., Kukor, J.J. and Merritt, R.W. (1980) The digestion of protein and carbohydrate by the stream detritivore, *Tipula abdominalis* (Diptera, Tipulidae). *Oecologia*, **46**, 360–4.

Master, L. (1990) The imperiled status of North American aquatic animals. *Biodivers. Net. News*, **3**, 5–7.

Matthews, W.J. (1985) Critical current speeds and microhabitats of the benthic fishes *Percina roanoka* and *Etheostoma flabellare. Environ. Biol. Fish.*, **12**, 303–8.

Matthews, W.J. and Hill, L.G. (1980) Habitat partitioning in the fish community of a southwestern river. *Southwestern Nat.*, **25**, 51–66.

Matthews, W.J. and Zimmerman, E.G. (1990) Potential effects of global warming on native fishes of the southern Great Plains and the southwest. *Fisheries*, **15**, 26–31.

Mathur, D., Bason, W.H., Purdy, E.J., Jr. and Silver, C.A. (1985) A critique of the instream flow incremental methodology. *Can. J. Fish. Aqua. Sci.*, **42**, 825–31.

May, R.M. (1984) An overview: real and apparent patterns in community structure, in *Ecological Communities: Conceptual Issues and the Evidence*, (eds D.R. Strong, Jr., D. Simberloff, L.G. Abele and A.B. Thistle). Princeton University Press, Princeton, NJ, pp. 3–16.

Mayer, M.S. and Likens, G.E. (1987) The importance of algae in a shaded headwater stream as food for an abundant caddisfly (Trichoptera). *J. N. Am. Benthol. Soc*, **6**, 262–9.

McAuliffe, J.R. (1983) Competition, colonization patterns, and disturbance in stream benthic communities, in *Stream Ecology: Application and Testing of General Ecological Theory*, (eds J.R. Barnes and G.W. Minshall). Plenum Press, New York, pp. 137–56.

McAuliffe, J.R. (1984a) Resource depression by a stream herbivore: effects on distributions and abundances of other grazers. *Oikos*, **42**, 327–33.

McAuliffe, J.R. (1984b) Competition for space, disturbance, and the structure of a benthic stream community. *Ecology*, **65**, 894–908.

McCullough, D.A., Minshall, G.W. and Cushing, C.E. (1979) Bioenergetics of a stream 'collector' organism, *Tricorythodes minutus* (Insects: Ephemeroptera). *Limnol. Oceanogr.*, **24**, 45–58.

McDiffett, W.F. (1970) The transformation of energy by a stream detritivore, *Pteronarcys scotti. Ecology*, **51**, 399–420.

McDiffett, W.F., Carr, A.E. and Young, D.L. (1972) An estimate of primary productivity in a Pennsylvania trout stream using a diurnal oxygen curve technique. *Am. Midl. Nat.*, **87**, 564–70.

McDowall, R.M. (1990) *New Zealand Freshwater Fishes: a Natural History and Guide*, Heinemann, Auckland, New Zealand.

McDowell, W.H. and Fisher, S.G. (1976) Autumnal processing of dissolved organic matter in a small woodland stream ecosystem. *Ecology*, **57**, 561–9.

McFadden, J.T. and Cooper, E.L. (1962) An ecological comparison of six populations of brown trout *(Salmo trutta). Trans. Am. Fish. Soc.*, **91**, 53–62.

McIntire, C.D. (1966) Some factors affecting respiration of periphyton communities in lotic environments. *Ecology*, **47**, 918–30.

McIntire, C.D. (1968) Structural characteristics of benthic algal communities in laboratory streams. *Ecology*, **49**, 520–37.

McIntire, C.D. (1973) Periphyton dynamics in laboratory streams: a simulation model and its implications. *Ecol. Monogr.*, **43**, 399–420.

McIntire, C.D. (1975) Periphyton assemblages in laboratory streams, in *River Ecology*, (ed. B.A. Whitton). Blackwell Scientific Publications, Oxford, pp. 403–30.

McIntire, C.D. and Colby, J.A. (1978) A heirarchical model of lotic ecosystems. *Ecol. Monogr.*, **48**, 167–90.

McIntire, C.D. and Phinney, H.K. (1965) Laboratory studies of periphyton production and community metabolism in lotic environments. *Ecol. Monogr.*, **35**, 237–58.

McLay, C.L. (1970) A theory concerning the distance travelled by animals entering the drift of a stream. *J. Fish. Res. Board Can.*, **27**, 359–70.

McMahon, R.F. (1975) Growth, reproduction and bioenergetic variation in three natural populations of a freshwater limpet *Laevapex fuscus* (C B Adams). *Proc. Malacol. Soc. London*, **41**, 331–42.

McShaffrey, D. and McCafferty, W.P. (1986) Feeding behavior of *Stenonema interpunctatum* (Ephemeroptera: Heptageniidae). *J. N. Am. Benthol. Soc.*, **5**, 200–10.

Meffe, G.K. (1984) Effects of abiotic disturbance on coexistence of predator-prey fish species. *Ecology*, **65**, 1525–34.

Meier, P.G. and Bartholomae, P.G. (1980) Diel periodicity in the feeding activity of *Potamanthus myops* (Ephemeroptera). *Arch. Hydrobiol.*, **88**, 1–8.

Melillo, J.M., Aber, J.D. and Muratore, J.F. (1982) Nitrogen and lignin control of hardwood leaf litter decomposition dynamics. *Ecology*, **63**, 621–6.

Melillo, J.M., Naiman, R.J., Aber, J.D. and Eshleman, K.N. (1983) The influence of substrate quality and stream size on wood decomposition dynamics. *Oecologia*, **58**, 281–5.

Mendelson, J. (1975) Feeding relationships among

species of *Notropis* (Pisces: Cyprinidae)in a Wisconsin stream. *Ecol. Monogr.*, **45**, 199–230.

Merritt, R.W. and Cummins, K.W. (1984) *An Introduction to the Aquatic Insects of North America*, 2nd edn, Kendall/Hunt, Iowa.

Merritt, R.W., Ross, D.H. and Larson, G.J. (1982) Influence of stream temperature and seston on the growth and production of overwintering larval black flies (Diptera: Simuliidae). *Ecology*, **63**, 1322–31.

Metz, J.P. (1974) Die Invertebratendrift an der Oberflache eines Voralpenflusses und ihre selektive Ausnutzung durch die Regenbogenforellen (*Salmo gairdneri*). *Oecologia*, **14**, 247–67.

Meybeck, M. (1976) Total mineral dissolved transport by world major rivers. *Hydrolog. Sci. Bull.*, International Association of Scientific Hydrology, **21**, 265–84.

Meybeck. M. (1982) Carbon, nitrogen, and phosphorus transport by world rivers. *Am. J. Sci.*, **282**, 401–50.

Meyer, J.L. (1979) The role of sediments and bryophytes in phosphorus dynamics in a headwater stream ecosystem. *Limnol. Oceanogr.*, **24**, 365–75.

Meyer, J.L. (1986) Dissolved organic carbon dynamics in two subtropical blackwater rivers. *Arch. Hydrobiol.*, **108**, 119–34.

Meyer, J.L. (1990) A blackwater perspective on riverine ecosystems. *Bioscience*, **40**, 643–51.

Meyer, J.L. and Edwards, R.T. (1990) Ecosystem metabolism and turnover of organic carbon along a blackwater river continuum. *Ecology*, **71**, 668–77.

Meyer, J.L., Edwards, R.T. and Risley, R. (1987) Bacterial growth on dissolved organic carbon from a blackwater stream. *Microb. Ecol.*, **13**, 13–29.

Meyer, J.L. and Johnson, C. (1983) The influence of elevated nitrate concentration on rate of leaf decomposition in a stream. *Freshwater Biol.*, **13**, 177–83.

Meyer, J.L. and Likens, G.E. (1979) Transport and transformation of phosphorus in a forest stream ecosystem. *Ecology*, **60**, 1255–69.

Meyer, J.L., Likens, G.E. and Sloane, J. (1981) Phosphorus, nitrogen, and organic carbon in a headwater stream. *Arch. Hydrobiol.*, **91**, 28–44.

Meyer, J.L. and O'Hop, J. (1983) Leaf-shredding insects as a source of dissolved organic carbon in headwater streams. *Am. Midl. Nat.*, **109**, 175–83.

Meyer, J.L. and Pulliam, W.M. (1992) Modification of terrestrial–aquatic interactions by a changing climate, in *Global Climate Change and Freshwater Ecosystems*, (eds P. Firth and S.G. Fisher). Springer-Verlag, New York, pp. 177–91.

Meyer, J.L. and Tate, C.M. (1983) The effects of watershed disturbance on dissolved organic carbon dynamics of a stream. *Ecology*, **64**, 33–44.

Miller, J. and Georgian, T. (1992) Estimation of fine particulate transport in streams using pollen as a seston analog. *J. N. Am. Benthol. Soc.*, **11**, 172–80.

Miller, R.R., Williams, J.D. and Williams, J.E. (1989) Extinctions of North American fishes during the past century. *Fisheries*, **14**, 22–38.

Milliman, J.D. (1990) Fluvial sediment in coastal seas: flux and fate. *Nature Resour.*, **26**, 12–22.

Minckley, W.L. (1963) The ecology of a spring stream: Doe Run, Meade County, Kentucky. *Wildl. Monogr. Chestertown*, **11**(124).

Minckley, W.L. (1964) Upstream movements of *Gammarus* (Amphipoda) in Doe Run, Meade County, Kentucky. *Ecology*, **45**, 195–7.

Minshall, G.W. (1967) Role of allochthonous detritus in the trophic structure of a woodland springbrook community. *Ecology*, **48**, 139–49.

Minshall, G.W. (1978) Autotrophy in stream ecosystems. *Bioscience*, **28**, 767–71.

Minshall, G.W. (1984) Aquatic insect–substratum relationships, in *The Ecology of Aquatic Insects*, (eds V.H. Resh and D.M. Rosenberg), Praeger Scientific, New York, pp. 358–400.

Minshall, G.W., Cummins, K.W., Petersen, R.C. *et al.* (1985) Developments in stream ecosystem theory. *Can. J. Fish. Aquat. Sci.*, **42**, 1045–55.

Minshall, G.W. and Minshall, J.N. (1977) Microdistribution of benthic invertebrates in a Rocky Mountain (U.S.A.) stream. *Hydrobiologia*, **55**, 231–49.

Minshall, G.W., Petersen, R.C., Cummins, K.W. *et al.* (1983). Interbiome comparisons of stream ecosystem dynamics. *Ecol. Monogr.*, **53**, 1–25.

Minshall, G.W., Petersen, R.C., Bott, T.L. *et al.* (1992) Stream ecosystem dynamics of the Salmon River, Idaho: an 8th-order system. *J. N. Am. Benthol. Soc.*, **11**, 111–37.

Moeller, J.R., Minshall, G.W., Cummins, K.W. *et al.* (1979) Transport of dissolved organic carbon in streams of differing physiographic characteristics. *Organic Geochem.*, **1**, 139–50.

Molles, M.C., Jr. and Pietruszka, R.D. (1983) Mechanisms of prey selection by predaceous stoneflies: roles of prey morphology, behavior and predator hunger. *Oecologia*, **57**, 25–31.

Monk, D.C. (1976) The distribution of cellulase in freshwater invertebrates of different feeding habits. *Freshwater Ecol.*, **6**, 471–5.

Moon, H.P. (1940) An investigation of the movements of freshwater invertebrate faunas. *J. Anim. Ecol.*, **9**, 76–83.

Moore, J.W. (1972) Composition and structure of algal communities in a tributary stream of Lake Ontario. *Can. J. Bot.*, **50**, 1663–74.

Moore, J.W. (1975) The role of algae in the diet of *Asellus aquaticus* L. and *Gammarus pulex* L. *Ecology*, **44**, 719–30.

Moore, J.W. (1977) Some factors affecting algal densities in a eutrophic farmland stream. *Oecologia*, **29**, 257–67.

Morehardt, J.E. (1986) *Instream Flow Methodologies*, Electric Power Research Institute Report, EPRIEA-4819, Palo Alto, CA.

Morgan, M.J. and Godin, J.G.J. (1985) Antipredator benefits of schooling behaviour in a cyprinodontid fish, the banded killifish (*Fundulus diaphanus*). *Z. Tierpsychol.*, **70**, 236–46.

Morisawa, M. (1968) *Streams: their Dynamics and Morphology*, McGraw Hill, New York.

Morris, A.W., Bale, A.J. and Howland, R.J.M. (1982) Chemical variability in the Tamar Estuary, South-west England. *Estuar. Coast. Shelf Sci.*, **14**, 649–61.

Morris, W.M. (1955) A new concept of flow in rough conduits. *Trans. Am. Soc. Civ. Eng.*, **120**, 373–98.

Mottram, J.C. (1932) The living drift of rivers. *Trans. Newbury District Field Club*, **6**, 195–8.

Moyle, P.B. and Baltz, D.M. (1985) Microhabitat use by an assemblage of California stream fishes: developing criteria for instream flow determinations. *Trans. Am. Fish. Soc.*, **114**, 695–704.

Moyle, P.B. and Cech, J.J., Jr. (1982) *Fishes: An Introduction to Ichthyology*, Prentice-Hall, Englewood Cliffs, NJ.

Moyle, P.B. and Li, H.W. (1979) Community ecology and predator–prey relations in warmwater streams, in *Predator–prey Systems in Fisheries Management*, (ed. H. Clepper), Sports Fishing Institute, Washington DC, pp. 171–80.

Moyle, P.B., Li, H.W. and Barton, B.A. (1986) The Frankenstein effect: Impact of introduced fishes on native fishes in North America, in *Fish Culture in Fisheries Management*, (ed. R.H. Stroud), American Fisheries Society, Bethesda, MD, pp. 415–26.

Moyle, P.B and Sato, G.M. (1991) On the design of preserves to protect native fishes, in *Battle against Extinction: Native Fish Management in the American West*, (eds W.L. Minckley and J.E. Deacon), The University of Arizona Press, Tucson, AZ, pp. 155–70.

Moyle, P.B. and Senanayake, F.R. (1984) Resource partitioning among the fishes of rainforest streams in Sri Lanka. *J. Zool., London*, **202**, 195–223.

Moyle, P.B. and Vondracek, B. (1985) Persistence and structure of the fish assemblage in a small California stream. *Ecology*, **66**, 1–13.

Moyle, P.B. and Williams, J.E. (1990) Biodiversity loss in the temperate zone: Decline of the native fish fauna of California. *Conserv. Biol.*, **4**, 275–84.

Mulholland, P.J., Elwood, J.W., Newbold, J.D. and Ferrin, L.A. (1985a) Effect of a leaf-shredding invertebrate on organic matter dynamics and phophorus spiralling in heterotropic laboratory streams. *Oceologia*, **66**, 199–206.

Mulholland, P.J., Newbold, J.D., Elwood, J.W., Ferrin, L.A. and Webster, J.R. (1985b) Phosphorus spiralling in a woodland stream: seasonal variations. *Ecology*, **66**, 1012–23.

Mulholland, P.J., Newbold, J.D., Elwood, J.W. and Hom, C.L. (1983) The effect of grazing intensity on phosphorus spiralling in autotrophic streams. *Oecologia*, **58**, 358–66.

Mulholland, P.J. and Watts, J.A. (1982) Transport of organic carbon to the oceans by the rivers of North America: a synthesis of existing data. *Tellus*, **34**, 176–86.

Müller, K. (1954) Investigations on the organic drift in North Swedish streams. *Rep. Inst. Freshwater Res. Drottningholm*, **35**, 133–48.

Müller, K. (1963) Diurnal rhythm in 'organic drift' of *Gammarus pulex*. *Nature*, **198**, 806–7.

Müller, K. (1965) Field experiments on periodicity of freshwater invertebrates, in *Circadian Clocks*, (eds. J. Aschoff), North-Holland, Amsterdam, pp. 314–17.

Müller, K. (1966) Zur Periodik von *Gammarus pulex*. *Oikos*, **17**, 207–11.

Müller, K. (1973) Circadian rhythms of locomotor activity in aquatic organisms in the subarctic summer. *Aquilo Ser. Zool.*, **14**, 1–18.

Müller, K. (1982) The colonization cycle of freshwater insects. *Oecologia*, **52**, 202–7.

Müller, K. and Mendl, H. (1980) On the biology of the stonefly species *Leuctra digitata* in a northern Swedish coastal stream and its adjacent coastal area (Plecoptera: Leuctridae). *Entomol. Gen.*, **6**, 217–23.

Müller-Haeckel, A. (1971) Locomotive behavior of single celled algae of running waters. *Natur Mus.*, **10**, 167–72.

Munn, M.D. and Brusven, M.A. (1991) Benthic macroinvertebrate communities in nonregulated and regulated waters of the Clearwater River, Idaho, U.S.A. *Regulated Rivers: Res. Manage.*, **6**, 1–11.

Munn, N.L. and Meyer, J.L. (1988) Rapid flow through the sediments of a headwater stream in the southern Appalachians. *Freshwater Biol.*, **20**, 235–40.

Munn, N.L. and Meyer, J.L. (1990) Habitat-specific solute retention in two small streams: an intersite comparison. *Ecology*, **71**, 2069–82.

Munteanu, N. and Maly, E.J. (1981) The effect of current on the distribution of diatoms settling on submerged glass slides. *Hydrobiologia*, **78**, 273–82.

Muntz, W.R.A. (1982) Visual adaptations to different light environments in Amazonian fishes. *Rev. Can. Biol. Exp.*, **41**, 35–46.

Murphy, M.L. and Hall, J.D. (1981) Varied effects of clear-cut logging on predators and their habitat in small streams of the Cascade Mountains, Oregon. *Can. J. Fish. Aquat. Sci.*, **38**, 137–45.

Murphy, M.L., Hawkins, C.P. and Anderson, N.H. (1981) Effects of canopy modification and accumulated sediment on stream communities. *Trans. Am. Fish. Soc.*, **110**, 469–78.

Naiman, R.J. (1976) Primary production, standing stock, and export of organic matter in a Mohave

Desert thermal stream. *Limnol. Oceanogr.*, **21**, 60–73.
Naiman, R.J. (1982) Characteristics of sediment and organic carbon export from pristine boreal forest watersheds. *Can. J. Fish. Aquat. Sci.*, **39**, 1699–718.
Naiman, R.J. and Melillo, J.M. (1984) Nitrogen budget of a subarctic stream altered by beaver (*Castor canadensis*). *Oecologia*, **62**, 150–5.
Naiman, R.J., Melillo, J.M. and Hobbie, J.E. (1986) Ecosystem alteration of boreal forest streams by beaver (*Castor canadensis*). *Ecology*, **67**, 1254–69.
Naiman, R.J., Melillo, J.M., Lock, M.A., Ford, T.E. and Reice, S.R. (1987) Longitudinal patterns of ecosystem processes and community structure in a subarctic river continuum. *Ecology*, **68**, 1139–56.
Naiman, R.J. and Sedell, J.R. (1979) Characterization of particulate organic matter transported by some Cascade Mountain streams. *J. Fish. Res. Board Can.*, **36**, 17–31.
Naiman, R.J. and Sedell, J.R. (1980) Relationships between metabolic parameters and stream order in Oregon. *Can. J. Fish. Aquat. Sci.*, **37**, 834–47.
Neave, F. (1930) Migratory habits of the mayfly, *Blasturus cupidus* Say. *Ecology*, **11**, 568–76.
Needham, P.R. (1928) A quantitative study of the fish food supply in selected areas. *N.Y. Conserv. Dep. Suppl. Annu. Rep.*, **17**, 192–206.
Nehlsen, W., Williams, J.E. and Lichatowich, J.A. (1991) Pacific salmon at the crossroads: stocks of salmon at risk from California, Oregon, Idaho, and Washington. *Fisheries*, **16**, 4–21.
Neves, R.J. (1979) Movements of larval and adult *Pycnopsyche guttifer* (Walker) (Trichoptera: Limnephilidae) along Factory Brook, Massachusetts. *Am. Mid. Nat.*, **102**, 51–8.
Neves, R.J. (1991) Mollusks, in *Virginia's Endangered Species*, (ed. K. Terwilliger), The McDonald and Woodward Publishing Company, Blacksburg, VA, pp. 251–320.
Newbold, J.D. (1992) Cycles and spirals of nutrients, in *The Rivers Handbook – Volume One: Hydrological and Ecological Principles*, (eds P. Calow and G.E. Petts), Blackwell Scientific Publications, Oxford, pp. 379–408.
Newbold, J.D., Elwood, J.W., O'Neill, R.V. and Sheldon, A.L. (1983a) Phosphorus dynamics in a woodland stream ecosystem: A study of nutrient spiralling. *Ecology*, **64**, 1249–65.
Newbold, J.D., Elwood, J.W., Schulze, M.S., Stark, R.W. and Barmeier, J.C. (1983b) Continuous ammonium enrichment of a woodland stream: uptake kinetics, leaf decomposition, and nitrification. *Freshwater Biol.*, **13**, 193–204.
Newbold, J.D., Elwood, J.W., O'Neil, R.V. and Van Winkle, W. (1981) Measuring nutrient spiralling in streams. *Can. J. Fish. Aquat. Sci.*, **38**, 860–3.
Newbold, J.D., Elwood, J.W., O'Neill, R.V. and Van Winkle, W. (1982) Nutrient spiralling in streams: Implications for nutrient limitation and invertebrate activity. *Am. Nat.*, **120**, 628–52.
Newman, R.M. (1991) Herbivory and detritivory on freshwater macrophytes by invertebrates: a review. *J. N. Am. Benthol. Soc.*, **10**, 89–114.
Newman, R.M. and Waters, T.F. (1984) Size-selective predation on *Gammarus pseudolimnaeus* by trout and sculpins. *Ecology*, **65**, 1535–45.
Nickerson, M.A. and Mays, C.E. (1973) *The Hellbenders*, Milwaukee Public Museum Publications in Biology and Geology, Number 1.
Niemi, G.J., DeVore, P., Detenbeck, N. *et al.* (1990) Overview of case studies on recovery of aquatic ecosystems from disturbance. *Environ. Manage.*, **14**, 571–88.
Nilsson, C. (1986) The occurrence of lost and malformed legs in mayfly nymphs as a result of predator attacks. *Ann. Zool. Fenn.*, **23**, 57–60.
Nowell, A.R.M. and Jumars, P.A. (1984) Flow environments of aquatic benthos. *Annu. Rev. Ecol. System.*, **15**, 303–28.
O'Connell, T.R. and Campbell, R.S. (1953) The benthos of the Black River and Clearwater Lake, Missouri. *Univ. Mo. Stud.*, **26**(2), 25–41.
O'Keeffe, J.H. and De Moor, F.C. (1988) Changes in the physico-chemistry and benthic invertebrates of the Great Fish River, South Africa, following an interbasin transfer of water. *Regulated Rivers: Res. Manage.*, **2**, 39–55.
Odum, H.T. (1956) Primary production in flowing waters. *Limnol. Oceanogr.*, **1**, 102–17.
Odum, H.T. (1957) Trophic structure and productivity of Silver Springs. *Ecol. Monogr.*, **27**, 55–112.
Odum, W.E., Kirk, P.W. and Zieman, J.C. (1979) Non-protein nitrogen compounds associated with particles of vascular plant detritus. *Oikos*, **32**, 363–7.
Oemke, M.P. and Burton, T.M. (1986) Diatom colonization dynamics in a lotic ecosystem. *Hydrobiologia*, **139**, 153–66.
Økland, J. (1983) Factors regulating the distribution of fresh-water snails (Gastropoda) in Norway. *Malacologia*, **24**, 277–88.
Økland, J. and Økland, K.A. (1986) The effects of acid deposition on benthic animals in lakes and streams. *Experientia*, **42**, 471–86.
Omernik, J.M. (1977) Nonpoint source-stream nutrient level relationships: A nationwide study. EPA-600/3–77–105.
Omernik, J.M., Abernathy, A.R. and Male, L.M. (1981) Stream nutrient levels and proximity of agricultural and forest land to streams: some relationships. *J. Soil Water Conserv.*, **36**, 227–31.
Omerod, S.J. (1985) The diet of breeding Dippers *Cinclus cinclus* and their nestlings in the catchment of

the River Wye, mid-Wales: a preliminary study by faecal analysis. *Ibis*, **127**, 316–31.

Omerod, S.J., Boole, P., McCahon, C.P. *et al.* (1987) Short-term experimental acidification of a Welsh stream: comparing the biological effects of hydrogen ions and aluminium. *Freshwater Biol.*, **17**, 341–56.

Orth, D.J. (1987) Ecological considerations in the development and application of instream flow–habitat models. *Regulated Rivers: Res. Manage.*, **1**, 171–81.

Orth, D.J. and Leonard, P.M. (1990) Comparison of discharge methods and habitat optimization for recommending instream flows to protect fish habitat. *Regulated Rivers: Res. Manage.*, **5**, 129–38.

Osborne, L.L. and Kovacic, D.A. (1993) Riparian vegetated buffer strips in water-quality restoration and stream management. *Freshwater Biol.*, **29**, 243–58.

Osborne, L.L. and Wiley, M.J. (1988) Empirical relationships between land use/cover and stream water quality in an agricultural watershed. *J. Environ. Manage.*, **26**, 9–27.

Osborne, L.L., Wiley, M.J. and Larrimore, R.W. (1988) Assessment of the water surface profile model: Accuracy of predicted instream fish habitat conditions in low gradient, warmwater streams. *Regulated Rivers: Res. Manage.*, **2**, 619–31.

Osterkamp, W.R., Lane, L.J. and Foster, G.R. (1983) An analytical treatment of channel-morphology relations. *U.S. Geo. Sur. Prof. Pap.*, **1288**, 21 pp.

Ostrofsky, M. (1993) Effect of tannins on leaf processing and conditioning rates in aquatic ecosystems: an empirical approach. *Can. J. Fish. Aquat. Sci.*, **50**, 1176–80.

Oswood, M.W., Milner, A.M. and Irons, J.G., III (1992) Climate change and Alaskan rivers and streams, in *Global Climate Change and Freshwater Ecosystems*, (eds P. Firth and S.G. Fisher), Springer-Verlag, New York, pp. 192–210.

Otto, C. (1971) Growth and population movements of *Potamophylax cingulatus* (Trichoptera) larvae in a south Swedish stream. *Oikos*, **22**, 292–301.

Otto, C. (1993) Long-term risk sensitive foraging in *Rhyachophila nubila* (Trichoptera) larvae from two streams. *Oikos*, **68**, 67–74.

Otto, C. and Sjöström, P. (1983) Cerci as antipredatory attributes in stonefly nymphs. *Oikos*, **41**, 200–4.

Otto, C. and Svensson, B. (1976) Consequences of removal of pupae for a population of *Potamophylax cingulatus* (Trichoptera) in a south Swedish stream. *Oikos*, **27**, 40–3.

Otto, C. and Svensson, B. (1980) The significance of case material selection for the survival of caddis larvae. *J. Anim. Ecol.*, **49**, 855–65.

Otto, C. and Svensson, B. (1981) How do macrophytes growing in or close to water reduce their consumption by aquatic herbivores? *Hydrobiologia*, **78**, 107–12.

Owens, M. (1965). Some factors involved in the use of dissolved-oxygen distributions in streams to determine productivity. *Mem. Int. Ital. Idrobiol.*, **18** Suppl., 209–24.

Pace, M.L., Findlay, S.F. and Lints, D.L. (1988) Variability in Hudson River zooplankton: Similarities and differences with open water. *Trans. Am. Geophys. Union*, **69**, 1136.

Pace, M.L., Findlay, S.F. and Lints, D.L. (1992) Zooplankton in advective environments: the Hudson River community and a comparative analysis. *Can. J. Fish. Aquat. Sci.*, **49**, 1060–9.

Page, L.M. and Schemske, D.W. (1978) The effect of interspecific competition on the distribution and size of darters of the subgenus *Catonotus* (Percidae: *Etheostoma*). *Copeia*, **1978**, 406–12.

Palmer, M.A. (1990) Temporal and spatial dynamics of meiofauna within the hyporheic zone of Goose Creek, Virginia. *J. N. Am. Benthol. Soc.*, **9**, 17–25.

Palmer, M.A. (1992) Incorporating lotic meiofauna into our understanding of faunal transport processes. *Limnol. Oceanogr.*, **37**, 329–41.

Pandian, T.J. and Marian, M.P. (1986) An indirect procedure for the estimation of assimilation efficiency of aquatic insects. *Freshwater Biol.*, **16**, 93–8.

Parker, B.C., Samsel, G.L., Jr. and Prescott, G.W. (1973) Comparison of microhabitats of macroscopic subalpine stream algae. *Am. Midl. Nat.*, **90**, 143–53.

Patrick, R. (1948) Factors affecting the distribution of diatoms. *Bot. Rev.*, **14**, 473–524.

Patrick, R. (1961) A study of the numbers and kinds of species found in rivers in eastern United States. *Proc. Acad. Nat. Sci., Philadelphia*, **113**, 215–58.

Patrick, R. (1964) A discussion of the result of the Catherwood Expedition to the Peruvian headwaters of the Amazon. *Verh. Int. Ver. Theor. Ang. Limnol.*, **15**, 1084–90.

Patrick, R. (1975) Stream communities, in *Ecology and Evolution of Communities*, (eds M.L. Cody and J.M. Diamond), Belknap Press, Cambridge MA, pp. 445–59.

Pattee, E., Lascombe, C. and Delolme, R. (1973) Effects of temperature on the distribution of turbellarian triclads, in *Effects of Temperature on Ectothermic Organisms*, (ed. W. Weiser), Springer-Verlag, New York, pp. 201–8.

Paulson, L.J. and Baker, J.R. (1981) Nutrient interactions among reservoirs on the Colorado River, in *Proceedings of the Symposium on Surface Water Impoundments*, (ed. H.G. Stefan), American Society of Civil Engineers, New York, pp. 1647–56.

Payne, A.I. (1986) *The Ecology of Tropical Lakes and Rivers*, Wiley, New York.

Payne, R. (1959) *The Canal Builders; the Story of Canal Engineers through the Ages*, MacMillan, New York.

Peckarsky, B.L. (1980) Predator–prey interactions between stoneflies and mayflies: behavior observations. *Ecology*, **61**, 932–43.
Peckarsky, B.L. (1982) Aquatic insect predator–prey relations. *Bioscience*, **32**, 261–6.
Peckarsky, B.L. (1983) Biotic interactions or abiotic limitations? A model of lotic community structure, in *Dynamics of Lotic Ecosystems*, (eds T.D. Fontaine III and S.M. Bartell), Ann Arbor Science, Ann Arbor, MI, pp. 303–23.
Peckarsky, B.L. (1984) Predator–prey interactions among aquatic insects, in *The Ecology of Aquatic Insects*, (eds V.H. Resh and D.M. Rosenberg), Praeger Scientific, NY, pp. 196–254.
Peckarsky, B.L. (1985) Do predaceous stoneflies and siltation affect the structure of stream insect communities colonizing enclosures? *Can. J. Zool.*, **63**, 1519–30.
Peckarsky, B.L. (1987) Mayfly cerci as defense against stonefly predation: deflection and detection. *Oikos*, **48**, 161–70.
Peckarsky, B.L., Cowan, C.A., Penton, M.A. and Anderson, C. (1993) Sublethal consequences of stream-dwelling predatory stoneflies on mayfly growth and fecundity. *Ecology*, **74**, 1836–46.
Peckarsky, B.L. and Dodson, S.I. (1980a) An experimental analysis of biological factors contributing to stream community structure. *Ecology*, **61**, 1283–90.
Peckarsky, B.L. and Dodson, S.I. (1980) Do stonefly predators influence benthic distributions in streams? *Ecology*, **61**, 1275–82.
Peckarsky, B.L. and Penton, M.A. (1989) Mechanisms of prey selection by stream-dwelling stoneflies. *Ecology*, **70**, 1203–18.
Peierls, B.L., Caraco, N.F., Pace, M.L. and Cole, J.J. (1991) Human influence on river nitrogen. *Nature*, **350**, 386–7.
Pennak, R.W. (1985) The fresh-water invertebrate fauna: Problems and solutions for evolutionary success. *Am. Zool.*, **25**, 671–87.
Perrin, C.J., Bothwell, M.L. and Slaney, P.A. (1987) Experimental enrichment of a coastal stream in British Columbia: effects of organic and inorganic additions on autotrophic periphyton production. *Can. J. Fish. Aquat. Sci.*, **44**, 1247–56.
Perry, M.C. and Uhler, F.M. (1982) Food habits of diving ducks in the Carolinas. *Proc. Annu. Conf. Southeastern Assoc. Fish Wildl. Agencies*, **36**, 492–504.
Peters, G.T., Benfield, E.F. and Webster, J.R. (1989) Chemical composition and microbial activity of seston in a southern Appalachian headwater stream. *J. N. Am. Benthol. Soc.*, **8**, 74–84.
Petersen, B.J. and Frey, B. (1987) Stable isotopes in ecosystem studies. *Annu. Rev. Ecol. System.*, **18**, 293–320.
Petersen, R.C., Jr. (1992) The RCE: a riparian, channel and environmental inventory for small streams in the agricultural landscape. *Freshwater Biol.*, **27**, 295–306.
Petersen, R.C and Cummins, K.W. (1974) Leaf processing in a woodland stream. *Freshwater Biol.*, **4**, 343–68.
Petersen, R.C., Jr., Madsen, B.L., Wilzbach, M.A. *et al.* (1987) Stream management: Emerging global similarities. *Ambio*, **16**, 166–79.
Petersen, R.C., Jr. and Petersen, L.B.M. (1992) A building block model for stream restoration, in *River Conservation and Management*, (eds P. Boon, G. Petts and P. Calow), Wiley, London.
Peterson, B.J., Hobbie, J.E. and Corliss, T.L. (1983) A continuous-flow periphyton bioassay: tests of nutrient limitation in a tundra stream. *Limnol. Oceanogr.*, **28**, 583–91.
Peterson, B.J., Hobbie, J.E. and Corliss, T.L. (1986) Carbon flow in a tundra stream ecosystem. *Can. J. Fish. Aquat. Sci.*, **43**, 1259–70.
Peterson, B.J., Hobbie, J.E., Hershey, A.E. *et al.* (1985) Transformation of a tundra river from heterotrophy to autotrophy by addition of phosphorus. *Science*, **229**, 1383–6.
Petranka, J.W. (1984) Ontogeny of diet and feeding behavior of *Eurycea bislineata* larvae. *J. Herpetol.*, **18**, 48–55.
Petrere, M., Jr. (1989) River fisheries in Brazil: A review. *Regulated Rivers: Res. Manage.*, **4**, 1–16.
Petts, G.E. (1984) *Impounded Rivers*, John Wiley, Chichester.
Petts, G.E. (1989) Perspectives for ecological management of regulated rivers, in *Alternatives in Regulated River Management*, (eds J.A. Gore and G.E. Petts), CRC Press, Boca Raton, FL, pp. 3–24.
Philipson, G.N. (1978) The undulatory behaviour of larvae of *Hydropsyche pellucidula* Curtis and *Hydrosyphe siltalai* Döhler, in *Proc. Second Internat. Symp. Trichoptera*, (ed. M.I. Crichton), Dr. W. Junk, The Hague, pp. 241–7.
Pielou, E.C. (1975) *Ecological Diversity*, John Wiley, New York.
Pinder, L.V.C. (1985) Studies on Chironomidae in experimental recirculating stream systems. I. *Orthocladius (Euorthocladius) calvus* sp. nov. *Freshwater Biol.*, **15**, 235–41.
Plafkin, J.L., Barbour, M.T., Porter, K.D., Gross, S.K. and Hughes, R.M. (1989) Rapid bioassessment protocols for use in streams and rivers: benthic macroinvertebrates and fish. U.S. Environmental Protection Agency EPA/444/4-89-011, Washington D.C.
Platts, W.S., Megahan, W.F. and Minshall, G.W. (1983) Methods for evaluating stream, riparian and biotic conditions. U.S.D.A. Forest Service, General

Technical Report INT-138. Intermountain Forest and Range Experiment Station, Ogden, UT.

Ploskey, G.R. and Brown, A.V. (1980) Downstream drift of the mayfly *Baetis flavistriga* as a passive phenomenon. *Am. Midl. Nat.*, **104**, 405–9.

Poff, N.L. and Ward, J.V. (1989) Implications of streamflow variability and predictability for lotic community structure: a regional analysis of streamflow patterns. *Can. J. Fish. Aquat. Sci.*, **46**, 1805–18.

Poff, N.L. and Allan, J.D. (1994) Hydrologic correlates of functional organization in stream fish assemblages. *Ecology* (in press).

Polunin, N.V.C. (1982) Processes contributing to the decay of reed (*Phragmites australis*) litter in freshwater. *Arch. Hydrobiol.*, **94**, 155–62.

Polunin, N.V.C. (1984) The decomposition of emergent macrophytes in fresh water, in *Advances in Ecological Research, Volume 14*, (eds A. Macfadyen and E.D. Ford), Academic Press, New York, 115–66.

Pomeroy, L.R. and Weibe, W.J. (1988) Energetics of microbial food webs. *Hydrobiologia*, **159**, 7–18.

Porter, K.G. (1977) Enhancement of algal growth and productivity by grazing zooplankton. *Science*, **192**, 1332–4.

Postel, S. (1992) *Last Oasis, Facing Water Scarcity*, W.W. Norton, New York.

Power, M.E. (1983) Grazing responses of tropical freshwater fishes to different scales of variation in their food. *Environ. Biol. Fish.*, **9**, 103–15.

Power, M.E. (1984a) Habitat quality and the distribution of algae-grazing catfish in a Panamanian stream. *J. Anim. Ecol.*, **53**, 357–74.

Power, M.E. (1984b) Depth distributions of armored catfish: predator-induced resource avoidance? *Ecology*, **65**, 523–8.

Power, M.E. (1987) Predator avoidance by grazing fishes in temperate and tropical streams: Importance of stream depth and prey size, in *Predation: Direct and Indirect Impacts on Aquatic Communities*, (eds W.C. Kerfoot and A. Sih), University Press of New England, Hanover, NH, pp. 333–52.

Power, M.E. (1990) Effects of fish in river food webs. *Science*, **250**, 811–14.

Power, M.E. (1992) Habitat heterogeneity and the functional significance of fish in river food webs. *Ecology*, **73**, 1675–88.

Power, M.E. and Matthews, W.J. (1983) Algae-grazing minnows (*Campostoma anomalum*), piscivorous bass (*Micropterus* spp) and the distribution of attached algae in a small prairie-margin stream. *Oecologia*, **60**, 328–32.

Power, M.E. and Stewart, A.J. (1987) Disturbance and recovery of an algal assemblage following flooding in an Oklahoma stream. *Am. Midl. Nat.*, **117**, 333–45.

Power, M.E., Matthews, W.J. and Stewart, A.J. (1985) Grazing minnows, piscivorous bass, and stream algae: dynamics of a strong interaction. *Ecology*, **66**, 1448–56.

Prato, T., Shi, H.-Q., Rhew, R. and Brusven, M. (1989) Soil erosion and nonpoint-source pollution control in an Idaho watershed. *J. Soil Water Conserv.*, **44**, 323–8.

Pringle, C.M. (1990) Nutrient spatial heterogeneity: effects on community structure, physiognomy, and diversity of stream algae. *Ecology*, **71**, 905–20.

Pringle, C., Vellidis, G., Heliotis, F., Bandacu, D. and Cristofor, S. (1993) Environmental problems of the Danube delta. *Am. Sci.*, **81**, 350–61.

Pringle, C.M. and Bowers, J.A. (1984) An in situ substratum fertilization technique: diatom colonization on nutrient-enriched, sand substrata. *Can. J. Fish. Aquat. Sci.*, **41**, 1247–51.

Pritchard, G. (1966) On the morphology of the compound eyes of dragonflies (Odonata: Anisoptera) with special reference to their role in prey capture. *Proc. R. Entomol. Soc. London*, **41**, 1–8.

Prophet, C.W. and Ransom, J.D. (1974) Summer stream metabolism values for Cedar Creek, Kansas. *Southwestern Nat.*, **19**, 305–8.

Prowse, G.A. and Talling, J.F. (1958) The seasonal growth and succession of plankton algae in the White Nile. *Limnol. Oceanogr.*, **3**, 222–38.

Quinn, F. (1987) Interbasin water diversions: A Canadian perspective. *J. Soil Water Conserv.*, **42**, 389–93.

Rabeni, C.F. and Minshall, G.W. (1977) Factors affecting micro-distribution of stream benthos insects. *Oikos*, **29**, 33–43.

Rau, G.H. and Anderson, N.H. (1981) Use of $^{13}C/^{12}C$ to trace dissolved and particulate organic matter utilization by populations of an aquatic invertebrate. *Oecologia*, **48**, 19–21.

Rebsdorf, A., Thyssen, N. and Erlandsen, M. (1991) Regional and temporal variation in pH, alkalinity and carbon dioxide in Danish streams, related to soil type and land use. *Freshwater Biol.*, **25**, 419–35.

Redfield, A.C. (1958) The biological control of chemical factors in the environment. *Am. Sci.*, **46**, 205–21.

Redfield, A.C., Ketchum, B.H. and Richards, F.A. (1963) The influence of organisms on the composition of sea water, in *The Sea, Volume 2*, (ed. M.N. Hill), Interscience, New York, pp. 26–77.

Regier, H.A. and Meisner, J.D. (1990) Anticipated effects of climate change on freshwater fishes and their habitat. *Fisheries*, **15**, 10–15.

Reice, S.R. (1980) The role of substratum in benthic macroinvertebrate microdistribution and litter decomposition in a woodland stream. *Verh. Int. Ver. Theor. Ang. Limnol.*, **20**, 1396–400.

Reice, S.R. (1983) Predation and substratum: factors in lotic community structure, in *Dynamics of Lotic*

Ecosystems, (eds T.D. Fontaine III and S.M. Bartell), Ann Arbor Science, Ann Arbor, MI, p. 325–45.

Reice, S.R. and Edwards, R.L. (1986) The effects of vertebrate predation on macrobenthic communities with and without indigenous fish populations. *Can. J. Zool.*, **64**, 1930–6.

Reid, L.M. and Dunne, T. (1984) Sediment production from forest road surfaces. *Water Resour. Res.*, **20**, 1753–61.

Reif, C.B. (1939) The effect of stream condition on lake plankton. *Trans. Am. Microsc. Soc.*, **58**, 398–403.

Reynolds, J.D. (1988) Crayfish extinctions and crayfish plague in central Ireland. *Biol. Conserv.*, **45**, 279–85.

Reynoldson, T.B. and Bellamy, L.S. (1971) The establishment of interspecific competition in field populations, with an example of competition in action between *Polycelis nigra* (Mill.) and *P. tenuis* (Ijima) (Turbellaria, Tricladida). *Proc. Adv. Stud. Inst. Dyn. Numbers Pop.*, **1970**, 282–97.

Rice, D.L. (1982) The detritus nitrogen problem: new observations and perspectives from organic geochemistry. *Mar. Ecol. Prog. Ser.*, **9**, 153–62.

Richards, C. and Minshall, G.W. (1988). The influence of periphyton abundance on *Baetis bicaudatus* distribution and colonization in a small stream. *J. N. Am. Benthol. Soc.*, **7**, 77–86.

Richards, K. (1982) *Rivers. Form and Process in Alluvial Channels*, Methuen, London.

Richey, J.E., Perkins, M.A. and Goldman, C.R. (1975) Effects of kokanee salmon (*Onchorhynchus nerka*) decomposition on the ecology of a subalpine stream. *J. Fish. Res. Board Can.*, **32**, 817–20.

Richey, J.S., McDowell, W.H. and Likens, G.E. (1985) Nitrogen transformations in a small mountain stream. *Hydrobiologia*, **124**, 129–39.

Ricker, W.E. (1979) Growth rates and models, in *Fish Physiology, Volume 8*, (eds W.S. Hoar, D.J. Randall and J.R. Brett), Academic Press, New York, pp. 677–743.

Ricker, W.E. (1972) Hereditary and environmental factors affecting certain salmonid populations, in *The Stock Concept in Pacific Salmon*, (eds R.C. Simon and P.A. Larkin), University of British Columbia, Vancouver, Canada, pp. 19–160.

Ricklefs, R.E. (1987) Community diversity: Relative roles of local and regional processes. *Science*, **235**, 167–71.

Riley, G.A. (1970) Particulate organic matter in sea water. *Adv. Mar. Biol.*, **8**, 1–118.

Ringleberg, J. (1964) The positively phototactic reaction of *Daphnia magna* Straus. *Neth. J. Sea Res.*, **2**, 319–406.

Ringler, N.H. (1979a) Prey selection by drift-feeding brown trout (*Salmo trutta*). *J. Fish. Res. Board Can.*, **36**, 392–403.

Ringler, N. (1979b) Prey selection by benthic feeders, in *Predator–prey Systems in Fisheries Management*, (eds R. Stroud and H. Clepper), Sports Fishing Institute, Washington DC, pp. 219–29.

Ringler, N.H. (1983) Variation in foraging tactics of fishes, in *Predators and Prey in Fishes*, (eds D.L.G. Noakes *et al.*), Dr. W. Junk, The Hague, pp. 159–71.

Roback, S.S. (1968) Notes on the food of Tanypodinae larvae. *Entomol. News*, **80**, 13–18.

Roberts, T.R. (1972) Ecology of fishes in the Amazon and Congo basins. *Bull. Mus. Comp. Zool.*, Harvard, **143**, 117–47.

Robinson, C.T. and Rushforth, S.R. (1987) Effects of physical disturbance and canopy cover on attached diatom community structure in an Idaho stream. *Hydrobiologia*, **154**, 149–59.

Robinson, F.W. and Tash, J.C. (1979) Feeding by Arizona trout (*Salmo apache*) and brown trout (*Salmo trutta*) at different light intensities. *Environ. Biol. Fish.*, **4**, 363–8.

Rodgers, J.H., Jr., McKevitt, M.E., Hammerlund, D.O., Dickson, K.L. and Cairns, J., Jr. (1983) Primary production and decomposition of submergent and emergent aquatic plants of two Appalachian rivers, in *Dynamics of Lotic Ecosystems*, (eds T. Fontaine and S. Bartell), Ann Arbor Science, Michigan, pp. 283–301.

Rooke, J.B. (1984) The invertebrate fauna of four macrophytes in a lotic system. *Freshwater Biol.*, **14**, 507–13.

Roos, T. (1957) Studies on upstream migration in adult stream-dwelling insects. I. *Rep. Inst. Freshwater Res. Drottingholm*, **38**, 167–93.

Rosemond, A.D., Mulholland, P.J. and Elwood, J.W. (1993) Top-down and bottom-up control of stream periphyton: effects of nutrients and herbivores. *Ecology*, **74**, 1264–80.

Ross, H.H. (1967) The evolution and past dispersal of the Trichoptera. *Annu. Rev. Entomol.*, **12**, 169–206.

Ross, S.T. (1986) Resource partitioning in fish assemblages: A review of field studies. *Copeia*, 352–88.

Ross, S.T. (1991) Mechanisms structuring stream fish assemblages: are there lessons from introduced species? *Environ. Biol. Fish.*, **30**, 359–68.

Ross, S.T., Matthews, W.J. and Echelle, A.A. (1985) Persistence of stream fish assemblages: Effects of environmental change. *Am. Nat.*, **122**, 583–601.

Roth, N.E. (1994) Land use, riparian vegetation, and stream ecosystem integrity in an agricultural watershed. MS Thesis, The University of Michigan, 148 pp.

Rottman, R. (1977) Management of weedy lakes and ponds with grass carp. *Fisheries*, **2**, 8–14.

Round, F.E. (1964) The ecology of benthic algae, in *Algae and Man*, (ed. D.F. Jackson), Plenum Press, New York, pp. 138–84.

Round, F.E. (1981) *The Ecology of Algae*, Cambridge University Press, Cambridge.

Rounick, J.S. and Winterbourn, M.J. (1983) The formation, structure and utilization of stone surface organic layers in two New Zealand streams. *Freshwater Biol.*, **13**, 57–72.

Rubenstein, D.I. and Koehl, M.A. (1977) The mechanisms of filter-feeding: some theoretical considerations. *Am. Nat.*, **111**, 981–94.

Rundle, S.D. and Omerod, S.J. (1991) The influence of chemistry and habitat factors on the microcrustacea of some upland Welsh streams. *Freshwater Biol.*, **26**, 439–52.

Russell-Hunter, W.D. (1970) *Aquatic Productivity*, Macmillan, London.

Russell-Hunter, W.D., Appley, M.L., Burky, A.J. and Meadows, R.T. (1967) Interpopulational variations in calcium metabolism in the stream limpet, *Ferrissia rivularis* (Say). *Science*, **155**, 338–40.

Ruttner, F. (1926) Bermerkungen über den Sauerstoffgehalt der Gewasser und dessen respiratorischen Wert. *Naturwissenschaften*, **14**, 1237–39.

Rzóska, J., Brooks, A.J. and Prowse, G.A. (1952) Seasonal plankton development in the White and Blue Nile near Khartoum. *Verh. Int. Ver. Theor. Ang. Limnol.*, **12**, 327–34.

Sain, P., Robinson, J.B., Stammers, W.N., Kaushik, N.K. and Whitely, H.R. (1977) A laboratory study on the role of stream sediment in nitrogen loss from water. *J. Environ. Qual.*, **6**, 274–8.

Sale, P.F. (1977) Maintenance of high diversity in coral reef fish communities. *Am. Nat.*, **111**, 337–59.

Sanchez-Sierra, G. (1993) Outlook for hydropower in Latin America and the Caribbean. *Hydro Rev.*, (February 1993), 16–24.

Sattler, W. (1963) Über den Körperbau und Ethologie der Larve und Puppe von *Macronema* Pict. (Hydropsychidae), ein uls Larve sich von 'Mikro-Drift' ernährendes Tricopter aus dem Amazongebiet. *Arch. Hydrobiol.*, **59**, 26–60.

Saunders, M.J. and Eaton, J.W. (1976) A method for estimating the standing crop and nutrient content of the phytobenthos of stoney rivers. *Arch. Hydrobiol.*, **78**, 86–101.

Saunders, J.F. and Lewis, W.E. (1988) Zooplankton abundance and transport in a tropical white-water river. *Hydrobiologia*, **162**, 147–55.

Sazima, I. (1983) Scale eating in characoids and other fishes. *Environ. Biol. Fish.*, **9**, 87–101.

Scheich, H., Langner, G. Tidemann, C., Coles, R.B. and Guppy, A. (1986) Electroreception and electrolocation in platypus. *Nature*, **319**, 401–2.

Schindler, D.W., Kasian, S.E.M. and Hesslein, R.H. (1989) Biological impoverishment in lakes of the Midwestern and North-eastern United States from acid rain. *Environ. Sci. Technol.*, **23**, 573–80.

Schlesinger, W.H. and Melack, J.M. (1981) Transport of organic carbon in the world's rivers. *Tellus*, **33**, 172–87.

Schlosser, I.J. (1982) Fish community structure and function along two habitat gradients in a headwater stream. *Ecol. Monogr.*, **52**, 395–414.

Schlosser, I.J. and Toth, L.A. (1984) Niche relationships and population ecology of rainbow (*Etheostoma caeruleum*) and fantail (*E. flabellare*) darters in a temporally variable environment. *Oikos*, **42**, 229–38.

Schoener, T.W. (1974) Resource partitioning in ecological communities. *Science*, **185**, 27–39.

Schwarz, P. (1970) Autokologische Untersuchungen zum Lebenszyklus von Setipalpia-Arten (Plecoptera). *Arch. Hydrobiol.*, **67**, 103–40.

Scott, D. and Shirvell, C.S. (1987) A critique of the instream flow incremental methodology and observations on flow determination in New Zealand, in *Regulated Streams: Advances in Ecology*, (eds J.F. Craig and J.B. Kemper), Plenum Press, New York, pp. 27–43.

Scrivener, J.C. and Brownlee, M.J. (1989) Effects of forest harvesting on spawning gravel and incubation survival of chum (*Oncorhynchus keta*) and coho salmon (*O. kisutch*) in Carnation Creek, British Columbia. *Can. J. Fish. Aquat. Sci.*, **46**, 681–96.

Sedell, J.R. and Froggatt, J.L. (1984) Importance of streamside forests to large rivers: The isolation of the Willamette River, Oregon, U.S.A., from its floodplain by snagging and streamside forest removal. *Verh. Int. Ver. Theor. Ang. Limnol.*, **22**, 1828–34.

Sedell, J.R., Naiman, R.J., Cummins, K.W., Minshall, G.W. and Vannote, R.L. (1978) Transport of particulate organic material in streams as a function of physical processes. *Verh. Int. Ver. Theor. Ang. Limnol.*, **20**, 1366–75.

Sedell, J.R., Reeves, G.H., Hauer, F.R., Stanford, J.A. and Hawkins, C.P. (1990) Role of refugia in recovery from disturbances: modern fragmented and disconnected river systems. *Environ. Manage.*, **14**, 711–24.

Seegrist, D.W. and Gard, R. (1972) Effects of floods on trout in Sagehen Creek, California. *Trans. Am. Fish. Soc.*, **101**, 478–82.

Sepkoski, J.J., Jr. and Rex, M.A. (1974) Distribution of freshwater mussels: coastal rivers as biogeographic islands. *Syst. Zool.*, **23**, 165–88.

Sharma, B.R., Martin, M.M. and Shafer, J.A. (1984) Alkaline proteases from the gut-fluids of detritus-feeding larvae of the crane fly, *Tipula abdominalis* (Say) (Diptera, Tipulidae). *Insect Biochem.*, **14**, 37–44.

Shearer, C.A. and Lane, L.D. (1983) Comparison of three techniques for study of aquatic hyphomycete communities. *Mycologia*, **75**, 498–508.

Sheath, R.G. and Cole, K.M. (1992) Biogeography of

stream macroalgae in North America. *J. Phycol.*, **28**, 448–60.

Sheldon, A.L. (1972) Comparitive ecology of *Arcynopterx* and *Diura* (Plecoptera) in a California stream. *Arch. Hydrobiol.*, **69**, 521–46.

Shelford, V.E. (1918) Conditions of coexistence, in *Freshwater Biology*, (eds H.B. Ward and G.C. Whipple), John Wiley, New York, pp. 21–60.

Shepard, B.G., Hartman, G.F. and Wilson, W.J. (1986) Relationships between stream and intragravel temperatures in coastal drainages, and some implications for fisheries workers. *Can. J. Fish. Aquat. Sci.*, **43**, 1818–22.

Shepard, B.S. and Minshall, G.W. (1981) Nutritional value of lotic insect feces compared with allochthonous materials. *Arch. Hydrobiol.*, **90**, 467–88.

Sherman, B.J. and Phinney, H.K. (1971) Benthic algal communities of the Metolius River. *J. Phycol.*, **7**, 269–73.

Shireman, J.V. (1984) Control of aquatic weeds with exotic fishes, in *Distribution, Biology and Management of Exotic Fishes*, (eds W.R. Courtenay and J.R. Stauffer, Jr.), Johns Hopkins University Press, Baltimore, MD, pp. 302–12.

Shireman, J.V. and Smith, C.R. (1983) Synopsis of biological data on the grass carp *Ctenopharyngodon idella* (Cuvier and Valenciennes 1844). FAO Fish Synopsis 135, Rome: FAO.

Short, R.A., Canton, S.P. and Ward, J.V. (1980) Detrital processing and associated macroinvertebrates in a Colorado mountain stream. *Ecology*, **61**, 727–32.

Short, R.A. and Maslin, P.E. (1977) Processing of leaf litter by a stream detritivore: effect on nutrient availability to collectors. *Ecology*, **58**, 935–8.

Shortreed, K.R.S. and Stockner, J.G. (1983) Periphyton biomass and species composition in a coastal rainforest stream in British Columbia: effects of environmental changes caused by logging. *Can. J. Fish. Aquat. Sci.*, **40**, 1887–95.

Shreve, R.L. (1966) Statistical law of stream numbers. *J. Geol.*, **74**, 17–37.

Shreve, R.L. (1967) Infinite topologically random channel networks. *J. Geol.*, **74**, 178–86.

Siegfried, C.A. and Knight, A.W. (1976) Prey selection by a setipalpian stonefly nymph, *Acroneuria (Calineuria) californica* Banks (Plecoptera: Perlidae). *Ecology*, **57**, 603–8.

Sih, A. (1980) Optimal behavior: Can animals balance two conflicting demands? *Science*, **210**, 1041–3.

Sih, A. (1982a) Optimal patch use: Variation in selective pressure for efficient foraging. *Am. Nat.*, **120**, 666–85.

Sih, A. (1982b) Foraging strategies and the avoidance of predation by an aquatic insect, *Notonecta hoffmanni*. *Ecology*, **63**, 786–96.

Sih, A. (1987) Predators and prey lifestyles: An evolutionary and ecological overview, in *Predation: Direct and Indirect Impacts on Aquatic Communities*, (eds W.C. Kerfoot and A. Sih), University Press of New England, Hanover, pp. 203–24.

Silvester, N.R. and Sleigh, M.A. (1985) The forces on microorganisms at surfaces in flowing water. *Freshwater Biol.*, **15**, 433–48.

Sinsabaugh, R.L., Linkins, A.E. and Benfield, E.F. (1985) Cellulose digestion and assimilation by three leaf-shredding aquatic insects. *Ecology*, **66**, 1464–71.

Sinsabaugh, R.L., Repert, D., Weiland, T., Golladay, S.W. and Linkins, A.E. (1991) Exoenzyme accumulation in epilithic biofilms. *Hydrobiologia*, **222**, 29–37.

Sioli, H. (1964) General features of the limnology of Amazonia. *Verh. Int. Ver. Theor. Ang. Limnol.*, **15**, 1053–8.

Sioli, H. (1984) *The Amazon: Limnology and Landscape Ecology of a Mighty Tropical River and its Basin*, W. Junk, Dordrecht, 763 pp.

Sjöström, P. (1985) Hunting behavior of the perlid stonefly nymph *Dinocras cephalotes* (Plecoptera) under different light conditions. *Anim. Behav.*, **33**, 534–40.

Skinner, W.D. (1985) Night–day drift patterns and the size of larvae of two aquatic insects. *Hydrobiologia*, **124**, 283–5.

Small, M.J. and Sutton, M.C. (1986) A regional pH-alkalinity relationship. *Water Res.*, **20**, 335–43.

Smith, J.R. (1975) *Turbulence in Lakes and Rivers*. Freshwater Biological Association Scientific Publication 29, Ambleside, UK.

Smith, G.R., Stearley, R.F. and Badgley, C.E. (1988) Taphonomic bias in fish diversity from Cenozoic floodplain environments. *Palaeogr. Palaeoclimatol. Palaeoecol.*, **63**, 263–73.

Smith, S.D., Wellington, A.B., Nachlinger, J.L. and Fox, C.A. (1991) Functional responses of riparian vegetation to streamflow diversion in the eastern Sierra Nevada. *Ecol. Appl.*, **1**, 89–97.

Smock, L.A., Metzler, G.M. and Gladden, J.E. (1989) Role of debris dams in the structure and functioning of low-gradient headwater streams. *Ecology*, **70**, 764–75.

Søballe, D.M. and Bachmann, R.W. (1984) Influence of reservoir transit on riverine algal transport and abundance. *Can. J. Fish. Aquat. Sci.*, **41**, 1803–13.

Søballe D.M. and Kimmel, B.L. (1987) A large-scale comparison of factors influencing phytoplankton abundance in rivers, lakes, and impoundments. *Ecology*, **68**, 1943–54.

Söderström, O. (1987) Upstream movements of invertebrates in running waters. *Arch. Hydrobiol.*, **111**, 197–208.

Sollins, P., Glassman, C.A. and Dahm, C.N. (1985) Composition and possible origin of detrital material in streams. *Ecology*, **66**, 297–9.

Solon, B.M. and Stewart, K.W. (1972) Dispersal of algae and protozoa via the alimentary tracts of selected aquatic insects. *Environ. Entomol.*, **1**, 309–14.

Soluk, D.A. (1985) Macroinvertebrate abundance and production of Psammophilous Chironomidae in shifting sand areas of a lowland river. *Can. J. Fish. Aquat. Sci.*, **42**(7), 1296–302.

Stallard, R.F. and Edmond, J.M. (1983) Geochemistry of the Amazon 2: The influence of the geology and weathering environment on the dissolved load. *J. Geophys Res.*, **88**, 9671–88.

Stanford, J.A. and Ward, J.V. (1979) Stream regulation in North America, in *The Ecology of Regulated Rivers*, (eds J.V. Ward and J.A. Stanford), Plenum Publishing Corporation, New York, pp. 215–36.

Stanford, J.A. and Ward, J.V. (1986) The Colorado River system, in *The Ecology of River Systems*, (eds B.R. Davies and K.F. Walker), Dr. W. Junk Publishers, Dordrecht, The Netherlands, pp. 353–74.

Stanford, J.A. and Ward, J.V. (1989) Serial discontinuities of a Rocky Mountain river. I. Distribution and abundance of Plecoptera. *Regulated Rivers: Res. Manage.*, **3**, 169–75.

Starnes, L.B. and Starnes, W.C. (1981) Biology of the blackside dace *Phoxinus cumberlandensis*. *Am. Midl. Nat.*, **106**, 360–71.

Starnes, W.C. and Etnier, D.A. (1986) Drainage evolution and fish biogeography of the Tennessee and Cumberland rivers drainage realm, in *Zoogeography of North American Freshwater Fishes*, (eds C.H. Hocutt and E.O. Wiley), Wiley Interscience, New York, pp. 325–61.

Statzner, B. (1981) The relation between 'hydraulic stress' and microdistribution of benthic macroinvertebrates in a lowland running water system, the Schierenseebrooks (North Germany). *Arch. Hydrobiol.*, **91**, 192–218.

Statzner, B. (1988) Growth and Reynolds number of lotic invertebrates: a problem for adaptation of shape to drag. *Oikos*, **51**, 84–7.

Statzner, B., Gore, J.A. and Resh, V.H. (1988) Hydraulic stream ecology: observed patterns and potential applications. *J. N. Am. Benthol. Soc.*, **7**, 307–60.

Statzner, B. and Higler, B. (1985) Questions and comments on the river continum concept. *Can. J. Fish. Aquat. Sci.*, **42**, 1038–44.

Statzner, B. and Holm, T.F. (1982) Morphological adaptations of benthic invertebrates to stream flow– an old question studied by means of a new technique (Laser Doppler Anenometry). *Oecologia*, **53**, 290–2.

Statzner, B., and Mogel, R. (1985) An example showing that drift net catches of stream mayflies (*Baetis* spp., Ephemeroptera, Insecta) do not increase during periods of higher substrate surface densities of the larvae. *Verh. Int. Ver. Theor. Ang. Limnol.*, **22**, 3238–43.

Statzner, B. and Müller, R. (1989) Standard hemispheres as indicators of flow characteristics in lotic benthic research. *Freshwater Biol.*, **21**, 445–59.

Steedman, R.J. (1988) Modification and assessment of an index of biotic integrity to quantify stream quality in Southern Ontario. *Can. J. Fish. Aquat. Sci.*, **45**, 492–501.

Steedman, R.J. and Anderson, N.H. (1985) Life history and ecological role of the xylophagous aquatic beetle, *Lara avara* LeConte (Dryopoidea: Elmidae). *Freshwater Biol.*, **15**, 535–46.

Stein, R.A. (1979) Behavioral response of prey of fish predators, in *Predator–prey Systems in Fisheries Management*, (ed. H. Clepper), Sport Fishing Institute, Washington, DC, pp. 343–53.

Stein, R.A. and Magnuson, J.J. (1976) Behavioral response of crayfish to a fish predator. *Ecology*, **58**, 571–81.

Steinman, A.D. (1992) Does an increase in irradiance influence periphyton in a heavily-grazed woodland stream? *Oecologia*, **91**, 163–70.

Steinman, A.D. and Boston, H.L. (1993) The ecological role of aquatic macrophytes in a woodland stream. *J. N. Am. Benthol. Soc.*, **12**, 17–26.

Steinman, A.D. and McIntire, C.D. (1986) Effects of current velocity and light energy on the structure of periphyton asemblages in laboratory streams. *J. Phycol.*, **22**, 352–61.

Steinman, A.D. and McIntire, C.D. (1987) Effects of irradiance on the community structure and biomass of algal assemblages in laboratory systems. *Can. J. Fish. Aquat. Sci.*, **44**, 1640–8.

Steinman, A.D., McIntire, C.D., Gregory, S.V., Lamberti, G.A. and Ashkenas, L.R. (1987a) Effects of herbivore type and density on taxonomic structure and physiognomy of algal assemblages in laboratory streams. *J. N. Am. Benthol. Soc.*, **6**, 175–88.

Steinman, A.D., McIntire, C.D. and Lowry, R.R. (1987b) Effects of herbivore type and density on chemical composition of algal assemblages in laboratory streams. *J. N. Am. Benthol. Soc.*, **6**, 189–97.

Steinmann, P. (1908) Die Tierelt der Gebirgbäche. *Arch. Hydrobiol.*, **3**, 266–73.

Stevenson, R.J. (1990) Benthic algal community dynamics in a stream during and after a spate. *J. N. Am. Benthol. Soc.*, **9**, 277–88.

Stockner, J.G. and Shortreed, K.R.S. (1976) Autotrophic production in Carnation Creek, a coastal rainforest stream on Vancouver island, British Columbia. *J. Fish. Res. Board Can.*, **33**, 1553–63.

Stockner, J.G. and Shortreed, K.R.S. (1978) Enhancement of autotrophic production by nutrient addition in a coastal rainforest stream on Vancouver Island. *J. Fish. Res. Board Can.*, **35**, 28–34.

Stone, L.J. (1974) Effects of geology and nutrients on water-quality development. *J. Am. Water Works Assoc.*, **66**, 489–94.
Stout, R.J. (1989) Effects of condensed tannins on leaf processing in mid-latitude and tropical streams: a theoretical approach. *Can. J. Fish. Aquat. Sci.*, **46**, 1097–106.
Stout, R.J. and Vandermeer, J.H. (1975) Comparison of species richness for stream-inhabiting insects in tropical and mid-latitude streams. *Am. Nat.*, **109**, 263–80.
Strahler, A.N. (1952) Hypsometric (area-altitude) analysis of erosional topograph. *Bull. Geol. Soc. Am.*, **63**, 1117–42.
Strahler, A.N. (1964) Quantitative geomorphology of drainage basins and channel networks; section 4-2, in *Handbook of Applied Hydrology*, (ed. Ven te Chow), McGraw-Hill, New York.
Straskraba, M. (1965) The effects of fish on the number of invertebrates in ponds and streams. *Mitt. Int. Ver. Theor. Ang. Limnol.*, **13**, 106–27.
Strayer, D.L. (1991) Projected distribution of the zebra mussel, *Dreissena polymorpha*, in North America. *Can. J. Fish. Aquat. Sci.*, **48**, 1389–95.
Stream Solute Workshop, (1990) Concepts and methods for assessing solute dynamics in stream ecosystems. *J. N. Am. Benthol. Soc.*, **9**, 95–119.
Strong, D.R., Jr. (1973) Amphipod amplexus, the significance of ecotypic variation. *Ecology*, **54**, 1383–8.
Strong, D.R., Simberloff, D., Abele, L.G. and Thistle, A.B. (eds) (1984) *Ecological Communities, Conceptual Issues and the Evidence*, Princeton Univ. Press, Princeton, NJ.
Stumm, W. and Morgan, J.J. (1981) *Aquatic Chemistry*, 2nd edn, John Wiley, New York.
Suberkropp, K. and Arsuffi, T.L. (1984) Degradation, growth, and changes in palatability of leaves colonized by six aquatic hyphomycete species. *Mycologia*, **76**, 398–407.
Suberkropp, K.F., Godshalk, G.L. and Klug, M.J. (1976) Changes in the chemical composition of leaves during processing in a woodland stream. *Ecology*, **57**, 720–7.
Suberkropp, K. and Klug, M.J. (1976) Fungi and bacteria associated with leaves during processing in a woodland stream. *Ecology*, **57**, 707–19.
Suberkropp, K. and Klug, M.J. (1980) The maceration of deciduous leaf litter by aquatic hyphomycetes. *Can. J. Bot.*, **58**, 1025–31.
Sumner, W.T. and Fisher, S.G. (1979) Periphyton production in Fort River, Massachusetts. *Freshwater Biol.*, **9**, 205–12.
Sumner, W.T. and McIntire, C.D. (1982) Grazer-periphyton interactions in laboratory streams. *Arch. Hydrobiol.*, **93**, 135–57.
Sutcliffe, D.W. and Carrick, T.R. (1973) Studies on mountain streams in the English Lake District. I. pH, calcium and the distribution of invertebrates in the River Duddon. *Freshwater Biol.*, **3**, 437–62.
Sutcliffe, D.W. and Carrick, T.R. (1983) Relationships between chloride and major cations in precipitation and streamwaters in the Windermere catchment (English Lake District). *Freshwater Biol.*, **13**, 415–41.
Sutcliffe, D.W. and Carrick, T.R. (1988) Alkalinity and pH of tarns and streams in the English Lake District (Cumbria). *Freshwater Biol.*, **19**, 179–89.
Sutcliffe, D.W., Carrick, T.R. and Willoughby, L.G. (1981) Effects of diet, body size, age and temperature on growth rates in the amphipod *Gammarus pulex*. *Freshwater Biol.*, **11**, 183–214.
Svensson, B.W. (1974) Population movements of adult Trichoptera at a South Swedish stream. *Oikos*, **25**, 157–75.
Swank, W.T. and Caskey, W.H. (1982) Nitrate depletion in a second-order mountain stream. *J. Environ. Qual.*, **11**, 581–4.
Swanson, C.O. and Bachman, R.W. (1976) A model of algal exports in some Iowa streams. *Ecology*, **57**, 1076–80.
Swanson, F.J., Benda, L.E., Duncan, S.H. *et al.* (1987) Mass failures and other processes of sediment production in Pacific Northwest forest landscapes, in *Streamside Management: Forestry and Fishery Interactions. Contribution 57*, (eds E.O. Salo and T.W. Cundy), Institute of Forest Resources, University of Washington, Seattle, Washington, USA, pp. 9–38.
Sweeney, B.W. (1984) Factors influencing life-history patterns of aquatic insects, in *The Ecology of Aquatic Insects*, (eds V.H. Resh and D.M. Rosenberg), Praeger, New York, pp. 56–100.
Sweeney, B.W., Jackson, J.K., Newbold, J.D. and Funk, D.H. (1992) Climate change and the life histories and biogeography of aquatic insects in eastern North America, in *Global Climate Change and Freshwater Ecosystems*, (eds P. Firth and S.G. Fisher), Springer-Verlag, New York, pp. 143–76.
Sweeney, B.W. and Vannote, R.L. (1978) Size variation and the distribution of hemimetabolous aquatic insects: two thermal equilibrium hypotheses. *Science*, **200**, 444–6.
Sweeney, B.W. and Vannote, R.L. (1981) *Ephemerella* mayflies of White Clay Creek: Bioenergetic and ecological relationships among six coexisting species. *Ecology*, **62**, 1353–69.
Sze, P. (1981) A culture model for phytoplankton succession in the Potomac River near Washington D.C. (U.S.A). *Phycologia*, **20**, 285–91.
Tachet, H. (1977) Vibrations and predatory behaviour of *Plectrocnemia conspersa* larvae (Trichoptera) *Z. Tierpsychol.*, **45**, 61–74.
Taghon, G.L. and Jumars, P.A. (1984) Variable inges-

tion rate and its role in optimal foraging behavior of marine deposit feeders. *Ecology*, **65**, 549–58.

Talling, J.F. and Rzóska, J. (1967) The development of plankton in relation to hydrological regime in the Blue Nile. *J. Ecol.*, **55**, 637–62.

Tanaka, H. (1960) On the daily change of the drifting animals in stream, especially on the types of daily change observed in taxonomic groups of insects. (Japanese, English summary). *Bull. Freshwater Fish. Res. Lab.*, **9**, 13–24.

Tate, C.M. and Meyer, J.L. (1983) The influence of hydrologic conditions and successional state on dissolved organic carbon export from forested watersheds. *Ecology*, **64**, 25–32.

Taylor, J.N., Courtenay, W.R. and McCann, J.A. (1984) Known impacts of exotic fishes in the continental United States, in *Distribution, Biology, and Management of Exotic Fishes*, (eds W.R. Courtenay and Stauffer, J.R., Jr.), The Johns Hopkins University Press, Baltimore, MD, pp. 322–73.

Teague, S.A., Knight, A.W. and Teague, B.N. (1985) Stream microhabitat selectivity, resource partitioning, and niche shifts in grazing caddisfly larvae. *Hydrobiologia*, **128**, 3–12.

Teal, J.M. (1957) Community metabolism in a temperate cold spring. *Ecol. Monogr.*, **27**, 282–302.

Tennant, D.L. (1976) Instream flow regimes for fish, wildlife, recreation and related environmental resources. *Fisheries*, **1**(4), 6–10.

Tett, P., Gallegos, C., Kelly, M.G., Hornberger, G.M. and Cosby, B.J. (1978) Relationships among substrate, flow, and benthic microalgal pigment density in the Mechums River, Virginia. *Limnol. Oceanogr.*, **23**, 785–97.

Theinemann, A. (1954) Ein drittes biozonotisches Grundprinzip. *Arch. Hydrobiol.*, **49**, 421–2.

Thomas, N.A. and O'Connell, R.L. (1966) A method for measuring primary production by stream benthos. *Limnol. Oceanogr.*, **11**, 386–92.

Thomas, G.W. and Crutchfield, J.D. (1974) Nitrate-nitrogen and phosphorus contents of streams draining small agricultural watersheds in Kentucky. *J. Environ. Qual.*, **3**, 46–9.

Thompson, B.H. (1987) The use of algae as food by larval Simuliidae (Diptera) of Newfoundland streams. III. Growth of larvae reared on different algae and other foods. *Arch. Hydrobiol. Suppl.*, **76**(4), 459–66.

Thorp, J.H. (1986) Two distinct roles for predators in freshwater assemblages. *Oikos*, **47**, 75–82.

Thurman, E.M. (1985) *Organic Geochemistry of Natural Waters*, Martinus Nijhoff/Dr W. Junk Publishers, Dordrecht, The Netherlands, 497 pp.

Tippets, W.E. and Moyle, P.B. (1978) Epibenthic feeding by rainbow trout (*Salmo gairdneri*) in the McCloud River, California. *J. Anim. Ecol.*, **47**, 549–59.

Tokeshi, M. (1986) Resource utilization, overlap and temporal community dynamics: a null model analysis of an epiphytic chironomid community. *J. Anim. Ecol.*, **55**, 491–506.

Tokeshi, M. and Pinder, L.C.V. (1985) Microhabitats of stream invertebrates on two submersed macrophytes with contrasting leaf morphology. *Holarctic Ecol.*, **8**, 313–19.

Towns, D.R. (1981) Effects of artificial shading on periphyton and invertebrates in a New Zealand stream. *N.Z. J. Mar. Freshwater Res.*, **15**, 185–92.

Towns, D.R. (1983) Life history patterns of six sympatric species of Leptophlebiidae (Ephemeroptera) in a New Zealand stream and the role of interspecific competition in their evolution. *Hydrobiologia*, **99**, 37–50.

Townsend, C.R. (1989) The patch dynamics concept of stream community ecology. *J. N. Am. Benthol. Soc.*, **8**, 36–50.

Townsend, C.R. and Crowl, T.A. (1991) Fragmented population structure in a native New Zealand fish: an effect of introduced brown trout? *Oikos*, **61**, 347–54.

Townsend, C.R. and Hildrew, A.G. (1976) Field experiments on the drifting, colonization and continuous redistribution of stream benthos. *J. Anim. Ecol.*, **45**, 759–72.

Townsend, C.R. and Hildrew, A.G. (1978) Predation strategy and resource utilization by *Plectrocnemia conspersa* (Curtis) (Trichoptera: Polycentropodidae), in *Proceedings of the Second Annual Symposium on Trichoptera*, Junk, The Hague, Holland, pp. 283–91.

Townsend, C.R. and Hildrew, A.G. (1979) Resource partitioning by two freshwater invertebrate predators with contrasting foraging strategies. *J. Anim. Ecol.*, **48**, 909–20.

Townsend, C.R., Hildrew, A.G. and Francis, J. (1983) Community structure in some southern English streams: the influence of physicochemical factors. *Freshwater Biol.*, **13**, 521–44.

Townsend, C.R., Hildrew, A.G. and Schofield, K. (1987) Persistence of stream invertebrate communities in relation to environmental variability. *J. Anim. Ecol.*, **56**, 597–613.

Townsend, C.R. and McCarthy, T.K. (1980) On the defensive strategy of *Physa fontinalis* (L.), a freshwater pulmonate snail. *Oecologia*, **46**, 75–9.

Trautman, M.B. (1981) *The Fishes of Ohio*, Ohio State University Press, Columbus, OH.

Triska, F.J., Kennedy, V.C., Avanzino, R.J. and Reilly, B.N. (1983) Effect of simulated canopy cover on regulation of nitrate uptake and primary production by natural periphyton assemblages, in *Dynamics of Lotic Ecosystems*, (eds T. Fontaine and S. Bartell), Ann Arbor Science Publishers, Michigan, pp. 129–59.

Triska, F.J. and Oremland, R.S. (1981) Denitrification

associated with periphyton communities. *Appl. Environ. Microbiol.*, **42**, 745–8.

Triska, F.J., Sedell, J.R., Cromack, K., Jr., Gregory, S.V. and McCorison, F.M. (1984) Nitrogen budget for a small coniferous mountain stream. *Ecol. Monogr.*, **54**, 119–40.

Tsui, Ph.T.P. and Hubbard, M.D. (1979) Feeding habits of the predaceous nymphs of *Dolania americana* in northwestern Florida (Ephemeroptera: Behningiidae). *Hydrobiologia*, **67**, 119–24.

Turcotte, P. and Harper, P.P. (1982) Drift patterns in a high Andean stream. *Hydrobiologia*, **89**, 141–51.

Turner, R.E. and Rabalais, N.N. (1991) Changes in Mississippi River water quality this century. *Bioscience*, **41**, 140–7.

U.S. Fish and Wildlife Service (1981) Interim regional policy for New England stream flow recommendations. Memorandum from H.N. Larsen, Director, Region 5, U.S. Fish and Wildlife Service, Newton Corner, Massachusetts.

Ulfstrand, S. (1968) Benthic animal communities in Lapland streams. *Oikos*, Suppl **10**, pp. 1–120.

Vannote, R.L., Minshall, G.W. Cummins, K.W. *et al.* (1980) The river continuum concept. *Can. J. Fish. Aquat. Sci.*, **37**, 130–7.

Vannote, R.L. and Sweeney, B.W. (1980) Geographic analysis of thermal equilibria: A conceptual model for evaluating the effect of natural and modified thermal regimes on aquatic insect communities. *Am. Nat.*, **115**, 667–95.

Vaux, W.G. (1968) Intragravel flow and interchange of water in a stream bed. *Fish. Bull.*, **66**, 479–89.

Vermeij, G.J. and Covich, A.P. (1978) Coevolution of freshwater gastropods and their predators. *Am. Nat.*, **112**, 833–43.

Vincent, R.E. and Miller, W.H. (1969) Altitudinal distribution of brown trout and other fishes in a headwater tributary of the South Platte River, Colorado. *Ecology*, **50**, 464–6.

Vogel, S. (1981) *Life in Moving Fluids*, Princeton University Press, Princeton, NJ.

Walde, S.J. and Davies, R.W. (1984) Invertebrate predation and lotic prey communities: Evaluation of *in situ* enclosure/exclosure experiments. *Ecology*, **65**, 1206–13.

Wallace, J.B. (1975) The larval retreat and food of *Arctopsyche*; with phylogenetic notes on feeding adaptations in Hydropsychidae larvae (Trichoptera). *Ann. Entomol. Soc. Am.*, **68**, 167–73.

Wallace, J.B., Cuffney, T.F., Webster, J.R. *et al.* (1991) Export of fine organic particles from headwater streams: effects of season, extreme discharges, and invertebrate manipulation. *Limnol. Oceanogr.*, **36**, 670–82.

Wallace, J.B. and Gurtz, M.E. (1986) Response of *Baetis* mayflies (Ephemeroptera) to catchment logging. *Am. Midl. Nat.*, **115**, 25–41.

Wallace, J.B. and Malas, D. (1976) The fine structure of capture nets of larval Philopotamidae, with special reference on *Doliphilodes distinctus*. *Can. J. Zool.*, **54**, 1788–802.

Wallace, J.B. and Merritt, R.W. (1980) Filter-feeding ecology of aquatic insects. *Annu. Rev. Entomol.*, **25**, 103–32.

Wallace, J.B. and O'Hop, J. (1985) Life on a fast pad: waterlily leaf beetle impact on water lilies. *Ecology*, **66**, 1534–44.

Wallace, J.B., Ross, D.H. and Meyer, J.L. (1982) Seston and dissolved organic carbon dynamics in a southern Appalachian stream. *Ecology*, **63**, 824–38.

Wallace, J.B., Webster, J.R. and Woodall, W.R. (1977) The role of filter feeders in flowing waters. *Arch. Hydrobiol.*, **79**, 506–32.

Wallace, J.B., Webster, J.R. and Cuffney, T.F. (1982) Stream detritus dynamics: regulation by invertebrate consumers. *Oecologia*, **53**, 197–200.

Wallace, J.B., Woodall, W.R. and Sherberger, F.F. (1970) Breakdown of leaves by feeding of *Peltoperla maria* nymphs (Plecoptera: Peltoperlidae). *Ann. Entomol. Soc. Am.*, **63**, 562–7.

Walling, D.E. (1984) Dissolved loads and their measurement, in *Erosion and Sediment Yield: Some Methods of Measurement and Modelling*, (eds R.F. Hadley and D.E. Walling), Geo Books, Regency House, Norwich, UK, pp. 111–77.

Walling, D.E. and Webb, B.W. (1975) Spatial variation of river water quality: a survey of the River Exe. *Trans. Inst. British Geographers*, **65**, 155–69.

Wallis, P.M., Hynes, H.B.N. and Telang, S.A. (1981) The importance of groundwater in the transportation of allochthonous dissolved organic matter to the streams draining a small mountain basin. *Hydrobiologia*, **79**, 77–90.

Walton, O.E., Jr. (1980) Invertebrate drift from predator-prey associations. *Ecology*, **61**, 1486–97.

Walton, O.E., Jr., Reice, S.R. and Andrews, R.W. (1977) The effects of density, sediment particle size and velocity on drift of *Acroneuria abnormis* (Plecoptera). *Oikos*, **28**, 291–8.

Wankowski, J.W.J. (1979) Morphological limitations, prey size selectivity, and growth response of juvenile Atlantic salmon, *Salmo salar*. *J. Fish. Biol.*, **14**, 89–100.

Wankowski, J.W.J. (1981) Behavioral aspects of predation by juvenile Atlantic salmon (*Salmo salar* L.) on particulate, drifting prey. *Anim. Behav.*, **29**, 557–71.

Wankowski, J.W.J. and Thorpe, J.W. (1979) The role of food particle size in the growth of juvenile Atlantic salmon (*Salmo salar* L.). *J. Fish. Biol.*, **14**, 351–70.

Ward, G.M. and Aumen, N.G. (1986) Woody debris as a source of fine particulate organic matter in coniferous forest stream ecosystems. *Can. J. Fish. Aquat. Sci.*, **43**, 1635–42.

Ward, G.M. and Cummins, K.W. (1979) Effects of food quality on growth of a stream detritivore, *Paratendipes albimanus* (Meigen) (Diptera: Chironomidae). *Ecology*, **60**, 57–64.

Ward, G.M., Ward, A.K., Dahm, C.N. and Aumen, N.G. (1990) Origin and formation of organic and inorganic particles in aquatic systems, in *The Biology of Particles in Aquatic Systems*, (ed. R.S. Wotton), CRC Press, Boca Raton, FL, pp. 27–56.

Ward, G.M., Ward, A.K., Dahm, C.N. and Aumen, N.G. (1994) Origin and formation of organic and inorganic particles in aquatic systems. animals, in *Particulate and Dissolved Material in Aquatic Systems*, (ed. R.S. Wotton), Lewis Publishers, Chelsea, Michigan.

Ward, G.M. and Woods, D.R. (1986) Lignin and fiber content of FPOM generated by the shredders *Tipula abdominalis* (Diptera: Tipulidae) and *Tallaperla cornelia* (Needham and Smith) (Plecoptera: Peltoperlidae). *Arch. Hydrobiol.*, **107**, 545–62.

Ward, J.V. (1974) A temperature-stressed stream ecosystem below a hypolimnial release mountain reservoir. *Arch. Hydrobiol.*, **74**, 247–75.

Ward, J.V. and Dufford, R.G. (1979) Longitudinal and seasonal distribution of macroinvertebrates and epilithic algae in a Colorado springbrook-pond system. *Arch. Hydrobiol.*, **86**, 284–321.

Ward, J.V. and Stanford, J.A. (eds) (1979) *The Ecology of Regulated Rivers*, Plenum Publishing Corporation, New York.

Ward, J.V. and Stanford, J.A. (1987) The ecology of regualted streams: past accomplishments and directions for future research, in *Regulated Streams–Advances in Ecology*, (eds J.F. Craig and J.B. Kemper), Plenum, New York, pp. 391–401.

Ware, D.M. (1971) Predation by rainbow trout (*Salmo gairdneri*): The effect of experience. *J. Fish. Res. Board Can.*, **28**, 1847–52.

Ware, D.M. (1972) Predation by rainbow trout (*Salmo gairdneri*): The influence of hunger, prey density and prey size. *J. Fish. Res. Board Can.*, **29**, 1193–201.

Ware, D.M. (1973) Risk of epibenthic prey to predation by rainbow trout (*Salmo gairdneri*). *J. Fish. Res. Board Can.*, **30**, 787–97.

Waters, T.F. (1961) Standing crop and drift of stream bottom organisms. *Ecology*, **42**, 352–7.

Waters, T.F. (1962a) Diurnal periodicity in the drift of stream invertebrates. *Ecology*, **43**, 316–20.

Waters, T.F. (1962b) A method to estimate the production rate of a stream bottom invertebrate. *Trans. Am. Fish. Soc.*, **91**, 243–50.

Waters, T.F. (1965) Interpretation of invertebrate drift in streams. *Ecology*, **46**, 327–34.

Waters, T.F. (1972) The drift of stream insects. *Ann. Rev. Entomol.*, **17**, 253–72.

Waters, T.F. (1983) Replacement of brook trout by brown trout over 15 years in a Minnesota stream: production and abundance. *Trans. Am. Fish. Soc.*, **112**, 137–46.

Waters, T.F. and Hokenstrom, S.C. (1980) Annual production and drift of the stream amphipod *Gammarus pseudolimnaeus* in Valley Creek, Minnesota. *Limnol. Oceanogr.*, **25**, 700–10.

Watson, D.J. and Balon, E.K. (1984) Ecomorphological analysis of fish taxocenes in rainforest streams of northern Borneo. *J. Fish. Biol.*, **25**, 371–84.

Webb, P.W. (1988) Simple physical principles and vertebrate aquatic locomotion. *Am. Zool.*, **28**, 709–25.

Webb, P.W. (1989) Station-holding by three species of benthic fishes. *J. Exp. Biol.*, **145**, 303–20.

Webb, P.W. and Weihs, D. (1986) Functional locomotor morphology of early life history stages of fishes. *Trans. Am. Fish. Soc.*, **115**, 115–27.

Webster, J.R. (1983) The role of benthic macroinvertebrates in detritus dynamics of streams: a computer simulation. *Ecol. Monogr.*, **53**, 383–404.

Webster, J.R. and Benfield, E.F. (1986) Vascular plant breakdown in freshwater ecosystems. *Annu. Rev. Ecol. System.*, **17**, 567–94.

Webster, J.R. and Patten, B.C. (1979) Effects of watershed perturbation on stream potassium and calcium dynamics. *Ecol. Monogr.*, **49**, 51–72.

Webster, J.R., Benfield, E.F. and Cairns, J., Jr. (1979) Model predictions of effects of impoundment on particulate organic matter transport in a river system, in *The Ecology of Regulated Rivers*, (eds J.V. Ward and J.A. Stanford), Plenum, New York, pp. 339–64.

Webster, J.R., Benfield, E.F., Golladay, S.W. *et al.* (1987) Experimental studies of physical factors affecting seston transport in streams. *Limnol. Oceanogr.*, **32**, 848–63.

Webster, J.R., Benfield, E.F., Golladay, S.W. *et al.* (1988) Effects of watershed disturbance on stream seston characteristics, in *Forest Hydrology and Ecology at Coweeta*, (eds W.T. Swank and D.A. Crossley, Jr.), Springer-Verlag, New York, pp. 279–94.

Webster, J.R. and Golladay, S.W. (1984) Seston transport in streams at Coweeta Hydrologic Laboratory, North Carolina, U.S.A. *Verh. Int. Ver. Theor. Ang. Limnol.*, **22**, 1911–19.

Webster, J.R., Golladay, S.W., Benfield, E.F. *et al.* (1990) Effects of forest disturbance on particulate organic matter budgets of small streams. *J. N. Am. Benthol. Soc.*, **9**, 120–40.

Webster, J.R. and Patten, B.C. (1979) Effects of watershed perturbation on stream potassium and calcium dynamics. *Ecol. Monogr.*, **49**, 51–72.

Welcomme, R.L. (1979) *Fisheries Ecology of Floodplain Rivers*, Longman, London.
Welcomme, R.L. (1984) International transfers of inland fish species, in *Distribution, Biology and Management of Exotic Fishes*, (eds W.R. Courtenay and J.R. Stauffer, Jr.), The Johns Hopkins University Press, Baltimore, MD, pp. 22–40.
Wentworth, C.K. (1922) A scale of grade and class terms for clastic sediments. *J. Geol.*, **30**, 377–92.
Wesche, T.A. (1980) The WRRI trout cover rating method: development and application. Water Resources Series 78, Water Resources Research Institute, University of Wyoming, Laramie.
Wesche, T.A. (1985) Stream channel modifications and reclamation structures to enhance fish habitat, in *Restoration of Rivers and Streams*, (ed. J.A. Gore), Butterworth Publishers, Boston, pp. 103–64.
Westlake, D.F. (1975a) Macrophytes, in *River Ecology*, (ed. B.A. Whitton), University of California Press, Berkeley, CA, pp. 106–28.
Westlake, D.F. (1975b) Primary production of freshwater macrophytes, in *Photosynthesis and Productivity in Different Environments*, (ed. J.P. Cooper), Cambridge University Press, Cambridge, pp. 189–206.
Westlake, D.F. (1968) The ecology of aquatic weeds in relationship to their management. *Proc. 9th Br. Weed Control Conf.*, 372–9.
Westlake, D.F., Casey, H., Dawson, F.H. *et al.* (1972) The chalk-stream ecosystem, in *Productivity Problems of Freshwaters. IBP/UNESCO Symposium*, (eds Z. Kajak and A. Hillbrict-Ilkowska), PWN Polish Scientific Publishers, Warszawa, pp. 615–37.
Wetzel, R.G. (1975) Primary production, in *River Ecology*, (ed. B.A. Whitton), University of California Press, Berkeley, CA, pp. 230–47.
Wetzel, R. (1983) *Limnology*, 2nd edn, Saunders, New York.
Wetzel, R.G. and Manny, B.A. (1971) Secretion of dissolved organic carbon and nitrogen by aquatic macrophytes. *Verh. Int. Ver. Theor. Ang. Limnol.*, **18**, 162–70.
White, D.S., Elzinga, C.H., Hendricks, S.P. (1987) Temperature patterns within the hyporheic zone of a northern Michigan river. *J. N. Am. Benthol. Soc.*, **6**, 85–91.
Whitford, L.A. (1960) The current effect and growth of fresh-water algae. *Trans. Am. Microsc. Soc.*, **79**, 302–9.
Whitford, L.A. and Schumacher, G.J. (1964) Effect of a current on respiration and mineral uptake in *Spirogyra* and *Oedogonium*. *Ecology*, **45**, 168–70.
Whitton, B.A. (1975a) Algae, in *River Ecology*, (ed. B.A. Whitton), University of California Press, Berkeley, CA, pp. 81–105.
Whitton, B.A. (ed.) (1975b) *River Ecology*, University of California Press, Berkeley CA.
Wiggins, G.B. and Mackay, R.J. (1978) Some relationships between systematics and trophic ecology in Nearctic aquatic insects, with special reference to Trichoptera. *Ecology*, **59**, 1211–20.
Wigley, T.M.L. and Jones, P.D. (1985) Influences of precipitation changes and direct CO_2 effects on streamflow. *Nature*, **314**, 149–52.
Wiley, M.J. (1981) Interacting influences of density and preference on the emigration rates of some lotic chironomid larvae (Diptera: Chironomidae). *Ecology*, **62**, 426–38.
Wiley, M.J. and Kohler, S.L. (1980) Positioning changes of mayfly nymphs due to behavioral regulation of oxygen consumption. *Can. J. Zool.*, **58**, 618–22.
Wiley, M.J., Pescitelli, S.M. and Wike, L.D. (1986) The relationship between feeding preference and consumption rates in grass carp and grass carp × bighead carp hybrids. *J. Fish. Biol.*, **29**, 507–14.
Wiley, M.J. and Wike, L.D. (1986) Energy balances of diploid, triploid, and hybrid grass carp. *Trans. Am. Fish. Soc.*, **115**, 853–63.
Wiley, M.J., Osborne, L.L. and Larimore, R.W. (1990) Longitudinal structure of an agricultural prairie river system and its relationship to current stream ecosystem theory. *Can. J. Fish. Aquat. Sci.*, **47**, 373–84.
Wilhm, J.L. and Dorris, T.C. (1968) Biological parameters for water quality criteria. *Bioscience*, **18**, 477–80.
Williams, L.G. (1964) Possible relationships between plankton-diatom species numbers and water-quality estimates. *Ecology*, **45**, 809–23.
Williams, D.D. (1977) Movement of benthos during the recolonization of temporary streams. *Oikos*, **29**, 306–12.
Williams, D.D. (1981) Migrations and distributions of stream benthos, in *Perspective in Running Water Ecology*, (eds M.A. Lock and D.D. Williams), Plenum Press, NY, pp. 155–208.
Williams, W.D. (1987) Salinization of rivers and streams: an important environmental hazard. *Ambio*, **16**, 180–5.
Williams, D.D. and Hynes, H.B.N. (1976) The recolonization mechanism of stream benthos. *Oikos*, **27**, 265–72.
Williams, D.D. and Moore, K.A. (1985) The role of semiochemicals in benthic community relationships of the lotic amphipod *Gammarus pseudolimnaeus*: a laboratory analysis. *Oikos*, **44**, 280–6.
Williams, J.E., Johnson, J.E., Hendrickson, D.A. *et al.* (1989) Fishes of North America endangered, threatened, or of special concern. *Fisheries*, **14**, 2–22.
Willoughby, L.G. and Mappin, R.G. (1988) The distribution of *Ephemerella ignita* (Ephemeroptera) in

streams: the role of pH and food resources. *Freshwater Biol.*, **19**, 145–55.

Willoughby, L.G. and Sutcliffe, D.W. (1976) Experiments on feeding and growth of the amphipod *Gammarus pulex* (L.) related to its distribution in the River Duddon. *Freshwater Biol.*, **6**, 577–86.

Wilson, D.S., Leighton, M. and Leighton, D.R. (1978) Interference competition in a tropical riffle bug (Hemiptera: Veliidae). *Biotropica*, **10**, 302–6.

Wilzbach, M.A., Cummins, K.W. and Hall, J.D. (1986) Influence of habitat manipulations on interactions between cutthroat trout and invertebrate drift. *Ecology*, **67**, 898–911.

Winemiller, K.O. (1989) Ontogenetic diet shifts and resource partitioning among piscivorous fishes in the Venezuelan Ilanos. *Environ. Biol.*, **26**, 177–99.

Winemiller, K.O. (1991) Ecomorphological diversification in lowland freshwater fish assemblages from five biotic regions. *Ecol. Monogr.*, **61**, 343–65.

Winterbourn, M.J. and Collier, K.J. (1987) Distribution of benthic invertebrates in acid, brown water streams in the South Island of New Zealand. *Hydrobiologia*, **153**, 277–86.

Winterbourn, M.J., Cowie, B. and Rounick, J.S. (1984) Food resources and ingestion patterns of insects along a West Coast, South Island river system. *N. Z. J. Mar. Freshwater Res.*, **18**, 43–52.

Winterbourn, M.J., Rounick, J.S. and Cowie, B. (1981) Are New Zealand ecosystems really different? *N. Z. J. Mar. Freshwater Res.*, **15**, 321–8.

Wise, D.H. and Molles, M.C., Jr. (1979) Colonization of artificial substrates by stream insects: influence of substrate size and diversity. *Hydrobiologia*, **65**, 69–74.

Wolman, M.G. and Miller, J.P. (1960) Magnitude and frequency of forces in geomorphic process. *J. Geol.*, **68**, 54–74.

Wong, S.L. and Clark, B. (1976) Field determination of the critical nutrient concentrations for *Cladophora* in streams. *J. Fish. Res. Board Can.*, **33**, 85–92.

Wotton, R.S. (1978) The feeding-rate of *Metacnephia tredecimatum* larvae (Diptera: Simuliidae) in a Swedish lake outlet. *Oikos*, **30**, 121–5.

Wotton, R.S. (1980) Coprophagy as an economic feeding tactic in black fly larvae. *Oikos*, **34**, 282–6.

Wotton, R.S. (1994) Methods for capturing particles in benthic animals, in *The Biology of Particles*, (ed. R.S. Wotton), Lewis Publishers, Boca Raton, Florida, pp. 183–204.

Wright, J.C. and Mills, J.K. (1967) Productivity studies on the Madison River, Yellowstone National Park. *Limnol. Oceanogr.*, **12**, 568–77.

Wright, K.A., Sendek, K.H., Rice, R.M. and Thomas, R.B. (1990) Logging effects on streamflow: Storm runoff at Caspar Creek in Northwestern California. *Water Resour. Res.*, **26**, 1657–67.

Wuhrmann, K. and Eichenberger, E. (1975) Experiments on the effects of inorganic enrichment of rivers on periphyton primary production. *Verh. Int. Ver. Theor. Ang. Limnol.*, **19**, 2028–34.

Yasuno, M., Fukushima, S., Hasegawa, J. *et al.* (1982) Changes in the benthic faunas and flora after application of temephos to a stream on Mt. Tsukuba. *Hydrobiologia*, **89**, 205–14.

Yodzis, P. (1986) Competition, mortality and community structure, in *Community Ecology*, (eds J.M. Diamond and T.J. Case), Harper & Row, New York, pp. 430–55.

Yokoe, Y., and Yasumasu, I. (1964) The distribution of cellulases in invertebrates. *Comp. Biochem. Physiol.*, **13**, 323–38.

Yount, D. (1990) The eco-invaders. *EPA J.*, **16**, 51–3.

Zaret, T.M. and Rand, A.S. (1971) Competition in stream fishes: support for the competitive exclusion principle. *Ecology*, **52**, 336–42.

Zelinka, M. (1974) Die Eintagsfliegen (Ephemeroptera) in Forellenbachen der Beskiden. III. Der Einfluss des verschiedenen Fischbestandes. *Vest. Cesk. Spolecnesti Zool.*, **38**, 76–80.

Zimmerman, M.C. and Wissing, T.E. (1978) Effects of temperature on gut-loading and gut-clearing times of the burrowing mayfly, *Hexagenia limbata. Freshwater Biol.*, **8**, 269–77.

Zimmerman, R.C., Goodlet, J.C. and Comer, G.H. (1967) The influence of vegetation on channel form of small streams, in *Symposium on River Morphology*. Publication 75, International Association of Scientific Hydrology, Bern, Switzerland, pp. 255–75.

Species index

Page numbers appearing in **bold** refer to figures and page numbers appearing in *italic* refer to tables.

Subject index

Page numbers appearing in **bold** refer to figures and page numbers appearing in *italic* refer to tables.